The Chemical Biology of Sulfur

Chemical Biology

Editor-in-chief:
Kira J. Weissman, *Lorraine University, France*

Titles in the series:

1: High Throughput Screening Methods: Evolution and Refinement
2: Chemical Biology of Glycoproteins
3: Computational Tools for Chemical Biology
4: Mass Spectrometry in Chemical Biology: Evolving Applications
5: Mechanisms of Primary Energy Transduction in Biology
6: Cyclic Peptides: From Bioorganic Synthesis to Applications
7: DNA-targeting Molecules as Therapeutic Agents
8: Protein Crystallography: Challenges and Practical Solutions
9: Oxidative Folding of Proteins: Principles, Biological Regulation and Design
10: Chemical and Biological Synthesis: Enabling Approaches for Understanding Biology
11: Synthetic Glycomes
12: The Chemical Biology of Sulfur

How to obtain future titles on publication:
A standing order plan is available for this series. A standing order will bring delivery of each new volume immediately on publication.

For further information please contact:
Book Sales Department, Royal Society of Chemistry, Thomas Graham House, Science Park, Milton Road, Cambridge, CB4 0WF, UK
Telephone: +44 (0)1223 420066, Fax: +44 (0)1223 420247,
Email: booksales@rsc.org
Visit our website at www.rsc.org/books

The Chemical Biology of Sulfur

By

Christopher T. Walsh
Stanford University, USA
Email: cwalsh2@stanford.edu

ROYAL SOCIETY
OF **CHEMISTRY**

Chemical Biology No. 12

Print ISBN: 978-1-83916-144-5
PDF ISBN: 978-1-83916-184-1
EPUB ISBN: 978-1-83916-194-0
Print ISSN: 2055-1975
Electronic ISSN: 2055-1983

A catalogue record for this book is available from the British Library

The Royal Society of Chemistry is a charity, registered in England and Wales, Number 207890, and a company incorporated in England by Royal Charter (Registered No. RC000524), registered office: Burlington House, Piccadilly, London W1J 0BA, UK, Telephone: +44 (0) 20 7437 8656.

For further information see our web site at www.rsc.org

Printed in the United Kingdom by CPI Group (UK) Ltd, Croydon, CR0 4YY, UK

Preface

Sulfur has a bad reputation historically. From the biblical connection of solid elemental sulfur with fire and brimstone ("burning stone"), to the characteristically unpleasant "rotten egg" odor of gaseous hydrogen sulfide, to the butyl mercaptan (*n*-butyl thiol) defensive spray of skunks, human interaction with these three different macroscopic forms of sulfur and its compounds (S^0, H_2S, RSH) bolsters aversive behavior.

Yet, sulfur is absolutely essential to life in both its inorganic chemistry and organic chemistry. This holds whether one looks at coenzymes and cofactors, at enzyme catalytic strategies, at the roles of cysteine and methionine as 2 of 20 proteinogenic amino acids or the posttranslational modifications of cysteine side chains in proteomes.

Perched above oxygen in the second row of the periodic table, sulfur does indeed show some of the characteristic divalent bonding properties of oxygen. H_2O is paralleled by H_2S. Water is the universal liquid solvent for life. H_2S, with sulfur's much lower tendency to hydrogen bond, is a gas under physiological conditions and toxic at high concentrations.

Alcohols (ROH) are similarly mirrored by divalent RSH (thiols). However, while the pK_a for water, and other simple alcohols (dissociation, to H^+ and OH^-/OR^-) are in the range of 14, H_2S/RSH are 50% dissociated to HS^-/RS^- at around pH 8. These pK_a differences of six orders of magnitude arise from greater basicity and availability of the nucleophilic hydrosulfide and/or thiolate anions for chemical biology.

The distinguishing chemical feature of sulfur that enhances and conditions its many biological roles is its redox versatility. Unlike oxygen, which is restricted to divalency, sulfur compounds of prime biologic relevance exist not only with S in the formal [−2] oxidation state, but all the way to the [+6] oxidation state in sulfate dianions. Indeed, sulfate anions (S = [+6]) reach 28 mM concentration in global oceans and are by far the most abundant

Chemical Biology No. 12
The Chemical Biology of Sulfur
By Christopher T. Walsh

Published by the Royal Society of Chemistry, www.rsc.org

form of sulfur available to organisms. In this oxidation state the sulfur atom is electrophilic, at the opposite end of reactivity from the H_2S or RSH nucleophilic oxidation state.

This volume analyzes contemporary chemical biology roles of both inorganic and organic sulfur compounds, but it is worth noting that in the early evolution of anaerobic life, before the planet was substantially oxygenated, sulfate-reducing bacteria were early colonizers. They made, and still make, a living transferring metabolite-derived electrons to sulfate, reducing it by eight electrons all the way to H_2S. In so doing the H_2S seeps up through anoxic water levels to the oxic–anoxic interface where sulfide oxidizers take it back up, by either two electrons to elemental sulfur, or by eight electrons all the way to sulfate. The sulfide oxidizers have learned to use O_2 as terminal electron acceptor.

Analysis of the molecular logic, protein/cofactor nanomachinery and spatial compartmentalization of sulfate reducers and sulfide oxidizing bacteria give insights into how early organisms solved many of the metabolic challenges of energy generation and the formation of compounds suitable for cellular work.

Sulfur in sulfate ions is electrophilic and $-SO_3^-$ groups are transferable to cosubstrate nucleophiles, such as tyrosyl side chains in membrane receptor proteins. However, it is the nucleophilicity of reduced sulfur in inorganic HS^- and cysteine as RS^- that is one of the two key chemical properties that dominate the metabolic diaspora of organosulfur metabolites in both one-electron and two-electron reaction manifolds.

The two preferred embodiments of broadly used thiol metabolite frameworks are the cysteamine terminus of coenzyme A and the cysteinyl thiol side chain embedded in the tripeptide glutathione. Coenzyme A (CoA) is the progenitor to acyl S–CoA thioesters that serve as primary cellular acyl transfer currencies They are kinetically stable enough to serve as freely diffusing molecules in cells but thermodynamically activated to drive acyl transfer equilibria, to enable acylation of a wide variety of cellular nucleophiles.

One could have imagined a world where acyl glutathiones served that role but instead glutathione (GSH) in its reduced thiol form is a surveillance and scavenging reagent with three main cellular purposes. One is to capture inorganic reactive oxygen species (*e.g.* hydrogen peroxide). A second role is covalent interception of organic electrophiles, including many xenobiotic molecules (*e.g.* Tylenol) that are ingested purposefully or otherwise. The third role is keeping intracellular proteins in the reduced dithiol forms as glutathione becomes oxidized to the GS–SG disulfide. These three roles of reduced GSH are so important that cells spend dihydronicotinamide-adenine dinucleotide phosphate (NADPH) molecules to re-reduce GS–SG back to two GSH molecules, such that more than 95% of cellular glutathione remains as the thiol protectant form.

Before noting the many roles of cysteine residues in the cellular collections of all proteins expressed in certain cells, tissue or organisms—the various proteomes—I note that the second major organosulfur metabolite is the other proteinogenic sulfur amino acid methionine. Methionine has its sulfur atom as a thioether rather than thiol linkage.

That thioether is generally muted in its sulfur-based chemical biology with the exception of one magnificent enzymatic transformation. This is the enzymatic conversion to *S*-adenosylmethionine (SAM), with the sulfur now permanently converted to positively charged trivalent sulfonium ion. Among the various master inventions of the molecules of life, this may be one of the most central–the transformation of a second-rate thioether nucleophile to a first-rate electron sink.

SAM is justly celebrated as the universal donor of any of the three substituents (adenosyl, aminobutyryl and methyl) attached to the sulfonium sulfur as electrophilic fragments. However, it is the predominant flux of $[CH_3^+]$ transfers, to any number of low molecular weight and high molecular weight cellular partners (including proteins, RNA, DNA, lipids and polysaccharides) that puts SAM in a class with ATP as the most widely used cellular metabolite in any cell cycle.

In recent decades bioinformatic analysis has turned up more than 100 000 candidates of a second, distinct class of SAM-dependent proteins, radical SAM enzymes, where the SAM is coordinated to a 4Fe–4S cluster. Iron–sulfur clusters are probably the most ancient of the iron-based nanomachines in protein active sites and are obligate one-electron donors, doubtless from pre-oxygenation planetary eons. These 10^5 open reading frames imply largely unexplored radical SAM chemistry. They use one electron input from the Fe–S cluster to the SAM to fragment it and generate a 5′ deoxyadenosyl radical. This $dA^\bullet$, in each case yet examined, then unleashes a wide range of (often novel) one-electron chemistry pathways in organisms. Thus, *S*-adenosylmenthionine and its embedded trivalent sulfur cation double up for both two-electron transfer and one-electron transfer chemistry.

The third part of this examination of sulfur chemical biology turns to the role of cysteine side chains in proteomes. In some senses the protein-based reactions recapitulate some of the chemical biology of free cysteine, CoA–SH, and glutathione metabolic reactions. In other senses, the protein-incorporated cysteine residues populate many of the catalytic strategies of enzymes and represent a large swath of posttranslational chemistries to provide distinct protein functions and localizations.

Many of the primary metabolic pathway transformations involve reactions in which electrophilic groups get transferred to cosubstrate nucleophiles. These include acyl transfers, phosphoryl transfers, sulfuryl transfers, alkyl transfers and glycosyl transfers. All but the glycosyl transfer category are mimicked at the proteome level *via* cysteine residue thiolates.

The covalent *S*-cysteinyl protein intermediates can be metastable intermediates in catalytic cycles or stable posttranslational modifications. Thus, acyl-*S* enzymes operate in more than 100 "cysteinyl" proteases, in ubiquitylation cascades, and in glutaminase catalytic domains. *S*-Phosphoenyzmes are intermediates in all the protein tyrosine phosphatase catalytic cycles and in the resuscitative functions of sulfiredoxin. The *S*-alkylation intermediates in thymidylate synthase and DNA methyltransferase are more reminiscent of glutathione behavior. A distinct set of alkylations that are stable occur in posttranslational *S*-prenylation of cysteine residues in dozens of lipidated proteins.

Among the remarkable facets of cysteine thiolate side chain chemistry in proteomes are the attacks on immediate upstream peptide carbonyl groups to yield tetrahedral addition adducts that can have remarkably distinct fates. Among them are the autocleavage and cholesterol esterification of the hedgehog signaling protein, the refashioning of peptide backbones into cyclic thiazoles and the evolutionarily intriguing acyl transfer reactions of protein splicing.

In summary, the chemical biology of sulfur permeates the logic and reactivity of many of the key molecules of life. The versatility of sulfur in offering several redox states to organisms for divergent chemistries lies at the core of its broad utility. From inorganic to organic frameworks, from sulfur nucleophiles to sulfur electrophiles, from one-electron to two-electron reaction manifolds, sulfur sits at distinct major metabolic intersections.

Two of four key sulfur-containing molecules, acetyl CoA, and *S*-adenosylmethionine, combine kinetic stability and thermodynamic activation, for acyl group and alkyl group transfers, respectively. The other two, adenylyl-5′-phosphosulfate and Fe–S clusters, represent the intersections of inorganic and organic sulfur chemistry, one as a mixed anhydride, the other as a universal one-electron redox cluster. As an ensemble, these four molecules give insight into how organisms have evolved to unlock the many attributes of sulfur chemical biology necessary for life.

I note that each of parts I, II and III could easily fill a volume of their own as each chapter could be substantially expanded by treating existing information in more scholarly depth. These could include the role of biological sulfur reduction and oxidation in global sulfur (and carbon) cycles, the many redox niches for multiple forms of iron–sulfur clusters in different cellular organelles, or the logic of ubiquitin cascades in control of protein locations and lifetimes, just to pick three of many topics in sulfur chemical biology.

Instead, this volume focuses on the chemistry available to and exhibited by molecular forms of sulfur that are central to life. For the inorganic metabolic universe of sulfur, these are largely simple combinations of oxygenated sulfur scaffolds, at varying redox states from inorganic sulfate to inorganic sulfide ions. Analogously, the connection between inorganic and organic forms of sulfur in biology is the enzyme-mediated transit of hydrogen sulfide to cysteine as ground zero for the subsequent organic chemical biology of sulfur.

The compact organic framework of the amino acid cysteine is itself a marvel of reactivity design and invention, with three different functional groups at each of the three carbons (carboxyl, amine and thiol). The approach of Part II centers around the chemistry available to the thiol of cysteine and, subsidiarily, of coenzyme A and glutathione, and then the thioether of the amino acid methionine, The one- and two-electron pathways from the trigonalized sulfonium ion in *S*-adenosylmethionine reveal new facets of cellular chemistry. Most notably, they introduce SAM as the methylation reagent for a plethora of low molecular weight and high molecular weight substrates, including thousands of side chains of protein members of proteomes.

For the proteomics discussion in Part III, which could wander into many different aspects of both enzyme catalyses and the arc of different protein posttranslational modifications. I have, instead, continued the close focus on the thiol group of cysteine. With the amine and carboxyl of cysteine residues tied up in peptide linkages, it is only the side chain thiol groups of cysteines that are free to react, along the same chemical principles as described in Part II. The thiol nucleophile as a protein bound reagent recapitulates its behavior as the free amino acid but also shows off some remarkable reactivity facets, such as protein splicing and conversion of peptide bonds to embedded thiazoles.

Finally, one could write at length about the reach of chemical biology as an emerging interfacial discipline, and what is its appropriate subject matter. For some, it could encompass any investigation that brought reagents, methodologies or technologies to an understanding of sulfur reactivity, in any of its modes, even if the molecules under study are not themselves constituents of life. Instead, in this volume I have taken a narrower focus: to discuss the chemistry of the various forms of the actual sulfur metabolites, inorganic, organic, in low molecular weight scaffolds and in proteomes, and how that chemistry conditions, constrains and enables their biological roles.

Christopher T. Walsh

Dedication

This work is dedicated to:

Diana
Allison
Thomas
Sean

Chemical Biology No. 12
The Chemical Biology of Sulfur
By Christopher T. Walsh

Published by the Royal Society of Chemistry, www.rsc.org

Contents

Part I: Inorganic Sulfur Chemical Biology

Chapter 1 Introduction to Sulfur Chemical Biology 5

1.1 Introduction 5
1.2 Examples of the Multiple Oxidation States of Sulfur in Inorganic and Organic Compounds Relevant to Biology 9
1.3 Chemical Biology of Inorganic Sulfur Compounds 11
1.4 Organic Sulfur Metabolic Scaffolds 15
1.5 Oxygen *vs.* Sulfur Chemical Biology 19
References 21

Chapter 2 Sulfate Reducing Bacteria 23

2.1 The Eight-electron Traverse Between Inorganic Sulfate and Inorganic Sulfide 23
2.2 Sulfate Uptake by Transport Proteins 25
2.3 Microbial Conversion of Sulfate to Sulfite: Two Enzymatic Steps and Two Electron Inputs 26
2.3.1 ATP Sulfurylase 26
2.3.2 AMP-sulfate Reductase 27
2.4 Sulfate Reducers 27
2.4.1 An Alternative Adenylyl Sulfate Reductase Logic 31
2.5 Microbial Reduction of Sulfite to Hydrogen Sulfide: Net Six-electron Reductions 35
References 42

Chemical Biology No. 12
The Chemical Biology of Sulfur
By Christopher T. Walsh

Published by the Royal Society of Chemistry, www.rsc.org

Chapter 3 Sulfide Oxidizing Bacteria 45

3.1 From Sulfide by Two Electrons to Elemental Sulfur and by Eight Electrons to Sulfate 45

3.1.1 Two Electron Oxidation of H_2S to Elemental Sulfur and Linear Polysulfide Variants Thereof 46

3.2 Enzymatic Machinery for Conversion of S^0 to Sulfite and Sulfate 50

3.3 Assimilatory Sulfate Reduction: Transition to Organic Sulfur from Nascent Hydrogen Sulfide 54

3.4 Summary of Enzymatic Strategies for the Two Arms of Inorganic Sulfur Redox Cycles 55

References 57

Part II: Chemical Biology of Organosulfur Metabolites

Chapter 4 Organosulfur Metabolites in Biological Systems 65

4.1 Organosulfur Molecular Scaffolds in Biological Systems 65

4.2 Assimilation of Inorganic H_2S into the Organic Scaffold of L-Cysteine 72

4.3 Transsulfuration Pathway Between Cysteine and Homoserine in Mammalian Metabolism 76

4.3.1 The Four-carbon Scaffold of Homocysteine Derives from Methionine Metabolism 76

4.3.2 Enzymatic Transsulfuration: Cystathionine as Intermediate 78

4.3.3 Misprocessing of Cysteine to H_2S by the Two Enzymes of the Transsulfuration Pathway 83

4.3.4 Mercaptopyruvate Transferase 85

4.4 H_2S Signaling Role in Mammals 87

References 89

Chapter 5 The Nucleophilic Thiolate in Cysteine and Cysteamine Scaffolds 92

5.1 Coenzyme A: Thiol Chemical Biology *via* Cysteamine 93

5.1.1 Acyl Thioester Formation by CoASH and Acyl Phosphate and Acyl Adenylates 95

5.2 Double Headed Reactivity of Acyl Thioesters as Carbon Electrophiles and Carbon Nucleophiles 98
5.3 Glutathione and Other Low Molecular Weight Thiols in Biology 103
5.3.1 Reaction of Reduced Glutathione with Cellular Electrophiles 108
5.3.2 Geranylgeranyl Glutathione 112
5.3.3 Reaction of Reduced Glutathione with Xenobiotic Electrophiles 114
5.3.4 Cysteine *vs.* Glutathione as a Sulfur Nucleophile 115
5.4 *S*-Nitrosylation of Glutathione 117
5.5 Glutathione: Sulfenates and Disulfides: Thiol Oxidations 119
References 125

Chapter 6 Protein Cysteine Persulfides and Fe–S Clusters: Intersection of Organic and Inorganic Sulfur Biochemistry 129

6.1 Persulfide-forming Cysteine Lyases 131
6.1.1 β-Replacement *vs.* β-Elimination Cysteine Lyases: Six-electron *vs.* Eight-electron Sulfur Atom Products 131
6.1.2 The Differential Reaction of Persulfides *vs.* Persulfide Anions 133
6.2 Iron–Sulfur Cluster Assembly 134
6.3 Variant Fe–S Clusters 145
6.3.1 Sacrifice of Fe–S Clusters in Biotin and Lipoic Acid Sulfur Insertions 146
6.4 Sulfur as Electrophile in Sulfenic Acids, Persulfides and Disulfides 148
6.4.1 Persulfides as Sulfur Donors to Pyrimidines 148
6.4.2 Intersection of Protein Persulfides, Carrier Protein Acyl Persulfides and Protein Thiocarboxylates as Sulfur Donors 151
6.5 One More Look at the Versatility of Persulfides as Electrophilic or Nucleophilic Forms of Sulfur 154
References 158

Chapter 7 Cysteine to Taurine: Thiol to Sulfonic Acid 161

7.1 Cysteine *S*-Oxygenation 161
7.2 Hypotaurine to Taurine 167

7.3 Taurine Amide Conjugation 170
7.4 Cysteine to Sulfoxide to Sulfenate to Thiocarbonyl Group 170
References 174

Chapter 8 Methionine: Thioether Biological Chemistry 176

8.1 Introduction 176
8.2 Methionine Biosynthesis in Plants and Microbes 177
8.3 Methionine Biosynthesis from Homocysteine: The Last Step in the Pathway 177
8.4 *S*-Adenosylmethionine: Thioether to Trivalent Sulfonium Cation 182
8.5 SAM as an Electrophilic Group Transfer Substrate 183
8.5.1 Transfer of Electrophilic One-carbon $[CH_3]^+$ Fragment from SAM to Cosubstrate Nucleophiles 184
8.6 SAM-ylide Formation, Carboxy-SAM Formation and Carboxymethyl Transfers 191
References 196

Chapter 9 One-electron *vs.* Two-electron Reaction Manifolds in Sulfur Enzymology 198

9.1 Thiol Radicals: One-electron Enzyme Chemistry 198
9.1.1 Ribonucleotide Reductases: A Bevy of Thiol Radicals in the Active Site 199
9.1.2 Tandem Glycyl and Thiol Radical Intermediates in Enzyme Catalysis 201
9.1.3 β-Lactam Formation *via* Substrate Thiol Radicals 205
9.2 Radical SAM Enzymatic Logic 209
9.2.1 Two Convergent Routes to the 5′-Deoxyadenosyl Radical 209
9.3 The Scope of Radical SAM Reactions 211
9.3.1 Lysine Aminomutase 216
9.3.2 SAM to 5′-Deoxyadenosine and Methionine: HemN Decarboxylase 216
9.3.3 Consumption of Two SAM Molecules to Two Sets of Products 219
9.3.4 Likely Ancient Coenzymes from Radical SAM Pathways 219

9.4 Cannibalization of 2Fe–2S or 4Fe–4S Clusters in Biotin and Lipoic Acid Biosynthesis 221
9.5 Combined Methylthiolation *via* Two SAMs and a 4Fe–4S Sacrifice 228
References 231

Chapter 10 Sulfuryl Transferases and Sulfatases 233

10.1 Sulfotransferases 233
10.1.1 Nonprotein Substrates for Sulfotransferases 236
10.1.2 Sulfated Disaccharide Repeats of Glycosaminoglycans as Extracellular Ligands for Receptor Protein Tyrosine Phosphatases 239
10.2 Protein *vs.* Proteoglycan Substrates for Sulfuryl Transfer 241
10.2.1 Proteoglycan *O*- and *N*-Sulfation 242
10.3 Protein Sulfotransferases 243
10.4 Sulfatases 248
10.4.1 Metalloprotein Sulfatases 249
10.4.2 Formylglycine (Fgly)-Containing Sulfatases 249
10.5 Lysosomal Sulfatase Deficiencies 252
References 255

Chapter 11 C–S Bond Formations in Primary and Secondary Metabolism 258

11.1 C–S Bond-forming Strategies in Primary Metabolic Pathways 258
11.2 Some C–S Bond-forming Strategies in Conditional Plant and Microbial Biosynthetic Pathways 262
11.2.1 Disulfide to Methylthioacetal 262
11.2.2 Iterative Conjugate Additions by Cysteine-persulfide Anions Followed by Oxidative Cleavage of the Resultant Disulfides 266
11.2.3 Thioglycosides and Glucosinolates 266
11.2.4 Methanethiol from Methionine-γ-lyase Activity 272
References 272

Part III: Sulfur Side Chains in Proteomes

Chapter 12 Cysteinyl Proteomes I: Metal Coordination and Catalytic Intermediates 279

12.1 Cysteinyl Proteomes: Introduction 279
12.2 Coordination to Iron and Zinc 279
12.2.1 Three Forms of Iron in Proteins 280
12.2.2 Zinc as a Pervasive Super Acid, Organizing Ligand in Proteins 281
12.2.3 Zinc Finger Domains 283
12.3 Covalent Changes in Cysteinyl Proteomes 283
12.3.1 Acyl-*S*–Enzyme Intermediates: Cysteinyl Proteases 284
12.3.2 The Ubiquitinylated Proteome: Covalent Information-rich Acyl-*S*–Cys-Proteins 286
12.3.3 Deubiquitinases 290
12.3.4 Glutaminase Catalytic Logic 293
12.3.5 Fatty Acid, Polyketides and Non-ribosomal Peptide Assembly Lines 295
12.4 *S*–Phosphoryl–Enzyme Intermediates in Protein Tyrosine Phosphatase Action 298
12.5 Covalent Addition of Enzyme Thiolate Initiates Catalysis on Thymidylate and Cytosine 300
12.6 Use of an Enzyme Cysteine Active Site Nucleophile to Create a Persulfide S–S Bond 304
12.7 Thiohemiacetal Adducts on the Way to Acyl Thioesters 305
12.8 Summary of the Types of Substrate Electrophiles that Form Covalent Enzyme–Cysteinyl Adducts 307
References 310

Chapter 13 The Posttranslational Cysteinyl Proteome 313

13.1 Post-translation Lipidations of Cysteinyl Side Chains 314
13.1.1 Long Chain Acyl-thioesters *via* Acyl–*S*–Cys–Enzyme Intermediates 314
13.1.2 Protein Prenylations: Posttranslational Lipidation *via* Thioether Adducts 319
13.2 Cystine Disulfides 324
13.2.1 Keeping Cytoplasmic Cysteinyl Side Chains Reduced 325

13.2.2 Peroxiredoxins 327
13.2.3 Sulfiredoxin 330
13.3 Protein Machinery for Disulfide Formation in Oxidizing Compartments 334
13.4 Examples of Protein Cys–S–S–Cys Disulfide Bonds 339
13.5 Nitrosylation of the Cysteinyl Proteome 340
13.6 Cysteinyl Residue Modifications from Reactive Oxygen Species 342
13.6.1 Protein Sulfenates and Sulfenamides 342
13.6.2 Dioxygenation of Cys_2 Residues in Certain Proteins Is a Signal for Sequential Arginylation, Polyubiqitinylation and Rapid Degradation During $[O_2]$ Sensing 344
13.7 Cysteine Persulfide Residues 345
13.8 Methionine Residue *S*-Oxidations and Reversibility 347
13.8.1 Nonenzymatic Methionyl Residue *S*-Oxidation 347
13.8.2 Methionine Residue Enzymatic Sulfidation in Actin 349
References 353

Chapter 14 Cysteinyl Proteomes: Exotic Modifications 358

14.1 Thioethers 358
14.2 Attack by a Cysteine Side Chain on the Immediate Upstream Peptide Carbonyl Group 361
14.2.1 Autoproteolysis and Capture of a Thioester Intermediate by the Alcohol Group of Cholesterol 361
14.2.2 Thiazole and Oxazole Formation 364
14.2.3 Thiocarbonyl Group Formation 366
14.2.4 Protein Splicing 368
14.3 Cysteine–Tyrosine Thioether Automodification 374
14.4 Capturing the Nucleophilic Thiol Side Chains by Foreign Electrophiles in Cysteinyl Proteomes 374
14.4.1 Protein Serine Kinases with Conserved Non-catalytic Cysteine Residues Near the Active Sites 377
References 379

Part IV: Overview and Summary

Chapter 15 A Traverse Through Sulfur Chemical Biology **387**

15.1 Inorganic Sulfur Compounds Emphasize the Eight-electron Spread in Redox States of the Sulfur Atom 387
15.2 AMP–Sulfate and Phosphoadenosine Phosphosulfate: Sulfuryl Transfer and Two-electron Redox Capacity 390
15.3 Six-electron Reduction and Reoxidation Between Sulfite and Sulfide: Heme Connected to 4Fe–4S Cluster 394
15.4 Traverse from Inorganic Hydrogen Sulfide to Organic Thiol Group in Cysteine 396
15.5 Coenzyme A (CoASH) and Glutathione (GSH): Acyl Thioesters *vs.* Electrophile Scavenging 398
15.6 Cysteine to Fe–S Clusters: Role of Persulfide Side Chains and Electrophilic –SH 399
15.7 Methionine to *S*-Adenosylmethionine: Two-electron Transfers 399
15.8 Radical SAM Enzymes; Fe–S–SAM Coordination: Prelude to Radical Chemistry 401
15.9 The Broad Sweep of the Cysteinyl Proteomes: Covalent Cys–X Enzyme Intermediates and Covalent Posttranslational Modifications 402
15.10 The Versatility of Inorganic and Organic Sulfur Metabolites 405
References 405

Subject Index **406**

Part I: Inorganic Sulfur Chemical Biology

The centrality of sulfur and its electron transfer chemistry in the evolution of microbial life, first in anaerobic and then increasingly oxic microenvionments, is brought into high relief by the two extremes of inorganic forms of sulfur and the two kinds of bacteria that evolved to use them to get energy and make a living. The most oxidized form of biologically active sulfur is the ubiquitous dianion, sulfate (SO_4^{2-}), of the strong mineral acid sulfuric acid (H_2SO_4). The most reduced form, spewed out by volcanoes both above and below the surface of the oceans is H_2S, hydrogen sulfide, a gas under physiological conditions.

The traverse from sulfate to hydrogen sulfide is an eight-electron reduction. Or, on the way back up from H_2S to sulfate dianion, an eight-electron oxidation. The ability of sulfur to span those oxidation states in physiological contexts that are thermodynamically and kinetically accessible to organisms has illuminated many facets of the strategy for chemolithoautorophic growth by early organisms on the planet.

The sulfate reducers came first. Before ambient O_2, elaborated by cyanobacterial neighbors in photosynthesis, became high enough to be useful as an electron sink, sulfate served that role for sulfate-reducing anaerobic microbes. As electrons enter, sulfate is progressively reduced to sulfide ions, the electrons experience a thermodynamically favorable potential drop, coupling electron flow to anisotropic pumping of protons across biological membranes in which the electron transport chains are embedded. The resultant transmembrane potential gradient thus generated is coupled to back inflow of protons through the membrane adenosine triphosphate (ATP) synthase machinery. ATP is formed and the bacterial and archaeal sulfate-reducing cells can use the ATP to couple many otherwise unfavorable metabolic and biosynthetic equilibria, including DNA, RNA and protein biosynthesis.

The hydrogen sulfide gaseous product exits the sulfate-reducing cells. In water columns, the H_2S can diffuse up from anoxic regions to the anoxic–oxic interface where dissolved O_2 levels can begin to support the growth of aerobic microbes. O_2 reduction to 2 H_2O, a four-electron traverse, practiced by all aerobe electron transport respiratory chains is half the distance (four electrons *vs.* eight electrons) of the sulfate to sulfide traverse, but per electron transferred, the potential drop is larger. In the O_2 *vs.* the SO_4^{2-} systems, more energy is available and harvested. Aerobes are more efficient in energy capture metabolism.

Chapters 1–3 focus on the inorganic chemical biology of sulfur, largely the electron transfer reactions and attendant protein and cofactor machinery that organisms evolved to perform sulfate reduction and then sulfide oxidation and capture energy in those biologic processes. They illuminate the differing redox states of sulfur from its electrophilic +6 formal oxidation state in sulfate to the −2 oxidation state of the nucleophilic monoanion of H_2S, the hydrosulfide anion (HS^-), readily available at physiological pH values. Along the route in each direction is the sulfite dianion, two electrons reduced from sulfate (six electrons up from sulfide). Also, remarkably the electron transfers on the way down and up pass through the equivalent of atomic sulfur in its S^0 oxidation state.

Chapter 1 introduces several features of sulfur inorganic chemistry, including its remarkable ability to exist as stable molecular entities in the −2, 0, +2, +4 and + 6 oxidation states. In some respects, at the −2 oxidation state, sulfur behaves with some similarity to oxygen: H_2O *vs.* H_2S. However, in the +6 oxidation state with participation of d level atomic orbitals in hybrid molecular orbitals, sulfur is remarkably distinct from oxygen. The sulfate dianion has more in common with the phosphate dianion. Both central atoms, phosphorus and sulfur, are electrophilic, and the mixed anhydride adenosine-5-phosphosulfate, is a metabolic donor of sulfuryl groups ($-SO_3^-$) to cosubstrate nucleophiles.

Chapter 2 examines the underlying chemical logic of sulfate reducers and the protein and cofactor machinery that performs the eight-electron reductive traverse in one two-electron and then one six-electron step. While the focus is on the chemical biology of inorganic sulfur, metabolic strategy and chemical logic cannot be disentangled from the redox active organic and inorganic coenzymes that organisms had to have invented for the metabolic pathways to proceed. In the sulfate reducers, the first two-electron step requires the coenzyme form of vitamin B2, riboflavin, and introduces flavin–sulfur chemistry. The six-electron sulfite reductase machinery brings two additional forms of sulfur into play. One is the thiolate side chains of the most central of the organosulfur metabolites, cysteine (Cys), that is largely the subject of Part II of his volume. The second is a four iron–four sulfur cluster, a one-electron transfer redox cube, that sits at the interface of inorganic sulfide and organic cysteine thiol chemical biology. Throw in one siroheme tetrapyrrolic iron macrocycle and the enzyme then has a three-component redox machinery [4Fe–4S cluster, Cys thiolate and siroheme] for transfer in of six electrons one at a time, to sulfite axially coordinated to the iron in the siroheme cofactor. Sulfite reductase is, integrated over all biology, one of the most remarkable protein nanomachines that organisms have invented for coordinated electron transfers to drive energy metabolism.

Chapter 3 takes up the logic and strategy that microbes have invented to go in the other direction from sulfide (HS^-, not S^{2-}, is the predominant species of dissolved sulfate in physiologic media) to sulfate ions. Two unusual enzymes, the sulfide quinone oxidoreductase, performing the first two-electron oxidative traverse (formally: S^{2-} to S^0) has an intriguing catalyst design. It again recruits a flavin coenzyme, forms a distinct kind of flavin–sulfur covalent adduct from that in chapter 2, and generates a nascent sulfane product (a six-electron sulfur) inserted as a middle sulfur atom in a Cys–S–S–S–Cys trisulfide link. Additional catalytic cycles can create polysulfide chains at the oxidation level of atomic sulfur.

The second unusual catalyst design is in the last two-electron oxidative step, oxidation of sulfite dianion ($S = +4$) to the sulfate dianion ($S = +6$), carried out by sulfite oxidase. The key coenzyme in this active site is a tricyclic organic pterin scaffold with an ene-dithiolate moiety that coordinates a molybdenum^{VI} ion (molybdopterin). The active site Mo is the redox active electron transfer agent. As sulfite is oxidized by two electrons to sulfate *via* a molybdate covalent sulfate ester intermediate, the Mo^{VI} is reduced to the Mo^{IV} oxidation state. It is then reoxidized in two one-electron steps *via* a heme iron cofactor to reset the oxidase for the next catalytic cycle.

Part I thus delves into both the enzyme chemistry of the several forms of sulfur—sulfate, sulfite, atomic sulfur and sulfide—that make up the known players in inorganic sulfur chemical biology. These metabolic transformations show the chemical virtuosity of sulfur and set the stage for the crossover to organosulfur metabolism by the single route from H_2S to cysteine.

CHAPTER 1

Introduction to Sulfur Chemical Biology

1.1 Introduction

The chemical biology of living organisms is typically cast in terms of the properties of the four major elements, (tetravalent) carbon, (monovalent) hydrogen, (trivalent) nitrogen and (divalent) oxygen and the variety of their compounds and embedded functional groups therein that comprise the metabolites of cells. Metabolomics, the science of metabolic transformations across cells, tissues or the whole organism, follows the rules and reactivity of the organic chemistries open to those functionalized molecular scaffolds.

The fifth element that is essential for life is sulfur, a second-row element. Directly above oxygen in the periodic table, sulfur consequently has similarities to oxygen chemistry, but dramatic differences, arising in substantial part from its electronic configuration with 3p electrons ($1s^2 2s^2 2p^6 3s^2 3p^4$) and empty 3d orbitals (Figure 1.1A). Sulfur is believed to form in the hot cores of stars by fusion of silicon and helium atoms. Unlike atomic oxygen ("O") which is not a stable compound under physiological conditions, elemental sulfur, S^0, is a solid, with various allotropes.[1] Indeed, elemental sulfur is an active metabolite in bacterial and archaeal reduction–oxidation (redox) pathways.

Among the most remarkable distinctions between sulfur and oxygen is the range of accessible oxidation states available for sulfur biology (chemed.chem.purdue.edu/chemed/bodnergroup/index.html). While oxygen is restricted to divalent chemistry [*e.g.* water (H_2O) and dioxygen (O_2)], sulfur is protean by comparison in its compounds. Although H_2S has analogies to H_2O in the formal oxidation state of −2, as a divalent hydride, H_2S is a gas at room temperature and pressure in contrast to liquid H_2O because of the lessened hydrogen bonding tendency in sulfur *vs.* oxygen.[2] The tendency of

Chemical Biology No. 12
The Chemical Biology of Sulfur
By Christopher T. Walsh

Published by the Royal Society of Chemistry, www.rsc.org

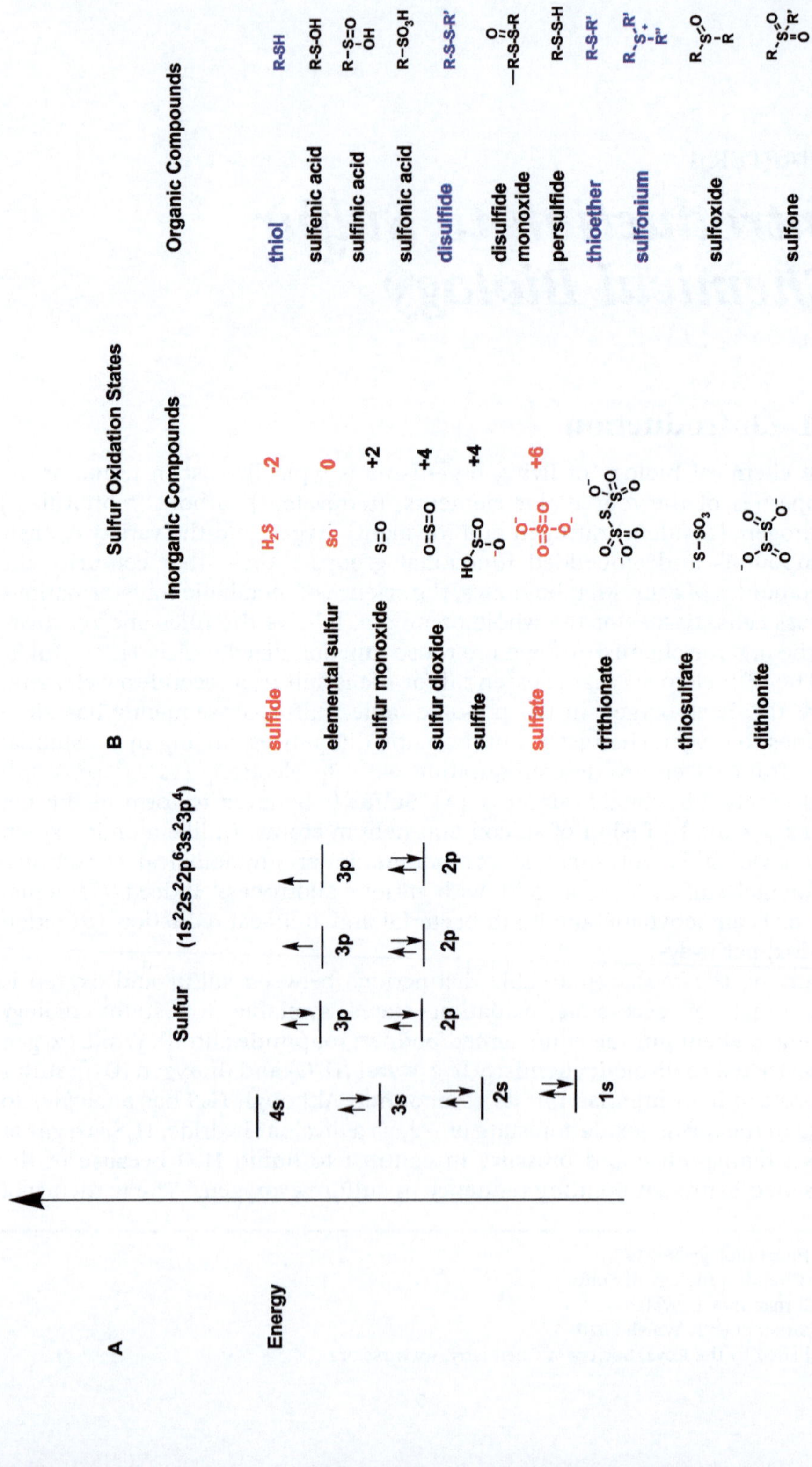

Figure 1.1 (A) Electronic configuration of sulfur: ($1s^2 2s^2 2p^6 3s^2 3p^4$); (B) Sulfur oxidation states in some biologically relevant compounds.

sulfur to form two bonds to partner atoms (*e.g.* H_2S) is consistent with the two partially filled 3p orbitals (Figure 1.1A).

Most notably, though, sulfur makes covalent bonds to partner atoms at oxidation states of −2, +2, +4 and +6. Most abundant in the earth's crust is the dianion of sulfuric acid[3] (H_2SO_4) in which S has an oxidation state of +6 (Figure 1.1B and 1.2) Sulfur monoxide has sulfur in the formal +2 oxidation state while sulfur dioxide has sulfur in the +4 oxidation state. Figure 1.1B also shows the S–S bonding in three of the more common forms of inorganic oxygenated sulfur ions, dithionate, trithionate and thiosulfate.

One comparison of the biological roles of phosphorus and sulfur can be made between phosphate and sulfate dianions. Both are water soluble forms of the P and S atoms, in their most oxidized biologic states. Phosphate groups, of course, are pervasive in biology, from small molecules, such as glucose-6-phosphate, to phosphoenzyme intermediates in signaling pathways to the covalent phosphodiester backbone linkages in RNA and DNA. In contrast, sulfate ions are much more restricted in their biological distribution. Yet, phosphorus chemical biology is restricted largely to phosphorus in the +5 oxidation state. It is not readily reduced to low-valent phosphorus metabolites or reoxidized back up to phosphate.[4] It would not be suitable for evolution of electron transport machinery in organisms.

The availability of sulfur to access so many distinct oxidation states between −2 and +6 allows it to occupy a quite distinct and complementary set of metabolic niches from phosphate.

On the organic side of Figure 1.1B, thiols, sulfenic acids and thioethers are examples of sulfur in the −2 oxidation state.[5,6] We shall note in subsequent chapters that thiols and thioethers have nucleophilic sulfur atoms while in sulfenic acids, the S atom is electrophilic. The sulfur atoms in disulfides are electrophilic while the formally comparable persulfide can react either as a sulfur nucleophile or electrophile.[7] Methionine with its thioether side chain offers the possibility of alkylation to form the trivalent sulfonium cation in the metabolite *S*-adenosylmethionine, thereby activating the three substituents at that trigonal sulfur for transfer to cosubstrate nucleophiles.

The flexibility of sulfur to populate such a broad range of oxidation states is a function of rehybridization of atomic valence orbitals to hybrid

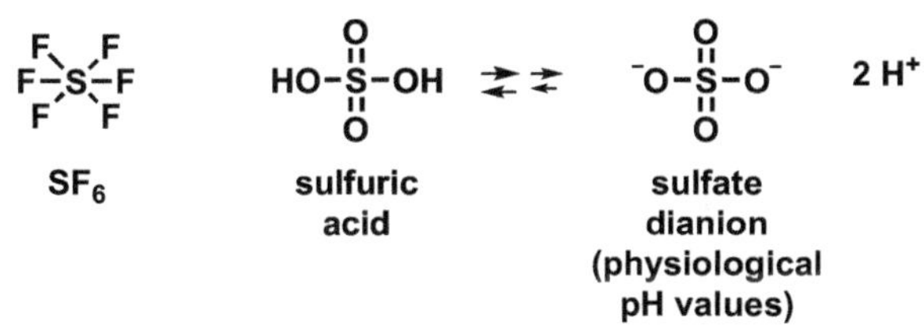

Figure 1.2 The ability of sulfur to rehybridize atomic orbitals to sp^3d^2 molecular orbitals allows sulfur to form hexavalent compounds. For example, SF_6 has six single S–F bonds. A hexavalent sulfur compound of biological importance is the SO_4^{2-} dianion.

molecular orbitals involving 3s, 3p and 3d atomic orbitals (https://mysite.du.edu/~jcalvert/phys/sulphur.htm, 2003). For example, the abiotic compound SF_6 (Figure 1.2) uses a set of equivalent sp^3d^2 molecular orbitals to generate the bonding pattern for hexavalent sulfur. The dramatic polyvalency of sulfur makes it much more versatile than oxygen in both its inorganic and organic chemistry in the molecules of life in distinct metabolic niches.

Sulfur makes up approximately 3% of the earth's mass, most of it in the core (an estimated 0.07% of the earth's crust),[8] but both elemental sulfur (S^0, the yellow solid that can take on several allotropic forms) and gaseous H_2S are emitted by volcanoes and undersea smoker vents (Figure 1.3). Solid sulfur as yellow chunks is still harvested from the immediate environs of Indonesian volcanoes. (sulfur liquifies at 159 °C). In the bible, sulfur and its attendant hydrogen sulfide with its characteristic odor (detectable at 0.02 ppm,[8,9] often characterized as the smell of "rotten eggs") was labeled "brimstone" as in the phrase "fire and brimstone" (brimstone is literally "burning stone": solid yellow sulfur burns in air to release malodorous sulfur dioxide – the odor of brimstone). Purportedly, the large deposits of sulfur in the Italian region of Campania that includes Mount Vesuvius, gave rise to the Latin name *sulphur,* now preferably sulfur, in many regions of the world.

During the two billion or so years when the planetary atmosphere was largely oxygen-free, sulfur in those low oxidation states was abundant. The sulfide at −2 oxidation state was subject to complexation with ferrous iron (also stable under anoxic conditions). A particularly abundant iron–sulfur compound is pyrite, FeS_2 ("fools gold").[10] The bonding in pyrite is approximated by $Fe^{2+}S_2^{2-}$, such that pyrite is more an iron persulfide than a ferrous sulfide. Other ionic forms of low-valent sulfur of historical import in ores are cinnabar (HgS), the source of mercury and of historical use in gold mining, lead sulfide (PbS) known as galena, as well as zinc sulfide (ZnS).

A B

Figure 1.3 (A) Elemental sulfur is a yellow solid, and has been mined continuously for the past two millennia. (B) Underwater black smoker vents release both H_2S and elemental sulfur.

As oxygen levels rose (from the global metabolic production of O_2 by photosynthesizing cyanobacteria), the weathering of rocks and flow of dissolved sulfur compounds, first into rivers, and then into global oceans was accompanied by biological and abiological oxidation of low-valent sulfur molecules. Among the various oxygenated forms of sulfur, inorganic sulfates (Figure 1.3) are the most oxidized and predominant forms. Some 10^{22} g of sulfur, largely as sulfate are estimated to be in the earth's crust.[9] These days extensive mining of sulfur results in approximately 10^{12} grams (approximately 4–5 million tons!) being mined each year, largely for use as sulfuric acid.[3] The sulfuric acid is used for many purposes, the largest of which are generation of sulfate and phosphate fertilizers, gunpowder, "vulcanization" of rubber tires and pesticides.[11]

The average concentration of sulfate in global oceans is about 28 mM, corresponding to a total of 1.3×10^{21} grams of dissolved dianion[3] ["With an estimated flux of 1.0×10^{14} g per year into (and out of) the ocean, the oceanic sulfate reservoir has a long response/turnover time" (approximately 13 Myr)]. The dissolved divalent calcium cation (Ca^{++}) is contemporaneously around 10 mM. During the multiple geological periods of warming and evaporation of seas, calcium sulfate as the dihydrate or pentahydrate ($CaSO_4 \cdot 2H_2O$ or $CaSO_4 \cdot 5H_2O$) precipitated out first (before the more abundant NaCl), as the material known as gypsum, accounting for huge inventories of gypsum beds and salt domes in sedimentary rock.

1.2 Examples of the Multiple Oxidation States of Sulfur in Inorganic and Organic Compounds Relevant to Biology

Examinations of abiotic sulfur chemistry have turned up a plethora of useful compounds and revealed multiple stable oxidation states of sulfur.[12] (Figure 1.1B) These include sulfur at its most reduced, −2 oxidation state (H_2S), the −1 oxidation state (Na_2S_2), elemental sulfur (S = zero) in various allotropes with S_0 and S_8 as abundant forms,[1] the +1 oxidation state (S_2Cl_2), +4 (SO_2, HSO_3^-) and +6 (SF_6, H_2SO_4) (Figure 1.2). The many oxidation states of sulfur indicate multiple ways to form energetically favorable bonds to ligands, consistent with the utilization of 3s and 3p electrons and the rehybridization of up to two of the empty 3d orbitals for bonding.

As the oxygen levels rose during biologically mediated planetary oxygenation over eons, multiple oxygen-containing sulfur compounds were formed and still occur contemporaneously (Figure 1.1B). They include thiosulfate ($S_2O_3^{2-}$) with an average sulfur oxidation state of +2, dithionite, often as the sodium salt ($Na_2S_2O_4$), with an average S oxidation state of +3, sulfur dioxide (SO_2 used in wine storage as an antimicrobial agent) that has sulfur in the [+4] oxidation state. Hydration of SO_2 yields the sulfite dianion (SO_3^{2-}), also common as a wine preservative, that is still at the [+4] oxidation state. Finally, both sulfur trioxide and its hydrated

SO_4^{2-} (sulfate) show sulfur at the [+6] oxidation state. More complex oxyanions of sulfur include trithionate ($S_3O_6^-$), with three sulfurs linked in a row[12] (Figure 1.4). Some of these oxygenated sulfur molecules are biological metabolites or alternate substrates in microbial biogeochemical cycles as noted in Chapters 2 and 3.

Metabolic reduction of readily available sulfate by sulfate-reducing microbes in anaerobic microenvironments gives rise to H_2S. The pH for 50%

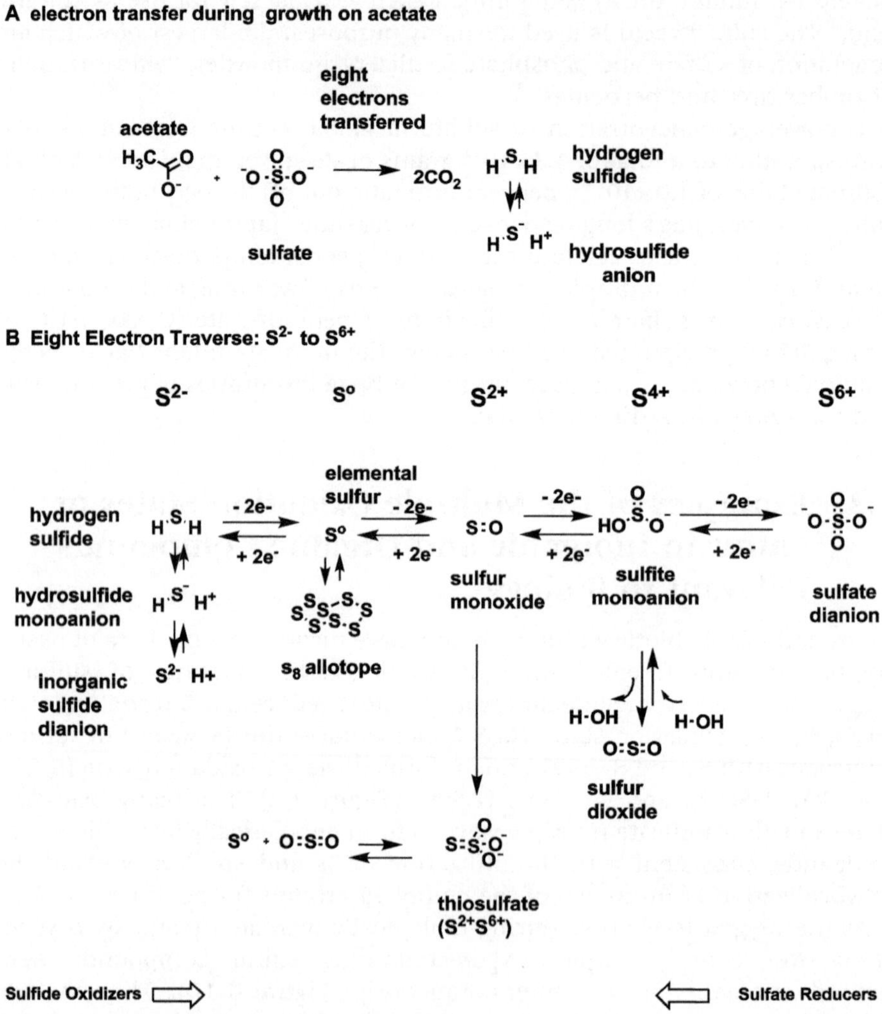

Figure 1.4 Sulfur can undergo an eight-electron traverse from sulfate where the sulfur atom is in the [+6] oxidation state to H_2S, where sulfur is in the [−2] oxidation state. (A) Chemotrophic sulfate-reducing bacteria can gain energy from oxidizing organic substrates such as acetate anion to two molecules of CO_2. (B) The eight electrons removed flow into sulfate, reducing it by the corresponding eight electrons to H_2S.

dissociation to the hydrosulfide anion (HS^-) is pH 6.3 (=the pKa), so hydrosulfide anion as a nucleophile is present wherever there is dissolved hydrogen sulfide at neutral pH. In Part II we will delve into the biological route to organification of sulfide *via* the nucleophilicity of HS^-, but only note here that the amino acid cysteine with its thiol (RSH) side chain is the metabolic entry point for inorganic sulfide to organosulfur metabolites.

Also shown in Figure 1.1B in the organosulfur metabolite column are a set of sulfur derivatives and oxidation states that are found in differing amounts in organic metabolites.[12] In addition to the reduced thiol form, the oxidatively dimerized disulfide grouping is common in proteins. The second major sulfur-containing amino acid is methionine. As shown in Figure 1.9B, it contains sulfur in a disubstituted thioether state. Most notably the downstream metabolite *S*-adenosylmethionine (SAM) has a trivalent sulfonium cation functional group. That sulfonium atom enables both two-electron and one-electron reaction manifolds that make SAM a remarkably versatile substrate and coenzyme across a broad swath of chemical biology transformations.[13,14] We will delve into the full array of organosulfur metabolites in Part II and Part III.

1.3 Chemical Biology of Inorganic Sulfur Compounds

The redox versatility of sulfur (−2 to +6, an eight-electron traverse) and the multiple forms of inorganic oxygenated sulfur compounds (Figure 1.1B) are the basis and signposts of biogeochemical cycling[3] of sulfur between sulfate [+6] and sulfide [−2]. This cycle involves an enormous tonnage of sulfur cycling every year on a planetary scale. Estimates of 120 trillion grams per year of total sulfur on the move correspond to about 120 million tons each year. Sulfate-reducing bacteria and sulfide-oxidizing bacteria are the major biological actors, as described in the next two chapters.

Dissimilatory sulfate reduction is the term typically applied to metabolic conversion of sulfate ions to sulfide ions when sulfate acts as the terminal electron acceptor for the enzymatic oxidation of reduced carbon substrates.[15] Thus, as sulfate is reduced microbiologically to sulfide, organic metabolites are mineralized to CO_2 molecules (Figure 1.4). Sulfate is serving in lieu of the role of O_2 in human mitochondrial electron transport chains.

This eight-electron redox traverse, that can be formalized as four sets of two electron inputs (Figure 1.4), provides a route to gain energy from oxidation of those substrates. Sulfate-reducing bacteria can be chemotrophic, removing electrons from such simple metabolites as the reduced organic acid anions, acetate, propionate, butyrate and succinate.[16] The sulfate-reducers can also be lithotrophic, oxidizing the inorganic substrate H_2 *via* hydrogenases acting as oxidative catalysts. Both the organic acids and the H_2 gas can be internally generated by the sulfate reducers themselves or can be available in mixed microbial producer colonies.[17] While one thinks of sulfate reducers in typical anaerobic terrestrial and marine microhabitats (marsh lands and water depths in the largely anoxic layers of the Black Sea) (<0.5 mg dissolved O_2 l^{-1} at depths below 50 meters, as prime exhibit; www.blackseascene.net), some 50% of healthy humans are colonized with *Desulfovibrio piger*.[18] These sulfate-reducing *vibrio* microbes in the anaerobic milieu of the distal colon are undoubtedly thriving in the neighborhood of primary fermenters, living on both the short chain fatty acids and H_2 as energy sources from their neighbors.

Four types of bacteria, including mesophiles and thermophiles (*e.g. Thermodesulfovibrios*) have the genes and encoded proteins to practice the above dissimilatory reduction as does one thermophilic archaeal species, *Archeoglobus.* The most well-studied mesophilic sulfate-reducing bacteria are in the δ-proteobacterial lineages, including the moderately aerotolerant *Desulfovibrios*. They can reduce O_2 as a protectant strategy.[19]

Because sulfate-reducing microbes carry out anaerobic metabolism and transfer electrons to sulfate rather than O_2, it was anticipated, and has been validated, that sulfate reducers may have been ancient and early in the tree of life. Sulfur isotope fractionation data can date fossil samples, reflecting the kinetic isotope effects of biological redox cycling, preferentially using the common ^{32}S isotope (95.02% abundance) over the minor ^{34}S isotope (4.21%). Over many biological redox cycles depletion of the heavy sulfur isotope in rock samples can be observed in gypsum ($CaSO_4{\cdot}5H_2O$) deposits in rocks in Australia that are estimated to be 3.4 billion years old.[9] This was a period in earth's evolution where O_2 levels in the atmosphere were still miniscule so there was little or no competition from aerobic organisms.

The thermodynamics of electron transfer to sulfate for net eight-electron reduction to sulfide are favorable,[20] albeit only 5–10% as favorable as dumping electrons into O_2. In the absence of O_2, sulfate reducers evolved to eke out a living. (Most contemporary sulfate-reducing bacteria are killed rapidly by exposure to air, probably *via* iron-initiated oxygen reductive radical chemistry). The reduction potential drop through which electrons fall from sulfate to sulfide is approximately −220 mV *vs.* −820 mV potential drop for the more favorable reduction of O_2 to H_2O. Insights into the chemical logic and protein machinery for how the sulfate reducers worked out an ancient strategy for coupling electron transfer to sulfate for gain of enough energy to sustain growth and reproduction can give signposts into the subsequent more efficient strategies of membrane electron transport.

The most abundant form of sulfur on the planet in its early evolution, before significant oxygenation of the atmosphere, was not posited to be the oxidized sulfate dianions. Most likely H_2S and elemental sulfur (S^0) were most abundant in the highly reducing atmosphere. Some microbes live anaerobically in the presence of elemental sulfur (S^0) deposits and carry out the reduction of S^0 to H_2S, just a two-electron traverse. We will examine the redox logic of those bacteria in Chapter 3.

Finally, there are bacteria that drive the sulfur redox equilibria in the opposite direction, from sulfide (S^{2-}) all the way to sulfate ($SO_4{}^{2-}$). These are lithotrophs, taking electrons out of inorganic rather than organic substrates. Those electrons removed in sulfur species oxidations are ultimately passed to O_2 as the electron acceptor. Thus, the sulfide oxidizers are aerobic microbes, in contrast to the anaerobic sulfate reducers (Figure 1.5). It is likely that they evolved as O_2 levels in the atmosphere (and dissolved in aqueous solutions) rose, long after the appearance of the sulfate-reducing microbes.

Inorganic sulfate, as the dianion (SO_4^{2-}), became the prevalent form of sulfur. Weathering of rocks and river inflows to seas and oceans led over time to the contemporary concentration of sulfate at 28 mM.[21] This provides an abundant substrate supply to sulfate reducers provided they can congregate in anoxic niches and microenvironments. We shall also note in Part II that the high levels of dissolved sulfate anions in sea water enable marine organisms, such as many algal species, to use sulfate in its activated AMP–sulfate form as the source of many sulfate groups in polysulfated carbohydrates that are signature molecules in marine algal cell walls.

These sulfate levels support both the sulfate-reducing bacteria and archaea that still occupy anaerobic niches, most notably in anoxic regions of water columns. Sulfate reducers provide the hydrogen sulfide that can sustain sulfide oxidizers provided they two microbial forms can coexist on opposite sides of oxic–anoxic interfaces. As H_2S seeps up to the anoxic–oxic interface in lakes and seas, sulfide-oxidizing microbes use it as substrate. They oxidize it either by two electrons to elemental sulfur or by eight

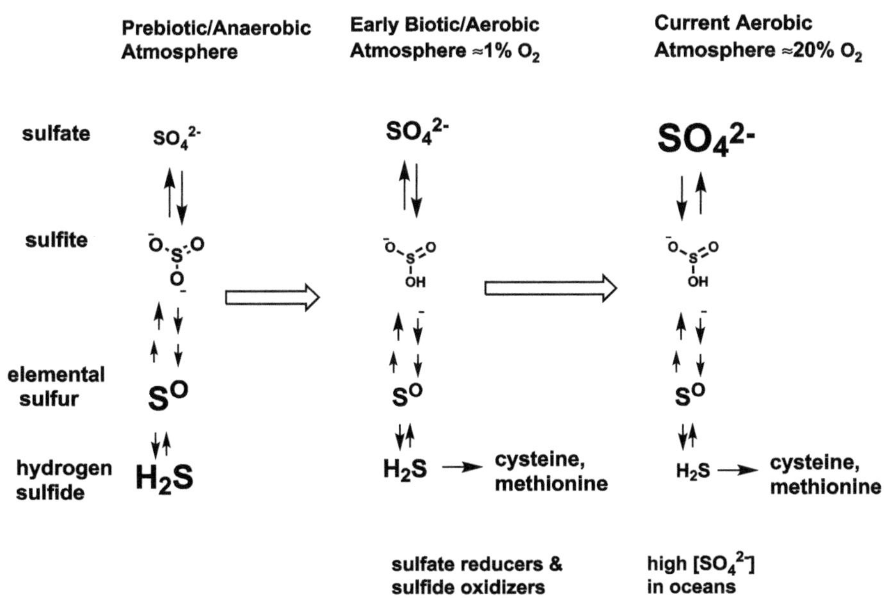

Figure 1.5 Schematic of shift in relative amounts of sulfate, sulfite, elemental sulfur, and hydrogen sulfide before and after global oxygenation of the planet.

electrons back to sulfate, dumping those electrons into O_2 as the terminal metabolic electron acceptor. Thus, both sulfate reducers and sulfide oxidizing microbes exist in spatially separated anoxic–oxic microenvirons and run the inorganic sulfur redox pathways in either direction (Figure 1.6; also see Figure 2.2).

Assimilatory sulfate reduction[15] is carried out by single-cell eukaryotes, such as fungi, and also by multicellular plants. They do not use sulfate ions as terminal electron acceptors in energy metabolism. Instead, they have evolved to use O_2 with a higher energy yield from organic substrate oxidations. They reduce enough sulfate to sulfide to provide sufficient HS^- to function in substrate quantities for the enzyme *O*-acetyl-serine sulfhydrylase, also known as cysteine synthase. We will examine the logic and mechanism of this key enzyme that converts inorganic sulfide to the organic thiol of the amino acid cysteine in Part II. Meanwhile in Part I, where we focus on the inorganic biochemistry of sulfate ions, sulfide ions, elemental sulfur and intermediates in between, we will observe the key role of cysteine residue thiolate side chains in many of the pathway enzymes. From cysteine as the organosulfur entry point, a host of other sulfur-containing organic scaffolds, including methionine, coenzyme A and glutathione are constructed.

On average, about 140 grams of organic sulfur are present in a 70 kg human. The recommended daily intake requirement for a 70 kg human adult is about 900 mg of sulfur. As a calibration a peanut contains about 35 mg of sulfur.

Globally, over some 7 billion humans on the planet, the total sulfur abundance in people corresponds to approximately 10^{11}–10^{12} grams of

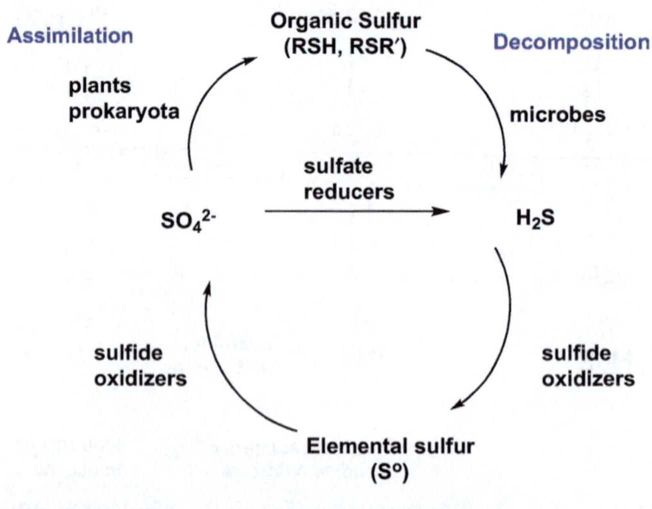

Figure 1.6 Simplified sulfur metabolic cycle: metabolic assimilation, dissimilation and interconversion of organic forms of sulfur [such as thiols (RSH)] and inorganic forms represented by S^0 elemental sulfur.

organic sulfur (approximately 10^5–10^6 tons!). The connection between inorganic sulfide and organic sulfur in the form of cysteine (and subsequent organic metabolites) also happens in reverse. Some of the sulfur atoms incorporated into the amino acid cysteine get transformed to persulfides (R–S–SH) (Figure 1.1B) and then are transiently released as inorganic sulfide at the specific enzyme sites of two-iron–two-sulfur and four-iron–four-sulfur cofactor constructions (Figure 1.7).

1.4 Organic Sulfur Metabolic Scaffolds

The common functional groups in sulfur chemical biology are organic thiols, thioethers, disulfides, persulfides and trivalent sulfonium atoms (Figure 1.1B). We will see that the role of thiol sulfur as a biological nucleophile and of the trivalent sulfonium atom in *S*-adenosylmethionine as a methyl donor are so pervasive that it is difficult to imagine life without these molecules and their chemistry.[13] Additionally, thiol-containing metabolites, such as cysteine, get oxygenated once to the sulfenate and twice to the sulfinate on the way to the carbon–sulfur bond in the sulfonate (SO_3^-) of taurine. In addition to enzymatic transfer of sulfur at the sulfide [$S = -2$] level, enzymatic sulfuryl transfers [$-SO_3^-$: $S = +6$] occur from the mixed sulfuric–phosphoric anhydride side chain of 3′-phosphoadenosine-5′-phosphosulfate (PAPS) to yield *O*-sulfate and *N*-sulfate linkages in oligosaccharide frameworks of heparan and the sulfated disaccharide repeat units of algal cell walls[22] (Figure 1.8).

In the universe of organosulfur metabolite interconversions, especially at the oxidation state of thiol groups, *e.g.* in glutathione, free cysteine and cysteinyl side chains in proteins, both one-electron and two-electron reaction pathways can be operant[23] (Figure 1.9). While we will focus on the nucleophiliciity of thiolate anions and the electophilicity of the SAM trivalent sulfur cation in two-electron reaction manifolds, those are not exclusive modes of reactivity, as will be discussed in Chapter 9.

Indeed, thiol radical chemistry (R–SH to RS•) is widespread in many metabolic contexts. Famously, ribonucleotide reductase from different sources, uses one-electron pathways involving up to three cysteinyl

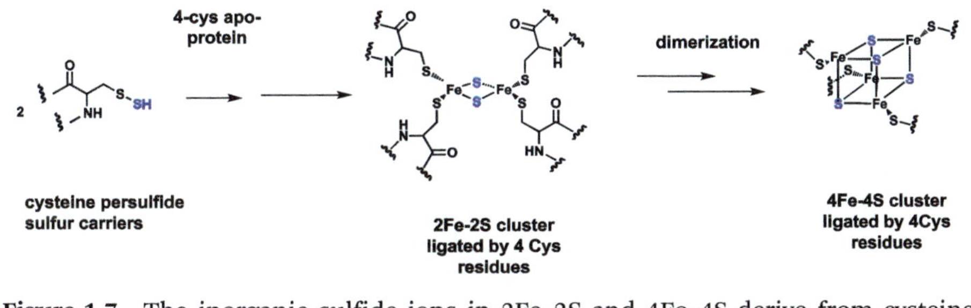

Figure 1.7 The inorganic sulfide ions in 2Fe–2S and 4Fe–4S derive from cysteine persulfide side chains in proteins that act as –SH chaperones.

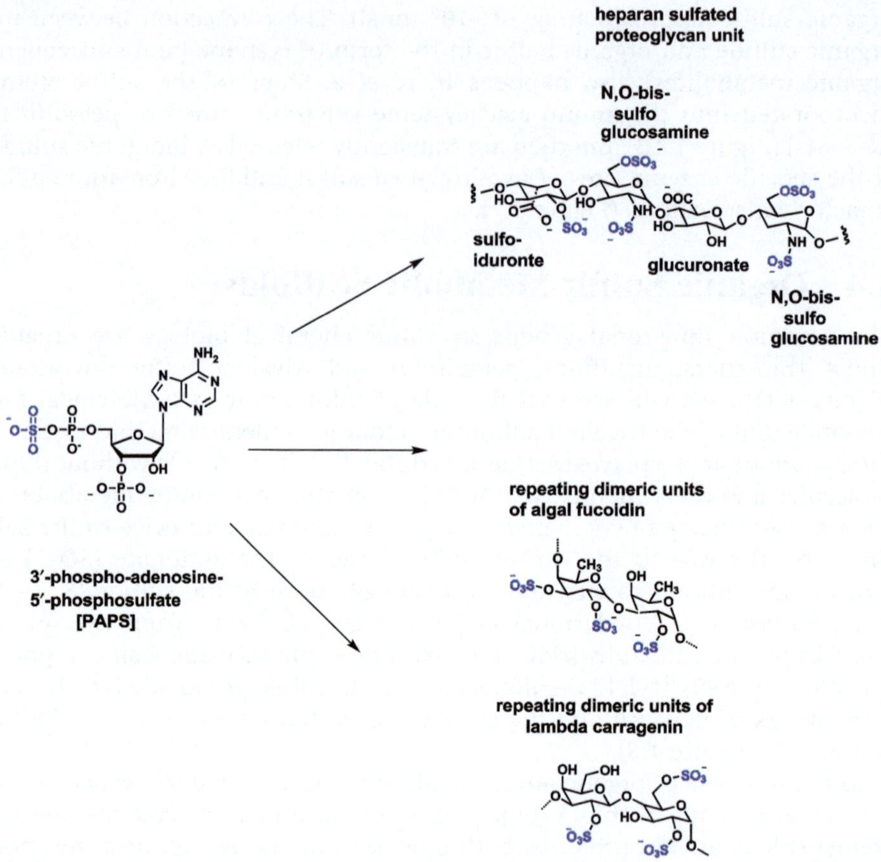

Figure 1.8 PAPS (3′-phosphoadenosine-5′-phosphosulfate) is the universal sulfuryl ($-SO_3^-$) donor; here, transferring the electrophilic sulfuryl (sulfo) group to hydroxyl group oxygen nucleophiles to create anionic sulfated polysaccharides.

radicals in its active site in each catalytic cycle.[24] The tripeptide glutathione (γ-glutamyl-cysteinyl-glycine) can act as a surveillance and detoxification metabolite both against organic electrophiles by two-electron pathways and reactive oxygen and nitrogen species by one- or two-electron modalities.

Part III of this volume turns to the role of cysteine and methionine residues in proteomes. With the exception of reversible *S*-oxygenation of methionyl side chains in actin, the sulfur proteomics universe is dominated by cysteine residues. Thiolates can function as catalytic nucleophiles in a range of enzyme families, generating acyl-*S* enzymes (proteases, ubiquitin cascades, glutaminases), *S*-phosphoryl cysteinyl enzymes in protein tyrosine phosphatase catalytic cycles and *S*-alkylated pyridines during thymidylate synthase and DNA methyl transferase catalyses.

Cysteinyl residues in proteomes are also favorite targets for covalent posttranslational modifications. These include two types of lipidations:

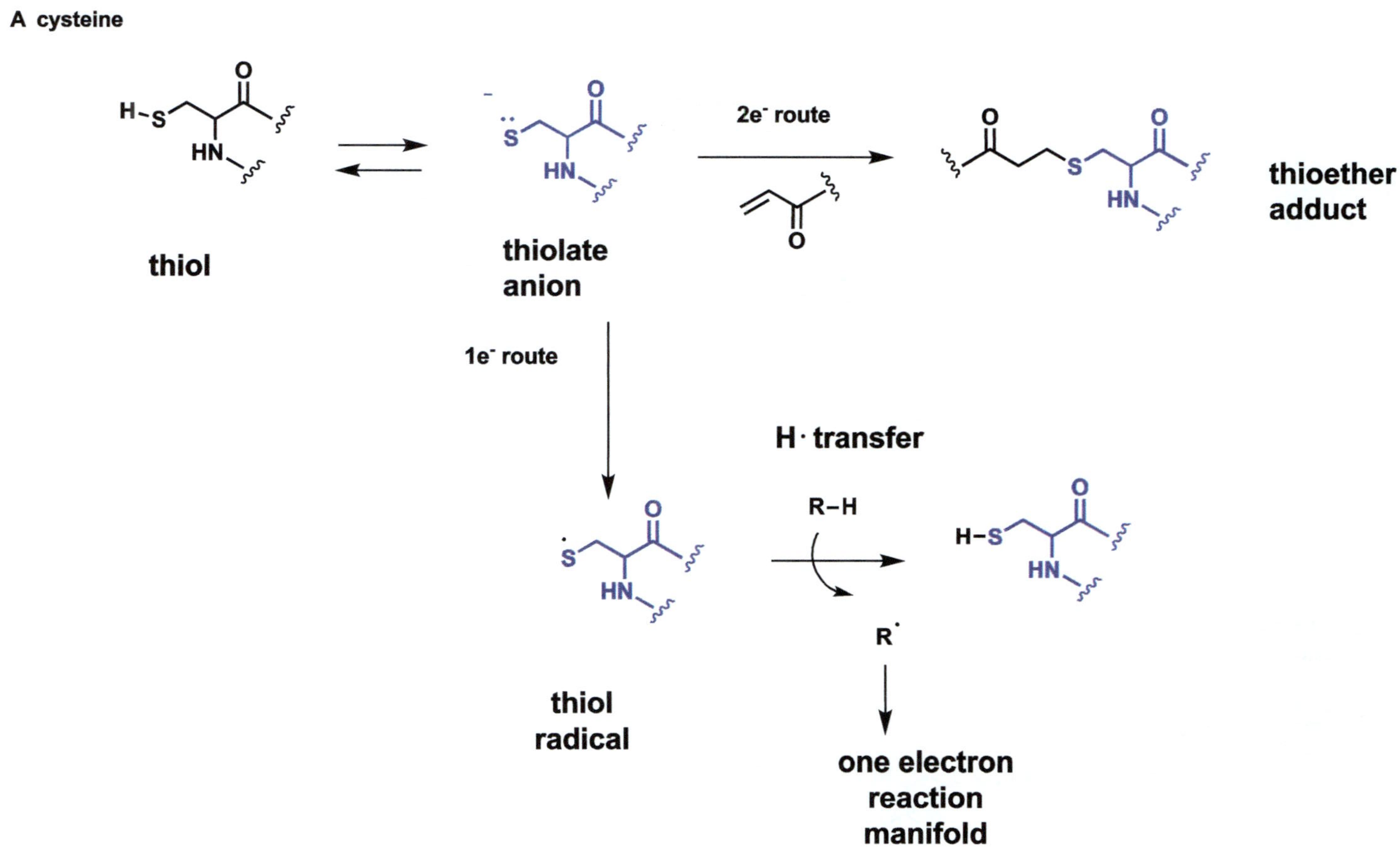

Figure 1.9 (A) Cysteine and (B) Methionine: characteristic reactions by one-electron and two-electron reaction pathways. The enzymatic adenosylation of methionine sulfur to *S*-adenosylmethionine creates the most prolific methyl transfer substrate (two-electron route) as well as the initiator of a host of radical reactions (one-electron-radical SAM enzyme family).

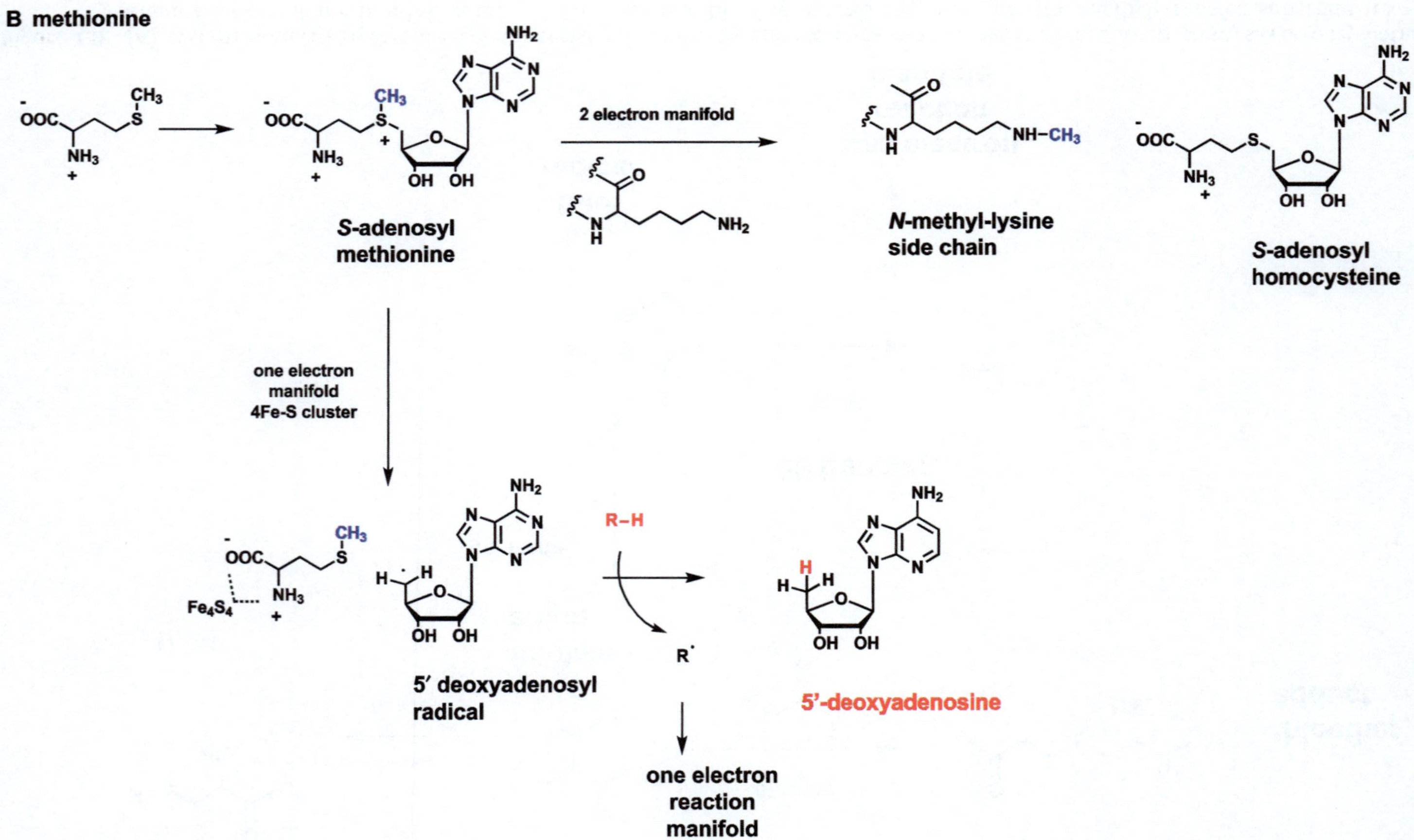

Figure 1.9 Continued.

palmitoylations and prenylations. They also encompass cystine disulfide bond formations, the common fate of cysteinyl residues in proteins passing through cellular secretory pathways. The hundreds of 2Fe–2S and 4Fe–4S clusters found as ancient redox elements in proteins are built from sulfur atoms carried as persulfide groups (protein–Cys–S–SH) on chaperone proteins. Less widely distributed, but remarkable, reactions of cysteine residues in proteins include lanthionine residue formation, peptide backbone autocleavage reactions, protein splicing and heterocyclic thiazole ring formations.

1.5 Oxygen *vs.* Sulfur Chemical Biology

We close out this introductory chapter with one more set of comparisons between the oxygen functional groups in primary metabolic pathways and their sulfur homologs. The parallel between divalent oxygen and divalent sulfur functional groups in biological metabolism is extensive. Yet, the distinct redox states available to sulfur and the profound changes in nucleophilicity–electrophilicity between the $S = -2$ and $S = +6$ oxidation states dramatically alter the chemical biology of sulfur.

One way to compare the roles of oxygen and sulfur, its next higher homolog in the periodic table, for their anticipated chemical biology roles is depicted in Table 1.1. Table 1.1A lists and compares oxygen functional groups that occur in molecules of primary metabolism. We note that many of them—alcohols, aldehydes and ketones, acids, amides, esters (think acetylcholine the neuromuscular transmitter molecule)—are key functional groups in the thousands of primary metabolites in prokaryotic and eukaryotic cells. In addition to oxygenated organic (carbon) compounds, the inorganic class of oxygenated molecules with central chemical biology are, of course, the phosphate monoesters and diesters on one hand (think ATP) and, on the other hand, inorganic O_2 and H_2O, and the three one-electron-reduced intermediates between them.

In reductive oxygen metabolism, the four-electron traverse from O_2 to two molecules of water, practiced as the key energy-conserving pathway in all aerobic cells, there are transiently formed superoxide anions (one-electron reduced), hydrogen peroxide (two-electron reduced) and hydroxyl radicals (three-electron reduced) between O_2 and H_2O. The peroxide can be inorganic or can arise from lipid autoxidations as alkyl peroxides (ROOH). Essentially all stable oxygenated molecules in biological processes are restricted to oxygen in a divalent state, albeit oxygen atoms can undergo the redox transfers noted (alcohol to aldehyde to acid that are essential in hydrogen or electron removal in energy metabolism).

Consistent with the placement of sulfur directly above oxygen in the periodic table there are a host of compounds with divalent sulfur that have a rich and essential set of chemical biology roles. Part A of Table 1.1 shows a one to one correspondence of divalent oxygen *vs.* divalent sulfur functional groups: water *vs.* hydrogen sulfide, alcohol to thiol, ether to thioether,

Table 1.1 Comparison of Oxygen and Sulfur Functional Groups in Biology.

(A)

Divalent Oxygen Functional Group		Divalent Sulfur Homolog	
HOH	Water	HSH	Hydrogen sulfide
ROH	Alcohol	RSH	Thiol
ROR′	Ether	RSR′	Thioether
R-OOH	Peroxide	RSSH	Persulfide
R-OOR′	Dialkyl peroxide	RSSR′	Disulfide
RCHO	Aldehyde	RCH=S	Thioaldehyde
RCOR′	Ketone	RCR′=S	Thioketone
RCOOH	Carboxylic acid	RCOSH	Thiocarboxylic acid
RCOOR′	Oxoester	RCOSR′	Thioester
RCONHR″	Amide	RCSNHR″	Thioamide
$ROPO_3^{2-}$	Phosphate ester	$RSPO_3^{2-}$	Thiophosphate
		RSOH	Sulfenic acid

(B)

Stable Tetravalent and Hexavalent Compounds Relevant to Biology

Oxygen: None	Sulfur	
	RSO_2H	Sulfinic acid
	RSO_3H	Sulfonic acid
	$HS\text{–}SO_3^-$	Thiosulfate
	SO_4^{2-}	Sulfate
	SO_3^{2-}	Sulfite
	RSOR′	Sulfoxide
	RSO_2R'	Sulfone
	RSSOR′	Disulfide monoxide
	RSSOCR′	Acyl disulfide

aldehyde to thioaldehyde, carboxylic acid to thiocarboxylic acid, ester to thioester, phosphate ester to thiophosphate ester. There are also analogs between alkyl peroxides and dialkyl peroxides with persulfides and disulfides. The one to one correspondence of divalent oxygen and sulfur functional groups is a correlative property predictable from the periodic table. Parts II and III take up the roles of these specific sulfur functional group in inorganic and organic metabolism.

However, the distinct size and electronic configuration of sulfur with its d electrons *vs.* oxygen (s and p electrons only) impart distinct chemical properties to the oxygen *vs.* sulfur homologs, which in turn affect reactivity, roles and abundance of the oxygen *vs.* sulfur functional groups in biology. Much of this volume explores the role of the divalent sulfur compounds and their specific properties that allow them to function in distinct niches from the corresponding oxygen compounds.

Water *vs.* hydrogen sulfide; universal liquid solvent *vs.* toxic gas is one starting point. The dramatic increase in basicity and consequent nucleophilicity of thiols *vs.* alcohols is another. The thermodynamic activation of acyl thioesters is much greater than that of corresponding acyl oxoesters, accounting for the central role of acyl thioesters as cellular energy currencies.

Cells use hydrogen peroxide sparingly; alkyl peroxides are toxic side products of reductive oxygen chemistry. On the other hand, persulfide groups in proteins are the progenitors of ubiquitous iron–sulfur redox clusters, and dialkyl disulfides, *e.g.* as cystine residues form the major crosslinking, stabilizing covalent bonds in most secreted and outer membrane proteins. In many, but not all, cases the oxidative traverse from a pair of protein thiols to the disulfide goes by way of the sulfenic aid. Furthermore, the water-soluble tripeptide thiol glutathione acts as cellular scavenger of organic and inorganic electrophiles, the latter role largely *via* the sulfenic acid intermediate.

While oxygen functional groups in organic metabolites are central to all metabolic pathways and much to be praised in their chemical biology, oxygen is restricted to divalent bonding but sulfur is not. Some of those higher-valent sulfur metabolites are listed in Table 1.1B.

The availability of dp hybrid molecular orbitals allows sulfur to expand its valence from 2 to 4 to 6, with oxidation states that vary from −2 in hydrogen sulfide (analogous to the −2 oxidation state of oxygen in water and other physiologically stable compounds) to +6 in the dianion of sulfuric acid, inorganic sulfate. Both the inorganic redox chemistry of sulfur, the eight-electron reversible traverse between sulfate dianion and sulfide anions, *and* the organic chemistry of sulfenates, sulfinates and sulfonates are all at play in sulfur chemical biology. At the [+6] oxidation state of sulfate, the central sulfur atom is electrophilic. In the [−2] oxidation state of hydrogen sulfide or organic thiols (*e.g.* cysteine and cystamine), the sulfur anion is powerfully nucleophilic. These differences in reactivities enable a wide range of discrete chemistries that are central to life. These are our starting precepts for exploring sulfur's chemical biology.

References

1. B. Meyer, Elemental Sulfur, *Chem. Rev.*, 1976, **76**, 367–390.
2. H. Kimura, N. Shibuya and Y. Kimura, Hydrogen sulfide is a signaling molecule and a cytoprotectant, *Antioxid. Redox Signaling*, 2012, **17**, 45–57.
3. D. A. Fike, A. Bradley and C. V. Rose, Rethinking the ancient sulfur cycle, *Annu. Rev. Earth Planet. Sci.*, 2015, **43**, 593–622.
4. F. H. Westheimer, Why nature chose phosphates, *Science*, 1987, **235**, 1173–1178.
5. J. F. Honek, Biological chemistry of the carbon-sulfur bond, *Can. J. Chem.*, 2015, **93**, 1051–1060.
6. L. B. Poole, The basics of thiols and cysteines in redox biology and chemsitry, *Free Radic. Biol. Med.*, 2015, 148–157.
7. E. G. Mueller, Trafficking in persulfides: delivering sulfur in biosynthetic pathways, *Nat. Chem. Biol.*, 2006, **2**, 185–194.
8. P. Brimblecombe, The Global Sulfur Cycle, *Encyclopedia of Atmospheric Sciences*, 2003, 646–679.

9. P. Brimblecombe, The global sulfur cycle, in *Treatis on Geochemistry*, Elsevier, Amsterdam, **vol. 10**, 2014.
10. C. S. Hurlbut and C. Klein, Manual of Minerology, in *Manual of Minerology*, John Wiley & Sons, NY, 1985, pp. 285–286.
11. P. J. Chenier, *Survey of Industrial Chemistry*, Springer, Heidelberg, 2002.
12. J. I. Toohey and A. Copper, Thiosulfoxide (sulfane) sulfur: new chemistry and new regulatory roles in biology, *Molecules*, 2014, **19**, 12789–12813.
13. M. Fontcave, M. Attila and E. Mulliez, S-adenosyl methionine: nothing goes to waste, *Trends Biochem. Sci.*, 2004, **29**, 243–249.
14. W. Loenen, S-adenosylmethionine: jack of all trades and master of everything?, *Biochem. Soc. Trans.*, 2006, **34**, 330–333.
15. H. D. Peck Jr., Enzymatic basis for dissimilatory and assimilatory sulfate reduction, *J. Bacteriol.*, 1961, **82**, 933–939.
16. G. Muyzer and A. J. Stams, The ecology and biotechnology of sulfate-reducing bacteria, *Nat. Rev. Microbiol.*, 2008, **6**, 441–454.
17. B. M. Brahmacharimayum, M. P. Mohanty and P. Ghosh, Theoretical and practical aspects of biological sulfate reduction: a review, *Global NEST J.*, 2019, **2**, 222–244.
18. J. Loubinoux, J.-P. Bronowicki, I. A. C. Pereira, J.-L. Mougenel and A. E. Le Faou, *FEMS Microbiol. Ecol.*, 2002, **40**, 107–112.
19. H. Cypionka, Oxygen respiration by desulfovibrio species, *Annu. Rev. Microbiol.*, 2000, **54**, 827–848.
20. K. L. Keller, B. Raoo-Giles, E. S. Semkiw, I. Porat, S. D. Brown and J. D. Wall, New model for electron flow for sulfate reduction in Desulovibrio alaskensis G20, *Appl. Enivon. Microbiol.*, 2014, **80**, 855–868.
21. D. E. Canfield and J. Farquahar, Animal evolution, bioturbation, and the sulfate concentration of the oceans, *Proc. Natl. Acad. Sci. U. S. A.*, 2009, **106**, 8123–82127.
22. M. Negishi, L. Pederses, E. Petrochenko, S. Shevtsov, A. Gorokhov, Y. Kakuta and L. G. Pedersen, Structure and Function of Sulfotransferases, *Arch. Biochem. Biophys.*, 2001, **390**, 149–157.
23. P. A. Frey, A. Hegeman and G. H. Reed, Free radical mechanisms in enzymology, *Chem. Rev.*, 2006, **106**, 3302–3316.
24. S. Y. Reece and M. Seyedsayamdost, Long-range proton-coupled electron trasfer in the E.coli class Ia ribonucleotide reductase, *Essays Biochem.*, 2017, **61**, 281–292.

CHAPTER 2

Sulfate Reducing Bacteria

2.1 The Eight-electron Traverse Between Inorganic Sulfate and Inorganic Sulfide

The earliest incarnations of the versatility of sulfur chemical biology are found in bacteria (and a few archaeal species) that take advantage of the multiple accessible oxidation states of sulfur in aqueous solutions.[1] At the most reduced end of the sulfur spectrum we have noted the gas hydrogen sulfide, with sulfur at its most nucleophilic, in a −2 oxidation state (Figure 2.1). Two electrons oxidized from sulfide (S^{2-}) is elemental sulfur (S^0),[2] a rare case of a (naked) chemical element participating in biology. At the other end of the sulfur redox spectrum is the contemporaneously most abundant, fully oxidized sulfate dianion (SO_4^{2-}). Two electrons down from sulfate is the sulfite dianion (SO_3^{2-}), actually a ubiquitous metabolite across prokaryotes and eukaryotes, with commercial application as an antimicrobial preservative, along with its dehydrated congener sulfur dioxide gas (SO_2), in many red wines. Sulfite at physiological pH is in equilibrium with the monoanionic bisulfite anion ($HOSO_2^-$).

Between sulfite and elemental sulfur is a gap of two electrons. As shown in Figure 2.1 that intermediate, with S^{2+} formal oxidation state, would be sulfur monoxide, a known but not very stable species. In the enzyme sulfite reductase the sulfur monoxide equivalent is ligated at the top axial position of a siroheme iron atom[3] and not known to be a fee species. In the section below on the mechanism of sulfite reductase we will note that exogenous thiosulfate (Figures 1.1 and 2.1) will serve as a reducible substrate.[4] Of the two sulfur atoms in thiosulfate, one is at the +2 oxidation state, the other at the +6 sulfate oxidation state, itself an

Chemical Biology No. 12
The Chemical Biology of Sulfur
By Christopher T. Walsh

Published by the Royal Society of Chemistry, www.rsc.org

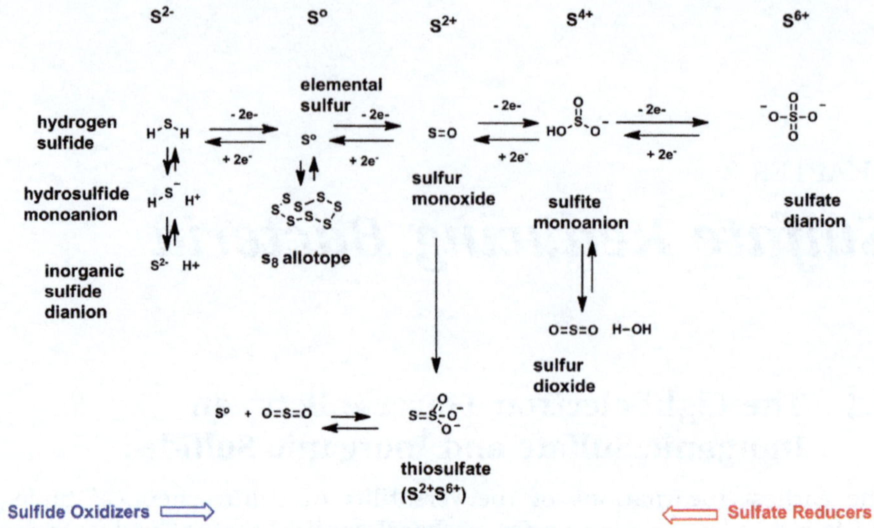

Figure 2.1 Eight electron traverse between hydrogen sulfide and sulfate. Molecular species between sulfide and sulfate separated by two electrons include elemental sulfur S^0, sulfur monoxide or equivalent and sulfite.

unusual situation, but a reminder of sulfur's redox versatility. One can imagine a formal loss of sulfite (SO_3^-) in the reductase active site to get to the S^{2+} atom as "on pathway" species. With the scheme for the eight-electron traverse of Figure 2.1, we observe that sulfate-reducing bacteria traverse from right to left, while sulfide-oxidizing bacteria (Chapter 3) traverse from left to right.

Sergei Winogradsky, a 19th century scientist sometimes termed the father of sulfur microbiology,[5] devised a simple vertical column for enrichment of both types of sulfur-utilizing bacteria (Figure 2.2).

In the top oxygenated layer of water one can find aerobic microbes, including bacteria that use O_2 as terminal electron acceptor for metabolic electrons, and also cyanobacteria with their armamentarium of pigments that protect their DNA from light. One layer down are the sufide or $sulfur^0$ oxidizers where there is still enough dissolved O_2 to serve as a terminal electron acceptor, while enough H_2S has diffused up from below to serve as oxidizable substrate.

Further down in O_2-depeleted, anoxic aqueous regions are anaerobes such as *Desulfovibrios* and *Clostridia.* The *Desulfovibrios* (Figure 2.2B) are Gram-negative bacteria that are classic sulfate reducers, producing the H_2S that can seep back up as substrate for the various forms of sulfite-oxidizers. This water column is reproduced in bodies of water such as the Black Sea where the deep layers are effectively anoxic. In addition to anoxic regions of lakes and seas, sulfate reducers are found in oil fields and natural gas reservoirs where they generate H_2S gas.

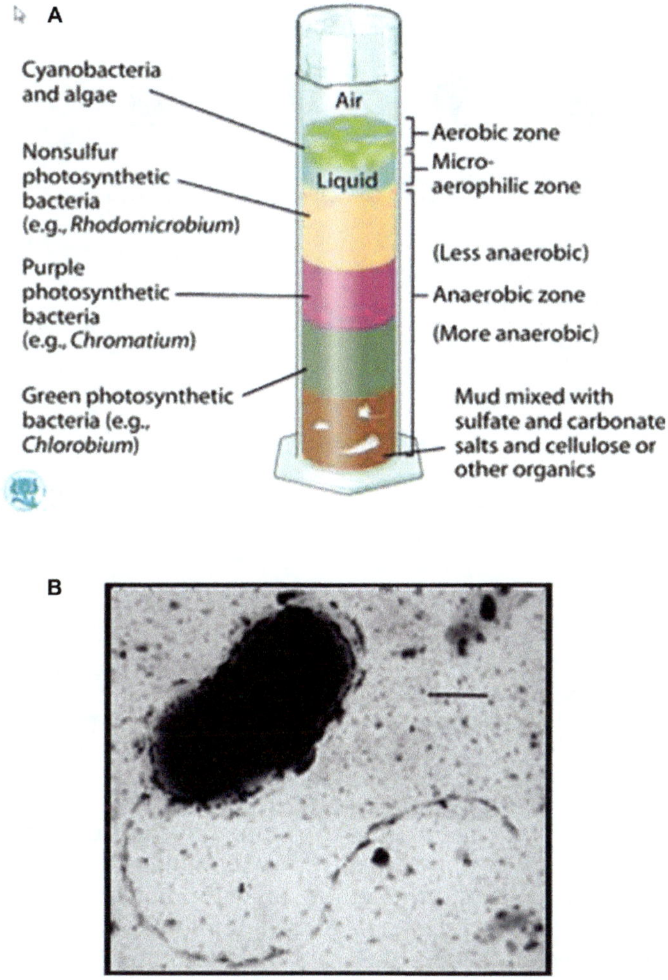

Figure 2.2 (A) Version of a Winogradsky column. Freshwater microbes will grow in stratified layers of a water column dependent on incident light and dissolved [O_2]. (B) *Desulfovibrio vulgaris.*

2.2 Sulfate Uptake by Transport Proteins

The first step in intracellular metabolic reduction of sulfate is its transit from the external environment into the microbial (or eukaryotic cell). Sulfate transporters, transmembrane proteins, are ubiquitously distributed in cells. They have apparently evolved several times to solve the problem of getting the doubly anionic sulfate anion across hydrophobic membranes[6] (Figure 2.3A). Some forms use ATP to provide energy for membrane carrier proteins. Others form channels and either pump cations (protons, sodium

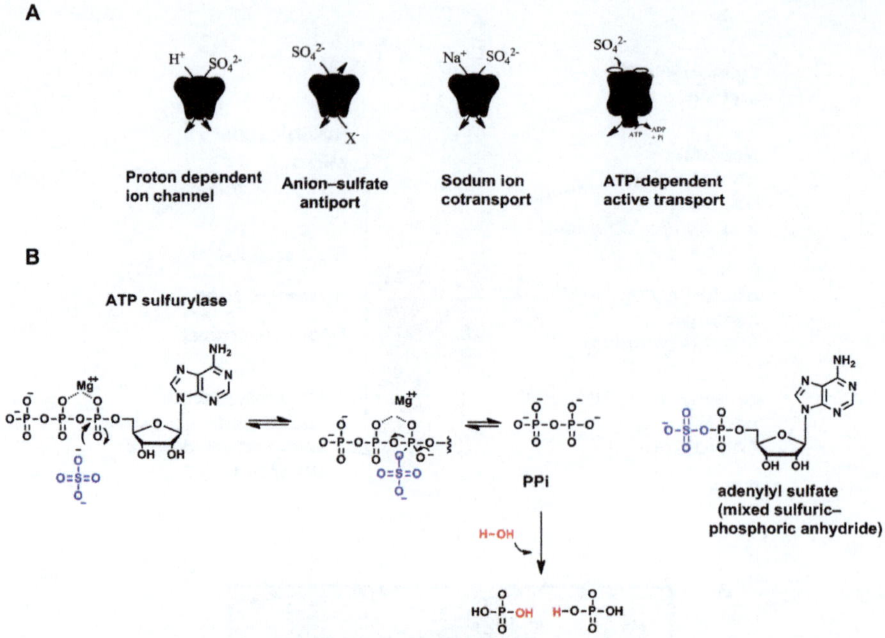

Figure 2.3 (A) Four major types of sulfate transport proteins, including carrier and ion channel strategies. (B) ATP sulfurylase generates the mixed sulfuric–phosphoric anhydride side chain in product adenosine 5′-phosphosulfate (APS) by attack of a sulfate oxygen anion on the electrophilic Pγ of ATP.

ions) in *sym*-port or sulfate *in trans* with anions such as bicarbonate or short chain carboxylates, in the opposite direction to manage the electrochemical gradients.[7] We start discussion assuming one or more transport routes have accumulated sufficient intracellular sulfate dianion concentrations to run the enzymatic conversions described later in this chapter.

2.3 Microbial Conversion of Sulfate to Sulfite: Two Enzymatic Steps and Two Electron Inputs

2.3.1 ATP Sulfurylase

The first two-electron step, reduction of sulfate dianions to sulfite dianion, is uphill energetically and thus unfavorable.[8–10] Microbes (and eukaryotes that came later) get around this energy barrier by converting inorganic sulfate to the mixed phosphoric sulfuric anhydride *via* action of the enzyme ATP sulfurylase. As shown in Figure 2.3B, in the sulfurylase active site, one of the oxygen anions of a bound sulfate molecule attacks the electrophilic α-phosphorus atom in the side chain of the Mg–ATP cosubstrate. The resultant pentacovalent phosphorane adduct can decompose in the forward direction with P–O bond cleavage, releasing inorganic pyrophosphate (PPi).

This action replaces the initial phosphoric anhydride bond between Pα and Pβ with the mixed anhydride between Pα and OSO_3^-. The mechanistically coupled cleavage of ATP drives the otherwise unfavorable equilibrium to accumulate adenylyl sulfate.

The key chemical feature of the product AMP-sulfate (adenylyl sulfate) is that it is a *mixed phosphoric–sulfuric anhydride*, thermodynamically activated but kinetically sufficiently stable in cells to serve as a diffusible substrate.[11] The kinetic stability arises in substantial part from the phalanx of two negative charges shielding the side chain from adventitious attack by a water molecule. Anhydride linkages are typically approximately 7 kcal $mole^{-1}$ uphill from the corresponding hydrolysis products (here AMP and inorganic sulfate), corresponding to a 10^5 push in the equilibrium constant for anhydride cleavage. Thus, the sulfurylase spends an ATP to create a form of sulfate activated for transfer and reduction.

2.3.2 AMP-sulfate Reductase

In the mixed sulfuric-phosphoric anhydride side chain of AMP–O–SO_3^- (APS) the S–O–P linkage is set up for S–O bond cleavage on attack by a nucleophile at the electrophilic sulfur atom. The transferring unit is the –SO_3^{2-} as an electrophilic fragment. That is the job of the next enzyme, AMP-sulfate reductase[12] (Figure 2.4).

2.4 Sulfate Reducers

The reductase enzyme from the sulfate reducing bacterium *Desulfovibrio vulgaris* is an αβ heterodimer. The larger α subunit contains one equivalent of flavin adenine dinucleotide (FAD) a tightly bound coenzyme form of vitamin B_2 riboflavin.[13] The smaller β-subunit has folds characteristic for iron–sulfur cluster proteins and has two of the 4Fe–4S clusters noted briefly in Figure 1.7 and further described in Chapter 9. The presence of Fe–S clusters typically signifies one-electron redox shuttling to or from a specific substrate. The FAD coenzyme can act either in a two-electron transfer mode or two separate one-electron transfer modes. Single crystals of the *D. vulgaris* enzyme have been grown, the AMP-sulfate bound and the structure analyzed by X-ray as shown in Figure 2.4B, right panel. Also, X-ray analysis of an N5–FAD–SO_3 intermediate is shown in the left panel of that figure. The structure of the *Archeoglobus fulgidus* reductase at 1.6 Ångstrom resolution has also been determined.[14]

The covalent N5–flavin–sulfite adduct is consistent with other evidence indicating that the adduct is on the pathway and arises by capture of the transferring –SO_3^- group of bound adenylyl sulfate by N-5 of the reduced form of the flavin ($FADH_2$). This is shown in Figure 2.4A, with formation of an N5–S nitrogen–sulfur bond. That flavin–sulfite adduct is prone to intramolecular decomposition by participation of the N1-anion of the flavin (Its pK_a is approximately 7, so that anion is kinetically available to participate

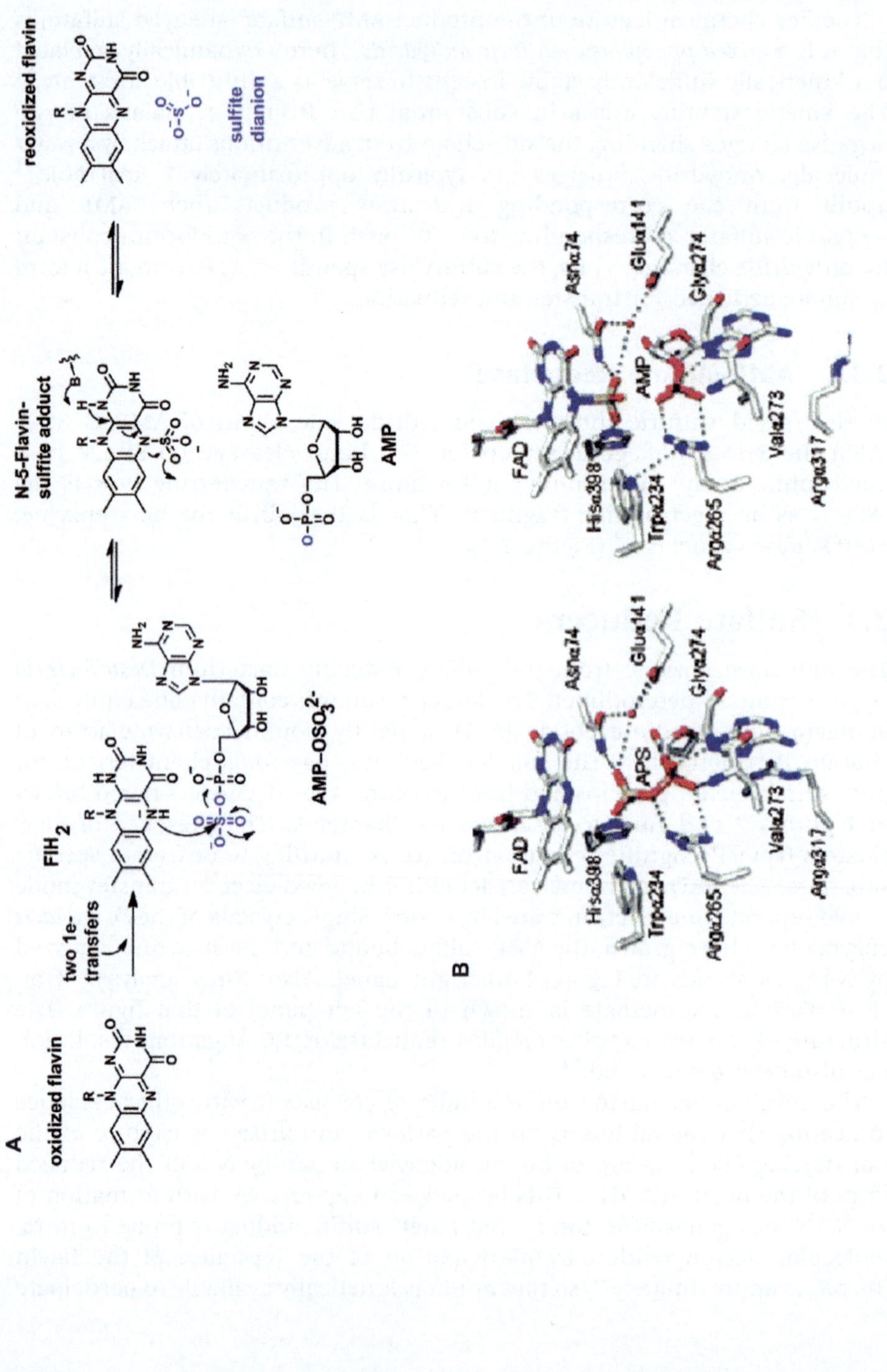
A
oxidized flavin
two 1e- transfers
FlH2
AMP–OSO3 2-
N-5-Flavin- sulfite adduct
AMP
reoxidized flavin
sulfite dianion
B
FAD
APS
Asnα74
Gluα141
Glyα274
Hisα398
Trpα234
Argα265
Valα273
Argα317

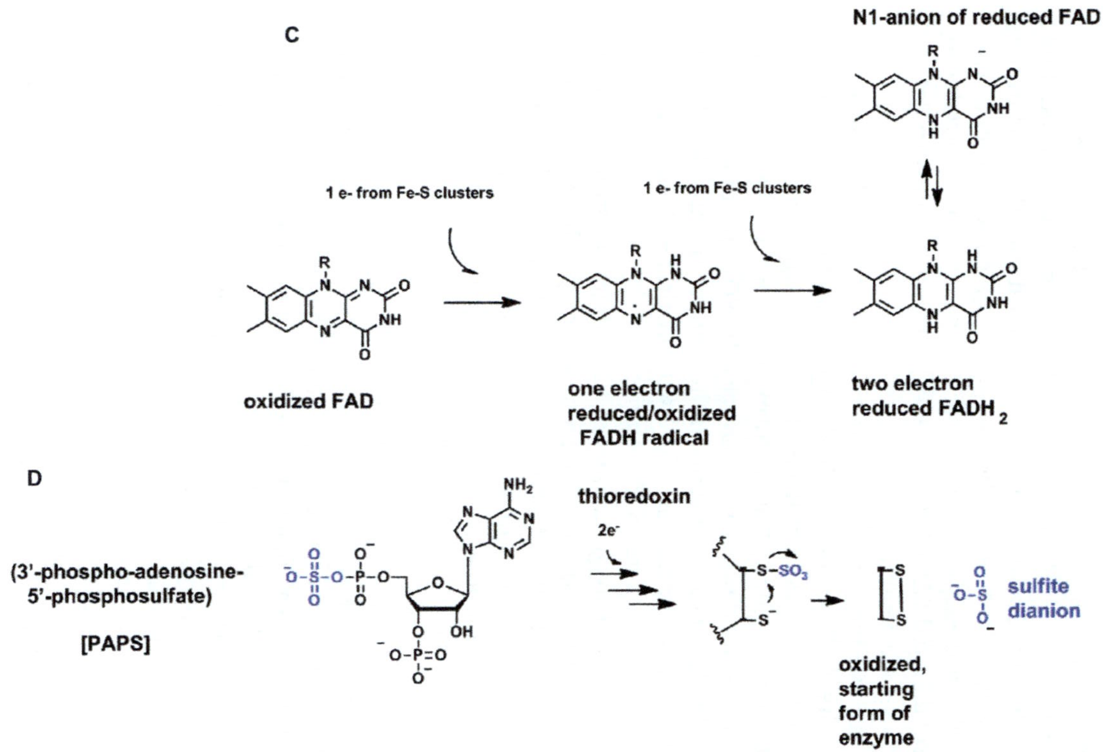

Figure 2.4 Bacterial adenylyl sulfate reductase. (A) Reaction mechanism for APS reductase involving bound FAD reduction to $FADH_2$ by external reductant to generate the N5–$FADH_2$ nucleophile to attack the electrophilic sulfuric anhydride portion of APS. The FAD–N5–sulfite adduct decomposes in the forward direction with intramolecular participation of the N1–C10a electron pair as shown. (B) X-ray structure of sulfate-reducer APS reductase showing FAD and bound sulfite in the left panel and bound AMP in the right panel (C). The external electron donor substrate funnels electrons one at a time through a 4Fe–4S cluster to the FAD. (D). The one electron at a time input mechanism means the FAD is reduced in two steps, *via* the one electron-reduced flavin semiquinone anion. (B) Reproduced from ref. 9 with permission from Royal Society of Chemistry.

as shown to reoxidize the FAD and expel the SO_3^- sulfite anion product). This unraveling of the flavin–sulfite adduct is the step completing the net two-electron reduction of sulfate to sulfite. The $FADH_2$ form of the enzyme has been reoxidized by two electrons to FAD.

We started the mechanistic analysis in mid catalytic cycle, where the nucleophilic N-5 of $FADH_2$ attacks the electrophilic sulfuric–phosphoric anhydride group of bound adenylyl sulfate. To get to that two-electron reduced state the two 4Fe–4S clusters are proposed to be one-electron conduits from an external electron donor through the β-subunit into the FAD in the α-subunit (Figure 2.4C). Necessarily, the FAD will undergo reduction in two consecutive one-electron steps and proceed *via* the one-electron-reduced semiquinone form denoted $FADH^{\bullet}$ (Figure 2.4C). Only in the dihydro $FADH_2$ form is the flavin N-5 nucleophilic.

Overall, there are two half reactions to the *D. gigas* bacterial AMP–SO_4^{2-} reductase. In the first half reaction a physiological external electron donor, as yet unidentified, must pass electrons in one at a time *via* the 4Fe–4S electron conduits to generate the $FADH_2$ form of the enzyme. In the second half reaction N-5 of nucleophilic $FADH_2$ attacks the SO_3^- portion of bound AMP–SO_4^{2-} to yield the covalent flavin–sulfite adduct. Intramolecular expulsion of sulfite as the flavin reoxidizes to FAD constitutes the two-electron reduction of the activated sulfate to free sulfite ion.

Note that one of the four oxygen atoms of the sulfate substrate has been removed as product sulfite is formed. That removed oxygen has left as part of the AMP, a low energy route for S–O bond cleavage. Still, the equilibrium for its transformation lies on the side of AMP–sulfate; the sulfate reducing bacterial cells use an inorganic pyrophosphatase to cleave the PPi coproduct hydrolytically to two molecules of organic phosphate to draw the equilibrium towards sulfite formation.

This first step in successful "dumping" of two electrons into sulfate to form a sulfite ion took some strategic and mechanistic chemical ingenuity on the part of these contemporary descendants of ancient, early microbes. They had to evolve chemical machinery to form the thermodynamically activated, kinetically stable, mixed sulfuric–phosphoric anhydride. In parallel, they had to learn how to use the two-electron reduced coenzyme form of riboflavin (vitamin B_2) as a nucleophile towards the sulfur atom in the mixed sulfuric–phosphoric anhydride, effecting a net sulfuryl transfer to the flavin coenzyme. Third, the organisms had to transfer the two electrons in the $FADH_2$ tricyclic ring system to the sulfur atom as the covalent adduct was decomposed in the product direction. (Note that this happens as the N5–S bond cleaves: the two electrons in that bond depart with the sulfur atom in the nascent sulfite ion product.)

Also, to accumulate usable amounts of product sulfite to continue dumping electrons into its sulfur framework, the microbes couple hydrolysis of PPi to get another approximately 10^5 kick in forward direction equilibrium. One should marvel at the chemical ingenuity and the reducibility of the sulfur atom in this initial stage of harvesting energy from [$S = +6$] reduction.

2.4.1 An Alternative Adenylyl Sulfate Reductase Logic

For green plants that reduce sulfate to sulfide for *assimilatory* incorporation into the amino acid cysteine, a topic described more extensively below, the same thermodynamic strategy of spending an ATP to convert sulfate to AMP–O–SO_3 is involved.[15] (Sometimes a second ATP is cleaved, this time to ADP as the migrating γ-phosphoryl group is attacked by the 3′-OH of the ribose group of APS to form 3′-phosphoadenosine-5′-phosphosulfate (PAPS; Figure 2.4D), but PAPS is used mostly for nonreductive transfers of SO_3^- as noted in Chapter 10.)

Studies on the mechanism of the plant reductase, and homologs from fungi, and some bacteria including mycobacteria, indicate the absence of FAD. There is one 4Fe–4S cluster and a conserved pair of cysteines that are thought to start each catalytic cycle as a disulfide. The external donor of two electrons is the 11 kDa protein thioredoxin (TRX).[16] As its name indicates, it engages in thiol disulfide interchange chemistry with client proteins, typically starting as the dithiol form of TRX and ending as the two-electron oxidized TRX disulfide. Because the dithiol form of TRX is the form that can donate two electrons, cells spend NADPH to convert the disulfide form of TRX back to the dithiol form (Figure 2.5). This is analogous to the logic that cells employ to keep glutathione in the reduced dithiol form by action of NADPH-dependent glutathione reductase[17] (Chapter 5) rather than the oxidized disulfide form.

The plant type of APS reductase also has a cysteine residue in the thiol oxidation state in the C-terminal portion of the enzyme, a domain that has homology to TRX, indicating that a core APS reductase and a TRX domain became fused at some point in evolution. It is that C-terminal thiolate side chain that appears to be the enzyme nucleophile that attacks the electrophilic SO_3^- mixed anhydride group of bound adenylyl sulfate substrate (Figure 2.6). The result is a covalent adduct between the enzyme cysteinyl residue and the itinerant SO_3^-, a *thiosulfate* linkage. In Part III we will examine the cysteinyl proteome and note here only that this thiosulfate protein linkage is, to date, unique among known cysteine side chain protein modifications.

The thiosulfate group is intriguing for the two very different oxidation states of the sulfur atoms in the S–S bond. The thiol sulfur is in the formal [−2] oxidation state while the SO_3^- sulfur is still in the [+6] oxidation state. One can argue that the next step in the catalytic cycle is the one that constitutes the two-electron reduction of that sulfur from [+6] to [+4].

That step is presumed to be attack of one of the TRX–SH_2 thiolates on the $S = -2$ side of the thiosulfate–enzyme adduct. As depicted in Figure 2.6, that leads to a TRX–reductase mixed disulfide and release of the product SO_3^-. Both electrons in the S–S bond of the enzyme S–SO_3^- have departed with the sulfite anion product. That is the two–electron reduction step. Meanwhile the mixed TRX–enzyme disulfide can be resolved by attack of the other thiol in TRX to release the enzyme thiol, ready for the next catalytic cycle and to produce the two-electron oxidized form of TRX as a coproduct. The role of the Fe–S cluster and the conserved disulfide in the plant, fungal and

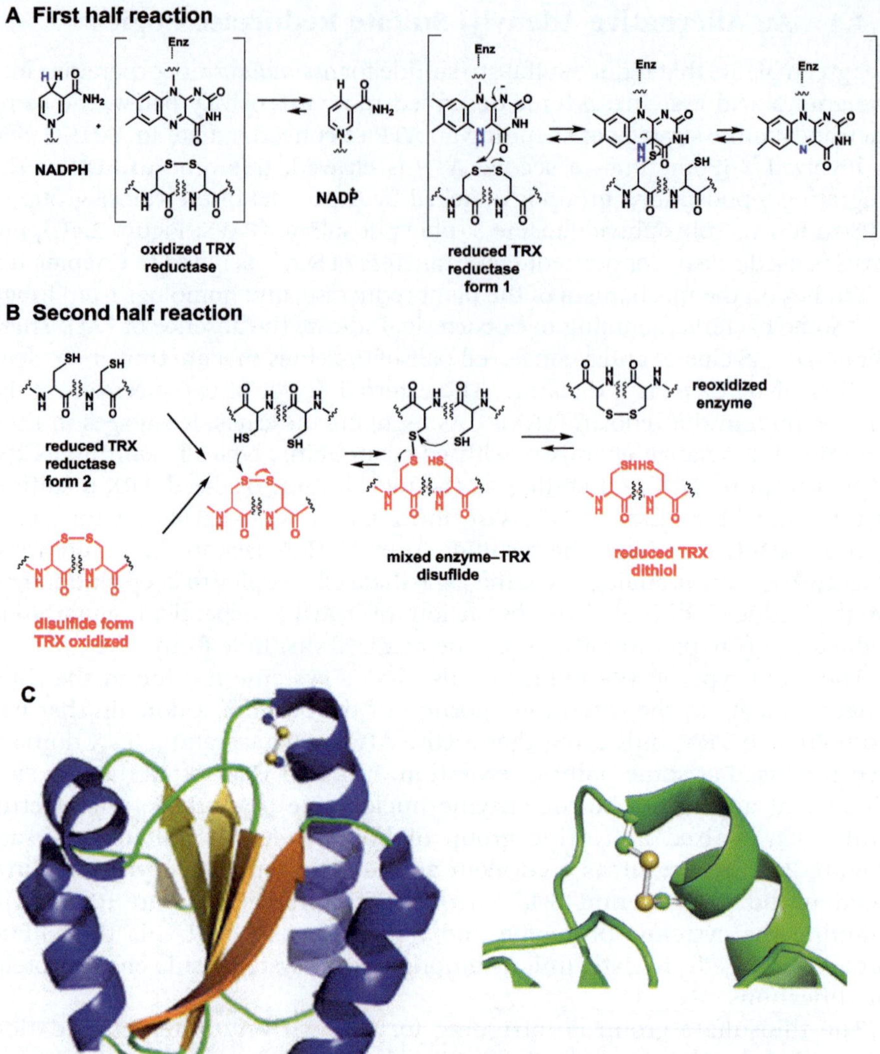

Figure 2.5 Thioredoxin as oxidant of a client protein dithiol to disulfide. (A) Enzymatic machinery for utilizing a hydride equivalent from NADPH to reduce the TRX active site disulfide to dithiol *via* a FAD intermediary: a key role for the thioflavin adduct. (B). Reduction of the oxidized, disulfide form of thioredoxin by the EH_2 form of thioredoxin reductase. Cells spend NADPH to keep the protein thioredoxin in reduced form. (C) Cartoon representation of thioredoxin architecture with redox active disulfide/dithiol (left) with close up view of the disulfide bond in the TRX_{ox} form (right).

bacterial APS reductases are as yet undefined. The conserved Cys–Cys disulfide in the enzyme could be the immediate reductant of the S–SO_3^- adduct, and only then would TRX act.

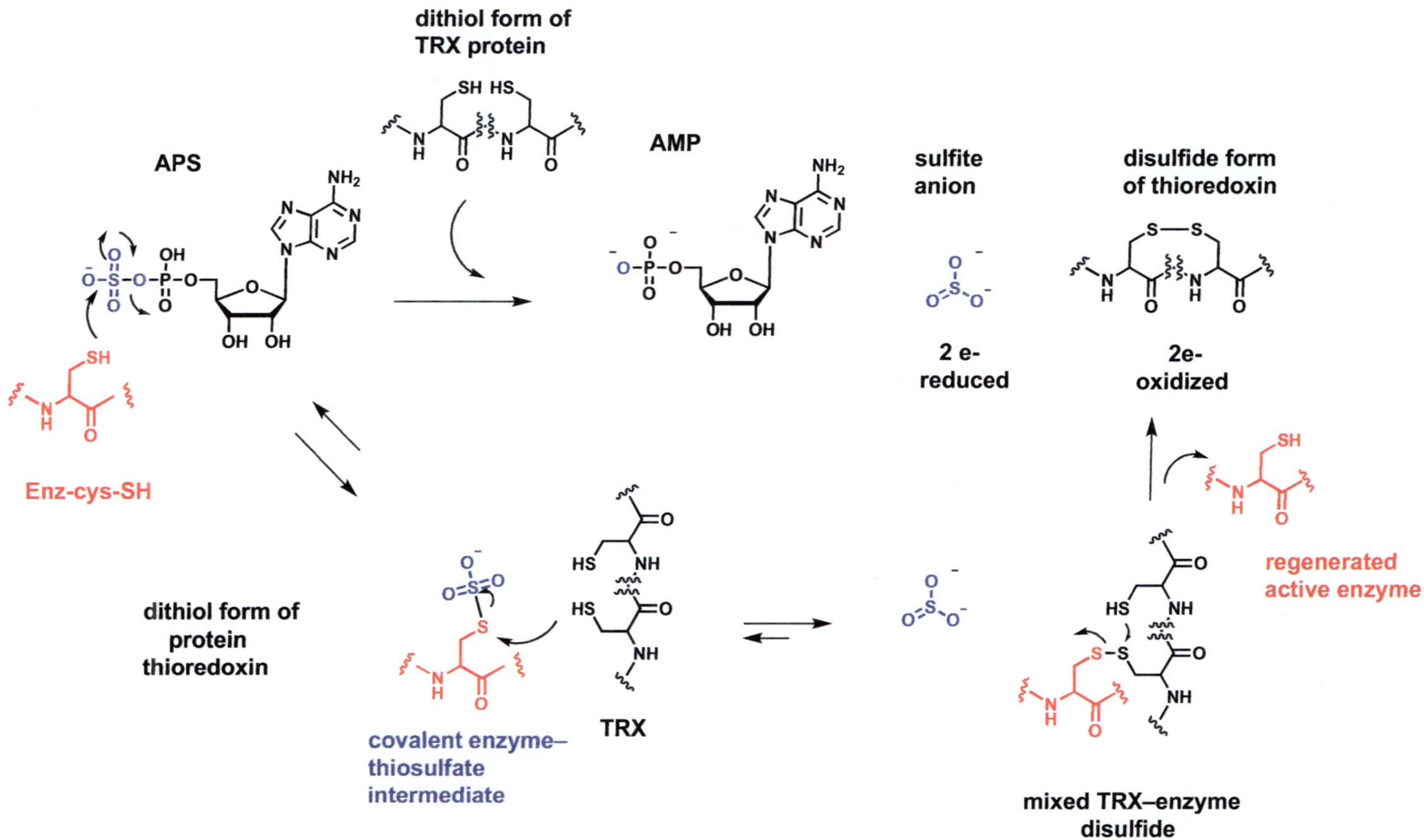

Figure 2.6 Plant APS reductase mechanism involves a covalent enzyme Cys–S–SO_3^- thiosulfate intermediate and then an enzyme Cys–S–S–Cys disulfide that is reduced by the dithiol form of thioredoxin cosubstrate.

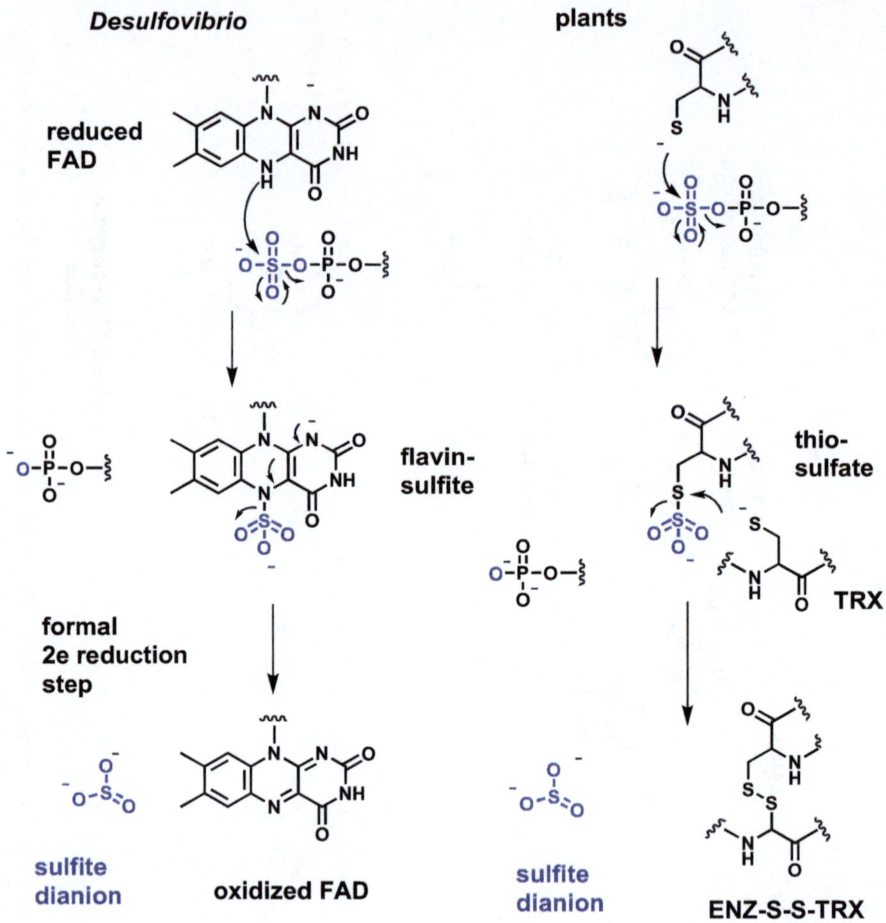

Figure 2.7 Two parallel distinct mechanistic strategies for adenylyl sulfate reductases.

A comparison of the microbial AMP–sulfate reductase logic and the independently evolved plant catalytic logic reveals two strategies for two-electron transfer steps to the electrophilic sulfur atom in that mixed sulfuric–phosphoric anhydride. In the presumably more primordial electron transfer protein nanomachinery of the sulfate reducing bacteria, the ancient 4Fe–4S clusters are electron conduits to a vitamin B_2-based flavin coenzyme. The sulfur atom gets reduced from the flavin–N5–sulfite adduct. In the alternate plant logic that coenzyme–SO_3^- covalent adduct has been replaced by an enzyme Cys–S–SO_3^- adduct. These are two distinct intersections between inorganic sulfur chemical biology and organic sulfur chemical biology (Figure 2.7).

Both adducts are *sui generis* and are not observed elsewhere in metabolism of contemporary organisms. Both rely on being presented with mixed sulfuric–phosphoric anhydride as a "sulfate" equivalent donor. The sulfate-reducing *D. gigas* uses $FADH_2$ as a nitrogen nucleophile. The plant nanomachine uses a side chain thiolate as a nucleophile. In each case thiol–disulfide chemistry is then part of the resolution phase of the catalytic cycle.

2.5 Microbial Reduction of Sulfite to Hydrogen Sulfide: Net Six-electron Reductions

Now the task facing sulfate reducing bacteria that take nascent sulfite ions on to organic sulfide gets even more challenging. Sulfate reducing bacteria and comparable archaea channel a large flux of electrons derived from co-oxidation of reduced metabolites into inorganic sulfide[1] (Figure 1.4).

Dissimilatory sulfite reductases (dsrs)[18] couple that flux to proton pumping across the cell membrane to create electrochemical gradients that are then used to make ATP, the central metabolic energy currency of cells.[4,19] It is the electrons that are valuable, transmuted into transmembrane proton gradients that are the primary energy currency from electron transfers (Figure 2.8A).

The bacteria dump the great bulk of the gaseous hydrogen sulfide as waste product into extracellular milieus.[20] They do funnel a small fraction into cysteine and the organosulfur metabolic universe. In contrast, in plants and fungi that practice *assimilatory* sulfite reduction, it is not the dumping of electrons into sulfite and on to sulfide that is the main purpose. Instead, it is now the actual reduced product hydrogen sulfide that is vectored into organic molecules, mainly *via* L-cysteine synthesis.

The core catalytic inventory of assimilatory and dissimilatory reductases appears at first glance to be similar, in terms of an four iron–four sulfur cluster connected by a bridging cysteine to a tetrapyrrolic-iron-containing siroheme[21] (Figure 2.8B).

However, the dissimilatory reductases have additional subunits not found in the assimilatory only organisms.[22,23] One of these subunits is thought to make a protein Cys–SSH persulfide as the final intermediate in the reduction process. A set of up to four membrane protein components are also encoded and may be involved in setting up the proton pumping and electrochemical gradient formation (Figure 2.8C). Given that dissimilatory sulfate reducers are among the most ancient microbial lineages from anaerobic antiquity, full understanding of the logic and array of the multiprotein components may give insights into early biological solutions for harvesting energy from membrane flow of electrons down potential drops.[8,24]

Six of the total of eight electrons funneled into sulfate on the way to inorganic sulfide (S^{2-}) occur at the stage of the multi-subunit metalloenzyme sulfite reductase, containing the 4Fe–4S unit (Figure 1.7) linked to siroheme *via* an intervening thiolate side chain of a cysteinyl residue from the enzyme core. Siroheme is in the tetrapyrrole macrocycle subfamily of isobacteriochlorins (Figure 2.8). Its redox potential may be tuned for iterative one-electron transfers into sulfite, which sits as its axial top ligand to the siroheme iron atom in the equatorial plane of the pyrrole nitrogens.[3,23]

The purified dissimilatory sulfite reductases, such as the well-characterized desulfoviridin $\alpha_2\beta_2$ heterotetramer, release partially reduced intermediates *in vitro,* indicating that they may need the missing dsrC subunit to couple the last two electron reductive steps to sulfide with membrane components dsrMKJOP and proton pumping[23] (Figure 2.8A). On the other hand, the cognate sulfite reductase from assimilatory organisms,

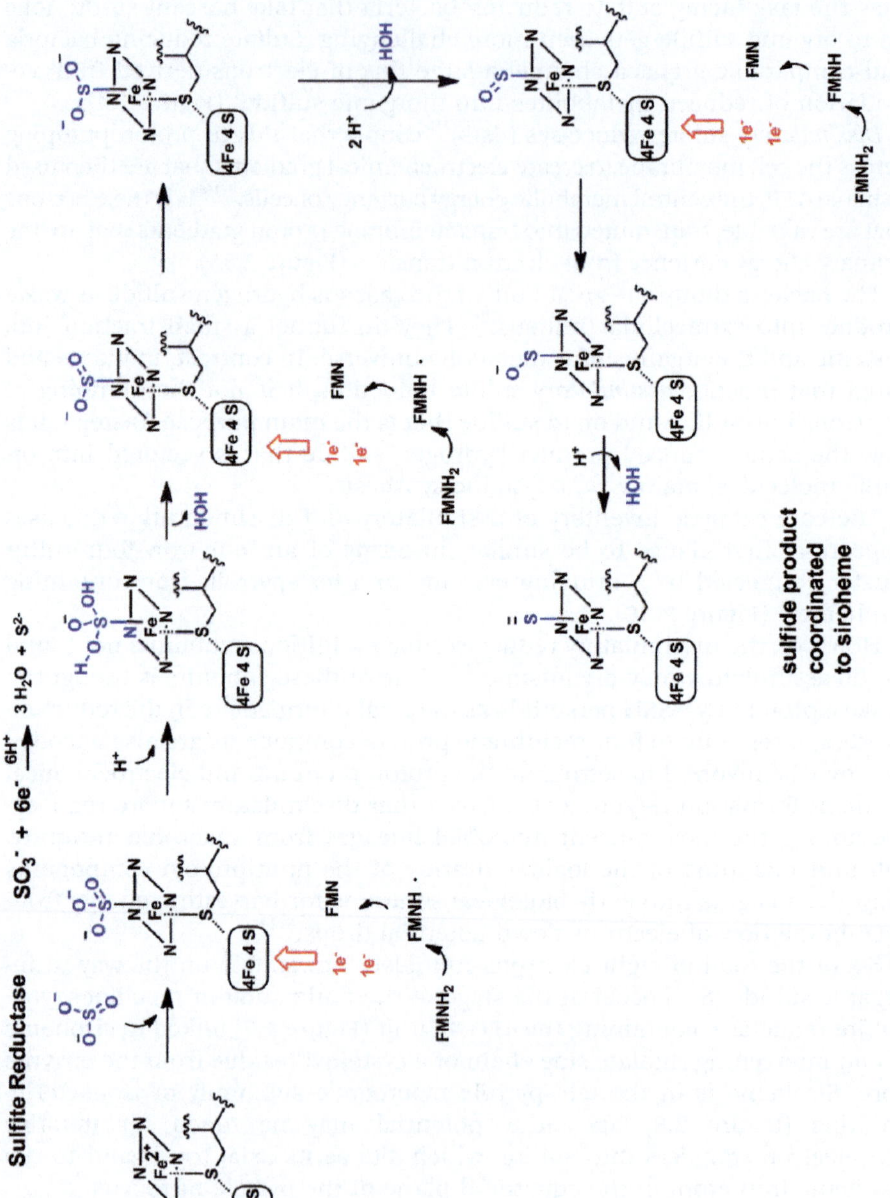
A
Sulfite Reductase
$SO_3^- + 6e^- \xrightarrow{6H^+} 3\,H_2O + S^{2-}$
4Fe 4S
FMNH2
FMNH ·
FMN
1e⁻
H⁺
HOH
2 H⁺
sulfide product
coordinated
to siroheme

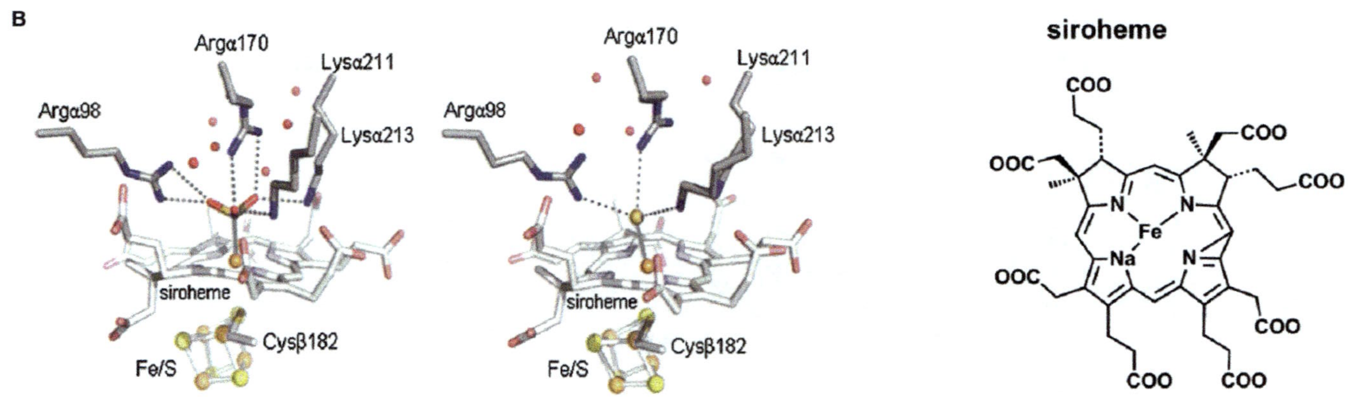

Figure 2.8 (A) Six-electron reduction of sulfite ion to sulfide dianion while coordinated to the siroheme group of sulfite reductase. (B) Left, schematic of dSIR–sulfite. Centre, schematic of dSIR–sulfide complex. Both from *A. fulgidus*, bound to siroheme in the sulfite reductase active site. Right, siroheme is an octacarboxy form of the heme tetrapyrrole macrocycle. (C). Multiple ancillary subunits couple the active site, containing the 4F–4S siroheme subunit to membrane associated subunits, allowing flow of the six electrons into the membrane respiratory chain of the sulfate reducing bacteria.
(B) Reproduced from ref. 9 with permission from Royal Society of Chemistry. (C) Reproduced from ref. 28 with permission from American Society for Microbiology, Copyright 2005.

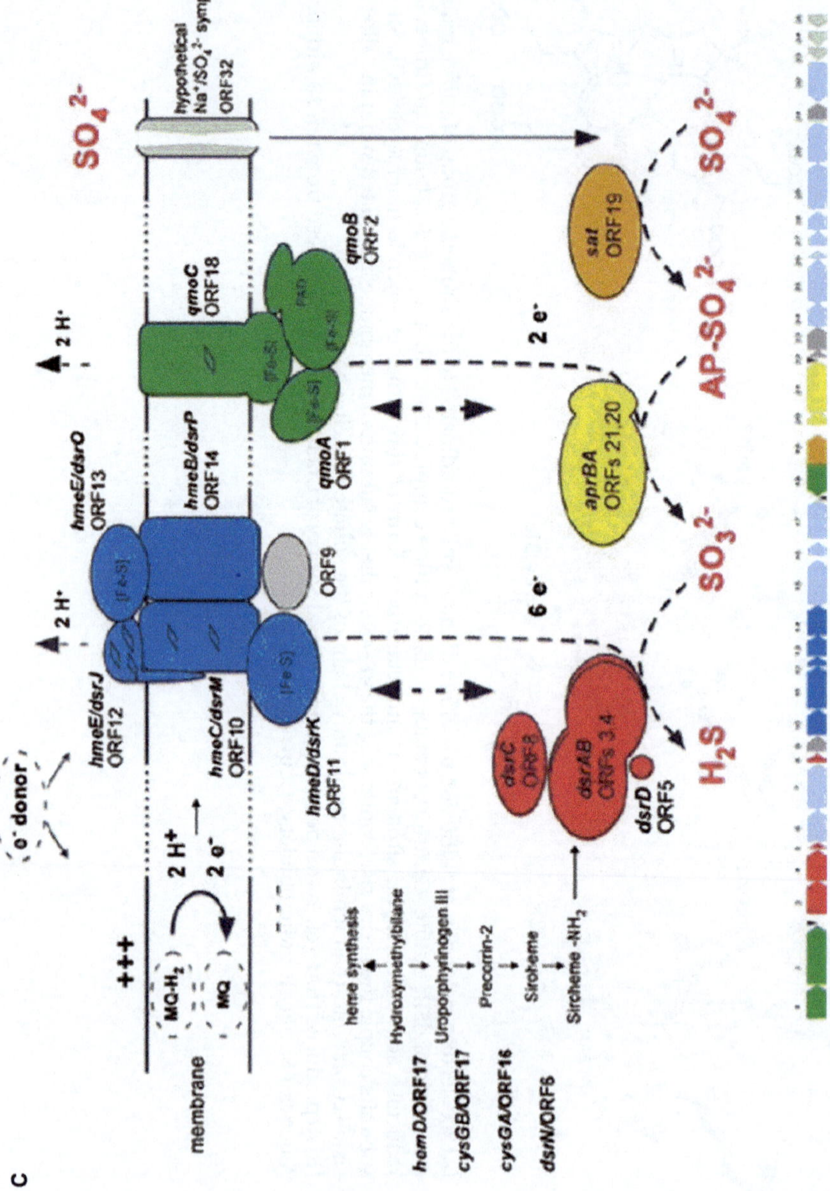

Figure 2.8 Continued.

including plants, reduces sulfite all the way to sulfide, consistent with the catalytic machinery separately from the membrane proton-pumping, energy-yielding functions.

Electron input to either form of sulfite reductase is probably derived ultimately from metabolites such as NADPH perhaps *via* flavin and Fe–S redox active clusters. Figure 2.9 shows such a schematic.

Presumably, the six electrons are fed one electron at a time *via* the 4Fe–4S redox cluster into the siroheme and then out to the top axial sulfite ligand. (More information on Fe–S clusters is given in Part II, Chapter 9 at the interface of inorganic sulfide and organic thiol chemical biology.) The readily available, diffusible redox currency in the cell is NADH or its 2′-phospho homolog, NADPH. It is an obligate hydride donor, two electrons at once along with the hydrogen. Given that the 4Fe–4S cluster and the siroheme are obligate one-electron acceptors and donors then there must be a two- electron–one-electron step down redox switch between NADPH and iron. That is the role of FMN in the sequence of electron transfers: NADPH to FAD to FMN to Fe–S to siroheme (Figure 2.9A). This is a typical logic pattern seen in many other redox enzymes where a two to one step down electron switch is required to turn NAD(P)H hydride-reducing equivalents into one-electron packets to iron centers.[25]

The observed release of two-electron-reduced trithionate and four-electron-reduced thiosulfate are thought to arise from aberrant off-loading of siroheme-bound SO_2 and SO, respectively[3] (Figure 2.10), rather than their existence as obligate free intermediates in the six-electron reduction process. The addition of two more electrons to the putative SO–siroheme intermediate would generate sulfide in the assimilatory sulfite reductase catalytic heterotetramers.

The above description is at best schematic about the timing and reaction manifold of both electron transfer and net dehydration steps as the three S–O bonds in sulfite are cleaved without loss of intermediates. The 4Fe–4S siroheme couple is efficient architecturally for six-electron input and the axial siroheme Fe–S axial bond must persist through the net of six-electron transfers that pass from the bottom axial 4Fe–4S cluster through the central siroheme Fe to the coordinated sulfur atom becoming ever more reduced.

Sulfite reductase is both a remarkably optimized and engineered metalloprotein catalyst, with two of the three kinds of biologic iron scaffolds as redox mediators of electron flow (The third is mononuclear nonheme iron, typically with two histidine and one aspartate or glutamate as enzyme ligands to ferrous iron: Figure 2.9B (discussed further in Chapter 7)).

This enzyme is the central node in both dissimilatory and assimilatory sulfate redox energy metabolism. Thereby, it is the central player in global biological cycling of sulfur. If even 10% of the approximately 10^{12} grams of sulfate cycling in and out of global oceans per year is due to sulfate reducing microbes, that would amount to hundreds of thousands of tons of sulfite reduced yearly by this enzyme family.[1,26]

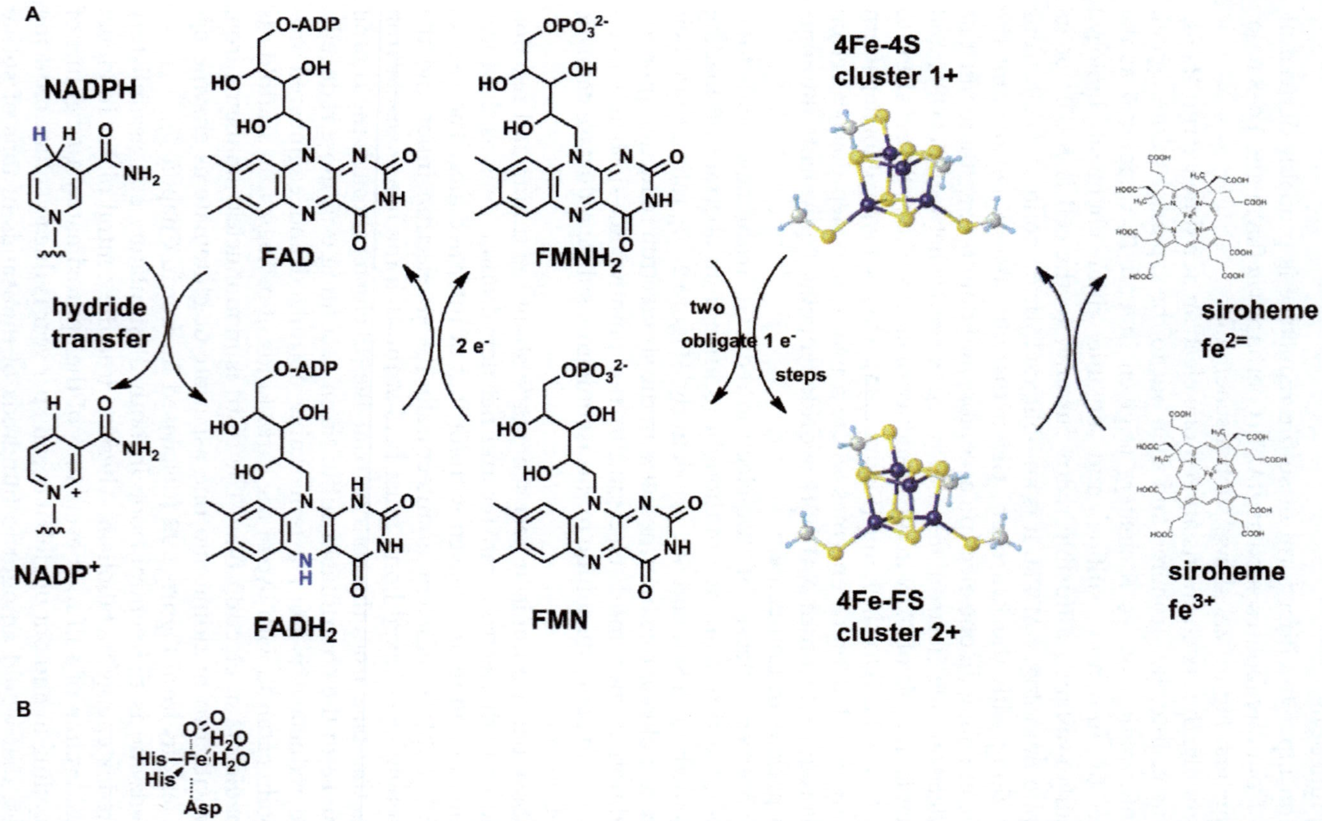

Figure 2.9 (A) Sulfite reductase electron flow: NADPH to FAD to FMN to 4Fe–4S cluster to siroheme to bound sulfite and reduced intermediates. (B) A typical mononuclear, nonheme iron coordination site. Nonheme mononuclear Fe^{2+} are the third type of redox-active iron center in enzymes.

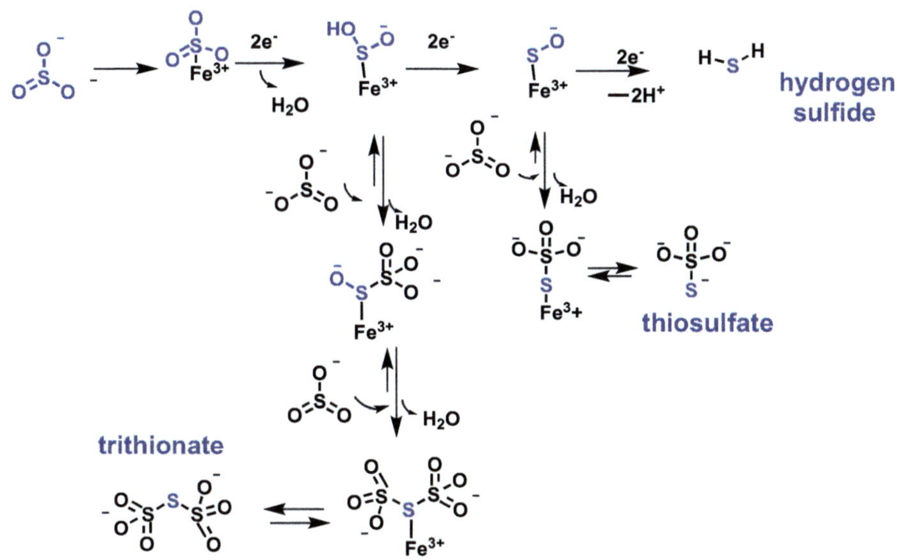

Figure 2.10 Leakage and utilization of trithionate and thiosulfate by sulfite reductase.

For the dissimilatory sulfite reductases, further consideration must be given to the logic and molecular strategy of sulfite reductases as membrane chain, terminal electron transfer catalysts. Still to be understood in detail is how the sulfite reductase catalytic subunits interact with membrane-bound subunits, to couple electron flow to net pumping of protons across the bacterial or archaeal membranes to generate the electrochemical gradients needed to drive membrane ATP synthases (Figures 2.8 and 2.11). The dsrC protein is thought to be a soluble partner that interfaces the dsrAB catalytic subunits with dsrJMOP proteins in the membrane.

There is evidence that the dsrC subunit can transfer the four-electron-reduced sulfur species, as an electrophile, from the siroheme tetrapyrrole to a cysteine residue of dsrC, generating a dsrC–Cys–S–SH persulfide.[22,23] Internal capture of that dsrC persulfide by another Cys–SH (Figure 2.11), could generate an enzyme trisulfide form to be carried to the membrane before final two-electron reductive release. Figure 2.11 B shows a schematic of the sulfite reductase active site and a blowup of the persulfide sulfur embedded within a cystine disulfide to yield an unusual linear enzyme *trisulfide*.[27] Alternatively, dsrX could convert the persulfide intermediate to free S^{2-} as the dianion of the final six-electron-reduced hydrogen sulfide while producing the disulfide form of dsrC. Reduction of dsrC back to the active dithiol form at the membrane would effectively couple the last two electrons removed to the membrane electron transport chain. The dithiol–disulfide strategy has analogies to the plant AMP sulfite reductase logic noted above.

The contemporary sulfate-reducing bacteria and archaea are descendants of some of the most ancient microbial lineages that developed and executed

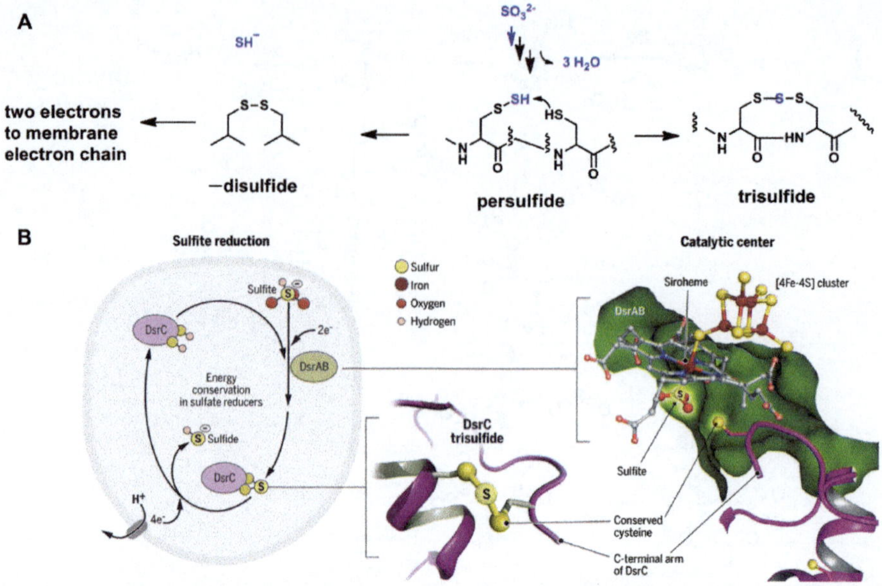

Figure 2.11 (A) Schematic for interaction of the DsrAB heterodimer with DsrC to couple electron transfer to sulfite and proton pumping across the bacterial membrane to create the transmembrane electrochemical potential that represents the energy capture strategy. (B). Atomic view of the DsrAB active site and a blowup of the enzyme region that can fix the persulfido sulfur as the middle sulfur in an enzyme trisulfide adduct.
(A) Reproduced with permission from ref. 27, courtesy of K. SUTLIFF/SCIENCE.

the principles of coupled electron transport and electrochemical gradients across membranes.[24] These catalytic nanomachines interface with membrane electron-accepting proteins to make ATP and ultimately other kinetically stable but thermodynamically activated molecules to perform cellular metabolic work.

References

1. D. A. Fike, A. Bradley and C. V. Rose, Rethinking the ancient sulfur cycle, *Annu. Rev. Earth Planet. Sci.*, 2015, **43**, 593–622.
2. B. Meyer, Elemental Sulfur, *Chem. Rev.*, 1976, **76**, 367–390.
3. B. R. Crane, L. Siegel and L. D. Getzoff, Probing the catalytic mechanism of sulfite reductase by X-ray crystallography: structures of the *E.coli* hemoprotein in complex with substrates, inhibitors, intermediates, and products, *Biochemistry*, 1997, **63**, 12120–12137.
4. K. Parey, E. Weerpana, P. M. H. Kroneck and U. Ermler, Reaction cycle of the dissimilatory sulfite reductase from *Archaeoglobus fulgidus*, *Biochemistry*, 2010, **49**, 8912–8921.

5. M. Dworkin and D. Gutnik, Sergei Winogradsky: a founder of modern microbiology and the first microbial ecologist, *FEMS Microbiol. Rev.*, 2012, **36**, 364–379.
6. H. Takahashi, P. Buchner, N. Yoshimoto, M. J. Hawkesford and S.-H. Shiu, Evolutionary relationships and functional diversity of plant sulfate transporters, *Front. Plant Sci.*, 2011, **2**, 119.
7. E. Aguilar-Barajas, C. Diaz-Perez, M. I. Ramiewz-Diaz, H. Riveros-Rosas and C. Cervantes, Bacterial transprt of sulfate, molybdate and related oxyanions, *Biometals*, 2011, **24**, 687–707.
8. Y. Shen and R. Buick, The antiquty of microbial sulfate reduction, *Earth-Sci. Rev.*, 2004, **64**, 243–272.
9. K. Parey, G. Fritz, U. Emmler and P. M. H. Kronek, Conserving energy-with sulfate around 100 °C-structure and mechanism of key metal enzymes in hyperthermophilic *Archaeoglobus fulgidus*, *Metallomics*, 2013, 203–220.
10. P. K. Demmer, W. Wynen and E. U. Dahl, Structural, biochemical and dissimilatory ATP sulfurylase from *Allochromatium vinosum*, *PLoS One*, 2013, 8.
11. C. T. Walsh, B. Tu and Y. Tang, Eight kinetically stable but thermodynamically activated molecules that power cell metabolism, *Chem. Rev.*, 2018, **118**, 1460–1494.
12. Y. Fu, J. Tang, G.-F. Yao, Z.-Q. Huang, Y.-H. Li, Z. Han, X.-Y. Chen, L.-Y. Hu, K.-D. Hu and H. Zhang, Central role of adenosine 5′-phosphosulfate reductase in the control of plant hydrogen sulfide metabolism, *Front. Plant Sci.*, 2018, 1404.
13. Y.-L. Chiang, Y.-C. Hsieh, J.-Y. Fang, E.-H. Liu and Y.-C. Huang, Crystal structure of an adenylylsulfate reductase from Desulfvirio, *J. Bacteriol.*, 2009, **191**, 7597–7608.
14. G. Fritz, A. Roth, A. Schiffer, T. Buchert, G. BOurenkov, H. D. Bartunik, H. Huber, K. O. Stetter, P. M. H. Kroneck and U. Ermler, Structure of adenylylsulfate reductase from the hyperthemophilic *Archaeoglobus fulgidus* at 1.6 Angstrom resolution, *Proc. Natl. Acad. Sci.*, 2002, 1836–1841.
15. J. A. Bick, J. Dennis, G. J. Zylstra, J. Nowack and T. Leustek, Identification of a new class of 5′-adenylylsulfate reductase from sulfate -assimilating bacteria, *J. Bacteriol.*, 2000, **182**, 135–142.
16. A. Holmgren, Thioredoxin and glutaredoxin systems, *J. Biol. Chem.*, 1989, **264**, 3900–3905.
17. M. Deponte, Glutathione catalysis and the mechanism of glutathione-dependent enzymes, *Biochim. Biophys. Acta*, 2013, **1830**, 3217–3266.
18. H. D. Peck Jr., Enzymatic basis for dissimilatory and assimilatory sulfate reduction, *J. Bacteriol.*, 1961, **82**, 933–939.
19. A. Schiffer, K. Parey, E. Warkentin, K. Diedrichs, H. Huber, K. O. Stetter, P. M. H. Kronek and U. Ermler, Structure of the dissimilatory sulfite reductase from the hyperthermophilic archaeon *Archaeoglobus fulgidus*, *J. Mol. Biol.*, 2008, **379**, 1063–1074.

20. G. Muyzer and A. Stams, The ecology and biotechnology of sulfate-reducing bacteria, *Nat. Rev. Microbiol.*, 2008, **6**, 441–454.
21. K. L. Keller, B. Rapp-Giles, E. S. Semkiw, I. Porat, S. D. Brown and J. D. Wall, New model for electron flow for sulfate reduction in *Desulovibrio alaskensis* G20, *Appl. Enivon. Microbiol.*, 2014, **80**, 855–868.
22. T. F. Oliveira, C. Vornheim, P. M. Matias, S. S. Venceslau, I. A. C. Pereira and M. Archer, The crystal structure of *Desulfovibrio vulgaris* dissimilatory sulfite reductase bound to DsrC provides novel insights into the mechanism of sulfate respiration, *J. Biol. Chem.*, 2001, **283**, 34141–34149.
23. T. F. Oliviera, C. Vornheim, P. Matias, S. Venceslau, I. C. Pereira and M. Archer, The crystal structure of *Desulfovibrio vulgaris* dissimilatory sulfite reductase bound to DsrC provides novel insight into the mechanism of sulfate respiartion, *J. Biol. Chem.*, 2008, **283**, 34141034149.
24. M. Wagner, A. Roger, J. L. Flax, G. A. Brussea and D. A. Stahl, Phylogeny of dissimilatory sulfide reductases supports an early origin in sulfate respiration, *J. Bacteriol.*, 1998, **180**, 2975–2982.
25. T. Wencewicz and C. Walsh, Flavoenzymes: versatile catalysts in natural product pathways, *Nat. Prod. Rep.*, 2012, **30**, 175–200.
26. B. M. Brahmacharimayum, M. P. Mohanty and P. Ghosh, Theoretical and practical aspects of biological sulfate reduction: a review, *Global NEST J.*, 2019, **2**, 222–244.
27. A. A. Santos, S. Venceslau, F. Grein, W. D. Leaitt and C. Dahl, A protein trisulfide couples dissimilatory sulfate reduction to energy conservation, *Science*, 2015, **350**, 1541–1545; accompanying commentary by G. Fritz, Kroneck PMH: sulfate to go, *Science* **350**, 1476–1477.
28. M. Mussman, M. Richter, T. Lombardot, A. Myeredierks, J. Kuever, M. Kube, F. Glockner and R. Amann, Clustetered genes related to sulfate respiration in unclultured prokaryotes support the theory of their co-comitant horizontal transfer, *J. Bacteriol.*, 2005, **187**, 7126–7137.

CHAPTER 3

Sulfide Oxidizing Bacteria

3.1 From Sulfide by Two Electrons to Elemental Sulfur and by Eight Electrons to Sulfate

As noted in Figure 2.2 as H_2S levels rise in water columns from action of sulfate reducers, the hydrogen sulfide gas can seep up to the oxic–anoxic interface where sulfide oxidizing bacteria, of the *Sulfobales* order (families *Thiobacilliacae, Acidithiobacilli, Aquifex* and others) can make a living by chemolithotrophic oxidation of the S^{2-} substrate.[1–4] These bacteria are often Gram-negative, rod-shaped and grow in filaments. Some bacteria can oxidize the H_2S or HS^- by only two electrons and so deposit elemental sulfur, as S^0 or in polysulfide S^n chain products. The polysulfides may be deposited as visible granules in the bacterial filaments. The two electrons removed end up either in O_2 or nitrate ion (NO_3^-) as terminal electron acceptors in those bacterial membrane electron transport chains (Figure 3.1A). Indeed, some sulfide oxidizers load up on nitrate ions in internal membranes to serve as terminal electron acceptors and the high intracellular vesicle allows identification of these bacteria morphologically. Among these bacteria are the *Chromatiacaea*, also called purple sulfur bacteria. Figure 3.1B shows a picture of purple sulfur bacteria from Yellowstone National Park.

However, most sulfide oxidizing bacteria have the full complement of enzymes to then take the elemental sulfur all the way up to sulfate, transferring the additional six electrons ultimately to O_2 as the membrane chain terminal acceptor[5–7] (Figure 3.1B). They use that energy, in ATP generation, ultimately to power CO_2 incorporation into reduced carbon metabolites. Thiobacilli live in neutral pH environments but some types of sulfide oxidizers grow in strongly acidic environments, including coal waste ponds. Some sulfide oxidizers are purple, some are green. Each makes a series of highly conjugated carotenoid lipids that can appear as orange, red or purple

Chemical Biology No. 12
The Chemical Biology of Sulfur
By Christopher T. Walsh

Published by the Royal Society of Chemistry, www.rsc.org

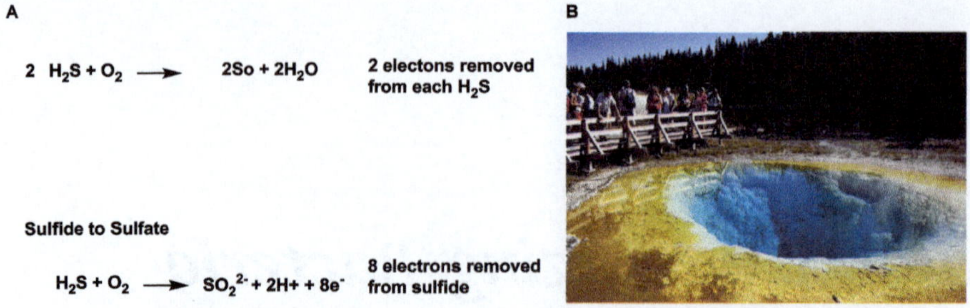

Figure 3.1 Sulfide oxidizing bacteria; (A) Reaction stoichiometries for sulfide oxidation to elemental sulfur and then on to sulfate. (B) Sinkhole in Yellowstone National Park. The purple area is the locale of purple sulfide oxidizing bacteria. The yellow area is elemental sulfur which is emitted as a gas and has condensed as a solid on the periphery of the fumarole.

depending on the number and placement of double bonds. The green sulfur oxidizers all have bacteriochlorophyll a. The carotenoids probably serve in part as photoprotectant molecules and perhaps peripheral photoantennae.

3.1.1 Two Electron Oxidation of H_2S to Elemental Sulfur and Linear Polysulfide Variants Thereof

The first enzymatic step, oxidation of sulfide to S^0 has been characterized in several bacteria as action of a sulfide quinone oxidoreductase[8,9] (Figure 3.2A). The enzyme has both a tightly bound flavin cofactor and active site disulfide to facilitate likely electron transfer routes.[10] The examination of inorganic sulfur redox enzymology, down and back in these two chapters, illuminates the mix of inorganic redox elements [iron-based tetrapyrroles, Fe–S clusters, and organic redox cofactors (flavins)] that were (had to have been) developed early in evolution and have continued as redox mainstays.

X-ray structures of sulfide-quinone oxidoreductases have shown not only the presence of a cystine disulfide in striking distance to the FAD coenzyme, but also product complexes with three to five sulfur[9,11] atoms bridging the two cysteine residues (Figure 3.2D). These observations support the mechanism shown in Figure 3.2B where the nucleophilic hydrosulfide (HS^-) anion form of substrate H_2S attacks the resting enzyme's active site disulfide, cleaving it to an enzyme–Cys–S–SH (persulfide) and the thiolate form of the other Cys residue. Attack of the terminal persulfide anion on the bridgehead carbon C4a of the oxidized FAD then produces the indicated unusual *persulfido-flavin adduct*. Now, attack of the other Cys thiolate anion on the proximal sulfur of that persulfido–FAD adduct generates reduced $FADH_2$ (the two electrons have been transferred out of substrate H_2S) and the S^0 product held between the two cysteine residues in an enzyme–trisulfide nascent product.[9]

The net two-electron transfer is completed in the second half reaction as $FADH_2$ transfers two electrons, perhaps one at a time *via* semiquinone species, to either a benzoquinone (*e.g.* Coenzyme Q10) or a naphthoquinine (*e.g.* menaquinone-7) electron-accepting cosubstrate. The bound FAD-oxidized coenzyme is thereby regenerated and the reduced hydroquinol form of the quinone product is released.

The enzyme apparently can catalyze multiple cycles of H_2S oxidation to S^0 equivalents without releasing the S^0 atoms from the oxidoreductase active site. Instead they can be stored in a growing polysulfide chain, like sulfur beads on a string. Figure 3.2C shows that two more catalytic cycles would lead to a trisulfide form of elemental sulfur bridging the two active site cysteines (net five sulfur atoms in a row: the S_3 trisulfide sandwiched between the two cysteine sulfurs), is a species seen in one of the X-ray structures.[9,10,12] Linear pentasulfide allotropes of elemental sulfur are also observed as displayed in Figure 3.2C. The size of the growing polysulfide bridge may reach a limiting volume that leads to release of polysulfide chains of different lengths.

The sulfide quinone oxidoreductase is a member of a family of FAD-dependent oxidoreductases utilizing an active site Cys–S–S–Cys disulfide as key redox element. This example demonstrates several features of sulfur chemical biology, including the nucleophilic properties of the hydrosulfide anion and the tendency of elemental sulfur (S^0) to form polysulfide allotropes. The enzyme-bound persulfide introduces the –S–SH group and reveals both the *nucleophilicity* of the $-S-S^-$ persulfide anion and the *electrophilicity* of that terminal sulfur when protonated as –S–SH for capture by thiolate anions. We will return to the chemical versatility of persulfide groups in Chapter 6 in the section on Fe–S cluster origins.

The flavin sulfur chemistry in the sulfide quinone oxidoreductase active site is the second example of flavin–sulfur covalent adducts, but with the sulfur at very different oxidation states (Figure 3.2C). In the AMP-sulfate reductase the reduced $FADH_2$ uses N-5 to attack the activated sulfuryl group (S^{+6}). Here, in the sulfide oxidoreductase it is the persulfide anion ($R-SS^-$) attacking the oxidized flavin at C4a. In the back reaction the C4a is a nucleophile, illustrating that both N5 and C4a of the flavin tricyclic ring system can be ports of electron entry and exit with sulfur at dramatically different oxidation states.[11]

Perhaps the most remarkable feature of the sulfide quinone oxidoreductase cycle is the accumulation of enzyme-bound trisulfide to pentasulfide chains at the oxidation states of elemental sulfur.[10,13] They may shed light on the dissimilatory sulfite reductase (dsr) subunit dsrC and its ability to generate bound –Cys–S–S–S–Cys species as nascent products. The details of how sulfane sulfur atoms are added to the growing polysulfide chain sandwiched between the two cysteine residues are not fully resolved (Figure 3.2C), but may be among the most surprising features of enzyme-catalyzed inorganic sulfur chemical biology. (A sulfane sulfur is defined as a sulfur with only sulfurs bonded to it.)

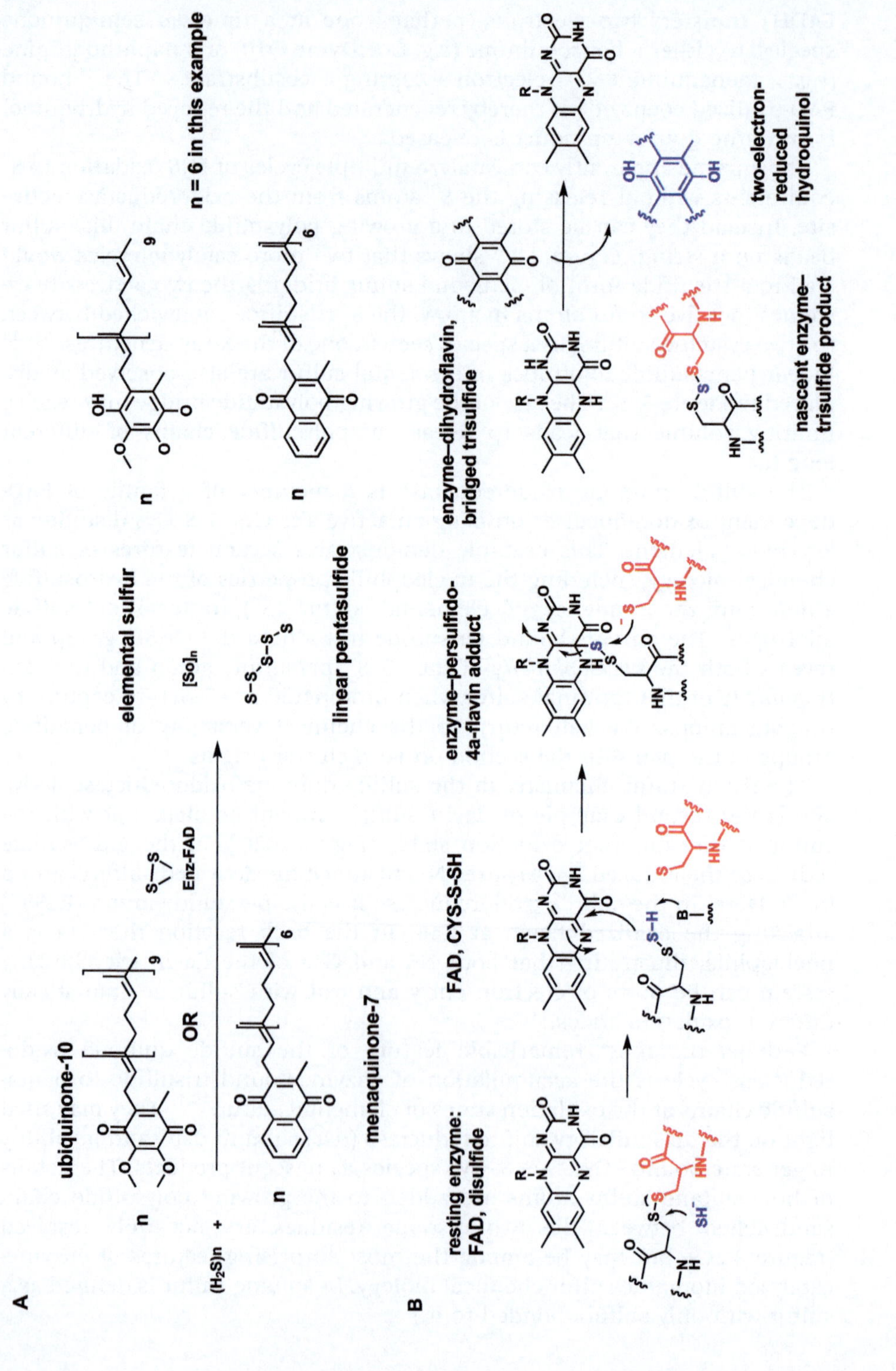
A
ubiquinone-10
elemental sulfur
n = 6 in this example
(H2S)n +
Enz-FAD
OR
[So]n
menaquinone-7
linear pentasulfide
B
resting enzyme:
FAD, disulfide
FAD, CYS-S-SH
enzyme–persulfido-
4a-flavin adduct
enzyme–dihydroflavin,
bridged trisulfide
two-electron-
reduced
hydroquinol
nascent enzyme–
trisulfide product

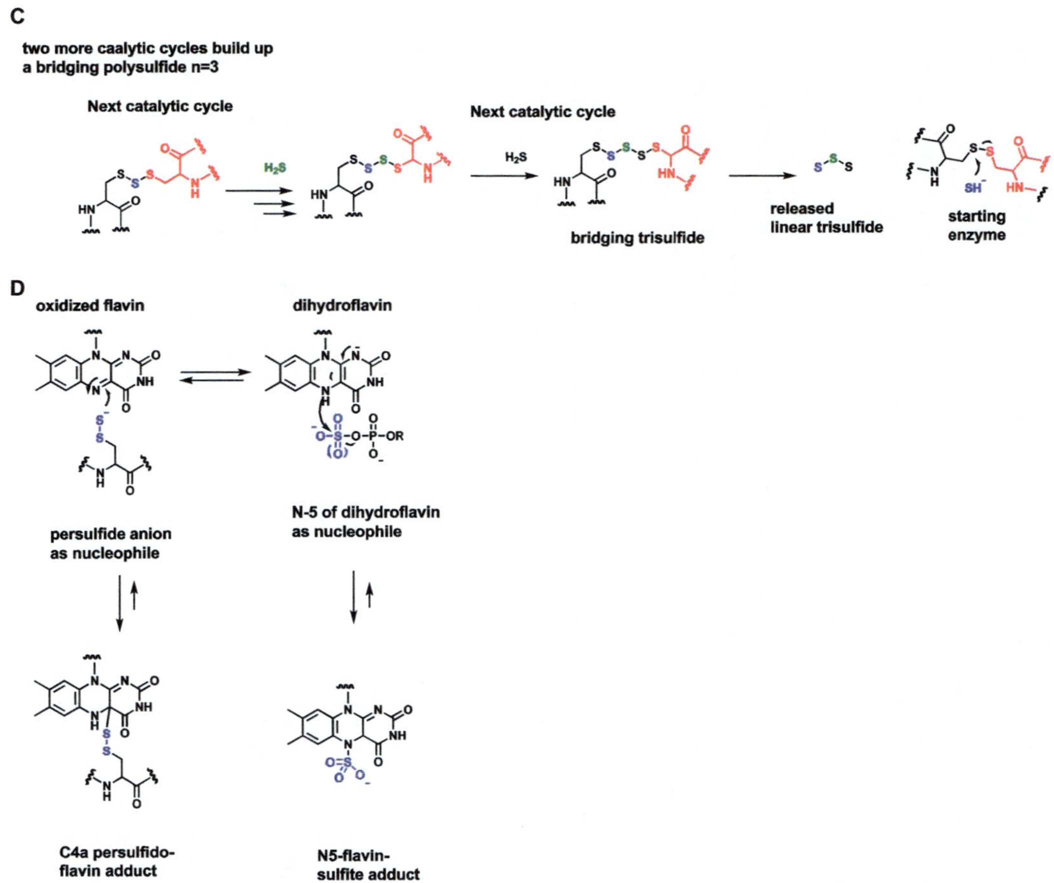

Figure 3.2 Sulfide quinone oxidoreductase. (A). Reaction stoichiometry for generation of elemental sulfur S^0 from inorganic sulfide. (B) Enzyme persulfide intermediate, precedes persulfide-4a–flavin adduct, precedes $FADH_2$ and enzyme trisulfide formation. (C) Two additional catalytic cycles build up a bridging trisulfide embedded between a pair of cysteine residues. (D) Comparison of FAD-N5–sulfite adduct with FAD-4a–persulfido enzyme adduct.

3.2 Enzymatic Machinery for Conversion of S^0 to Sulfite and Sulfate

The next four (to six) electrons removed from S^0 (or S^{2-}) to get to sulfite can be mediated by reverse reaction by homologs of the dissimilatory sulfite reductases[14] (Figures 2.8 and 2.9). The electron flow into or out of the 4Fe–4S and siroheme cofactor can apparently be run in either direction. Likewise, the final step of sulfite oxidation on to sulfate dianion can be a reverse of AMP–sulfate reductase noted in Section 2.3.2 (the reverse is in fact the favorable equilibrium direction), but is more typically the province of a remarkable molybdoenzyme sulfite oxidase, examined below. As the unique step in sulfide oxidizers *vs.* sulfate reducers, we focus on that sulfite oxidase.[14]

Sulfite oxidase (Figure 3.3) is found in bacteria and also in all eukaryotic mitochondria, presumably from an early symbiont. It contains a molybdopterin cofactor in its active site as well as a cytochrome b5 type heme, indicating a one-electron transfer route to the external acceptor, membrane-associated cytochrome c. The human sulfite oxidase is involved in terminal degradation of sulfur-containing amino acids. Defects in the encoding gene cause neuropathologies.

The molybdo cofactor derives biosynthetically from guanosine triphosphate (GTP) *via* expansion to a 6,6-fused bicyclic pterin ring and the related family is termed molybdopterins.[15] Figure 3.3 shows the coordination site around the molybdenum Mo^{VI} ion in three related enzyme catalysts: sulfite oxidase, xanthine oxidase and dimethyl sulfoxide reductase. Notable are the two thiol ligands in an *enedithiol* moiety of the molybdenum cofactor. Also, in sulfite oxidase and xanthine oxidase there is a double-bonded oxygen and a water ligand. An active site cysteine thiol ligand anchors the Mo^{VI} resting oxidation state in sulfite oxidase.

The b- and c-type cytochromes were initially named as pigments (chromes) in cells (cyto). They have different peripheral substituents around the central tetrapyrrole macrocycle, allowing for tuning of their one-electron redox potentials at the central Fe atom. This sets the directionality of electron flow from cytochrome b to the more positive potential of cytochrome c. They are both restricted ordinarily to one-electron transfer steps (Fe^{3+} to or from Fe^{2+}) and both the bottom and top axial ligands are typically provided by imidazole groups of histidine residues in their apoprotein partners, ruling out O_2 ligation and associated radical chemistry (Figure 3.3B).

The enzyme-bound b5 cytochrome has two histidine residue imidazole side chains as both top and bottom axial ligands to the heme as seen in Figure 3.3B. The external one-electron acceptor cytochrome c has a top axial histidine but a bottom axial methionine, which alters the reduction potential such that it functions as an electron acceptor, not a donor, from the reduced cytochrome b5 (Fe^{2+}).

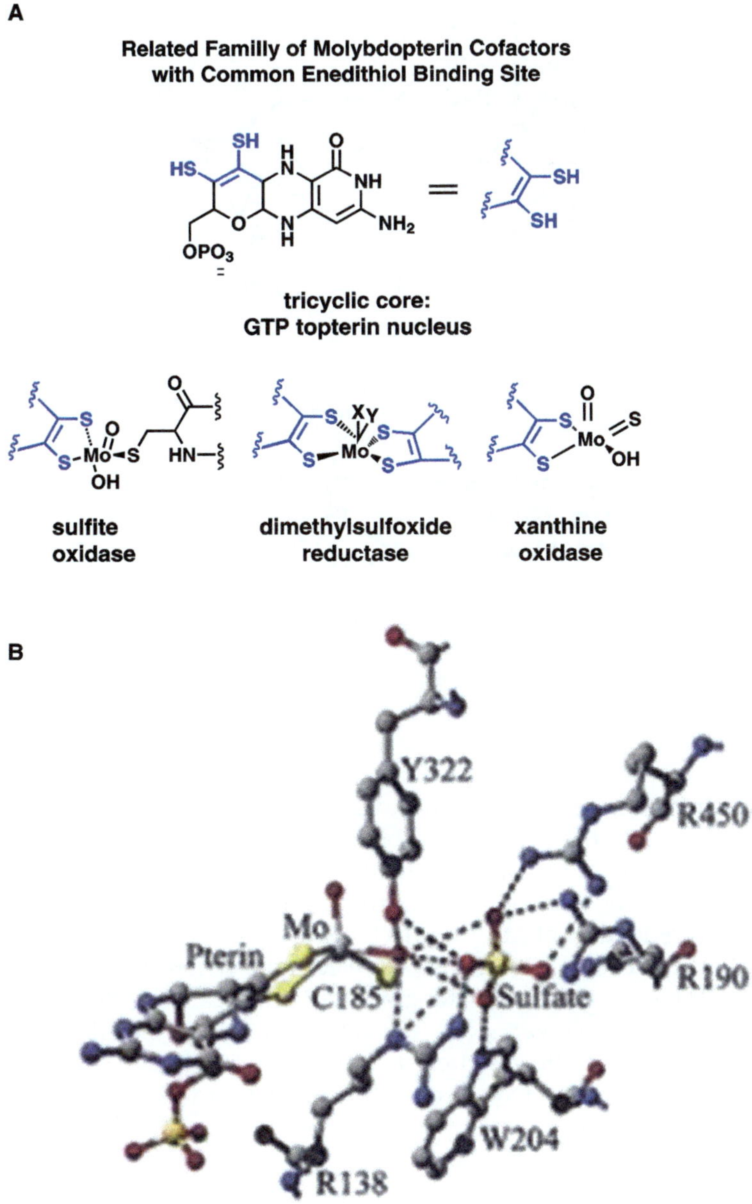

Figure 3.3 (A) Human sulfite oxidase is a molybdo-iron enzyme. The molybdopterin enedithiol group is the binding site for resting Mo^{VI} at the start of each catalytic cycle. A related family of molybdopterin enzymes has slightly different ligand sets around the molybdenum atom. (B) Molybdopterin–sulfate complex in sulfite oxidase (C). Top and bottom axial ligands from protein residues in cytochrome b5 and cytochrome c. (B) Reproduced from Proteopedia, courtesy of Kimberly Meyers.

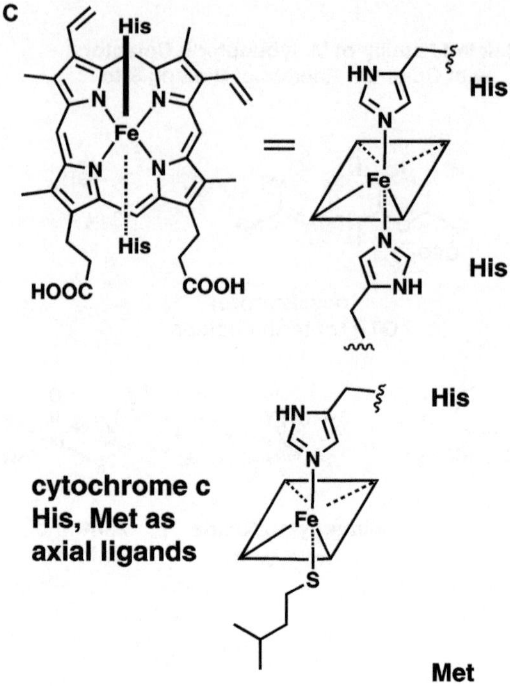

Figure 3.3 Continued.

The resting enzyme starts each catalytic cycle with the Mo in the +6 oxidation state and the cytochrome b5 in the Fe^{3+} state[16] (Figure 3.4). We will focus on electron transfer mechanism in the next paragraph but note that sulfite binding and net two-electron oxidation to sulfate product involves initial reduction of Mo^{VI} to Mo^{IV}.

The rest of the catalytic cycle involves electron transfer one at a time from Mo^{IV} to Fe^{III}. The first such transfer gives Mo^{V} and Fe^{II}. Now, this Fe^{II} version of the cytochrome b5 can pass one electron out to cosubstrate ferric cytochrome c. This reoxidizes the cytochrome b5 to the Fe^{III} state. Then, it can accept one more electron, from Mo^{V}, recreating the starting Mo^{VI} oxidation state and the Fe^{II} state. A second one-electron transfer out to ferric cytochrome c completes the catalytic cycle and has generated two molecules of Fe^{II}-cytochrome c as product.

Mechanistic investigations of sulfite reductase indicate that the nucleophilic sulfur of the sulfite substrate to be oxidized attacks the Mo=O double bond of the Mo^{VI} atom of the molybdopterin cofactor (Figure 3.4). The resultant adduct now has an Mo^{IV} oxidation state, signaling that sulfate oxidation has occurred by this point. Indeed, the intermediate is a Mo^{IV}-coordinated sulfate complex. Release by water addition creates the free sulfate product and the Mo^{IV} oxidation state that is presumed to be then reoxidized, one electron at a time, *via* an intermediate Mo^{V} oxidation state by

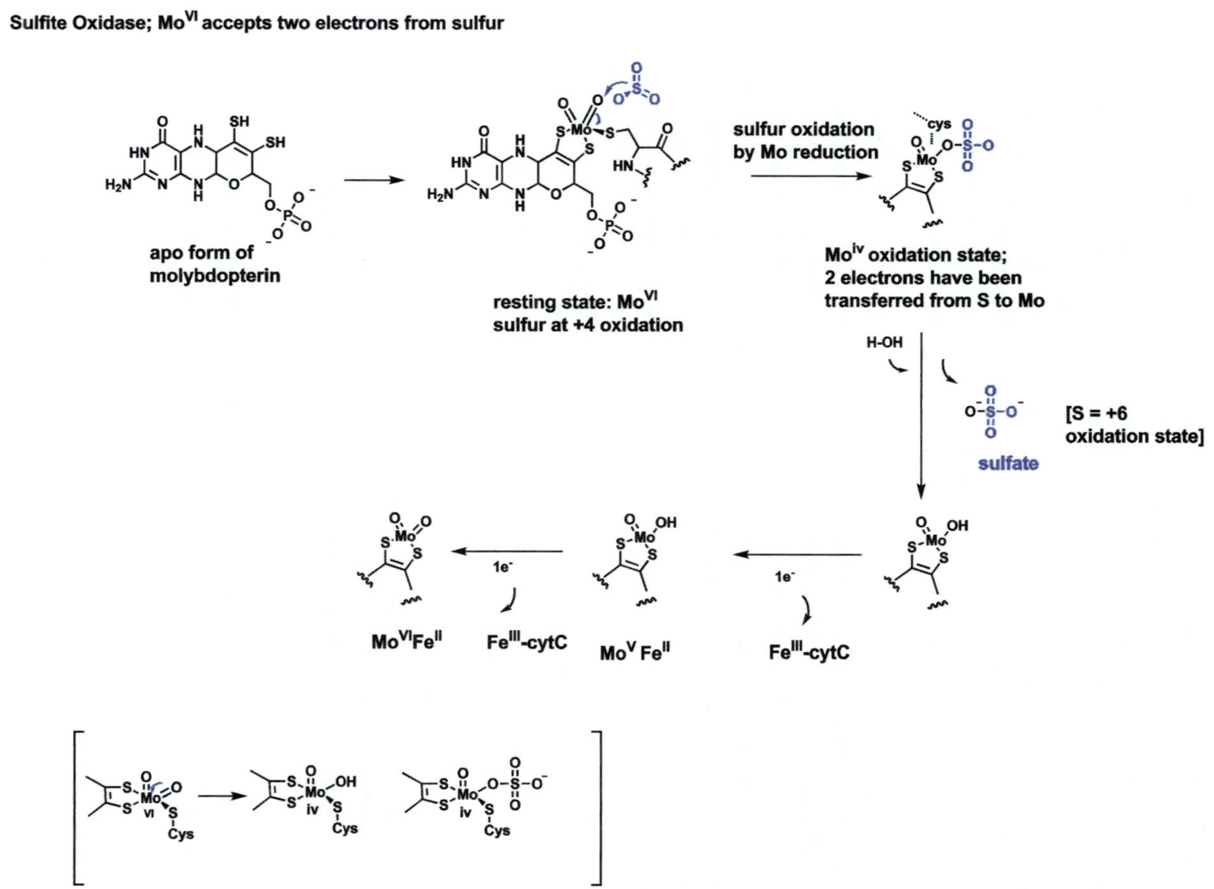

Figure 3.4 Mechanistic proposal for sulfite oxidase: attack of sulfite oxyanion on the molybdopterin–Mo=O to yield a coordinated sulfate ester that undergoes hydrolytic attack by water. The resultant Mo^{IV} species undergoes consecutive one-electron oxidations back to starting Mo^{VI}.

electron passage through the cytochrome b5 to exogenous cytochrome c as noted in the above paragraph and Figure 3.3A.

Examination of the Mo cofactor (Figures 3.3 and 3.4) structure reveals that the Mo atom is coordinated by two sulfurs in an enedithiol arrangement with a third nearby sulfur, from a cysteinyl side chain of the enzyme. The two oxo ligands complete the molybdenum^{VI} first coordination sphere. One of those oxo ligands is attacked by the nucleophilic sulfur atom of sulfite with flow of two electrons into Mo as the oxo double bond becomes a single bond and Mo^{VI} picks up two electrons to Mo^{IV} as shown (Figure 3.4). The resultant Mo^{IV} center has an oxidized sulfate ester as a ligand, readily exchangeable by a water molecule. That exchange releases the oxidized sulfate dianion product and leaves the Mo^{IV} oxidation state to undergo two single-electron transfers to the internal cytochrome c (Fe^{III} to Fe^{II} each time) to get back to the starting enzyme.

The biosynthesis of the molybdenum cofactor and the source of the two sulfurs has been studied extensively over several decades.[15] This turns out to be another pathway involving protein-bound cysteinyl–S–SH residues and will be considered in Part II. As noted in Figure 3.3 there are a small set of oxidases that contain the Mo cofactor or functional variants thereof, including the human enzyme xanthine oxidase, which is overly active in patients with gout symptomatology. Anti-gout drugs target and trap the intermediate Mo^{IV} state of that oxidase. This enzyme family is the major locus of molybdenum in human biology and reflects the affinity of thiol ligands for Mo^{VI} ions.[17] The dimethyl sulfoxide reduction to dimethyl sulfide is a widespread microbial process that generates the volatile thioether as product.

This presumptively ancient redox chemistry reveals how vitamin B_2, 4Fe–4S clusters, siroheme and molybdopterin had to have been biosynthesized and assembled in the key enzymes for the biological transit between inorganic sulfate and inorganic sulfide as dominant energy pathways for the approximately two billion years of early anaerobic life on the planet through to the last 600 million years of aerobic conditions. Plants in general practice assimilatory sulfate reduction but not dissimilatory sulfate reduction.

3.3 Assimilatory Sulfate Reduction: Transition to Organic Sulfur from Nascent Hydrogen Sulfide

A relatively small fraction of the inorganic sulfur, essentially exclusively at the level of inorganic sulfide ions, is converted biologically to organic sulfur. The main route for biological incorporation of inorganic sulfur to organic molecular frameworks occurs in bacteria as H_2S, from sulfate reduction as noted in Chapter 2, acts as a nucleophile in the reaction catalyzed by *O*-acetyl-L-serine sulfhydrylase (Figure 3.5) to generate L-cysteine.[18,19] We will examine the logic and mechanism of that enzymatic transformation in Part II of this volume. L-Cysteine is then metabolized to the other proteinogenic

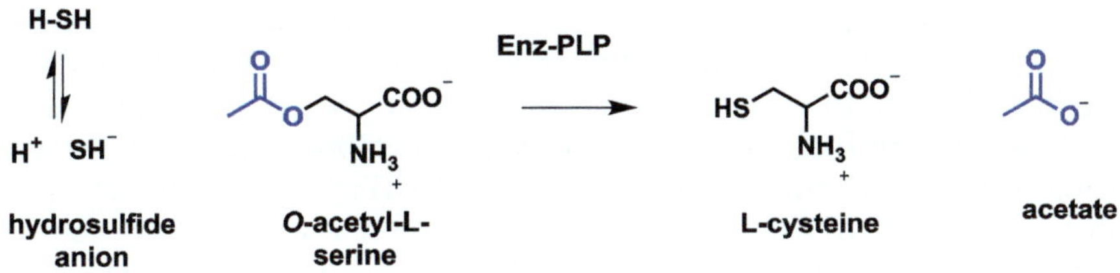

Figure 3.5 *O*-Acetyl serine sulfhydrylase: the pyridoxal-phosphate-containing enzyme that connects inorganic sulfide to the thiol framework of L-cysteine by net beta replacement.

amino acid L-methionine and the multiple roles of organic thiols and thioethers radiate from those two central sulfur amino acid metabolites.

Even though the focus in these two chapters has been on inorganic sulfate to sulfide metabolism and then back up, enabling two orthogonally acting sets of microbes to make livings on sulfur redox chemistry, it is also clear that many of the enzymes utilize cysteine–SH groups as essential participants in catalyses. Thus, the radiation of inorganic sulfur to the existing inventory of organosulfur molecular frameworks in microbes was effectively complete in order for the inorganic sulfur chemical biology to occur, as summarized in these two chapters.

3.4 Summary of Enzymatic Strategies for the Two Arms of Inorganic Sulfur Redox Cycles

In the evolution of early organisms on the earth to obtain energy from their surroundings, sulfur was more important than oxygen for more than a billion years. Although both oxygen and sulfur are in the same column of the periodic table of elements and so share some common properties, the ability of sulfur to exist in a variety of oxidation states from −2 (sulfide) to +6 (sulfate) allowed it to serve as both an electron donor in lithotrophic microbial energy metabolism and a terminal electron acceptor (elemental sulfur and sulfate) in chemotrophic metabolism. Sulfur the element (atomic sulfur) is a stable solid. Atomic oxygen is not stable, although its dimer, molecular oxygen, O_2, is and is now the dominant oxygen compound on the planet.

While O_2, when it finally evolved to high planetary levels from cyanobacterial photosynthesis, serves as a terminal electron acceptor for aerobic electron transport chains, it does not change its formal −2 oxidation state. In its redox versatility sulfur chemical biology is more akin to the multiple nitrogen redox states that are biologically accessible. Nitrogen, in its inorganic redox biochemistry, ranges from +5 in nitrate to −3 in NH_3. Indeed nitrate-reducing bacteria use this redox versatility to dump up to eight metabolically derived electrons to yield NH_3.

The differential positioning of microbes in water columns allows illustration of the spatial separation of the two roles of sulfur in microbial redox chemical biology. In the deeper anoxic regions, the sulfate reducers use SO_4^{2-} as the terminal electron acceptor for metabolic electrons removed during oxidation of various substrate molecules (Figure 2.2). Higher up the water column, often just above the anoxic–oxic interface zone one can find sulfide oxidizers. Now, the sulfur is the oxidizable substrate and O_2 is typically the terminal electron acceptor. This is a microcosm of both the evolution of anaerobic microbes to aerobes and the underlying chemical versatility of inorganic sulfur compounds to traverse eight redox states, with directionality controlled by partner molecules in the environment.

Sulfur is nucleophilic at the sulfide (−2) oxidation state and also at the +4 oxidation state in the sulfide anion. In the +6 oxidation state of sulfate the sulfur is only electrophilic. Sulfur in the low oxidation states also has high affinity for metal cations, most notably iron and molybdenum in redox biology as noted above, but also for Hg^{2+} and Zn^{2+} as will be discussed in the context of cysteine biology in Parts II and III. The ubiquitous iron–sulfur clusters in redox proteins throughout the kingdoms of life show the interplay between inorganic sulfide, organic thiolates and iron atoms.

The survey of inorganic sulfur redox biochemistry turned up two unusual covalent enzyme forms: a Cys–S–SO_3^- intermediate, and chains of polysulfides sandwiched between two cysteine residues (Figure 3.6).

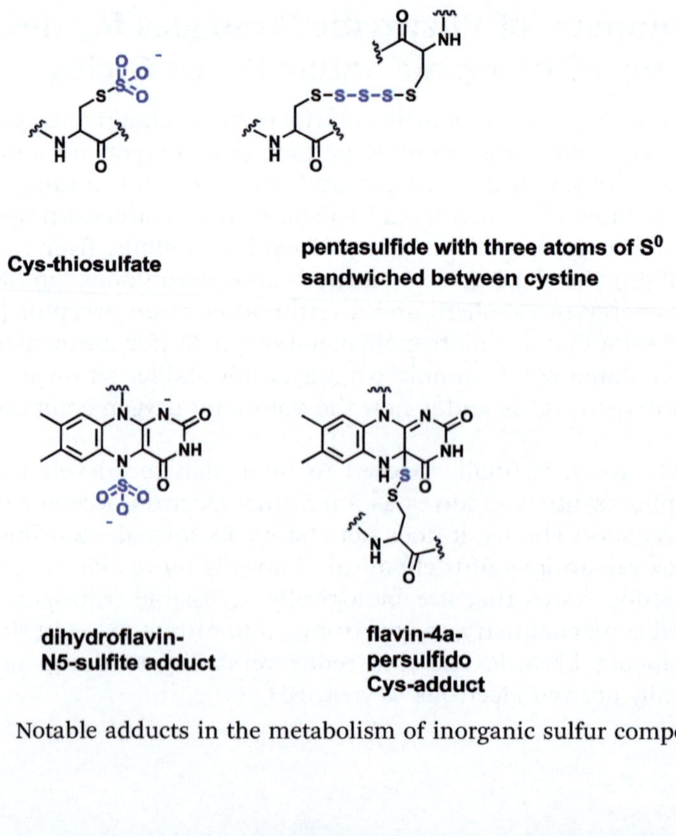

Figure 3.6 Notable adducts in the metabolism of inorganic sulfur compounds.

Additionally, two unusual flavin coenzyme adducts are the N5-sulfite adduct in sulfate-reducers' AMP sulfate reductase and the remarkable flavin-C4a-persulfido cysteinyl adduct (Figure 3.6) in the sulfide quinone oxido-reductase catalytic cycle. These four intermediates reveal the reach of both sulfur and flavin chemistry in inorganic sulfur redox metabolism.

Since the hydrogen sulfide precursor to Fe–S clusters in most organisms comes from cysteine *via* cysteine persulfides, that topic is also covered in Part II.

References

1. C. G. Friedrich, Physiology and genetics of sulfur-oxidizing bacteria, *Adv. Microb. Physiol.*, 1998, **39**, 235–289.
2. C. Cascio, Examples of sulfur-oxidizing Bacteria, education.seattlepi.com/examples-sulfuroxidizing-bacteria-4142.html, accessed June 5, 2019.
3. D. Pokorna and J. Zabranska, Sulfur-oxidizing bacteria in enironmental technology, *Biotechnol. Adv.*, 2015, **33**, 1246–1259.
4. W. Ghosh and B. Dam, Biochemistry and molecular biology of lithotrophicsulfur oxidation by taxonomically and ecologically diverse bacteria and archea, *FEMS Microbiol. Rev.*, 2009, **33**, 999–1043.
5. D. A. Fike, A. Bradley and C. V. Rose, Rethinking the ancient sulfur cycle, *Annu. Rev. Earth Planet. Sci.*, 2015, **43**, 593–622.
6. P. K. Moore and Y. Zhu, H_2S biosynthesis and catabolism: new insights from molecular studies, *Cell. Mol. Life Sci.*, 2017, **74**, 1391–1412.
7. P. Rose, P. Moore and Y. Z. Zhu, Hydrogen sulfide biosynthesis and catabolism: new insights from molecular studies, *Cell. Mol. Life Sci.*, 2017, **74**, 1391–1412.
8. U. Klapper and C. Dahl, Enzymology and molecular biology of prokaryotic sulfite oxidation, *FEMS Microbiol. Lett.*, 2001, **203**, 1–9.
9. M. M. Cherney, Y. Zhange, M. Solmonson, J. H. Weiner and M. N. G. James, Crystal structure of sulfide:quinone oxidoreductase from Acidithiobacillus ferrooxidans: Insights into sulfidotrophic respiration and detoxification, *J. Mol. Biol.*, 2010, **398**, 292–305.
10. J. A. Brito, F. Sousa, M. Stelter, T. M. Bandeiras, C. Vonrhein, M. Teixeira, M. M. Pereira and M. Archer, Structural and functional insights into sulfide quinone oxidoreductase, *Biochemistry*, 2009, **48**, 5613–5622.
11. M. R. Jackson, P. J. Loll and M. S. Jorns, X-ray strcture of human sulfide quinone oxidoreductase: insights into the mechanism of mitochondrial hydrogen sulfiide oxidation, *Structure*, 2019, **27**, 794–805.
12. N. Levine, Putting the spotlight on organic sulfur, *Science*, 2016, **354**, 418–419.
13. A. A. Santos, S. Venceslau, F. Grein, W. D. Leaitt and C. Dahl, A protein trisulfide couples dissimilatory sulfate reduction to energy conservation, *Science*, 2015, **350**, 1541–1545.

14. U. Kappler and C. Dahl, Enzymology and molecular biology of prokaryotic sulfite oxidation, *FEMS Microbiol. Lett.*, 2001, **203**(1), 1–9.
15. G. Schwarz and R. Mendel, Molybdenum cofactor biosynthesis and molybdenum enzymes, *Annu. Rev. Plant Biol.*, 2006, **57**, 623–647.
16. C. Feng, G. Tollin and J. H. Enemark, Sulfite oxidizing enzymes, *Biochim. Biophys. Acta*, 2007, **1774**, 527–539.
17. R. R. Mendel, The molybdenum cofactor, *J. Biol. Chem.*, 2013, **288**, 13165–13172.
18. W. Rabeh and P. Cook, Structure and mechanism of O-acetylserine sulfhydrylase, *J. Biol. Chem.*, 2004, **279**, 26803–26806.
19. B. Huang, M. Vatting and S. L. Roderick, The active site of O-acetylsulfhydrylase is the anchor point for bi-enzyme complex with serine acetyltransferase, *J. Bacteriol.*, 2005, **187**, 3201–3205.

Part II: Chemical Biology of Organosulfur Metabolites

Part II, comprising Chapters 4–11, focuses on the common organic scaffolds in which sulfur atoms are embedded in primary metabolism. The precepts of Part I obtain and some of the protein and cofactor machinery invented for dealing with inorganic sulfur also serves for organosulfur metabolite transformations. Most of the sulfur-based metabolic flux goes through the amino-acids cysteine and methionine and some key metabolites thereof (Figure 1).

The relevant sulfur oxidation states are the thiol side chain in cysteine, the thioether side chain in methionine and its cationic, trigonalized sulfonium derivative in *S*-adenosylmethionine (SAM). The fourth sulfur oxidation state of relevance in Part II is a return to sulfate in the form of the phosphoric–sulfuric mixed anhydride side chain of a 3′-phospho derivative of adenyl sulfate (from Chapter 2) as a sulfuryl donor to many kinds of cosubstrate nucleophilic oxygen and nitrogen groups.

The main bridge from inorganic sulfur to carbon–sulfur bond formation in organic metabolite scaffolds appears to be the fixation of sulfide (HS^- most likely) into the three-carbon amino acid framework of L-cysteine. That is the subject of Chapter 4. The coenzyme that dominates amino acid chemistry, the aldehyde form of vitamin B6-pyridoxal phosphate, is central to the armamentarium of cysteine-forming and cleaving enzymes. The metabolic chemistry of cysteine is dominated by the accessibility of the thiolate anion (pK_a approximately 8.5), the most powerful, freely diffusing small molecule nucleophile in cells.

In addition to the microbial and plant metabolic logic of directly replacing an *O*-acyl group in L-serine with the HS^- nucleophile, Chapter 4 examines the more circuitous mammalian route to cysteine. This is the transsulfuration pathway: net movement of the thiol at C4 of starting homocysteine to end up at C3 of cysteine by way of a seven-carbon covalent intermediate known as cystathionine. Again, PLP shows its coenzymatic virtuosity at the heart of two tandemly acting enzymes, cystathionine-β-synthase and then cystathionine-γ-lyase. Of the two thiol amino-acids, cysteine and its four-carbon homolog homocysteine, cysteine has arborized metabolically. Homocysteine has only one other key role to play, noted as a progenitor to methionine in Chapter 8.

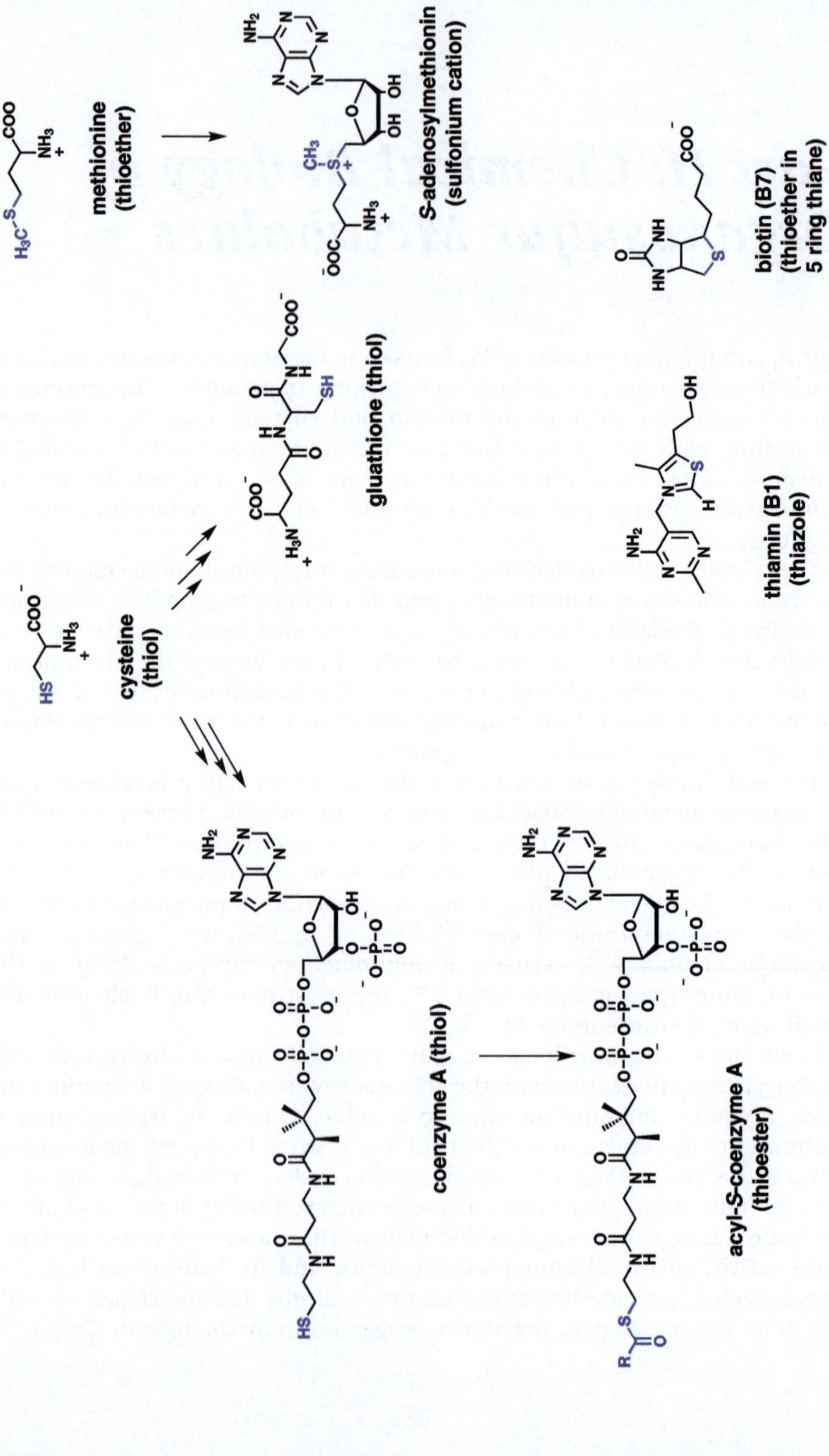
methionine
(thioether)
S-adenosylmethionin
(sulfonium cation)
cysteine
(thiol)
gluathione (thiol)
biotin (B7)
(thioether in
5 ring thiane)
thiamin (B1)
(thiazole)
coenzyme A (thiol)
acyl-S-coenzyme A
(thioester)

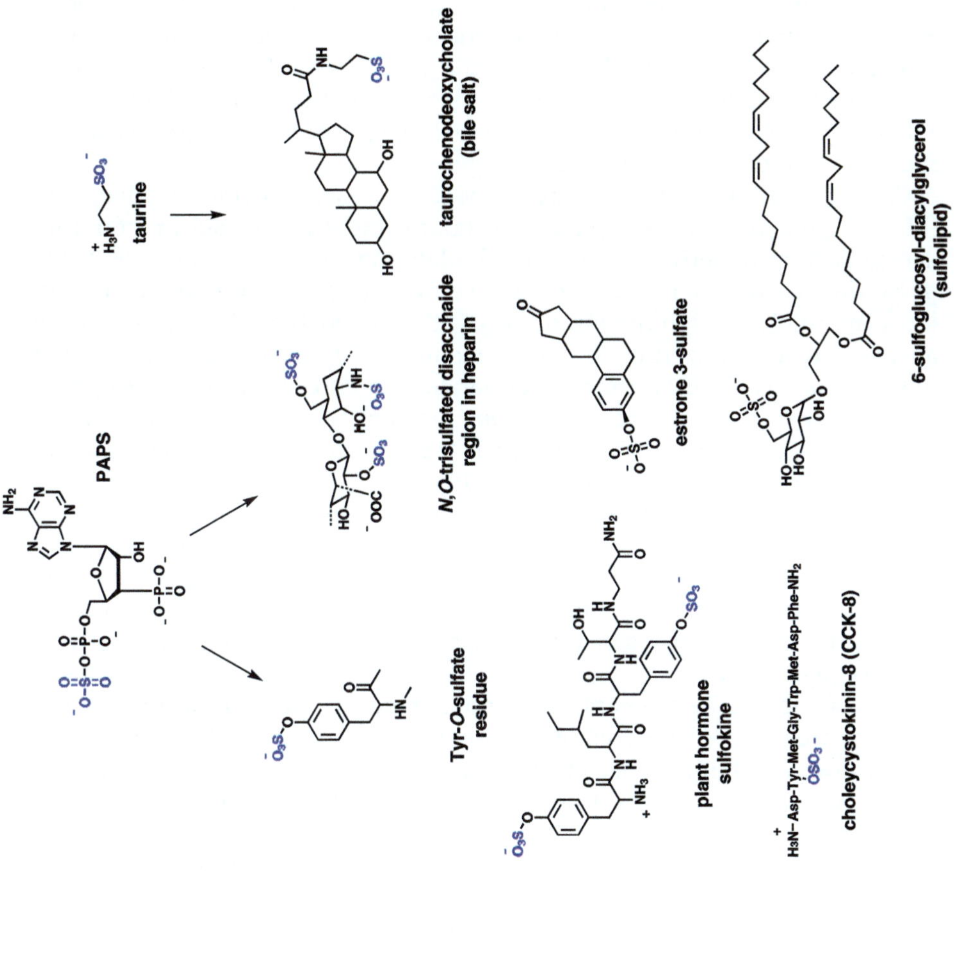

Figure 1 Examples of organosulfur metabolites.

Chapter 5 examines the scope of reactivity to which the cysteine-derived thiol is put in two cysteine-derived central metabolites: coenzyme A and glutathione. For vitamin B5-derived coenzyme A the main roles are to form acyl thioesters with a variety of activated carboxylic–phosphoric anhydrides, preserving the thermodynamic acyl group activation while enhancing kinetic stability. Acyl-*S*-CoA thioesters are the major acyl transfer currencies in electrophilic group transfers to many classes of cosubstrate nucleophiles, as well as activating C2 of those acyl groups as stabilized carbanion equivalents for C–C bond formations.

The second cysteine derivative with major metabolic role is the tripeptide glutathione with a cysteine residue embedded as the middle amino acid. Thiolate anion chemistry dominates the protective role of glutathione in scavenging reactive oxygen species (such as hydrogen peroxide), in covalently capturing and often detoxifying electrophilic organic groups and in the reduction of cystine disulfide bridges.

Chapter 6 starts with an unusual cysteine lyase, a pyridoxal phosphate (PLP)-enzyme, that cleaves cysteine to alanine and a sulfane sulfur atom that is captured by an active site cysteine thiolate to form the enzyme–Cys–S–SH persulfide adduct. It is this itinerant sulfur that is the ultimate source of inorganic sulfide ion in the hundreds of 2Fe–2S and 4Fe–4S clusters that are ancient one-electron transfer nanomachines. Of particular note is that PLP-enzymes can cleave cysteine either to this "six-electron" electrophilic sulfur in net β-replacement reactions or with different polarity, cleave cysteine to pyruvate, ammonia and HS^- in a β-elimination reaction manifold, an amazing chemical versatility.

Chapter 7 examines the purposeful modification of the thiolate sulfur of cysteine by sequential oxygen incorporation to the sulfenate, sulfinate, and finally the sulfonate oxidation state in which the SH of cysteine has become the trioxygenated sulfonic acid anion of taurine. This is both a pathway to control the amount of reactive free cysteine and a route to taurine for bile acid conjugations.

Two of the chapters, chapters 8 and 9, take up methionine first as the rather unreactive (compared with the cysteine thiol–thiolate) thioether-containing metabolite. That changes when the enzyme *S*-adenosylmethionine synthetase coaxes the thioether sulfur to attack C5′ of ATP to make *S*-adenosylmethionine (SAM). The permanently charged sulfonium cation activates all three carbon substituents for transfer as electrophilic groups to a range of cosubstrate nucleophilic atoms. Most dominant is the role of SAM as the premier methyl donor in cells and tissues, directly transferring a $[CH_3^+]$ equivalent to O, N, S and C nucleophiles.

Chapter 9, in contrast to the two-electron reaction paths described in the previous chapters, serves as a reminder that both thiols and the sulfonium center in SAM are also prone to one-electron transfer reaction manifolds under the aegis of specific enzyme catalysts. This introduces radicals in catalysis. While cysteinyl radicals are known species in some key enzymatic reactions, it is the one-electron chemical universe of SAM that is thought to populate more than 100 000 predicted enzymatic reaction cycles. In all those radical SM open reading frames, homolytic fragmentation of a SAM molecule coordinated to a 4Fe–4S cluster liberates the adenosyl moiety as a 5′-deoxyadenosyl carbon-centered radical (5′dA•) in the active site. 5′dA• is the robust one-electron oxidant that can create radicals in cosubstrates bound in the same enzyme. active site.

Chapter 10 returns to sulfur in the +6 oxidation state as sulfate and examines the molecular logic and catalytic protein machinery for sulfuryl group ($-SO_3^-$) transfer to both low molecular weight and high molecular weight (protein) substrates. The logic is entirely parallel to phosphoryl group ($-PO_3^{2-}$) transfers but of more limited biological scope. To continue the parallel to phosphate chemical biology, there are phosphatases for low molecular weight (*e.g.* glucose-6-phosphate) and for control of and hydrolytic removal of protein phosphate groups from protein side chains. Similarly, there are comparable sulfatase enzymes, with a novel mechanistic twist to the hydrolysis routes, involving an unusual formylglycine residue in sulfatase active sites. Failure to efficiently remove the $-SO_3^-$ group from sulfated proteins in lysosomes is the underlying cause of multiple human lysosomal storage diseases.

Chapter 11 is a short chapter focusing on how C–S bonds are formed in biological systems. It reprises the four major C–S bond-forming strategies in primary metabolism. Then, it examines a selected set of C–S bond-forming steps in secondary, natural product metabolic pathways to show similarities and distinctions in sulfur chemical biology.

The eight chapters in Part II are far from encyclopedic in tracking down every instance of thiolate, thioether, sulfonate or sulfate two-electron and one-electron chemical biology. Nor are they comprehensive explorations of the full scope of thiol and thioether oxidations. They are not meant to serve such roles.

Rather, they outline the major reactivity patterns of the central organosulfur low molecular weight metabolites—cysteine, homocysteine, methionine, *S*-adnosylmethionine (SAM), taurine and 3′-phosphoadenosine-5′-phosphosulfate (PAPS). These patterns arise from the one-electron and two-electron pathways available to the sulfur atoms embedded in those metabolite frameworks as they range between low valent, nucleophilic states ($S = -2$) to high valent electrophilic states ($S = +6$). Reactivity of the trivalent sulfonium cation in SAM, of the ancient one-electron Fe–S donors, of sulfane sulfur *vs.* HS^-, of RSH *vs.* RS–SH and RS–SR, and of the sulfuric–phosphoric anhydride moiety of adenosine phosphosulfate (APS) and PAPS, emphasize the richness of organosulfur chemical biology.

CHAPTER 4

Organosulfur Metabolites in Biological Systems

4.1 Organosulfur Molecular Scaffolds in Biological Systems

As we will note below in Section 4.2 and following on the skeletal outline of the transformation in Figure 3.5, there is one major metabolic pathway connector between the inorganic sulfur biologic universe and the broad sweep of organosulfur metabolites. The essential step is the enzymatic conversion of inorganic sulfide to the thiol group of cysteine. Essentially all organosulfur metabolites radiate from that three-carbon entry point[1,2] (Figure 4.1). The corresponding four-carbon thiol amino acid homocysteine is the bridge between cysteine and the thioether framework of the other sulfur-containing proteinogenic amino acid L-methionine. The catalyzed migration of the thiol group from the three-carbon cysteine to the four-carbon homocysteine scaffold occurs by consecutive action of three enzymes involved in the net *transsulfuration* process.

Methylation of the thiol group of homocysteine by two mechanistically distinct enzymatic variants creates the thioether linkage in methionine and opens up biological chemistry distinct from that of the thiol group. Most notable is the *S*-adenosylation by *S*-adenosylmethionine synthetase that trigonalizes the thioether sulfur, creating a permanent positive charge in the sulfonium ion sulfur center. The thermodynamic activation, combined with sufficient kinetic stability to hydrolysis,[3] allows *S*-adenosylmethionine (SAM) to function in two quite distinct metabolic niches.[4]

SAM is best known for its role as the premier biological methylating agent, transferring $[CH_3^+]$ equivalents to a huge variety of nucleophilic

Chemical Biology No. 12
The Chemical Biology of Sulfur
By Christopher T. Walsh

Published by the Royal Society of Chemistry, www.rsc.org

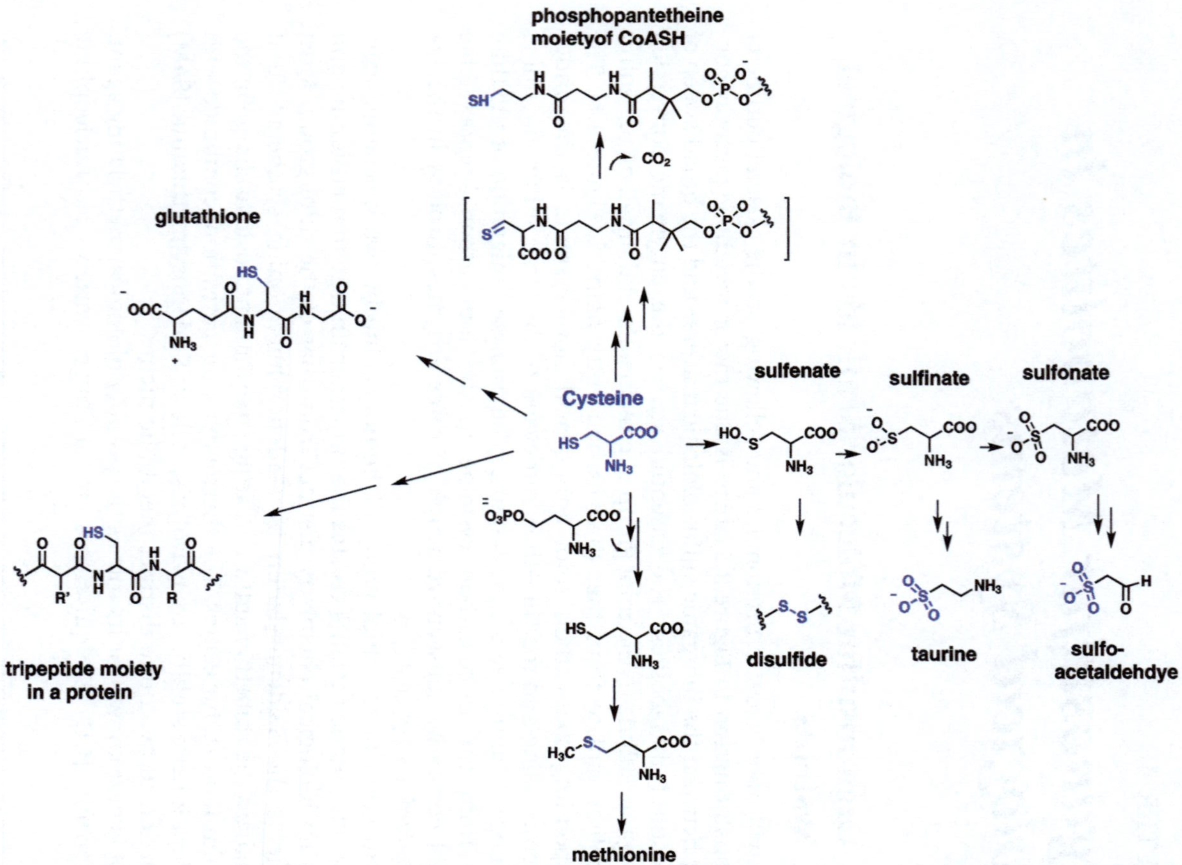

Figure 4.1 Cysteine is the point of entry from inorganic sulfide ions for diversification of sulfur-containing organic scaffolds in metabolism.

cosubstrates, large and small. In a parallel set of reaction manifolds, a large family of radical SAM enzymes carry out homolytic cleavage of the adenosyl-C5′–sulfur bond to create the 5′ deoxyadenosyl radical as initiator of a wide range of radical-based enzymatic transformations of substrates into products.

The thiol group of cysteine serves many cellular functions.[2] Two major metabolites are coenzyme A and the tripeptide glutathione. In coenzyme A (CoA) the three-carbon cysteine moiety has undergone biosynthetic decarboxylation and the terminal cysteamine thiol is the key to coenzyme functions. As a nucleophile the $CoAS^-$ thiolate attacks activated acyl-phosphate and acyl-AMPs to convert them from kinetically labile mixed anhydride metabolites to *acyl thioesters* that are kinetically stable but still thermodynamically activated.[3] Acetyl CoA, malonyl CoA, succinyl CoA, acetoacetyl CoA, hydroxymethylglutaryl CoA, along with myristoyl CoA and palmitoyl CoA, are examples of activated acyl group metabolites that are stable enough to serve as diffusible substrates (Figure 4.2).

The second activating function of the sulfur in acyl CoA thioesters is the lowering of energy barriers for deprotonation of C-2 of the acyl group to the C-2-carbanion. The negative charge on that carbon is delocalized onto the adjacent more basic sulfur as the thioester enolate and thereby provides stabilization. The kinetic acidity of the C–H bonds at C2 in acyl thioesters are higher than in the corresponding acyl oxoesters (Figure 4.3A). This gives an advantage (lower activation barrier) to C–C bond-forming steps from acyl groups of thioesters.

Analogously, the thermodynamic activation for acyl group transfer is more pronounced in the thioesters *vs.* the oxoesters. One measure is the equilibrium constant ($K_{eq.}$) for hydrolysis. The $K_{eq.}$ for acyl thioester hydrolysis is approximately 10^5, while the $K_{eq.}$ for the comparable acyl oxoester hydrolysis is approximately 10^2, reflecting a greater driving force for acyl transfer to water in the thioester case, a measure of thermodynamic group transfer potential[5] (Figure 4.3B).

The C-2-carbanions of acyl CoAs are the key nucleophiles in the first step of the citrate cycle and also the first two steps of the steroid biosynthetic pathway. The sulfur atom in acyl thioesters achieves an orthogonal activation at C-1 of the acyl group as carbon electrophile and at C-2 as carbon nucleophile[6] (discussed in detail in Chapter 5).

The conversion of cysteine to the tripeptide glutathione by two soluble, ATP lyases, generates the thiol-containing tripeptide in concentrations up to 10 mM in eukaryotic cells (discussed in the next chapter). The internal cysteine residue has its amine tied up in peptide linkage so *S*-acyl species do not rearrange to the more stable and irreversible *N*-acyl species as would happen in the free amino acid cysteine. The thiolate form of glutathione is then used as the major surveillance and/or detoxification molecule in many reactions where capture of cosubstrate electrophiles by the nucleophilic thiolate anion is the goal.

Figure 4.1 lists a set of other organosulfur functional groups found in cellular metabolites. Thee include oxygenated forms resulting from frequent autoxidation of cysteine, glutathione and cysteinyl residue thiol side chains in proteins to the monooxygenated sulfenic acid side chains.[7] (For the methionyl thioether functional group, the corresponding mono-oxidized grouping is the sulfoxide). In sulfenic acids the sulfur atom is now *electrophilic, not nucleophilic*. The most common fate of sulfenates is capture by another thiol, as the attacking thiolate anion, to create disulfide links, although isolated sulfenates can persist and have regulatory effects, *e.g.* on epidermal growth factor receptor (EGFR) function.[8,9] This is a major route of detoxification of reactive oxygen species by glutathione and also a common fate of dithiols in proteins in non-reducing microenvironments.

Further oxidation of cysteines to sulfinates and sulfonates are often accidental but actually purposeful in the biosynthesis of taurine as will be examined later in Part II. Correspondingly, methionine sulfoxides can go on irreversibly to methionine sulfone residues in proteins. *S*-Nitrosylation of

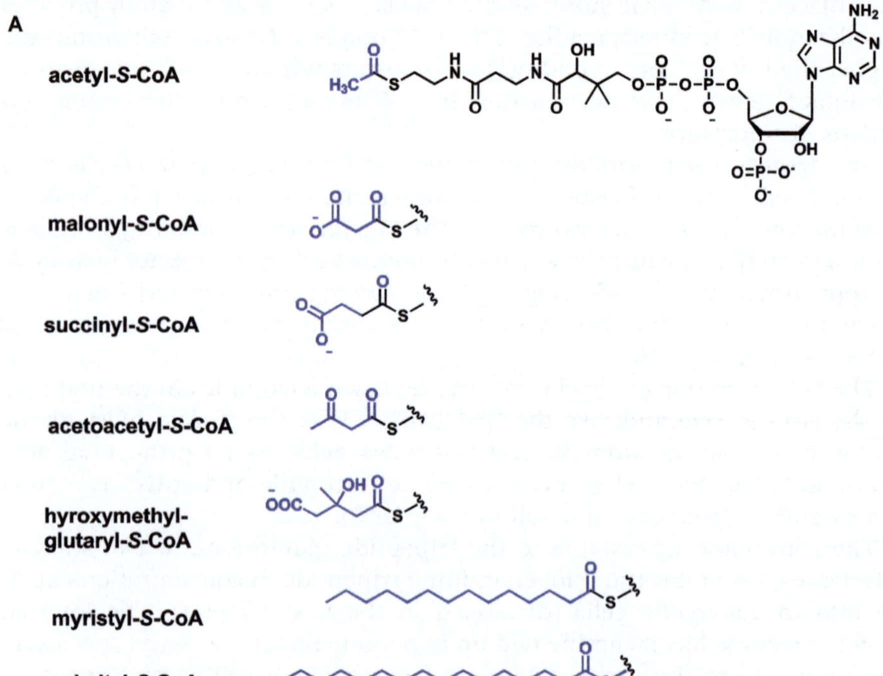

Figure 4.2 Acyl CoA metabolites; (A) some common acyl-*S*-CoA thioesters in metabolism. (B) The sulfur in the acyl thioester linkage activates C1 of the acyl group as an electrophile and C2 as a carbon nucleophile. The citrate synthase and acetoacetyl CoA thiolase enzymes demonstrate the synthetic utility of the carbanion equivalents in C–C bond formations.

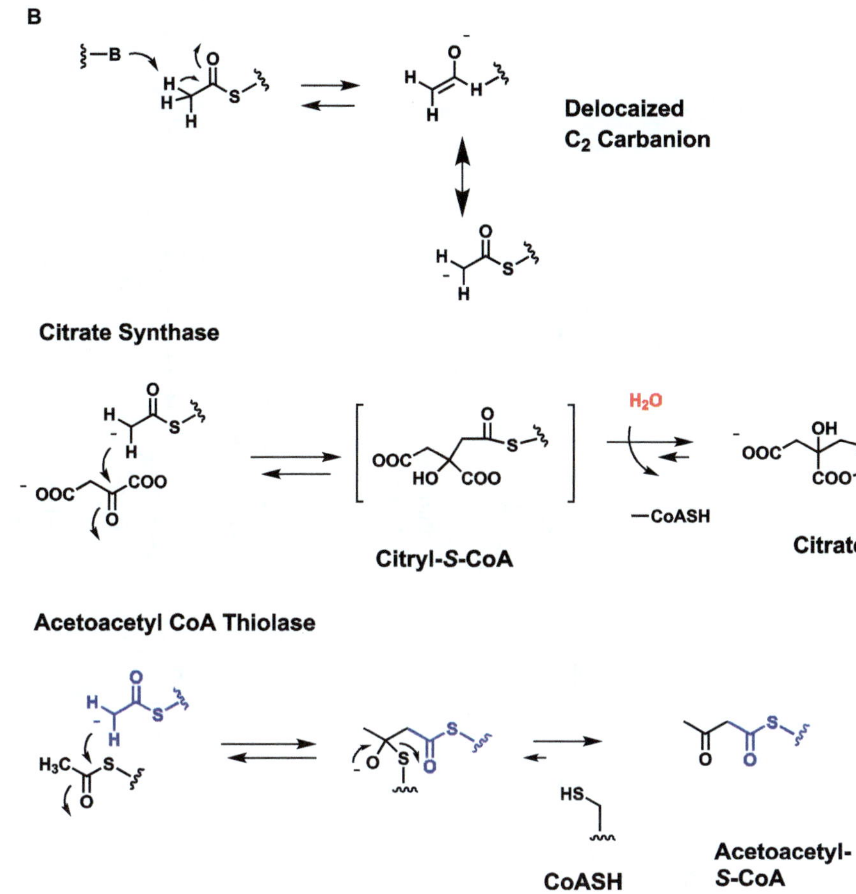

Figure 4.2 Continued.

glutathione and of cysteinyl residues in proteins is also an oxidizing outcome, thought, in some cases, to have signaling functions.

Perhaps the most surprising of the purposeful modifications of cysteine is the conversion of protein cysteinyl side chains to persulfide groups -Cys–S–SH.[10] The pK_a of the terminal SH is about two pH units (pK_a approximately 6.2) lower than that of the thiol so the persulfide anion (-Cys–S–S$^-$) is a major species at physiological pH ranges. Indeed, we have noted it as a nucleophile towards the bound flavin coenzyme in the sulfide quinone oxidoreductase mechanism in Chapter 3. However, when protonated the terminal SH of persulfides can be transferred as electrophilic sulfur species. Furthermore, the terminal sulfur of cysteinyl-persulfide residues is the progenitor to the inorganic sulfide ion that assembles with iron atoms into both 2Fe–2S and 4Fe–4S clusters, thought to be perhaps the most ancient of the iron-based one-electron-transfer cofactors in living cells (Figure 4.4).

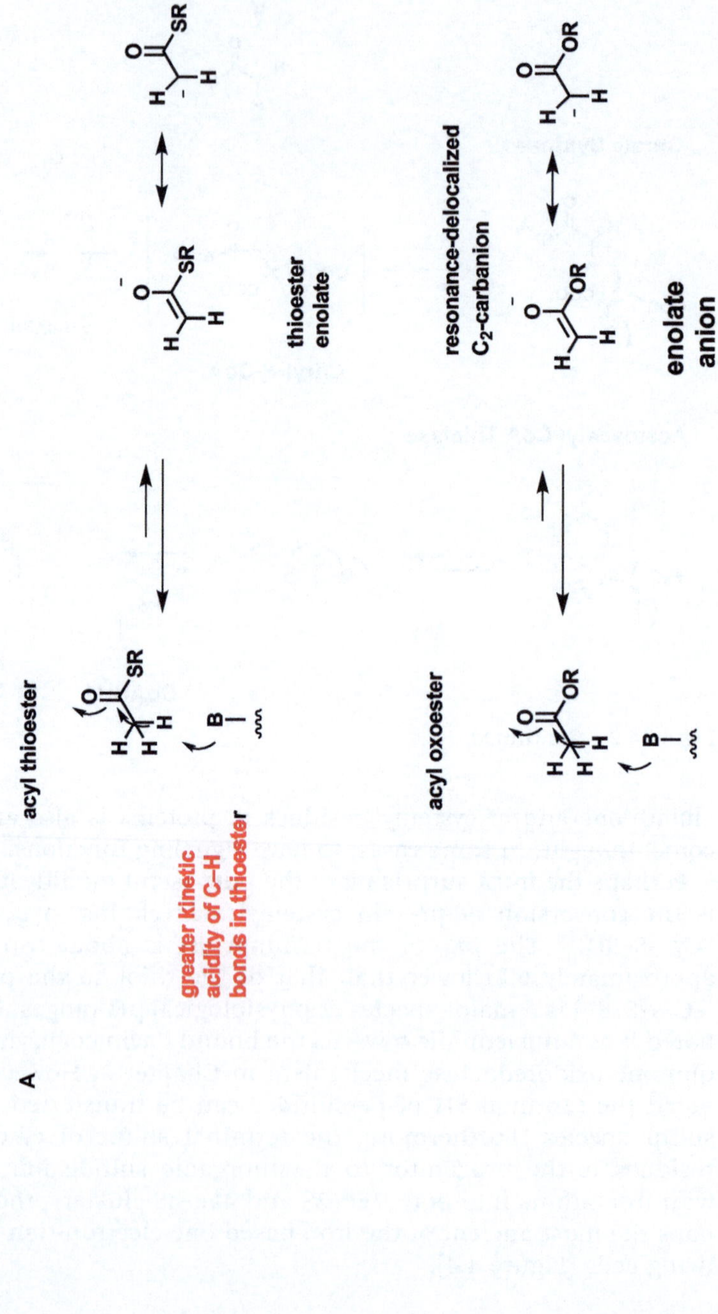
A
acyl thioester
greater kinetic
acidity of C–H
bonds in thioester
thioester
enolate
acyl oxoester
resonance-delocalized
C2-carbanion
enolate
anion

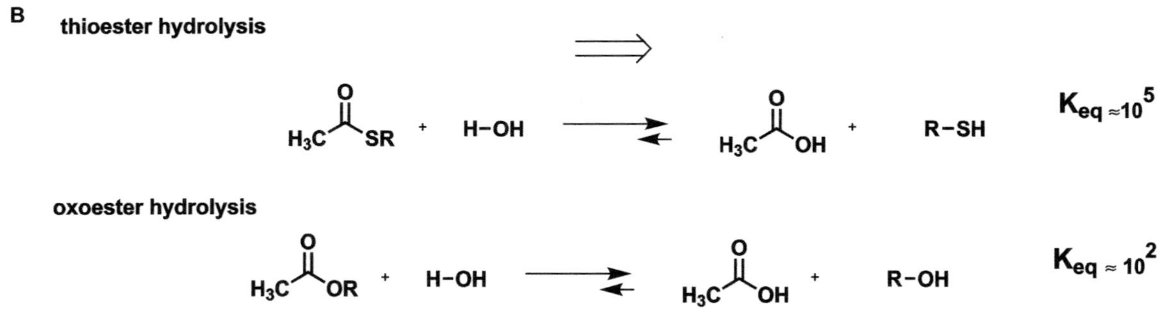

Figure 4.3 (A) Acyl thioesters *vs.* acyl oxoesters: thioester enolate *vs.* enolate as stabilized carbanion equivalents. (B) Difference in thermodynamic activation of acyl thioesters *vs.* acyl oxoesters revealed in equilibrium constants for hydrolysis.

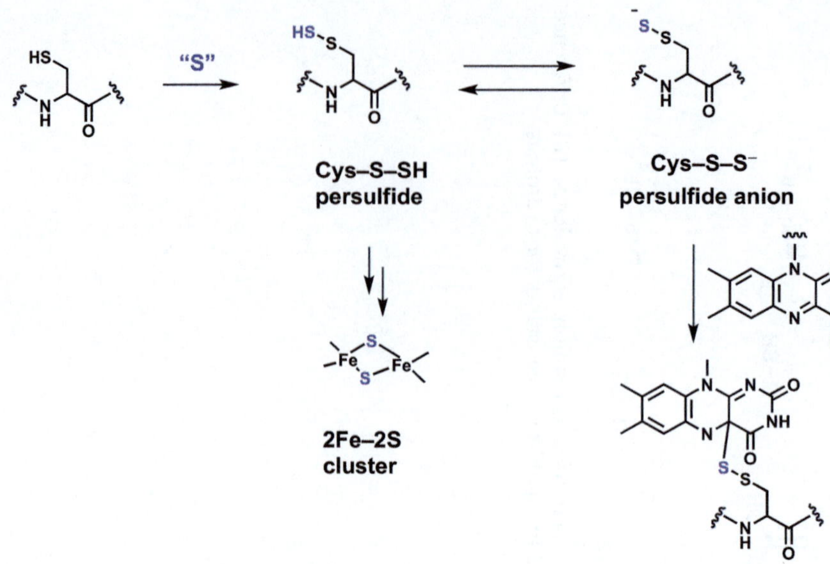

Figure 4.4 Cysteine persulfides: differential fate of protonated persulfide side chains *vs.* persulfide anions.

4.2 Assimilation of Inorganic H_2S into the Organic Scaffold of L-Cysteine

The major route for conversion of inorganic sulfur to the C–S bond in organic sulfur metabolites is catalyzed in plants and microbes by the enzyme *O*-acetyl serine sulfhydrylase. The two substrates are the hydrosulfide anion, in equilibrium with dissolved H_2S (pK_a = 7) and the amino acid L-serine that had been previously *O*-acetylated by an acetyl CoA-dependent *O*-transferase. The purpose of the *O*-acetylation is to set up the serine–OH as a lower energy leaving group in the next reaction, the sulfhydrylase reaction, also known as the cysteine synthase reaction.[11] In some organisms the transferase and sulfhydrylase form a functional complex to shepherd serine through to cysteine. We discuss the acyl group activating effect of the adjacent sulfur atom in such acyl thioesters in Chapter 5.

One quick way to think about leaving group energy barriers is to consider the pK_a of the protonated version of the potential leaving group. If unacetylated serine were the substrate, the pK_a of product HOH is approximately 14. For *O*-acetyl serine, with acetate as the leaving group, the pK_a of acetic acid is around 5. That is nine powers of 10 difference. If the ΔpK_a differences were all translated into the energy barrier for departure (it is not very likely that that is the only factor), that could correlate to a 10^9-fold rate advantage for cleaving C3–*O*–acetyl over C3–OH.

The *O*-acetyl serine lyase enzymes are members of the large family of pyridoxal-phosphate (PLP)-containing enzymes. This is the coenzymatically active form of vitamin B_6 (pyridoxine) and brings an aldehyde functional group to the catalytic inventory for cell metabolism.[6,12] The B_6-based PLP coenzyme typically binds tightly to its apoenzyme partner by both electrostatic interactions and most notably through a C=N imine linkage to the ε-NH_2 of a lysine side chain in the enzymes' active sites. Thus, PLP does not dissociate from its enzyme partner at the end of each catalytic cycle but stays tethered. The concentration of free PLP in most cells is negligible.

The aldehyde (PLP) to amine (*e.g.* of amino acids) linkage is termed an aldimine and reflects the resting stage of almost all PLP-containing enzymes (glycogen phosphorylase is a prominent outlier). The coenzymatic roles of PLP are synonymous with the interconversion of amino acid metabolites and participate in enzymatic reactions that cleave bonds at Cα-, Cβ- and Cγ of specific amino acid substrates.[13]

O-Acetylserine sulfhydrylase or cysteine synthase starts every catalytic cycle by forming the aldimine between the amino acid substrate and the PLP aldehyde carbonyl (Figure 4.5). Now, the C2–H is acidic, because its removal by an active site base yields the C2 carbanion that is stabilized by delocalization—the main role of the hetero-aromatic PLP coenzyme (sometimes termed an "electron sink"). That C2 carbanion electron pair can be used as neighboring group to assist in cleavage of the C3–O bond, resulting in elimination of the observed acetate product.[14,15]

The three-carbon serine-derived scaffold is then present as an enamino acid–PLP complex. In contrast to free enamines where the terminal carbon is nucleophilic, the terminal olefinic carbon in the enamino–PLP complex is electrophilic because the heteroaromatic PLP aldehyde attached to Cα is an electron sink.

At this stage the hydrosulfide anion (50% of the total dissolved H_2S at pH 7) acts as nucleophilic cosubstrate on C-3 of the enamino-PLP adduct. This is the C–S bond-forming step and constitutes the *organification* reaction that converts inorganic sulfur into carbon-based sulfur scaffolds.

Two steps remain for release of L-cysteine product. Protonation of the cysteinyl–PLP C2 carbanion precedes release of the free cysteine from the Cys–PLP aldimine nascent product as the active site lysine side chain ε-NH_2 adds back into the nascent cysteine=PLP aldimine product. The tetrahedral geminal diamine can collapse in the forward direction to liberate cysteine and regenerate the resting enzyme–Lys=PLP aldimine, ready for the next catalytic cycle.

In this key conduit from inorganic sulfide to the thiol group at C3 of cysteine there is no formal change in sulfur valency (−2). Both inorganic sulfide and the thiol of cysteine (and downstream homocysteine) are divalent and powerfully nucleophilic. Once embedded in the organic framework of cysteine by action of the *O*-acetylserine sulfhydrylase, the sulfur atom comes along as metabolism morphs the carbon scaffold of cysteine into many other metabolites described in the next sections. Some of

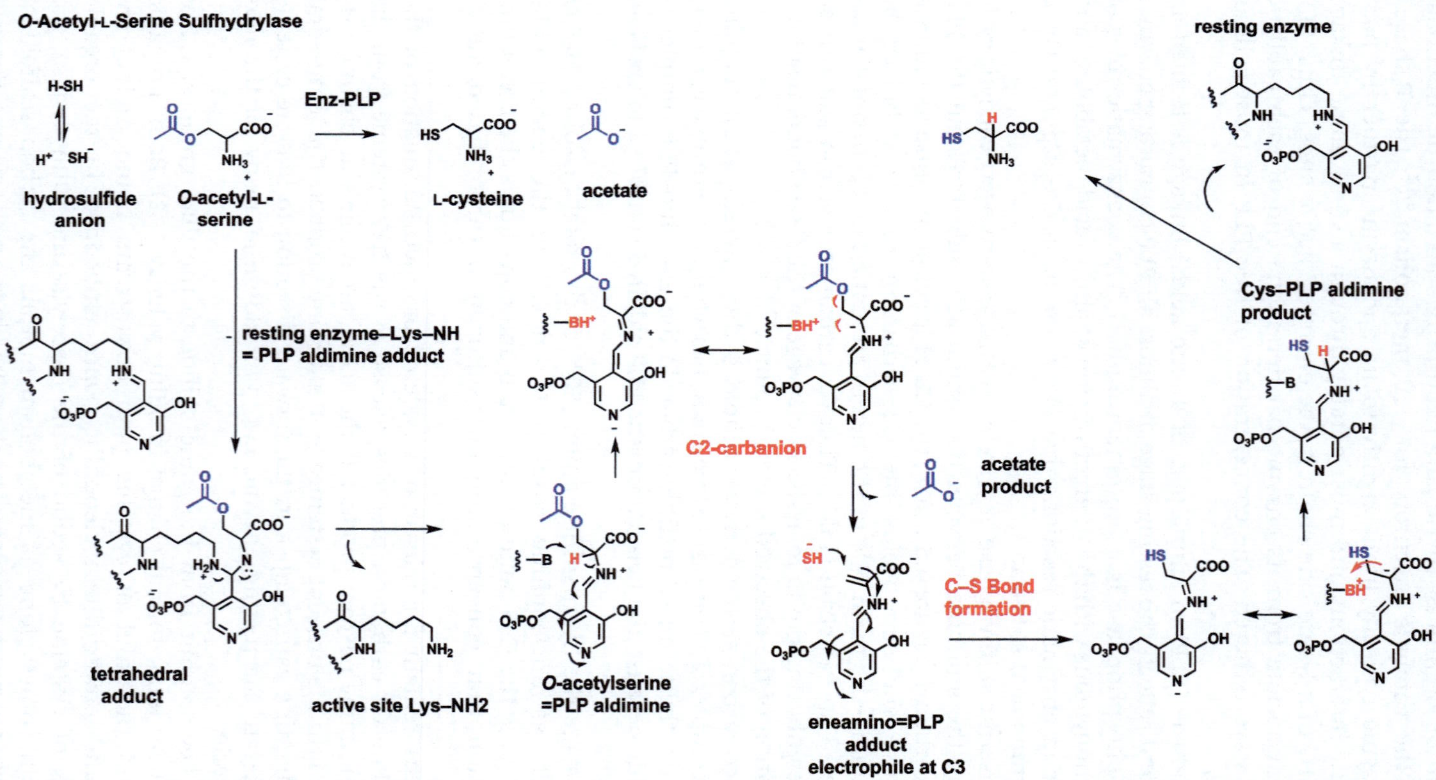

Figure 4.5 A detailed mechanistic scheme for the role of PLP as an electron sink, stabilizing substrate carbanion equivalents in amino acid–PLP aldimine adducts in enzyme active sites. *O*-Acetylserine sulfhydrylase catalyzes a net β-replacement of the departing *O*-acetyl group by the incoming HS in the cysteine product.

those enzymatic transformations do alter the oxidation state and connectivity of the sulfur atom.

The *O*-acetylserine to cysteine reaction is termed a β-replacement. In a net sense, the *O*-acetyl group is formally replaced by the incoming –SH group. This is the nomenclature for PLP–enzymes that leave the C2-amino group oxidation state unchanged between substrate and product. This nomenclature applies to *O*-acetylserine sulfhydrylase even though the actual mechanism is an elimination of the *O*-acetyl and then an addition of the –SH group, not a direct S_N2 replacement.

We will contrast this with β-elimination reactions (Figure 4.6) where C2 ends up as the oxidized keto acid rather than the amino group in the starting substrate.[6,13] One such example is serine deaminase where the PLP–enzyme carries out a net elimination of water across C2 and C3 of serine. The released product is the free enamino acid. That is the less stable tautomer of iminopyruvate. In aqueous solutions the imine is rapidly hydrolyzed to pyruvate and ammonia. The classic outcome of a β-eliminase is reduction of substrate amino acid at C3 and oxidation at C2. C3 becomes a methyl group, C2 becomes a ketone. The internal redox equation balances and no external electron transfer cosubstrate is required or consumed.

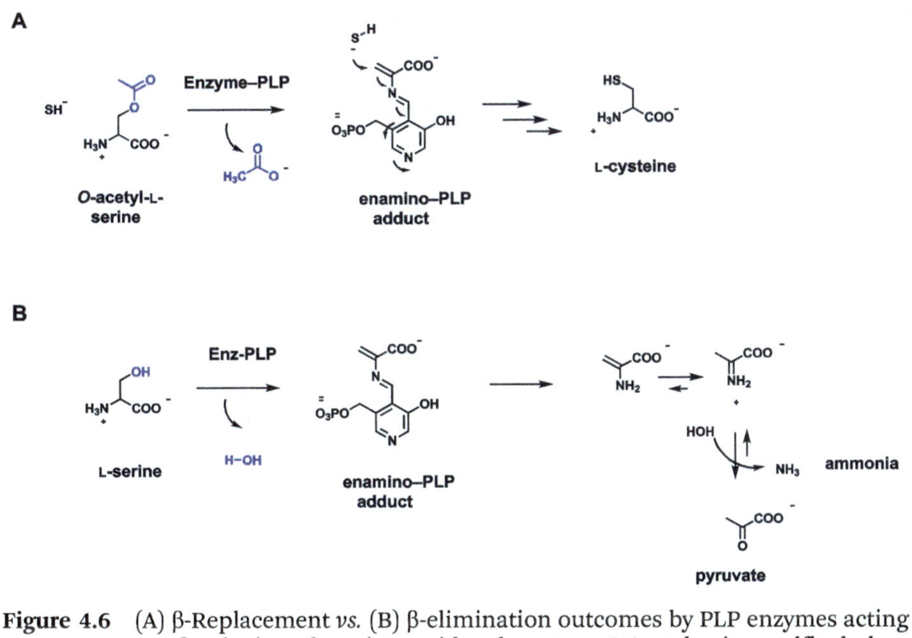

Figure 4.6 (A) β-Replacement *vs.* (B) β-elimination outcomes by PLP enzymes acting on β-substituted amino acid substrates. *O*-Acetylserine sulfhydrylase (=cysteine synthase) *vs.* serine deaminase. The β-elimination enzymes yield α-keto acid products as distinct from the α-amino acids in net replacement outcomes.

It is the kinetic fate of the enamino–PLP adduct that is at the heart of the underlying mechanistic logic (Figure 4.6). The bacterial enzyme from *Salmonella typhimurium* is an efficient catalyst, processing some 280 molecules of *O*-acetylserine to cysteine every second.[14] In each cycle, the loss of the *O*-acetyl group is rate-limiting, so HS^- addition is at least tenfold faster than the 280 s^{-1} rate.

4.3 Transsulfuration Pathway Between Cysteine and Homoserine in Mammalian Metabolism

4.3.1 The Four-carbon Scaffold of Homocysteine Derives from Methionine Metabolism

Humans do not have an enzyme equivalent to *O*-acetylserine sulfhydrylase. They obtain cysteine by a less direct pathway, ultimately from the four-carbon framework of homocysteine, itself a metabolite from methionine[16] (Figure 4.7). The thiol sulfur on the four-carbon framework is indirectly transferred to the three-carbon framework in serine. In turn methionine is one of the nine required amino acids that must be obtained dietarily: methionine is classified as an essential amino acid with a recommended daily intake of 19 mg kg^{-1}, corresponding to approximately1.4 g per day in a 150 lb adult. Ultimately then, it is dietary methionine sulfur that yields the sulfur atom in cysteine in humans.[17]

The homocysteine derives from methionine in three metabolic steps, of which two enzymes, *S*-adenosylmethionine synthetase and a host of SAM-utilizing methyltransferases, which are elaborated in Chapter 8, give *S*-adenosyl homocysteine (SAH). The third enzyme, SAH lyase carries out a C–C bond cleavage (Figure 4.8). It liberates the homocysteine chain from the C5′ ribose carbon by series of unobvious redox steps, proceeding through the 3′-keto-SAH, then elimination to homocysteine and the 4′,5′-enoyl-3′-ketoadenosine.[18] Conjugate addition of water to C5′ followed by re-reduction

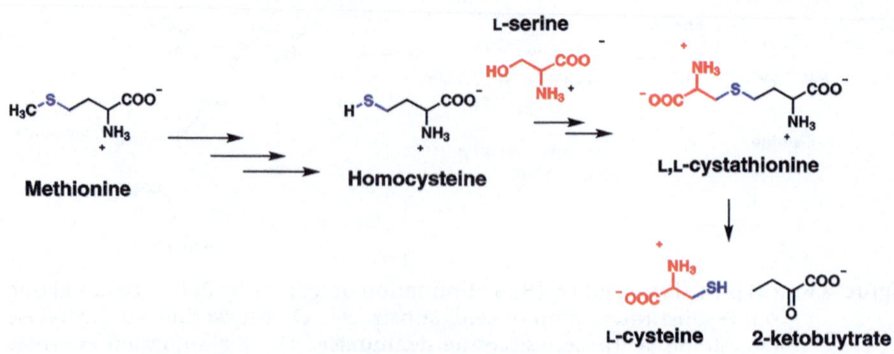

Figure 4.7 Cysteine biosynthesis in eukaryotes by a transsulfuration pathway from homocysteine and serine to cystathionine by β-replacement and then to cysteine and 2-ketobutyrate by γ-elimination.

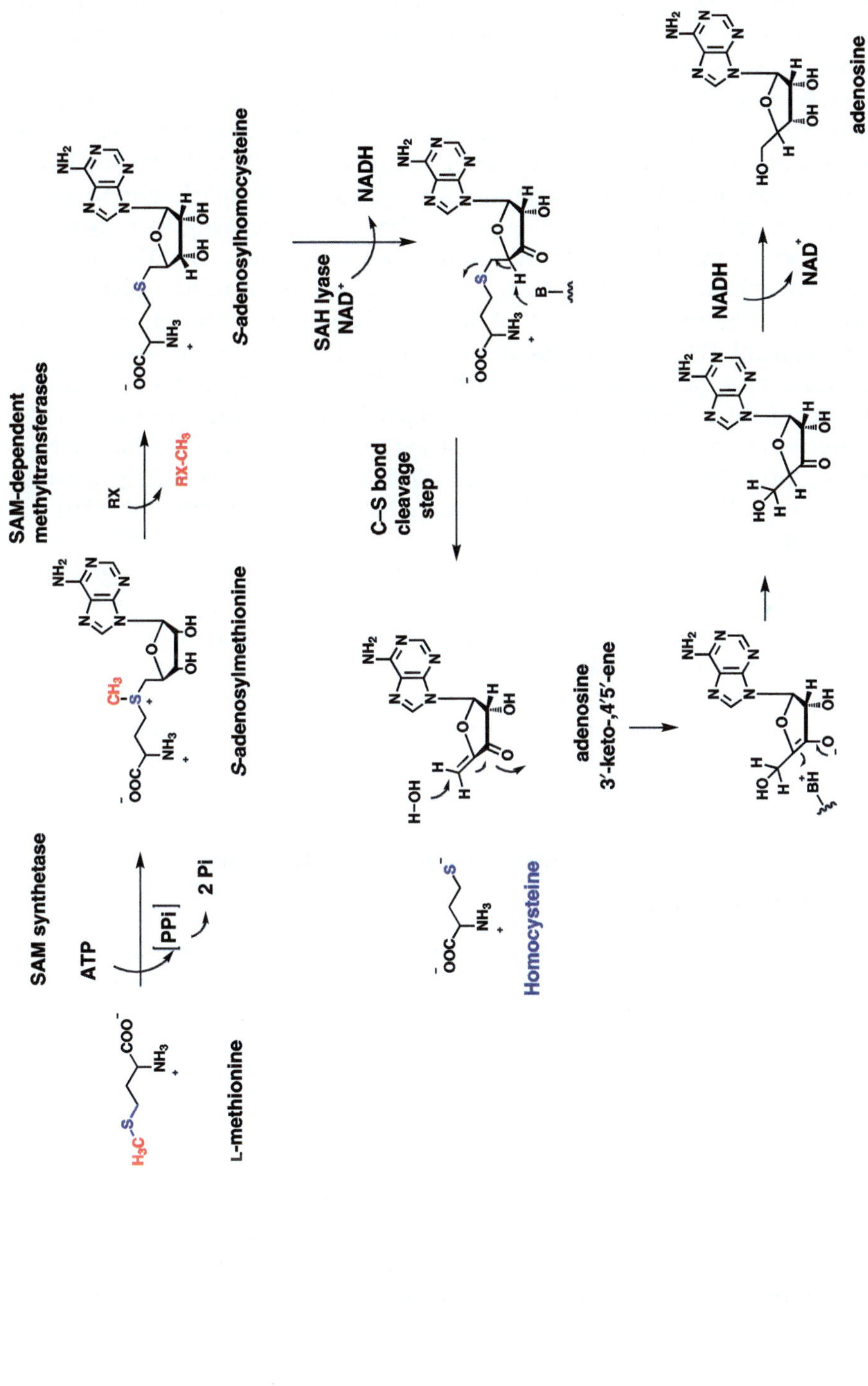

Figure 4.8 Three-enzyme pathway from methionine to homocysteine: SAM synthetase, SAM-dependent methyltransferases, SAH lyase.

of the 3′-keto to the alcohol results in the released product adenosine. This short three-enzyme pathway is effectively a route to retrieval of the homocysteine moiety from methionine and subsequently *S*-adenosylmethionine.

The sulfur traverse thus far is from ingested thioether methionine to the trivalent sulfonium cation of SAM, back to the thioether linkage in SAH, and then the C–S bond cleavage step that converts the thioether in SAH to the terminal thiol in homocysteine.[19]

Here, we note that C–S bond cleavage in the SAH lyase reaction was at first nonintuitive because there was no apparent low energy path to that C5′–S bond disconnection. It was only when Palmer and Abeles[18] purified the enzyme to homogeneity and detected an equivalent of tightly bound nicotinamide dinucleotide (NAD^+) coenzyme that the underlying logic became appreciated. Conversion of the ribose 3′-alcohol moiety to the 3′-ketone by hydride transfer to the bound NAD^+, generates NADH in mid-catalytic cycle (Figure 4.8). The newly generated 3′-ketone allows for carbanion formation at the adjacent C4′ by deprotonation because the negative charge can be delocalized into the ketone oxygen as the enolate anion. That C4′-carbanion is the requisite carbon nucleophile to expel homocysteine from C5′. This intramolecular participation by the C4 carbon nucleophile is the energy-lowering route for C5′–S bond cleavage. The thiolate is a low-energy leaving group, given that the pK_a for homocysteine to homocysteine thiolate is around pH 8.

In the process a 4′5′ C=C double bond forms that is now conjugated to the 3′-ketone. This is a classic αβ-unsaturated enone, electrophilic at the C5′ terminus and capturable by catalyzed water addition. At this juncture one has arrived at bound 3′-keto adenosine and NADH still in the active site. Hydride transfer back to the 3′-ketone, the reverse of the first step in the catalytic cycle, yields adenosine, which is released and the enzyme–NAD^+ complex is ready for the next round.

4.3.2 Enzymatic Transsulfuration: Cystathionine as Intermediate

The reason the homocysteine scaffold is worth the above three-enzyme retrieval strategy is that it then partners with serine in another three-enzyme process termed transsulfuration. The terminal thiol on the four-carbon framework of L-homocysteine is transferred to the three-carbon scaffold of L-cysteine (Figure 4.9). There is no low-energy mechanistic route to direct sulfur exchange between the two amino acids.

Instead, the kinetically accessible nucleophilic thiolate (RS^-) of homocysteine attacks an electrophilic serine-derived enamino–PLP adduct derived from L-serine. This parallels the strategy of HS^- addition to the same enamino–PLP adduct formed in the *O*-acetylsulfhydrylase reaction. This now yields a seven-carbon product, cystathionine. Then, a tandemly acting second PLP-containing enzyme cleaves the seven carbon thioether with the

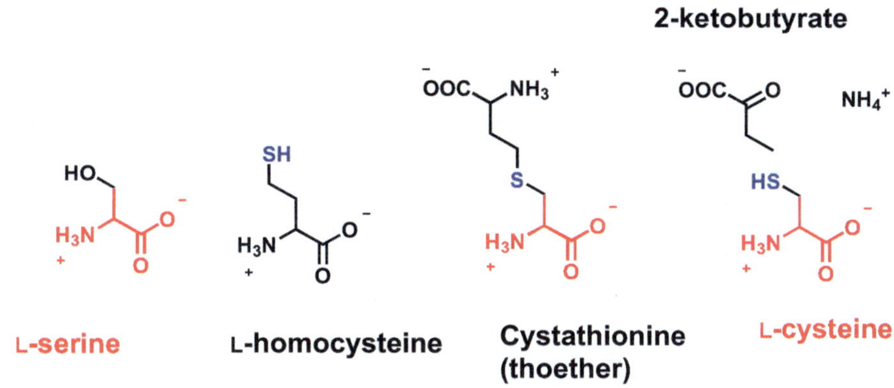

Figure 4.9 The five players in the transsulfuration pathway.

opposite regiochemistry and cysteine emerges. PLP-based regisoelective amino acid carbanions are the key enzyme-bound entities in this PLP-based reaction logic.[20]

4.3.2.1 Cystathionine-β-Synthase

The enzyme cystathionine β synthase recognizes L-serine as substrate and forms the L-serine–PLP aldimine adduct in the active site.[21,22] It then proceeds to generate the aminoacrylyl–PLP adduct by C2 carbanion formation and C3–OH expulsion, just as detailed earlier for the *O*-acetylserine sulfhydrylase (Figure 4.10A). The β-designation in the enzyme name indicates the locus C_β of the aminoacid–PLP adduct where the key chemical step occurs.

This aminoacrylyl–PLP is an electrophile at C3, capturable by the thiolate anion of cosubstrate. The cosubstrate that is bound productively is homocysteine, not H_2S or its HS^- monoanion. C–S bond formation (compare with the C–S bond cleavage strategy in SAH lyase described previously) creates the PLP adduct to the double-headed seven carbon bis-amino acid known as cystathionine. Transaldimination regenerates the resting enzyme–Lys=PLP aldimine and releases cystathionine. The reaction is a formal net replacement of the –OH at the beta carbon by the sulfur and its attached four-carbon homocysteine chain. This synthase is unique among PLP-dependent enzymes in also containing a tightly bound heme cofactor thought to act as a redox sensor to control catalytic activity.

Cystathionine has a very unusual framework. It has two amino acid halves, one a three-carbon half, the other a four-carbon half. In the three-carbon half the Cβ is attached to sulfur. In the four-carbon half it is Cγ (Figure 4.10). They are joined in the middle by a divalent sulfur atom in thioether linkage. In humans lacking this β-synthase, the substrate homocysteine piles up, exits cells and accumulates in the urine in a condition termed homocystinuria.

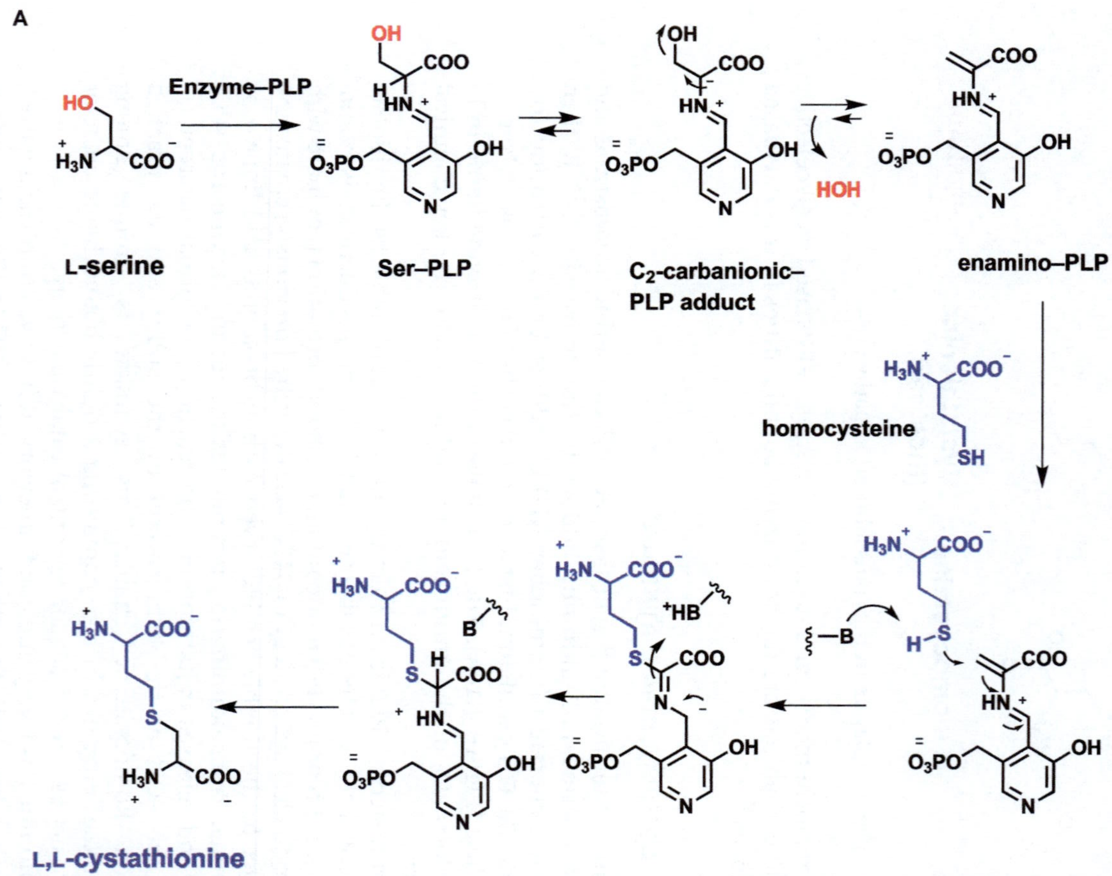

Figure 4.10 Consecutive action of two PLP enzymes makes and then fragments cystathionine. (A) Cystathionine β-synthase carries out a replacement of the C3–OH side chain of serine by the four-carbon framework of homocysteine. This involves a new sulfur–C3 bond as cystathionine is formed. (B) Cystathionine γ-lyase makes consecutive α-, then β-carbanion equivalents on the four-carbon arm of the cystathionine–PLP adduct on the way to elimination of L-cysteine as product. The four-carbon chain attached to PLP unravels as 2-ketobutyrate and NH_3.

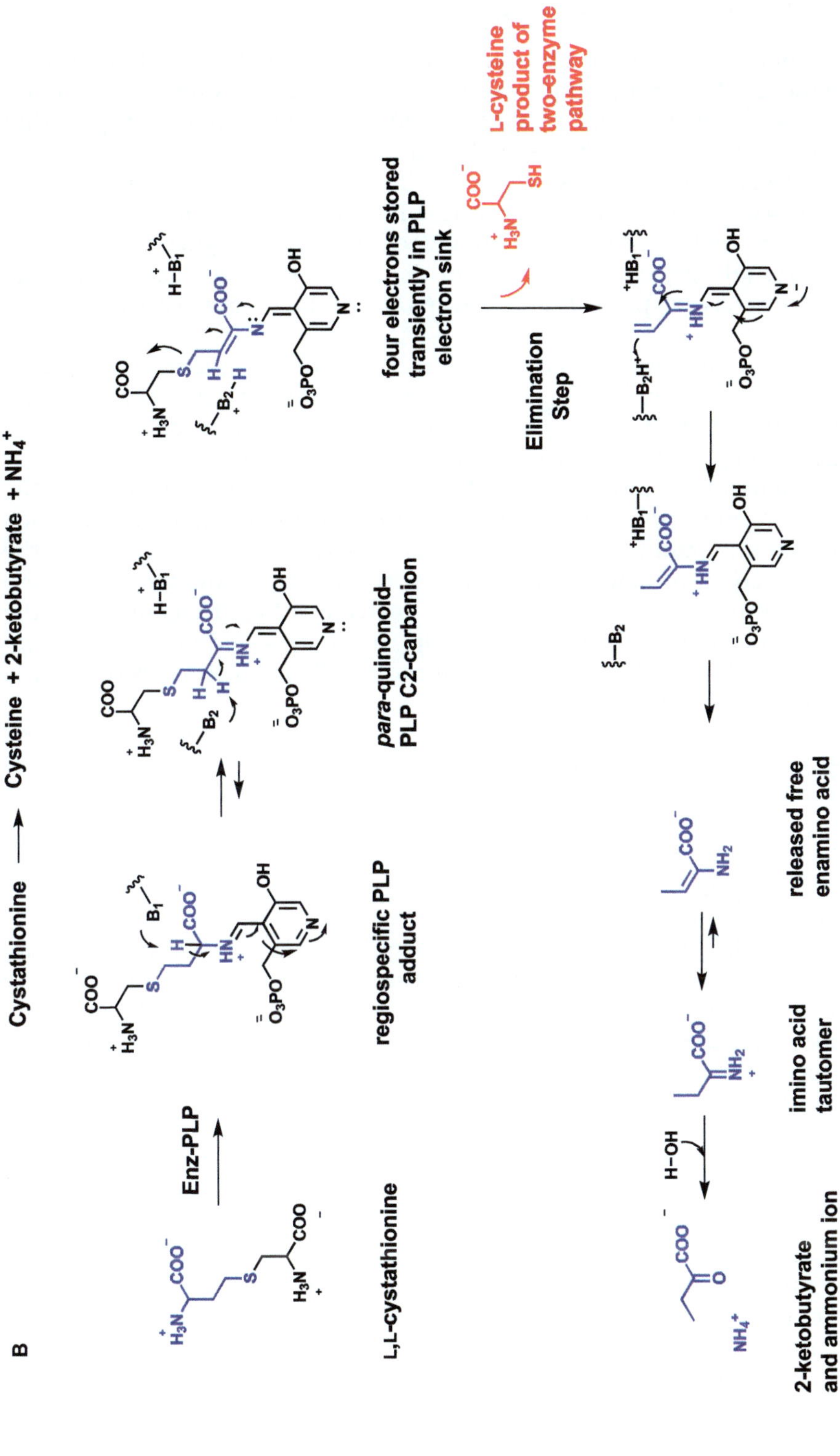

Figure 4.10 Continued.

4.3.2.2 Cystathionine-γ-lyase

To complete net sulfur transfer from the four-carbon scaffold (homocysteine) to the three-carbon scaffold (cysteine) requires not a β-eliminase but instead a γ-eliminase catalyst.[23] Again, a PLP-dependent enzyme that could generate and stabilize carbanion equivalents on the amino acid scaffold seems the likely strategy. In the case of such a γ-eliminase, the enzyme would have to get not just to a prototypic α-carbanion but also or instead to a stabilized β-carbanion–PLP equivalent. As Figure 4.10B depicts this takes advantage of the ability of the amino acid–PLP adduct to store four electrons (two pairs) transiently in the coenzyme as a "super" electron sink.[6]

The second enzyme in this transsulfuration pathway, cystathionine-γ-lyase is just such a PLP-containing catalyst (Figure 4.10B). It starts each catalytic cycle with a PLP–cystathionine aldimine adduct, regiospecifically on the amino group of the four-carbon arm, not the three-carbon arm. Subsequent C2-carbanion formation is a typical reaction sequence, here depicted as the fully conjugated *para*-quinonoid resonance contributor. What is an added and distinct feature of γ-elimination and γ-synthase PLP-enzymes is the ability to go on to make a C3-carbanion equivalent. In the indicated *para*-quinonoid form of the adduct the two methylene hydrogens on C3 of the homocysteinyl arm of substrate are adjacent to a C═N double bond. They are thereby kinetically acidic because the resultant C3 (beta) carbanion is stabilized by delocalization.

The electron counting for such a state is shown in Figure 4.10B and suggests that the PLP coenzyme can serve not only as a two-electron sink but also transiently as a four-electron sink. In a formal sense this is both a C2 and a C3 double carbanion species: a remarkable aspect of PLP chemical biology. It is this β (C3) carbanion that is used as the intramolecular neighboring group to cleave the adjacent C4–S bond as shown in Figure 4.10B.

This results in expulsion of the three-carbon arm with its terminal thiolate. This is the actual elimination step that forms the cysteine product. The remaining four-carbon PLP framework is unraveled sequentially by protonation at C3, then transaldimination to release the 2-amino-3-enoyl product fragment. On inspection, this is the enamine tautomer of the more stable C2-imino acid. In turn that hydrolyzes to 2-ketobutyrate and ammonia. As a γ-elimination catalyst, the Cγ (C4) has undergone a two-electron reduction while C2 (Cα) has undergone a two-electron oxidation. The stoichiometry balances internally: no external redox cosubstrate is required.

A reprise of the logic and strategy of this two-enzyme transsulfuration pathway reveals the essential role of PLP as a carbanion-generating coenzyme in the tandem cystathionine-β-synthase–cystathionine-γ-lyase pair of catalysts. The β-synthase makes stabilized α-carbanion–PLP adducts. Remarkably, the γ-lyase makes sequentially α-, then β-substrate–PLP carbanion equivalents.

The itinerant sulfur atom is shuffled from a four-carbon amino acid framework to a three-carbon amino acid framework *via* a seven-carbon thioether intermediate (cystathionine). The homocysteine input scaffold has been converted to a four-carbon α-keto acid. The three-carbon input scaffold serine ends up as the thiol-containing cysteine. What looks on paper to be a formal swap of the C3–OH to a C3–SH as L-serine goes to L-cysteine is more complex because of the chemical need for metabolic pathways to proceed by low energy reaction manifolds in order to proceed at rapid rates.

Cysteine, either formed directly from (*O*-acetyl)-L-serine by microbes and plants or indirectly in animals by the transsulfuration machinery, is then the central sulfur metabolite for many downstream pathways that use sulfur in its (−2) oxidation state as a powerful cellular nucleophile. The next higher, four-carbon homolog, homocysteine (again with sulfur in the thiol, −2 oxidation state) we have now seen is essential in the metabolism of the other sulfur-containing amino acid the thioether methionine. We deal further with homocysteine and methionine in Chapter 8. From inorganic H_2S to organic cysteine and homocysteine there is no formal change in the sulfur atom redox state.

4.3.3 Misprocessing of Cysteine to H_2S by the Two Enzymes of the Transsulfuration Pathway

We note, in passing, that each of the two PLP enzymes in the transsulfuration pathway above, cystathionine-β-synthase (normal substrate the four-carbon thiol amino acid homocysteine) and γ-cystathionase (normal substrate the thioether-linked seven-carbon, double-headed cystathionine) can process the three-carbon L-cysteine as an alternative, slow substrate.[24–26] In both those cases the enzymes catalyze net *β-elimination* of cysteine to pyruvate, ammonia and H_2S (Figure 4.11). This net production of hydrogen sulfide occurs in the absence of the cosubstrate homocysteine.[27]

Thus, the cystathionine-β-synthase acts only in β-elimination mode when it processes the three-carbon sulfur scaffold in cysteine as opposed to the replacement mode with serine as substrate (Figure 4.11). Presumably, when it processes L-serine in the absence of cosubstrate L-homocysteine, it may also run an abortive β-elimination route, by loss of the enamino–PLP adduct to premature release of the free enamino acid.[28] Over-oxidation of cysteine residues in the enzyme, perhaps *via* the built-in heme sensor, increases production of H_2S as a signaling molecule in the brain.[28]

Likewise, the second enzyme that normally acts on the seven-carbon cystathionine for sequential C2 and C3 carbanion formation, gets as far as the C2 carbanion when the three-carbon cysteine is the slow, alterative substrate and stops after a net β-elimination of H_2S. The three-carbon remainder scaffold is released as pyruvate and NH_3. Two additional routes to gaseous hydrogen sulfide from one of these two enzymes can arise from processing homocysteine with cysteine as the three-carbon cosubstrate instead of

A

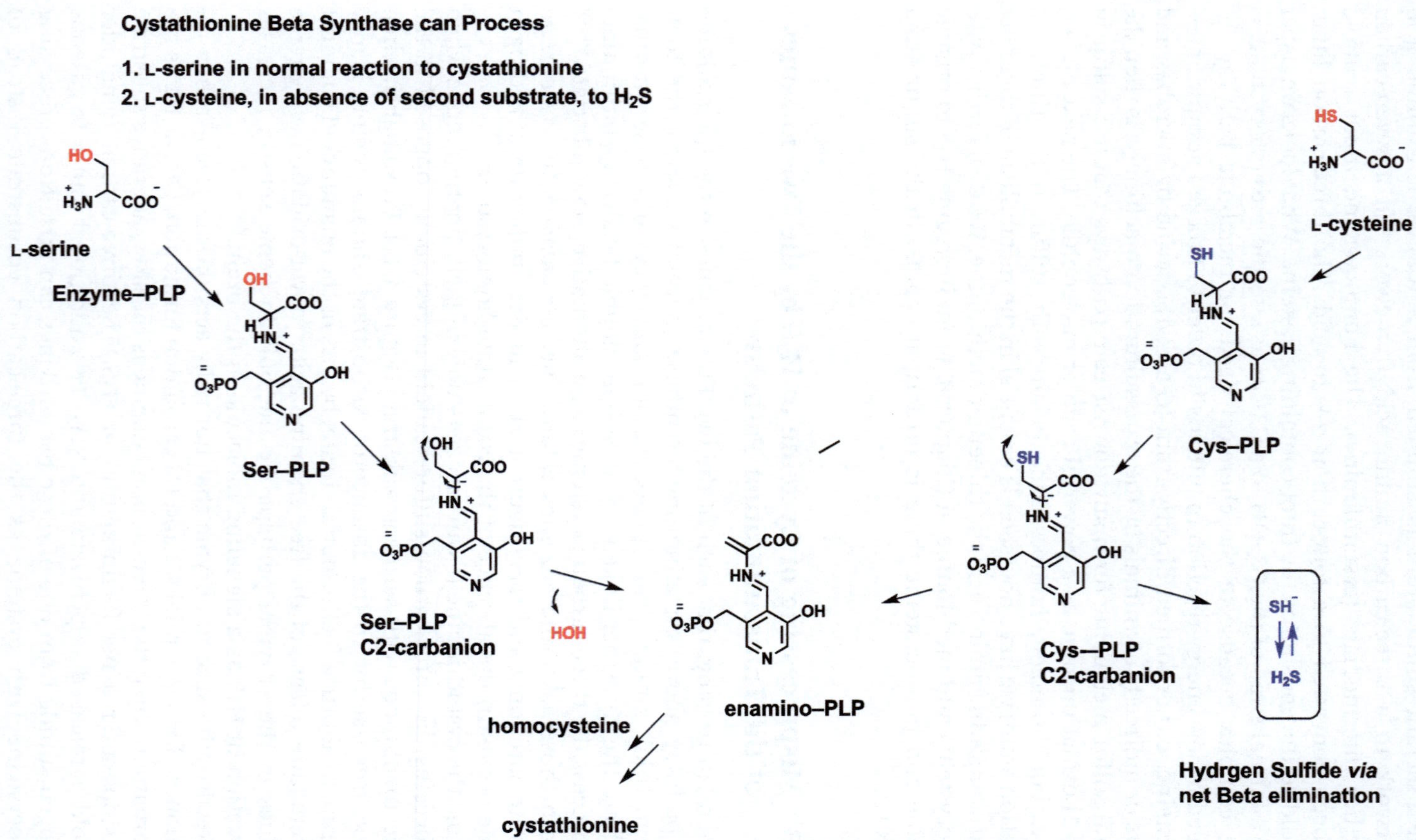

Figure 4.11 Cystathionine β-synthase normally processes serine and homocysteine to cystathionine. (A) It can take cysteine, in the absence of homocysteine, by β-elimination to pyruvate, ammonia and H_2S. This is one of the main routes to H_2S in mammalian tissues. (B). Cystathionine β-synthase and γ-lyase can process other substrates as indicated to also release H_2S.

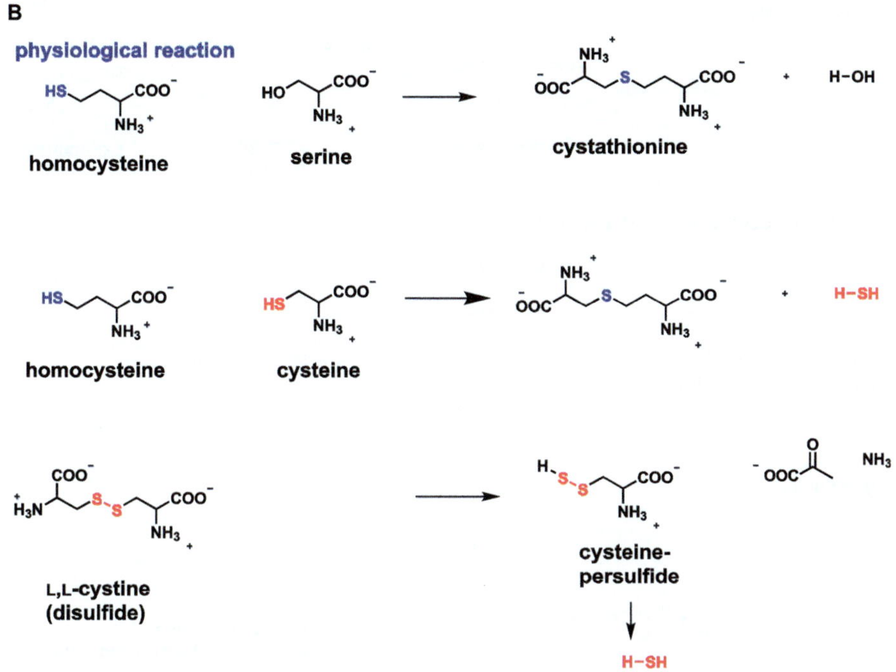

Figure 4.11 Continued.

serine.[26] The normal cystathionine product is favored but now H_2S instead of H_2O is the coproduct from the β-elimination step. A variant that leads into persulfide chemistry, discussed in further detail in Chapter 6, occurs when the disulfide L,L-cystine is processed for β-elimination.[29] Pyruvate and NH_3 are coproducts along with free cysteine persulfide (Cys–SSH). That can be reductively released to free H_2S.[30]

4.3.4 Mercaptopyruvate Transferase

A third enzyme that is involved in net cleavage of the C3–SH bond of cysteine is found in mitochondria.[12,13,31] This is mercaptopyruvate sulfotransferase and it acts when a PLP-dependent transaminase has converted L-cysteine to 3-mercaptopyruvate (3-thiopyruvate). As depicted in Figure 4.12, 2-ketoglutarate is the typical cosubstrate for the cysteine transaminase. As the enzyme name implies, the amino group is transferred from cysteine to the five-carbon framework of 2-ketoglutarate. The net reductive amination of ketoglutarate occurs *via* the intermediacy of an enzyme–pyridoxamine-P intermediate.[12,13]

Mercaptopyruvate sulfotransferse carries out an unusual mode of C3–SH bond cleavage. The thiolate side chain of cysteine 248 in the enzyme active site acts as nucleophile on the SH atom of mercaptopyruvate, transferring it as an *electrophilic sulfur atom*, a reversal in polarity from the normal

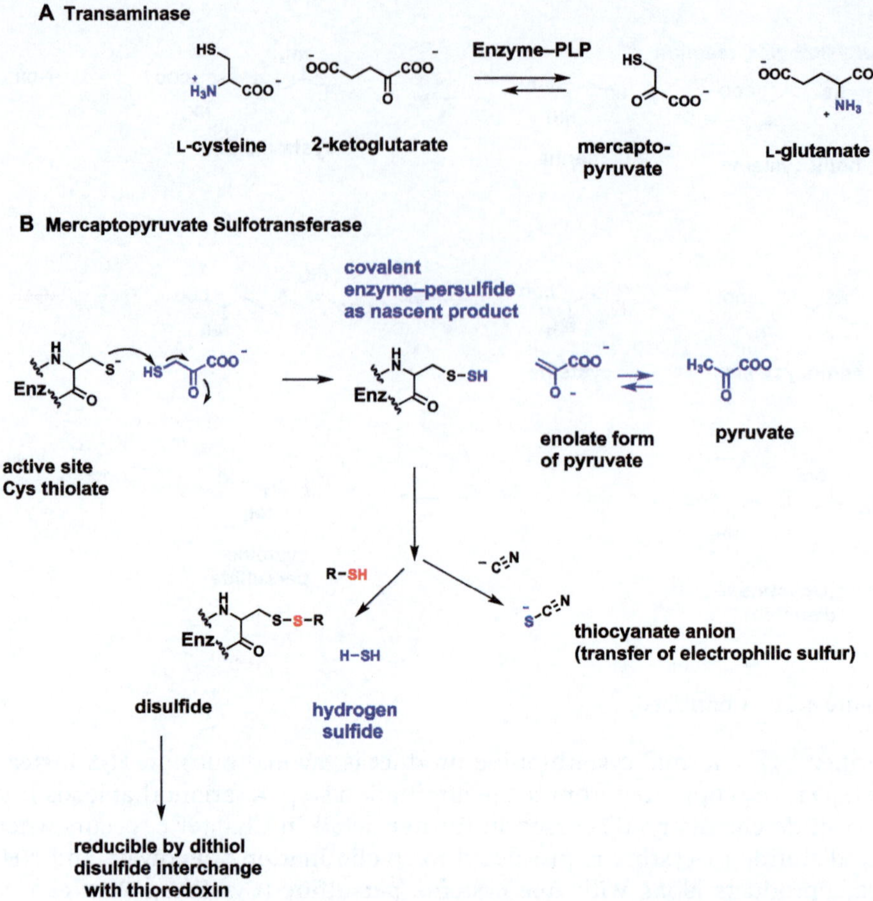

Figure 4.12 (A) Transaminase action on L-cysteine creates 3-mercaptopyruvate. This is a substrate for the enzyme mercaptopyruvate sulfotransferase (B) that forms a nascent enzyme–Cys–SSH persulfide adduct as product. Attack on the persulfide by an organic thiol (*e.g.* glutathione) would form the mixed disulfide while releasing H_2S for presumed signaling functions. Cyanide ion is a classic test for persulfides as the terminal sulfur behaves as electrophile and is released as a thiocyanate ion.

anticipated behavior of a thiol or thiolate group. The electrons in the C3–S bond stay behind on the three-carbon pyruvyl scaffold. The driving force for this *"ompolung"* (inversion of reactivity) may be that the electrons left at C3 can flow into the adjacent carbonyl to give the enolate anion as nascent product. This low-energy carbanion equivalent can then ketonize and protonate at C3.

Examination of the –SH group that has undergone transfer without its electron pair shows it becomes covalently attached to the thiolate side chain of Cys 248 in a persulfide (–SSH) linkage. The terminal S is also known as a sulfane sulfur and often behaves as an electrophile.[10,32] This mirrors the nascent persulfido linkage product seen above with L,L-cystine-processing by the cystathionine β-synthase (Figure 4.11B).

One of the identifying tests for this enzyme persulfide is its reactivity with cyanide ion to form the released thiocyanate anon (Figure 4.12). This must have arisen by reaction of cyanide as carbon nucleophile on the terminal sulfur of the enzyme Cys–SSH group with transfer of the sulfur as a six-electron, electrophilic sulfane. An alternative fate is for the nascent enzyme–SSH group to be reduced by the dithiol form of thioredoxin. This releases H_2S as thioredoxin is oxidized to the disulfide form. This reaction is a reminder that the conversion of a persulfide to a thiol is a net two-electron reduction.

We will examine persulfido chemical biology[10,32] further in Chapter 6, but for now have noted the several ways that mammalian enzymes can generate small quantities of H_2S and turn briefly to its role in mammalian signaling biology.

4.4 H_2S Signaling Role in Mammals

We started this chapter discussing how sulfide oxidizers have devised enzymes and catalytic strategies to oxidize H_2S to sulfate. This is also the case for higher organisms that act to keep the concentration of free H_2S at submicromolar levels. This involves the sulfide quinone oxidoreductase in mitochondrial membranes. Recall from Chapter 3 that this enzyme in sulfate oxidizing bacteria can accumulate persulfides attached to an active site cysteine.[33] In mitochondria the enzyme–SSH is captured by free glutathione (GSH) to produce the glutathione persulfide GSSH. This is the substrate for a nonheme iron oxygenase that ultimately converts the terminal sulfur of GSSH all the way to sulfite, in an oxygenase reaction not well studied. Then the mitochondrial sulfite oxidase, with its molybdopterin cofactor takes the sulfite to sulfate (Chapter 3).

In the past two decades a large body of evidence has accrued that H_2S, like carbon monoxide (CO) and nitrous oxide (NO), is a gaseous signaling molecule in mammalian physiology.[26,34] CO and NO bind tightly as axial ligands to different classes of heme proteins. NO is famously an activating ligand for the heme-containing guanylate cyclase that relaxes smooth muscle tone in vascular systems.

H_2S has at least two relevant reactivities.[35] One is similar to binding to heme iron in certain heme proteins. Inhibition of the respiratory chain's terminal enzyme, cytochrome c oxidase with both heme and copper atoms, occurs with metal coordination by any of the three gases CO, NO and H_2S.

Because the reservoir of protein persulfide groups can be in redox equilibrium with free H_2S,[10,32] it can be difficult to distinguish effects of free H_2S *vs.* the persulfide group modifications in specific target proteins. One route has been genetic knockouts of the H_2S-forming cystathionine synthase and lyase and the mercaptopyruvate sulfotransferase with pervasive phenotypic pathologies.

In mouse models, cytoprotective roles in a diverse set of organ systems, including vascular, endocrine, gastrointestinal and respiratory systems have been restored on addition of H_2S.[30,34] Persulfide groups in nuclear factor kappa B (NFκB), in P38 mitogen-activated protein (MAP) kinase, in the extracellular signal regulated kinase (ERK) kinase, in c-Jun N-terminal kinases (C-JNK), in nuclear factor erythroid-2-related factor 2 (NrF2), and the in histone deacetylase sirtuin 1 (SIRT-1) all have been shown to alter catalytic or other functions.[36] Nonetheless, there must be a fine line between signaling, cytoprotective properties and the toxicities seen at higher concentrations of H_2S.

One of the known deleterious effects of H_2S exposure is that hemoglobin (and myoglobin) can turn from their native red color to green-colored

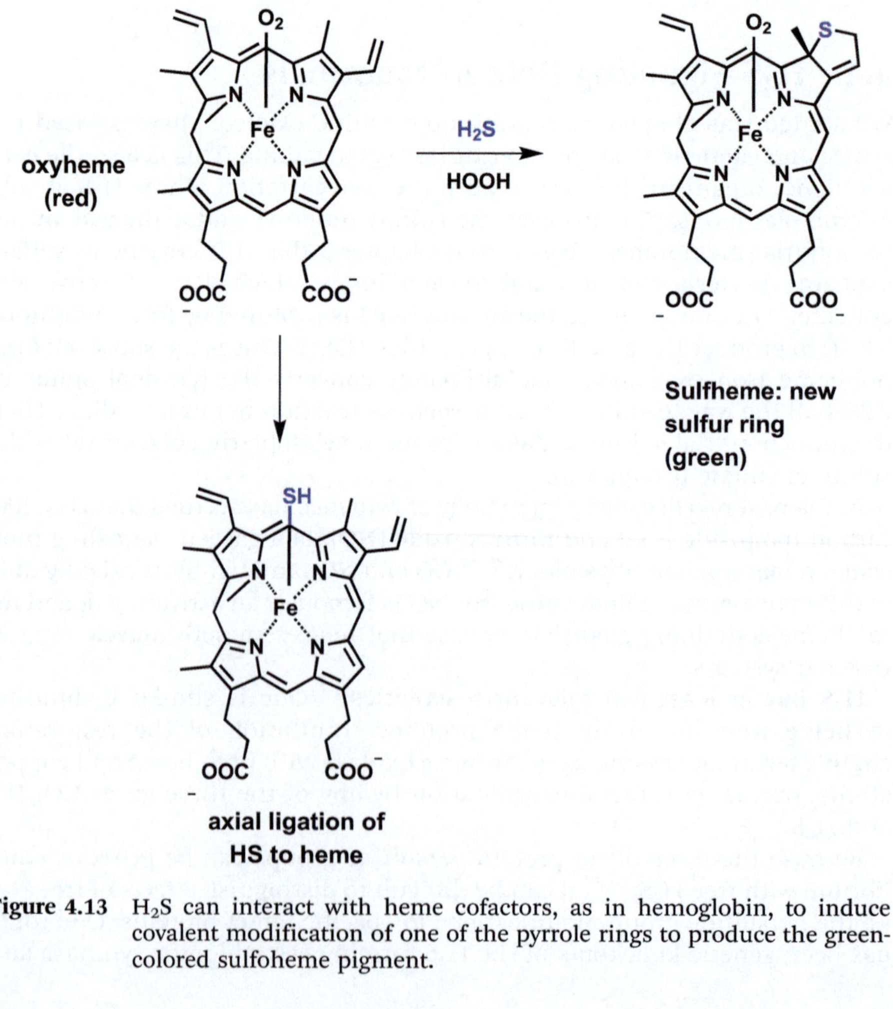

Figure 4.13 H_2S can interact with heme cofactors, as in hemoglobin, to induce covalent modification of one of the pyrrole rings to produce the green-colored sulfoheme pigment.

derivatives. It is known that the hydrosulfide anion has affinity for the central ferrous iron atom in heme. At first, it was assumed that some sequalae to such ligation was at work in the color change. O_2 can also compete with H_2S for axial ligand binding in the normal physiological function. Occasionally, the ligated O_2 can undergo one-electron reduction to superoxide anion and ferric heme (methemoglobin). It may be that some intersection of reactive oxygen species (ROS) chemistry and HS^- leads to the green-colored sulfoheme derivative (Figure 4.13). Structural analysis indicated that the HS atoms have added into the exocyclic vinyl group and also attached to the pyrrole double bond to create the new 5,5-fused ring system in the sulfoheme derivative.[37]

References

1. D. W. Bak, T. Bechtel, J. A. Falco and E. Weerapana, Cysteine reactivity across the subcellular universe, *Curr. Opin. Chem. Biol.*, 2019, **48**, 96–105.
2. L. B. Poole, The basics of thiols and cysteines in redox biology and chemsitry, *Free Radicals Biol. Med.*, 2015, 148–157.
3. C. T. Walsh, B. Tu and Y. Tang, Eight kinetically stable but thermodynamicallay activated molecules that power cell metabolism, *Chem. Rev.*, 2018, **118**, 1460–1494.
4. W. Loenen, S-adenosylmethionine: jack of all trades and master of everything? *Biochem. Soc. Trans.*, 2006, **34**, 330–333.
5. W. P. Jencks, *Catalysis in Chemistry and Enzymlogy*, McGraw-Hill, New York, 1969.
6. C. T. Walsh and Y. Tang, *The Chemical Biology of Human Vitamins*, Royal Society of Chemistry Press, London, 2018.
7. V. Gupta and K. Carroll, Sulfenic acid chemistry detection and cellular lifetime, *Biochim. Biophys. Acta*, 2014, **1840**, 847–875.
8. J. Yang, K. Carroll and D. C. Liebler, The expanding landscape of the thiol redox proteome, *Mol. Cell. Proteomics*, 2016, **15**, 1–11.
9. T. H. Truong, P. Ung, P. B. Palde, C. E. Paulsen, A. Schlessinger and K. S. Carroll, Molecular basis for redox activation of epidermal growth factor receptor kinase, *Cell Chem. Biol.*, 2016, **23**, 837–848.
10. J. I. Toohey and A. Cooper, Thiosulfoxide (sulfane) sulfur: new chemistry and new regulatory roles in biology, *Molecules*, 2014, **19**, 12789–12813.
11. B. Huang, M. Vatting and S. L. Roderick, The active site of O-acetylsulfhydrylase is the anchor point for bienezyme compex with serine acetyltransferase, *J. Bacteriol.*, 2005, **187**, 3201–3205.
12. E. E. Snell, From bacterial nutrition to enzyme structure: a personal journey, *Annu. Rev. Biochem.*, 1993, **62**, 1–27.
13. M. D. Toney, Controling reaction specificity in pyridoxal phosphate enzymes, *Biochim. Biophys. Acta*, 2011, **1814**, 1407–1418.

14. W. Rabeh and P. Cook, Structure and mechanism of O-acetylserine sulfhydrylase, *J. Biol. Chem.*, 2004, **279**, 26803–26806.
15. R. K. Yadav, M. Martinov, V. Vitvitsky, J. Seravalli, R. Wedmann, M. R. Filiopvic and R. Banerjee, Biosynthesis and reactivity of cysteine persulfides in signaling, *J. Am. Chem. Soc.*, 2016, **138**, 289–299.
16. S. Ravanel, B. Gakiere, D. Job and R. Douce, The specific features of methionine biosynthesis and metabolism in plants, *Proc. Natl. Acad. Sci. U. S. A.*, 1998, **95**, 7805–7812.
17. G. McBean, The transsufuration pathway'' a source of cysteine for glutathione in astrocyctes, *Amino Acids*, 2012, **42**, 199–205.
18. J. L. Palmer and R. Abeles, The mechanism of action of S-adenosylhomocysteinase, *J. Biol. Chem.*, 1979, **254**, 1217–1226.
19. S. M. Aitken, P. Lodha and D. J. K. Morneau, The enzymes of the transsulfuration pathways: Active-site characterizations, *Biochim. Biophys. Acta*, 2011, **1814**, 1511–5117.
20. J. F. Honek, Biological chemistry of the carbon-sulfur bond, *Can. J. Chem.*, 2015, **93**, 1051–1060.
21. M. Meier, M. Janosik, V. Kery, J. P. Kraus and P. Burkard, Structure of human cystathionine beta synthase: a unique pyridoxal-5′-phosphate heme protein, *EMBO J.*, 2001, **20**, 3910–3916.
22. R. Banerjee and C. Zou, Redox regulation aand reaction mechanism of human cystathionine beta synthase: A PLP dependent heme sensor protein, *Arch. Biochem. Biophys.*, 2005, **433**, 144–156.
23. J. P. Kraus, J. Hasek, V. Kozich, R. Collard, S. Venezia, B. Janosikova, J. Wang, S. P. Stabler, R. H. Allen, C. Jakobs, C. T. Finn, Y.-h. Chien, W.-l. Hwu, R. A. Hegele and S. H. Mudd, Cystathionine gamma lyase: clinical, metabolic, genetic and structural studies, *Mol. Genet. Metab.*, 2009, **97**, 250–259.
24. E. G. Mueller, Trafficking in persulfide delivering sulfur in biosynthetic pathways, *Nat. Chem. Biol.*, 2006, **2**, 185–194.
25. O. Kabil and R. Bannerjee, Redox biochemistry of hydrogen sulfide, *J. Biol. Chem.*, 2010, **285**, 21903–21907.
26. M. R. Filipovic, J. Zivanovic, B. Alvarez and R. Bannerjee, Chemical Biology of H_2S signaling through persulfidation, *Chem. Rev.*, 2018, **118**, 1253–1337.
27. T. Majtan, J. Krijt, J. Sokolova, M. Krizkova, A. Ralate, J. Kent, JF 3rd Gregory, V. Kozich and J. P. Kraus, Biogenesis of hydrogen sulfide and thioethers by cystathionine beta synthase, *Antioxid. Redox Signaling.*, 2018, **28**, 311–323.
28. K. Eto and H. Kimura, A novel enhancing mechanism of hydrogen sulfide-producing activity of cystathionine beta synthase, *J. Biol. Chem.*, 2002, **277**, 42680–42685.
29. D. Cavallini, C. Marco, B. Mondavi and B. G. Mori, The cleavage of cystine by cystathionase and the transsulfuration of hyotaurine, *Enzymologia*, 1960, **22**, 161–173.

30. H. Kimura, N. Shibuya and Y. Kimura, Hydrogen sulfide is a signaling molecule and a cytoprotectant, *Antioxid. Redox Signaling*, 2012, **17**, 45–57.
31. N. Nagahara, T. Okazaki and T. Nishio, Cytosolic mercaptopyruvate sulfur transferase is evolutionarily related to mitochondrial rhodanese, *J. Biol. Chem.*, 1995, **270**, 16230–16235.
32. G. Fritz and P. Kronek, Sulfane to go, *Science*, 2015, **350**, 1476.
33. A. A. Santos, S. Venceslau, F. Grein, W. D. Leaitt and C. Dahl, A protein trisulfide couples dissimilatory sulfate reduction to energy conservation, *Science*, 2015, **350**, 1541–1545.
34. Y. Kimura, S. Koike, N. Shibuya, D. Lefer, Y. Ogasawara and H. Kimura, 3-Mercaptopyruvate sulfur transfer produces potential redox regulators cysteine- and glutathione-persulfide (Cys-SSH and GSSH) together with signaling moecules H_2S_2, H_2S_3 nd H_2S, *Sci. Rep.*, 2017, **7**, 10459.
35. Z. Giuffre and J. Vicente, Bioxynthesis of hydrogen sulfide and interplay with other gaseous signaling agents, *Oxid. Med. Cell. Longevity*, 2018, 6290931.
36. P. Rose, P. Moore and Y. Z. Zhu, Hydrogen sulfide biosynthesis and catabolism: new insights from molecular studies, *Cell. Mol. Life Sci.*, 2017, **74**, 1391–1412.
37. R. Pietri, E. Román-Morales and J. Lopez-Garriga, Hydrogen sufide and hemeproteins: knowledge and mysteries, *Antioxid. Redox Signaling*, 2011, **15**, 393–404.

CHAPTER 5

The Nucleophilic Thiolate in Cysteine and Cysteamine Scaffolds

We have noted thus far that cysteine is the first organic metabolite in which inorganic sulfur, at the oxidation state of sulfide (S^{2-}, most probably as the monoprotonated HS^-) has crossed over from inorganic metabolism to organic metabolism.[1,2] Figure 5.1 notes two endpoint metabolites, coenzyme A and glutathione, that are the topics of this chapter, each reflecting the role of the thiol–thiolate group in these two central metabolites.

The pK_a of the thiol dissociation to the thiolate anion in cysteine is 8.2. Thus, at an intracellular (cytoplasmic) pH value of approximately 7.2, about 10% of the cysteine will be present as the nucleophilic thiolate anion (Figure 5.1). This is the strongest nucleophile in primary metabolic pathways and much of cysteine chemical biology revolves around this reactivity. Free cysteine is sufficiently reactive as the thiolate anion to be toxic at high concentrations so cells typically maintain free cysteine concentrations below millimolar values.[3] Instead, two major metabolites are formed from cysteine, keeping the nucleophilic thiol–thiolate function group but using it for largely distinct cellular functions. A third route to control cysteine concentrations is *S*-dioxygenation by cysteine dioxygenase,[4] examined in Chapter 7.

The first endpoint metabolite is coenzyme A, the second is the tripeptide glutathione (Figure 5.1). In the former metabolite the carboxylate moiety of cysteine has been trimmed away during biosynthesis of the coenzyme. Essentially all coenzyme A cellular functions involve two-electron chemical steps *via* the thiolate anion as a nucleophile. Glutathione in the reduced, thiol form (as opposed to the oxidized disulfide form) circulates in most cells

Chemical Biology No. 12
The Chemical Biology of Sulfur
By Christopher T. Walsh

Published by the Royal Society of Chemistry, www.rsc.org

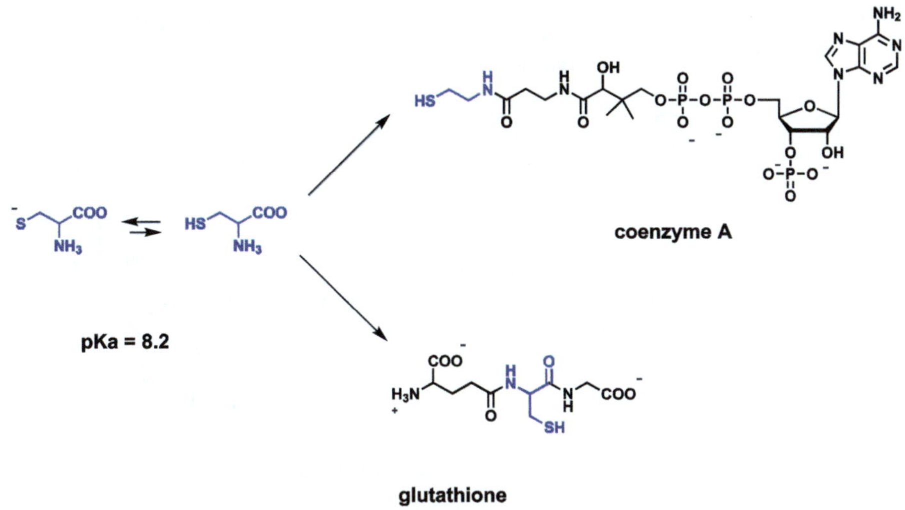

Figure 5.1 Cysteine is converted biosynthetically to two workhorse thiol frameworks, coenzyme A (CoASH) and the tripeptide glutathione (GSH), in cell metabolism.

as the most abundant thiol–thiolate molecular scaffold, at up to 100-fold higher levels than free L-cysteine. It can function coenzymatically and also as a consumable protective agent against reactive electrophiles. These include endogenous and exogenous organic molecules as well as reactive inorganic nitrogen and oxygen species. The thiolate of the cysteinyl residue in glutathione is key to essentially all of its chemical biology.

Entire books, large tomes packed with information, have been written separately about both coenzyme A and glutathione and interested readers are directed to the large amount of primary and review literature on both metabolites. The purpose here is to focus (narrowly) on the chemistry available to the cysteamine–cysteine thiolate moiety in these two molecules.

5.1 Coenzyme A: Thiol Chemical Biology *via* Cysteamine

The first of the two central, cysteine-derived metabolites, coenzyme A, is named for its historic and central roles in primary metabolism across all free-living organisms. Coenzyme A can be parsed structurally into three regions: the 3′5′-ADP moiety, connected to the 4-phosphopantetheine side chain, in which is embedded the terminal cysteamine moiety with the reactive thiol group[5] (Figures 5.1 and 5.2A).

Humans cannot biosynthesize the pantothenate internal moiety of coenzyme A. It is required in the diet as vitamin B5, but can then be further elaborated in ATP-consuming reactions to coenzyme A[5] (Figure 5.2B). The cysteamine moiety derives from enzymatic decarboxylation of a cysteinyl

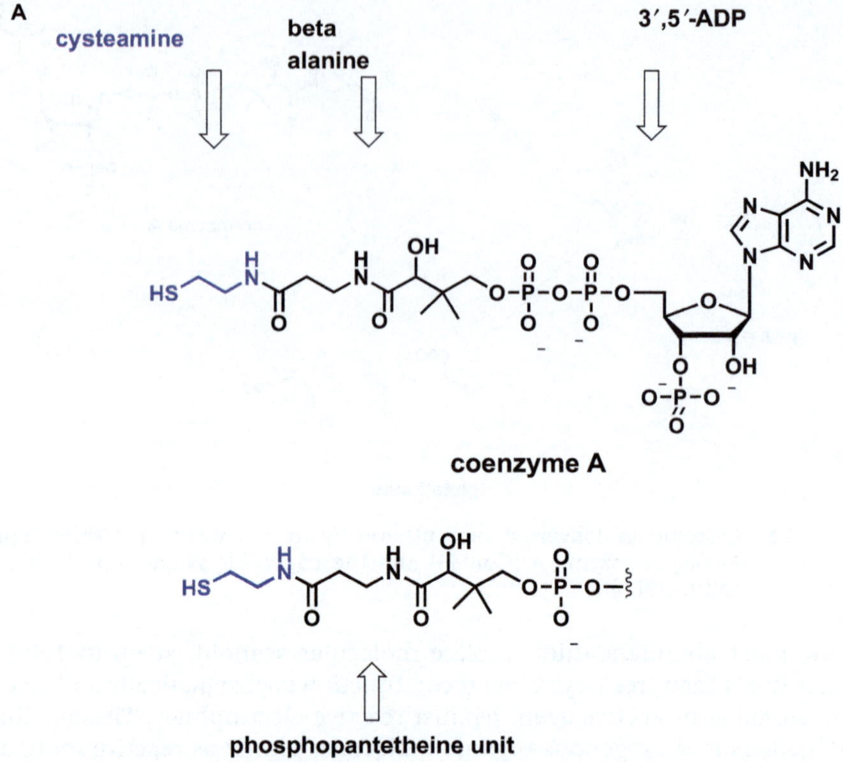

Figure 5.2 Coenzyme A. (A) Three regions of the coenzyme A molecule. (B) Pantothenate is vitamin B5 in human metabolism. Four nucleoside triphosphate-cleaving reactions are involved in human biosynthesis of CoASH from pantothenate. (C) The pantothenylcysteine decarboxylase contains a FAD cofactor, involved in initial oxidation of the cysteine side chain to an enethiol to set up an electron sink β to the carboxylate to lower the energy for decarboxylation to 4-phosphopantetheine.

unit, not as the free amino acid but as the phosphopantetheinyl cysteine precursor.[6,7] The decarboxylase utilizes a bound flavin adenine dinucleotide (FAD) coenzyme to reversibly create the -ene thiol as a beta electron sink to lower the energy barrier for the C–COO cleavage step. (Figure 5.2C).

A tautomer of the enethiol is the *thiol aldehyde*, an unusual oxidation state for the cysteine side chain. The underlying logic for this oxidation of thiol to thioaldehyde is that this establishes an electron sink β- to the cysteine carboxylate moiety. In turn, this provides a lower energy barrier to decarboxylation as the resulting carbanion can be delocalized into the thio aldehyde grouping.[8] Once the thioaldehyde has done its job in creating an electron sink to facilitate enzymatic decarboxylation, it gets re-reduced from $FADH_2$ prior to release from the enzyme active site. This regenerates the key cysteamine nucleophilic thiol form of the sulfur in the phosphopantetheine product. Examination of the overall stoichiometry reveals

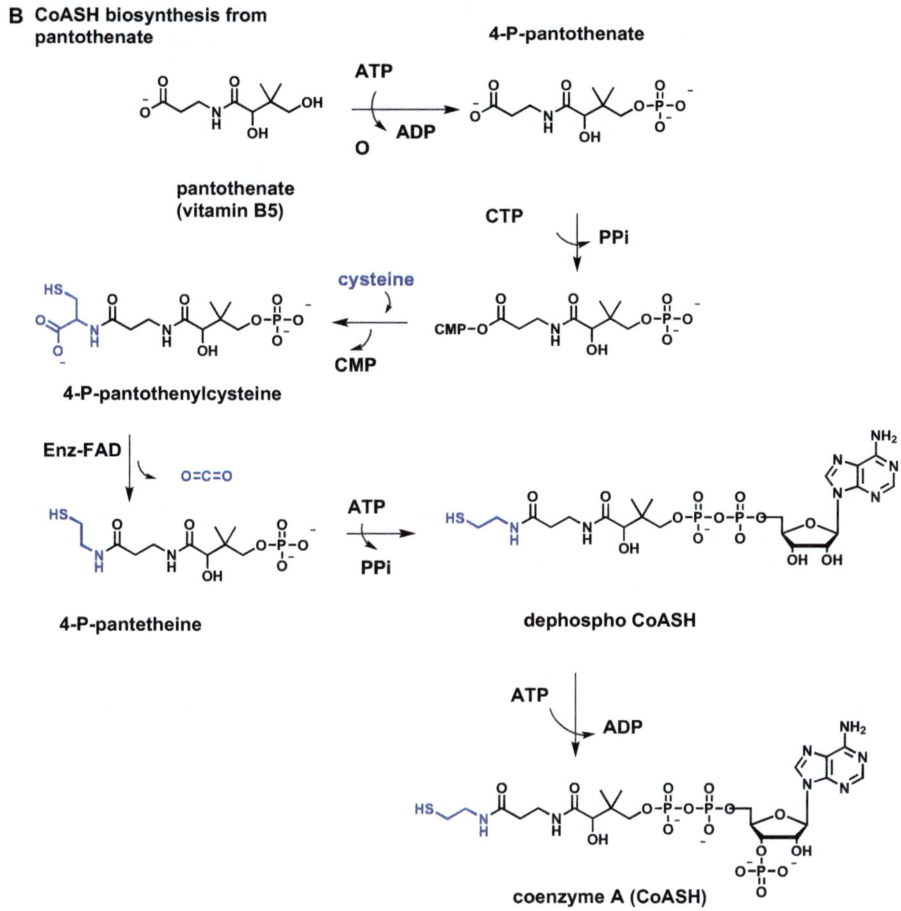

Figure 5.2 Continued.

that the thiol to *thioaldehyde* was a cryptic, reversible, yet mechanistically crucial operation.

5.1.1 Acyl Thioester Formation by CoASH and Acyl Phosphate and Acyl Adenylates

The thiolate anion form of the terminal cysteamine is the key nucleophile for reacting with kinetically labile acyl-adenosine monophosphate (AMP) mixed carboxylic acid anhydrides (Figure 5.3) to create acyl thioesters with release of AMP. The acyl thioesters are thermodynamically activated, for acyl transfer to cosubstrates, kinetically much more stable than the acyl-phosphoric anhydrides (half times of seconds to minutes), and suitable as diffusible packets of activated acyl groups throughout organelles and cells.[9]

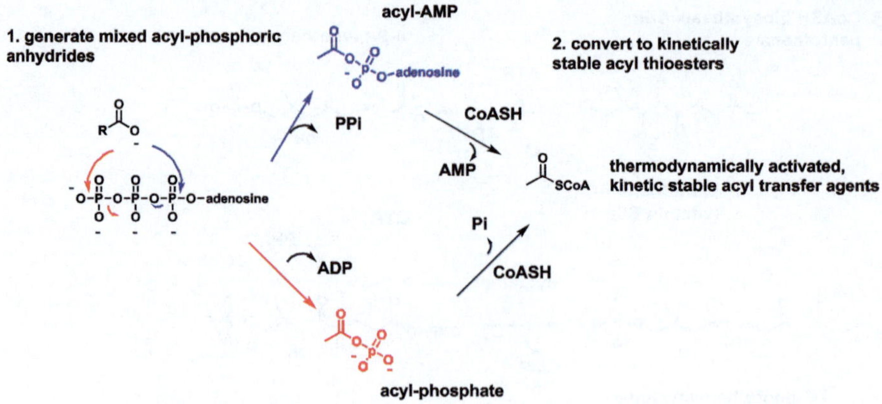

Figure 5.3 Carboxylic acid anions of metabolic acids are activated as mixed acyl-phosphoric anhydrides by ATP cleavage. These anhydrides, while thermodynamically activated for acyl transfer, are kinetically labile to hydrolysis. Conversion to the acyl-*S*-CoA thioesters preserves thermodynamic activation and dramatically enhances kinetic stability under physiological conditions such that they are utilized as diffusible packets of acyl transfer currency in cells.

That kinetic stability or thermodynamic activation for acyl group transfer underlies the widespread role of acetyl CoA and longer chain acyl CoAs for acyl transfer metabolism. Acetyl *S*-coenzyme A (acetyl CoA) is the paradigmatic cellular form of activated acetate. It is the acetyl donor in a vast range of enzyme-mediated acetyl group transfers to cosubstrate nucleophiles.

Four cellular substrates, of dozens that could be listed, in Figure 5.4A, are two with nitrogen nucleophiles, one with an oxygen nucleophile and one with a carbon nucleophile as the RX groups capturing the acetyl group from acetyl CoA. The glutamate *N*-acetylation creates an allosteric activator of one of the major carbamoyl phosphate synthetase isozymes for nitrogen cellular economy. The other nitrogen example is macromolecular, the acetylation of lysine side chains in proteins such as the N-terminal tail region of histone proteins. *O*-acetylation of choline creates the neurotransmitter acetylcholine, an oxoester. The last example shows a carbon nucleophile (resonance contributor) as attacking species on the activated acetyl group of acetyl CoA. This leads into the discussion of thioester enolate chemical biology described in the next section.

Other acyl groups of central importance in primary metabolic pathways that are ferried through the cell as acyl CoA species include succinyl CoA in the citrate cycle and malonyl CoA in fatty acid biosynthesis as chain elongation inventory. In every case they are activated thermodynamically for acyl group transfer to cosubstrate nucleophiles (Figure 5.4B).

Acyl thioesters are more activated thermodynamically than the corresponding oxoesters and thus have a wider role in acyl group transfer metabolism. (Two exceptions are the oxoester in the neurotransmitter [*O*]-acetylcholine

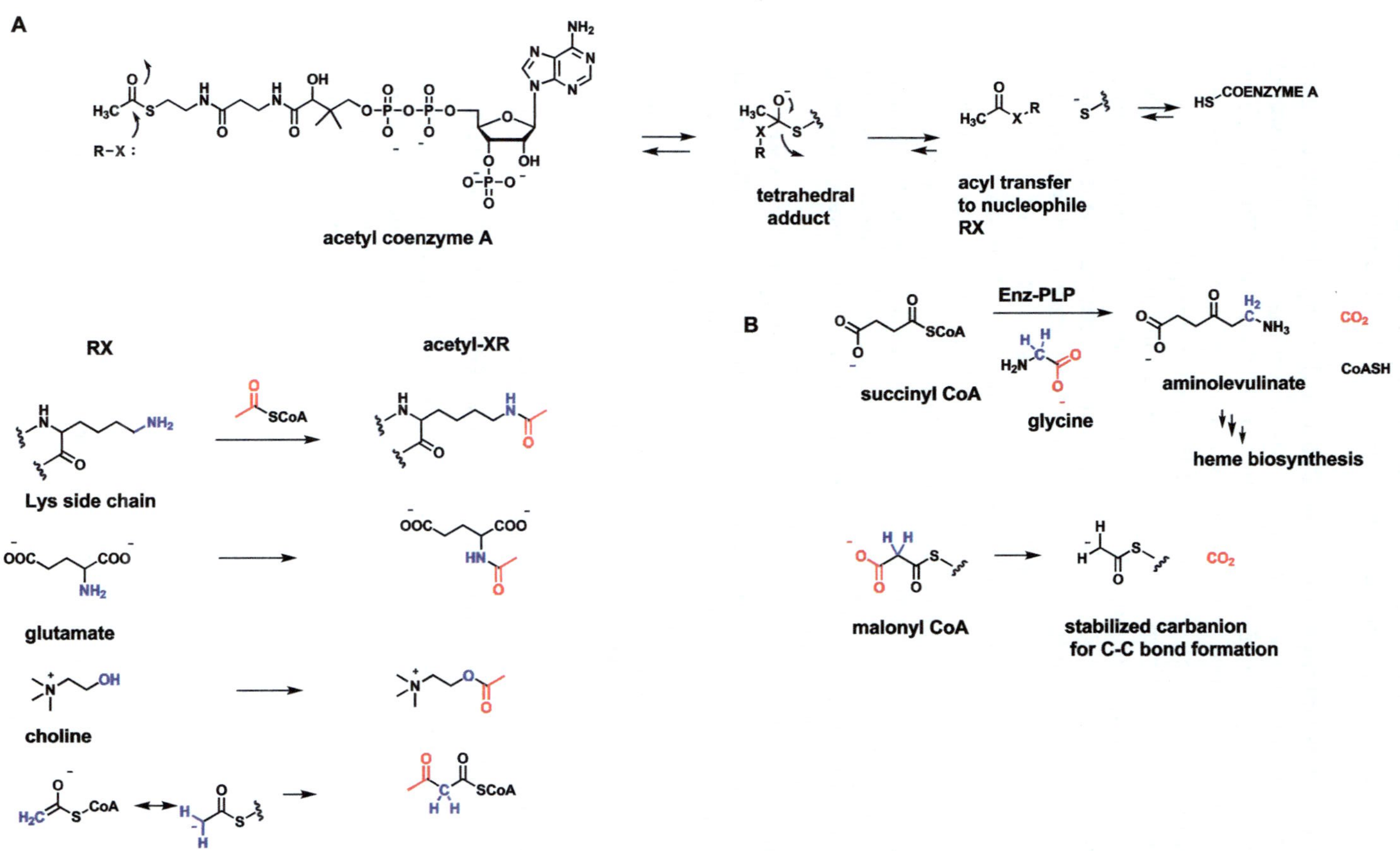

Figure 5.4 (A) Acyl transfers from acyl-*S*–CoAs to cosubstrate nucleophiles proceed *via* tetrahedral adducts. Acetyl transfer examples shown are lysine side chain protein acetylation, *N*-acetylglutamate, acetyl choline and acetoacetyl CoA. (B) The succinyl group from succinyl CoA is transferred to the C2 carbanion of a glycyl-PLP adduct in aminolevulinate formation. Also, the malonyl CoA C2 carbanion is a central nucleophile for fatty acid C–C bond formations during chain elongations.

(Figure 5.3) and the oxoester bonds in the approximately 60 aminoacyl-tRNA isoforms that deliver the aminoacyl monomers to the ribosome for protein biosynthesis).

The greater electrophilicity of the acyl carbonyl groups bonded to sulfur (thioester) than when bonded to oxygen (oxoester) may be due to lowered orbital overlap, at greater distance, of the 3p electrons of sulfur compared with the 2p electrons of oxygen. The carbonyl groups of the thioesters have more "ketone"-like character due to that lessened overlap.[8,10]

In some cases, notably in fatty acid synthases, and also polyketide synthases and non-ribosomal peptide synthase assembly lines, acyl thioester precepts on acyl transfer logic are followed, but not with freely diffusing acyl CoAs.[10] Instead, the phosphopantetheine ("business end") half of coenzyme A is transferred enzymatically to the serine-OH side chain of small acyl carrier proteins, as independently folding domains or free standing subunits[11] (Figure 5.5A). The thermodynamic activation and kinetic stability of acyl-*S*-pantetheinyl carrier proteins is equivalent to that of freely diffusing acyl-*S*-CoA thioesters but now they are tethered in fixed spots on the protein assembly lines. This covalent protein tethering, the avoidance of diffusible acyl thioester intermediates that could get away from multiprotein assembly lines, may aid in the efficiency of acyl chain build up.

In some polyketides such as the antifungal agent nystatin ("New York state-in"), some 18 phosphopantetheinyl way stations populate the protein assembly lines[12] (Figure 5.5B). The terminal thioesterase domain of the nystatin assembly line is an intramolecular cyclase, releasing the indicated lactone that is two oxidations and a glycosylation away from the mature antifungal natural product.

5.2 Double Headed Reactivity of Acyl Thioesters as Carbon Electrophiles and Carbon Nucleophiles

We have just examined the thermodynamic activation of the carbonyl group of acyl thioesters by the adjacent sulfur atom, powering the acyl group transfers to cosubstrate oxygen, sulfur and carbon nucleophiles. Equally important for reactions in primary metabolism is the effect the sulfur atom has on C2 of the attached acyl chains. The sulfur enhances the kinetic acidity of C2-hydrogens for removal as protons (Figures 4.3 and 5.6). This occurs because the resulting C2 carbanion negative charge is delocalized onto the adjacent oxygen to form the thioester enolate. Delocalization lowers the energy barrier to carbanion formation and, oxygen being more basic than carbon, the thioester enolate may be the major contributor to the resonance structure.[9]

The net effect is that in acyl thioesters the C2-acyl carbanions are kinetically accessible and function as carbon nucleophiles in C–C bond formations at key metabolic intersections. Typically, the carbon electrophiles with which they pair in C–C bond formation are carbonyl

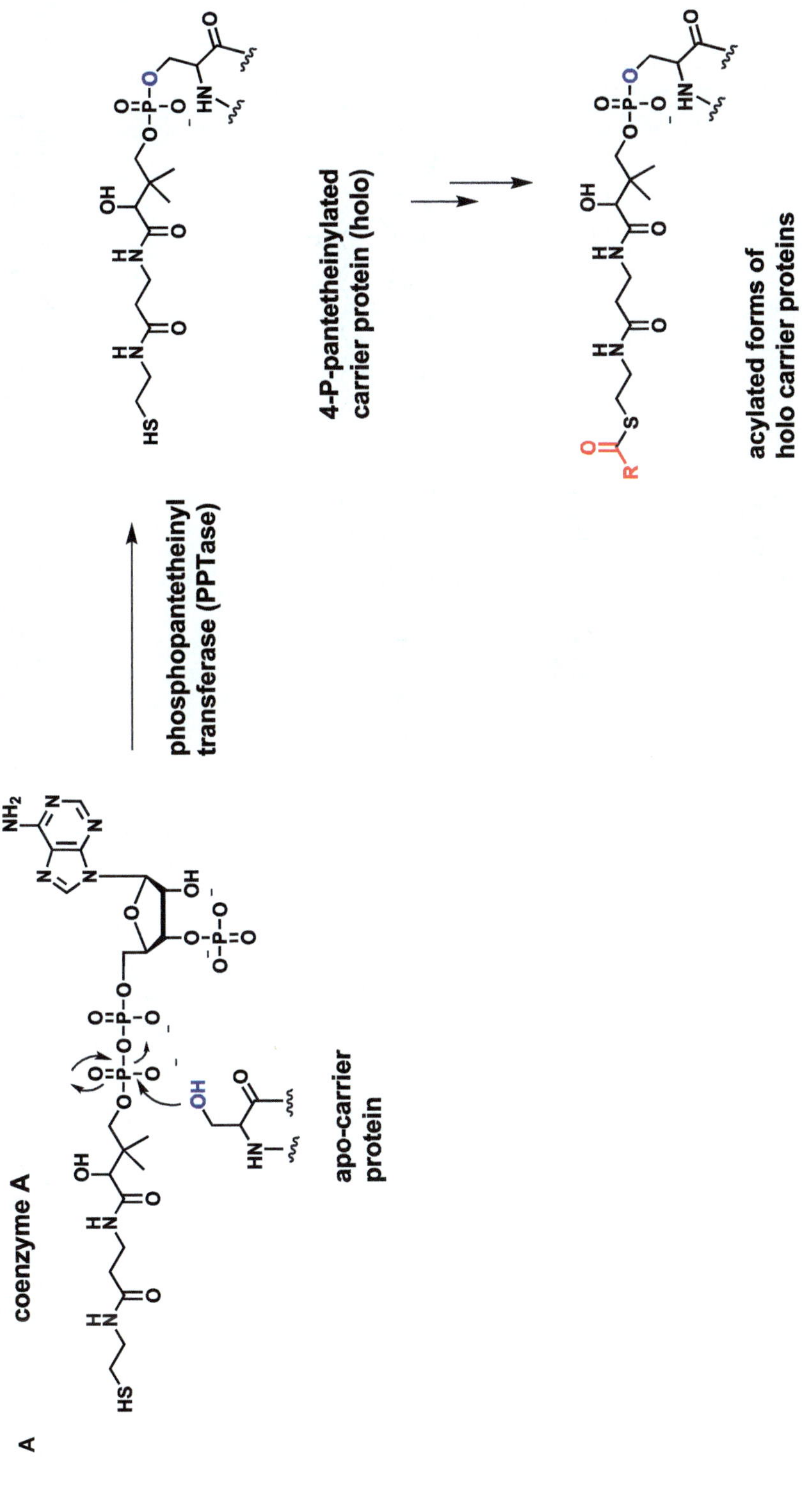

Figure 5.5 Continued on next page.

B

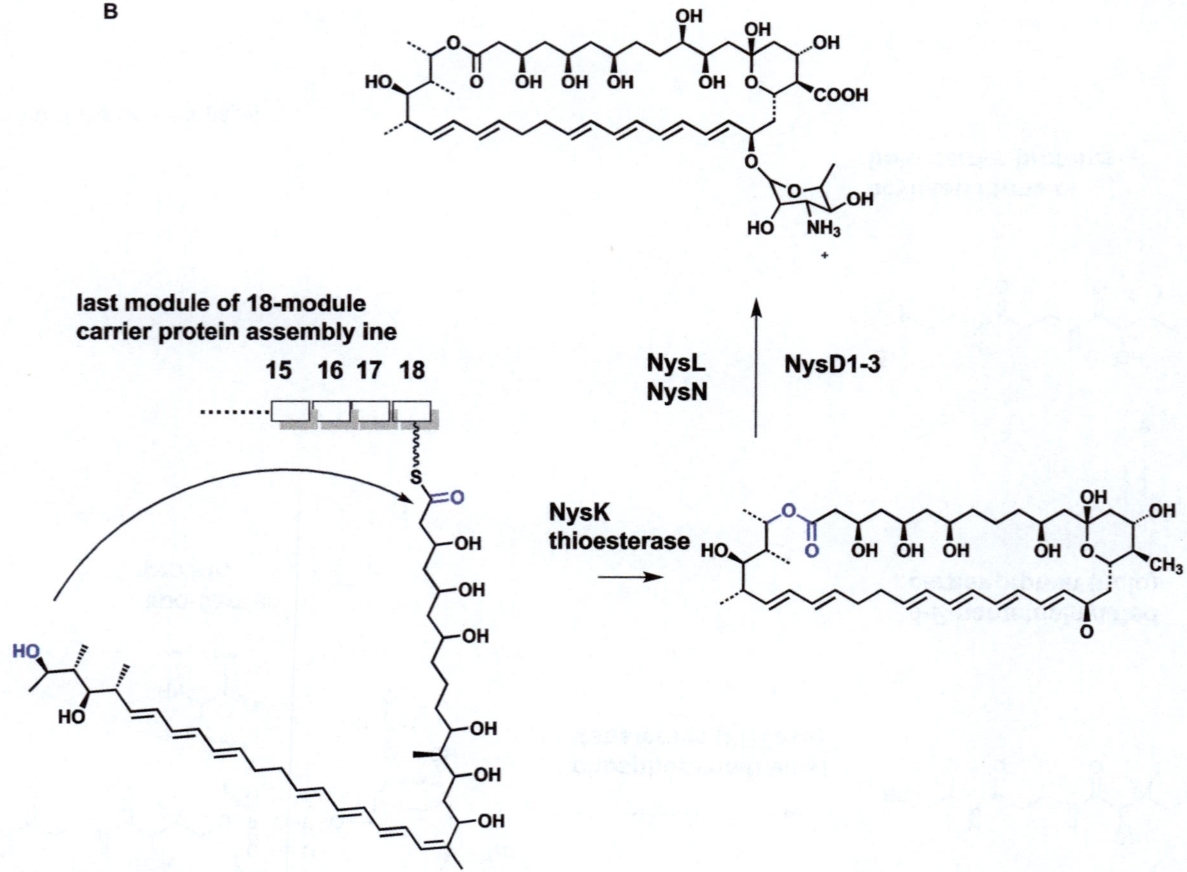

Figure 5.5 Phosphopantetheinylated carrier protein domains. (A) During biosynthesis of fatty acids, polyketides and non-ribosomal peptides, apo forms of 80–100 residue carrier proteins are posttranslationally phosphopantetheinylated from coenzyme A to yield the active, holo forms with the pantetheinyl terminal thiol installed as anchoring nucleophile for chain growth. (B) The biosynthesis of the antifungal natural product nystatin A involves iterative chain growth on a pantetheinylated carrier protein. Chain termination occurs *via* a cyclizing thioesterase to create the antifungal macrolactone scaffold.

carbons. Two such examples in primary metabolism are in the first step of the citric acid cycle, and the first two steps of steroid biosynthesis (Figure 5.6).

The citric acid cycle begins with catalysis by citrate synthase: attack by the C2-carbanion of acetyl CoA on the ketone carbonyl of oxaloacetate (Figure 5.6). The new C–C bond joins the former two-carbon acetyl group and the four-carbon oxaloacetate into a six-carbon product, citryl-*S*-CoA. To draw off the equilibrium in the forward direction, the synthase cleaves the thioester bond hydrolytically before nascent citryl CoA is released, getting a boost of about 10^5 in the forward direction of the equilibrium constant. The products are then citrate, with its tricarboxylate framework, and free CoASH. The thermodynamic activation of the acetyl thioester has been used to drive the biosynthetic equilibrium as C1 of the acetyl group is attacked as an electrophile by water. C2 of the acetyl thioester moiety has acted as a carbon nucleophile.

The adjacent, double headed, orthogonal reactivity of C1 and C2 of acetyl CoA is most aptly illustrated in the head to tail condensation of two molecules of acetyl CoA by the enzyme β-ketoacyl thiolase, named for its reaction in the reverse direction. Thiolase uses one acetyl CoA substrate molecule with acetyl C1 electrophile while it generates the C2 carbon nucleophile in the acetyl chain of the second acetyl CoA substrate (Figure 5.6). C–C bond formation ensues when the C2 carbanion of the first acetyl CoA_1 attacks the electrophilic C1-carbonyl group of the second acetyl CoA. Ketonizing elimination of the CoASH moiety yields the 3-keto group in the four-carbon acetoacetyl CoA product. Note that, again, the C–C bond formation involves hydrolysis of the thioester linkage, generating a 10^5 pull in the direction of condensation.

Even so, the overall equilibrium is not very far in the condensation direction. It is the next enzyme, hydroxymethylglutaryl CoA synthase, that effectively pulls the carbons into the steroid biosynthetic pathway. It uses a third molecule of acetyl CoA, generates its C2-carbanion–thioester enolate equivalent, and directs it to attack the ketone carbonyl of the acetoacetyl CoA cosubstrate (Figure 5.6). The synthase also hydrolyzes the nascent thioester of what had been the third acetyl CoA to utilize the thermodynamic driving force of hydrolysis to accumulate the branched six-carbon product. Note that the regiospecificity of attack on the acetoacetyl CoA ketone mandates a branched chain framework in the hydroxymethylglutaryl CoA product. This branch point persists in all the isoprenoid downstream metabolites (Figure 5.6).

The first two enzymes of steroid biosynthesis take three molecules of acetyl CoA and form two C–C bonds. The carbon nucleophiles are the C2 carbanions of the acetyl group, attacking first a thioester carbonyl and second a ketone carbonyl on the cosubstrate molecules. Two of the three thioester bonds have been hydrolytically cleaved to drive the equilibria in the desired biosynthetic direction to form the branched six-carbon hydroxymethylglutaryl chain. The third thioester will subsequently be reduced by

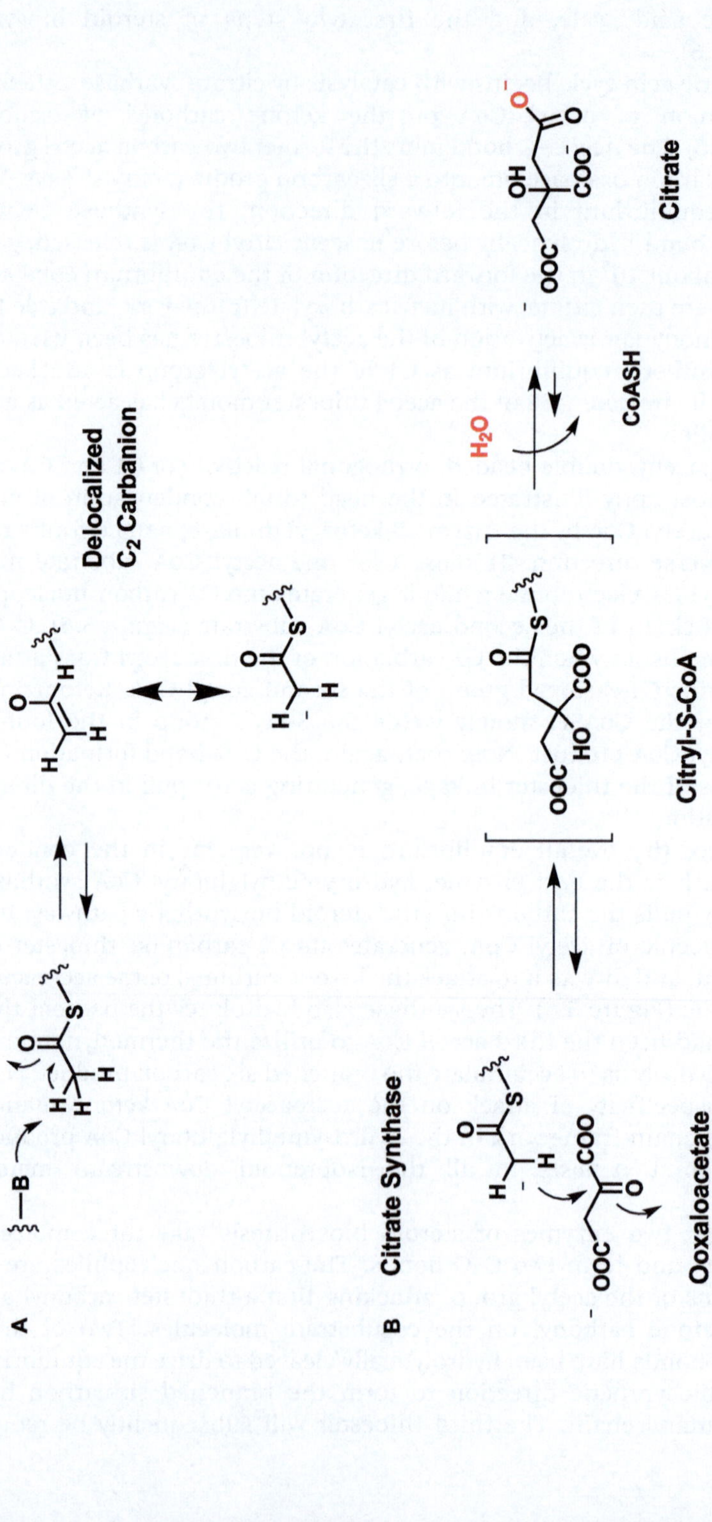

Figure 5.6 Continued on next page.

C Tandem Action of Thiolase and HMGCoA Synthase

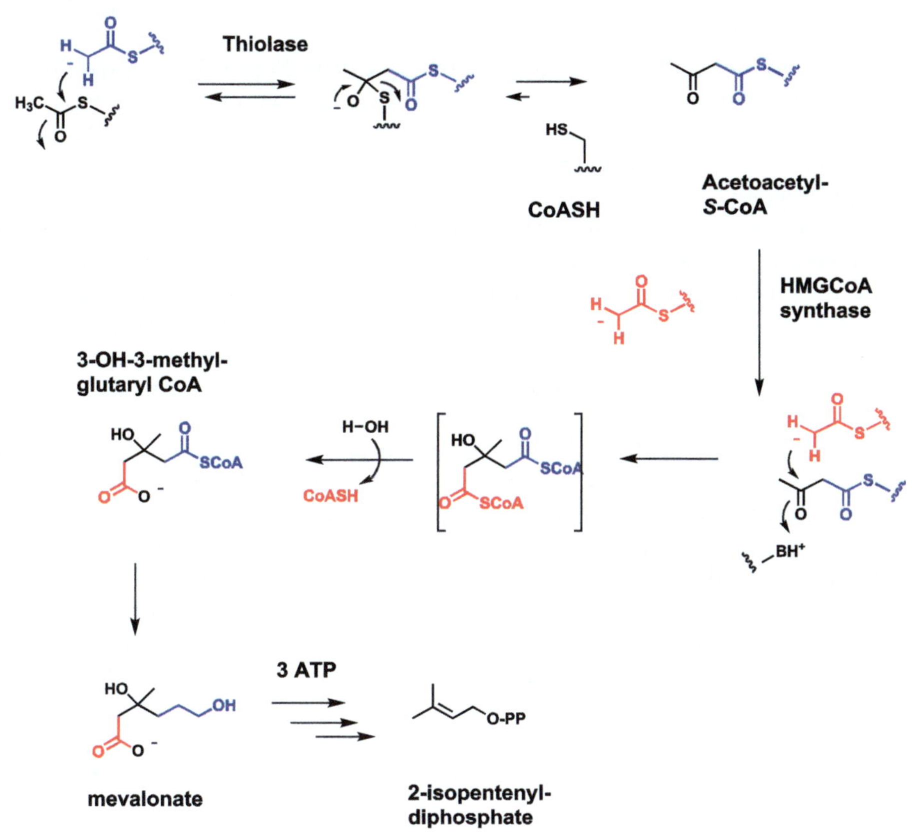

Figure 5.6 (A) The C2 carbon of acyl thioesters is setup for low-energy, delocalized carbanion formation and then serves as a carbon nucleophile in reactions such as those (B) catalyzed by citrate synthase and (C) the first two enzymes of the steroid biosynthetic pathway.

four electrons to the alcohol in the downstream metabolite mevalonate (Figure 5.6). (None of the activation chemical properties of all three thioester sulfurs go unutilized.)

5.3 Glutathione and Other Low Molecular Weight Thiols in Biology

In principle the reactive thiolate side chain of cysteine itself or of coenzyme A could be used to scavenge electrophilic molecules that could be toxic. These include xenobiotic molecules such as drugs (acetominophen) or toxins and also reactive oxygen species. In fact, neither free cysteine nor coenzyme A are used substantially by cells and tissues as nucleophilic thiol or thiolate scavenging and protective agents. That role falls principally to the cysteine-containing tripeptide glutathione (γ-glutamyl-cysteinyl-glycine) (Figure 5.1) which can be maintained at 50–100-fold higher levels in cells (particularly in hepatocytes) than free cysteine.

Glutathione (GSH) can accumulate to up to 10mM in cells, many-fold over free cysteine concentrations.[13,14] It is built up in two enzymatic steps from the three free amino acids, each cleaving a cosubstrate ATP molecule to ADP and Pi.[15,16] The first enzyme, γ-glutamyl-cysteine synthetase, makes a γ-glutamyl–PO_3^{2-} bound intermediate and captures it with the free amino group of cysteine to generate the dipeptide linkage. Then, glutathione (GSH) synthetase, make the dipeptidyl-acyl-phosphate mixed anhydride and captures it with the amine of glycine[17] (Figure 5.7A). The unusual γ-glutamyl peptide linkage rather than a typical α-glutamyl linkage contributes to resistance of GSH to normal cellular peptidase action. The name glutathione is a fusion of glutamate and thione, which is an historical term for a sulfur containing compound.

Glutathione is widely distributed in fungi, plants and animals. In fact, the dithiol form of GSH is the electron donor in plants for AMP–$SO4^{2-}$ reductase, discussed in Chapter 2 as the first committed step in sulfate reduction. GSH is present in mitochondria [we noted its role in removing the thiane sulfur from the enzyme persulfide in the sulfide quinone oxidoreductase catalytic cycle (Chapter 3)]. In liver, which is the recipient of first-pass blood flow from ingested foods and xenobiotics, GSH is particularly prominent.[18] Although as much as 20% of cellular GSH pools are consumed per day, more than 95% of the hepatic glutathione is in the reduced GSH state, *vs.* the oxidized GS–SG disulfide, thanks to the high activity of glutathione reductase; the mechanism is described below.

Glutathione is also present in plant chloroplasts and in some aerobic bacteria, such as cyanobacteria. However, the thiol-containing tripeptide is absent in many Gram-positive bacteria and anaerobes, consistent with the hypothesis that it may have evolved in aerobic organisms to deal with radical oxygen species.

Although anaerobic bacteria and Gram-positive bacteria usually lack the capacity to biosynthesize glutathione they do produce alternate low molecular weight thiols as protectant molecules.[19] For example a simple structural variant, glutathione amide (Figure 5.8) is present in purple sulfur bacteria, members of the anaerobic sulfate reducers.[20] The corresponding glutathione amide persulfide is also found in those bacteria, where the sulfane terminal sulfur in the persulfide is thought to be ferried to S^0 globules formed as the end-product of its sulfide to elemental sulfur redox metabolism.

In some analogy, green sulfur bacteria in the *Chlorobium* genus have *N*-methylbacillothiol.[21] As the name indicates bacillothiol was first detected in Gram-positive bacilli (Figure 5.8). A cysteinyl moiety with the requisite thiolate nucleophile is present. In the bacillothiol framework, the cysteine is

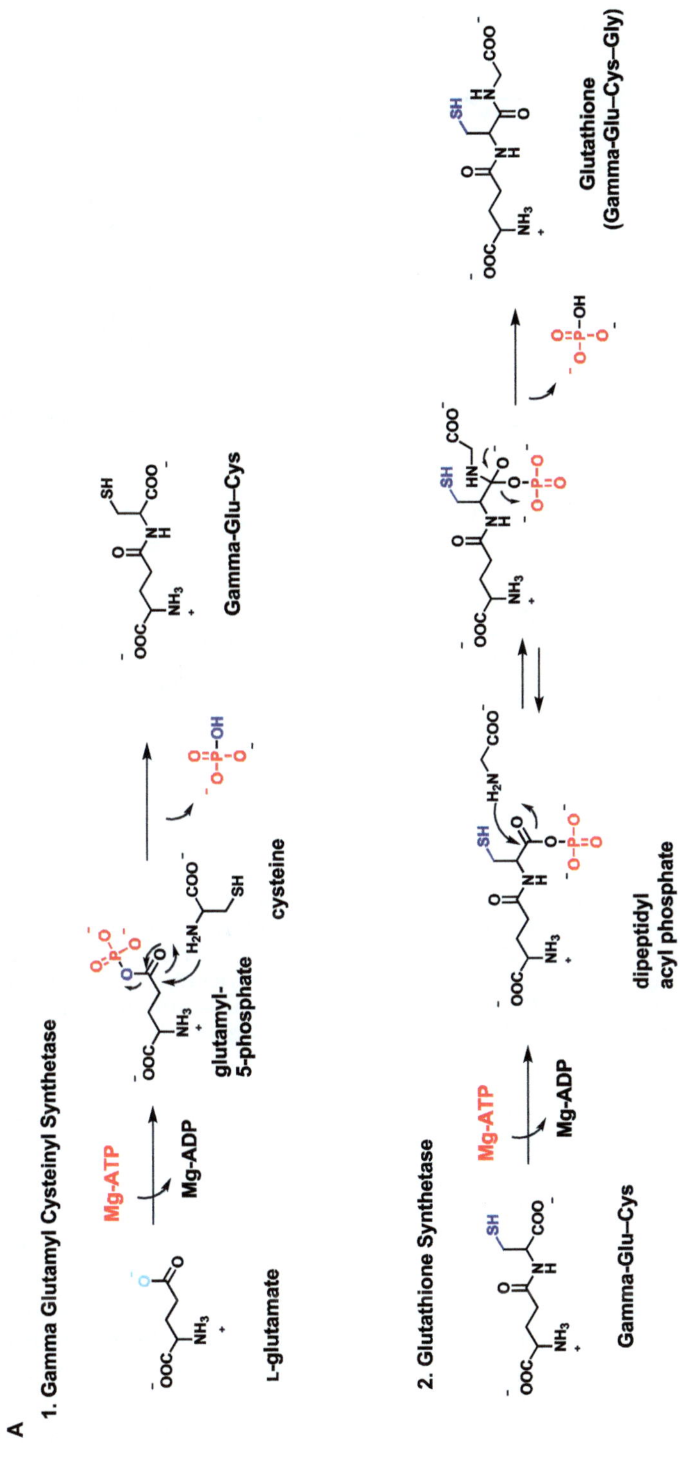

Figure 5.7 Continued on next page.

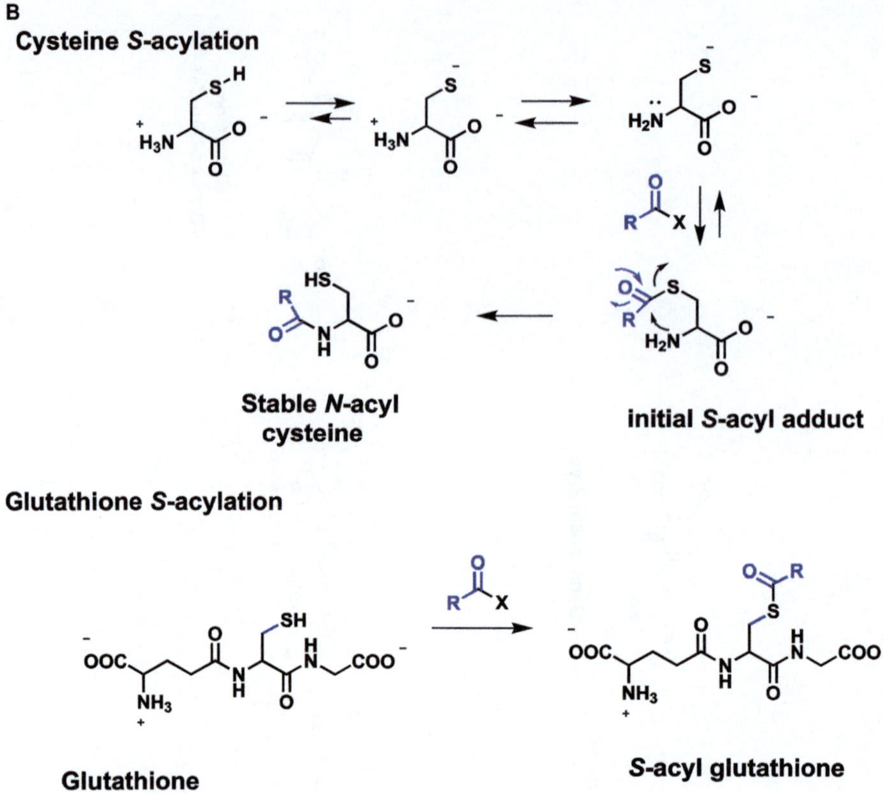

Figure 5.7 Glutathione biosynthesis and *S*- *vs.* *N*-acylations. (A). The cysteine-containing tripeptide glutathione has a non-standard γ-glutamyl-cysteinyl amide bond, signaling its non-ribosomal origin. Two enzymes of the ATP grasp family, act in tandem, first to make γ-Glu–Cys and then to add glycine to give γ-Glu–Cys–Gly = glutathione (GSH) *via* acyl phosphate intermediates. (B) Free cysteine, when *S*-acylated, then undergoes a thermodynamically favored S to N-shift (relevant to protein splicing in Chapter 14). Glutathione has the cysteine amine tied up in a peptide linkage and so does *not* undergo *S*- to *N*-acyl shifts.

in amide linkage to glucosamine and the sugar C1 is in acetal linkage to L-malate, a primary metabolite of the citrate cycle.[22]

Mycobacteria make the related mycothiol as corresponding circulating thiol metabolite.[23] Now the cysteinyl moiety is *N*-acetylated, in closer analogy to the internal cysteine residue of glutathione. The malate group is replaced with an inositol ring.

Finally, a separate class of microbial low molecular weight metabolites thought to be glutathione surrogates is the fungal metabolite ergothioneine, with an unusual C=S substituent on a modified histidine residue. The evolution of low molecular weight thiols in anaerobes presumably predated widespread oxygen radical species, but there were still organic electrophilic

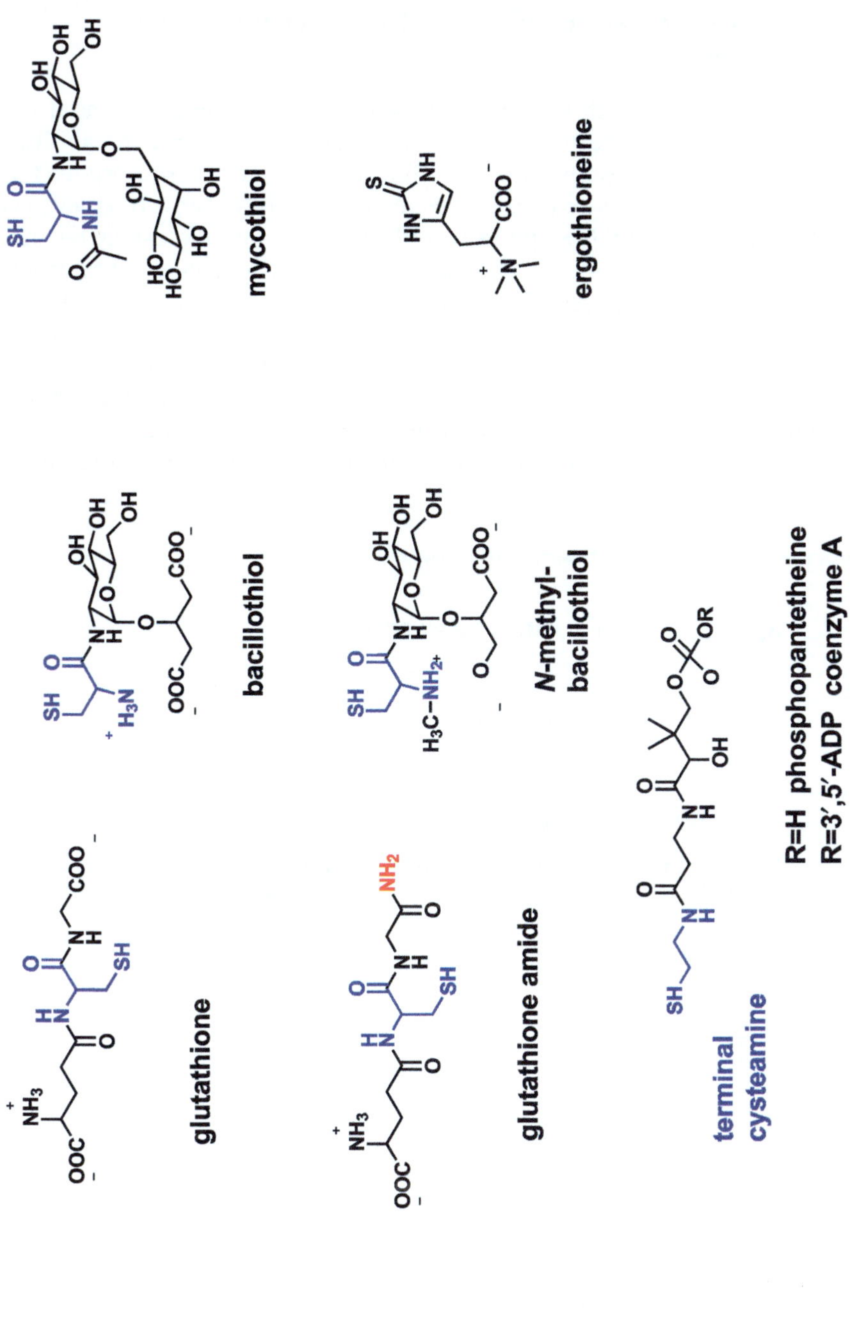

Figure 5.8 Low molecular weight thiols in different organisms as glutathione replacements.

metabolites, such as quinones, to be detoxified and inorganic metal ions to be chelated.

5.3.1 Reaction of Reduced Glutathione with Cellular Electrophiles

While we will highlight three examples below where GSH serves as a co-substrate towards three quite distinct endogenous electrophilic metabolites by two-electron reaction manifolds, one-electron chemistry is important. For instance, GSH acts to regenerate the one-electron oxidized forms of vitamin C (ascorbate) and vitamin E (tocopherol), and those react preferentially by one-electron redox routes[24] (Figure 5.9). Vitamin C is water-soluble and scavenges radicals in the aqueous phase of cells. Vitamin E is membrane-soluble and is the first-line reagent for scavenging lipid peroxide radicals generated by autoxidation. In both cases the thiolate anion (GS^-) or reduced glutathione (GSH) is a one-electron reductant to replenish the active forms of those two protectant vitamins. The $GS^\bullet$ radicals can couple with other radicals and/or react with O_2 for one-electron transfer to form GSH and superoxide anion

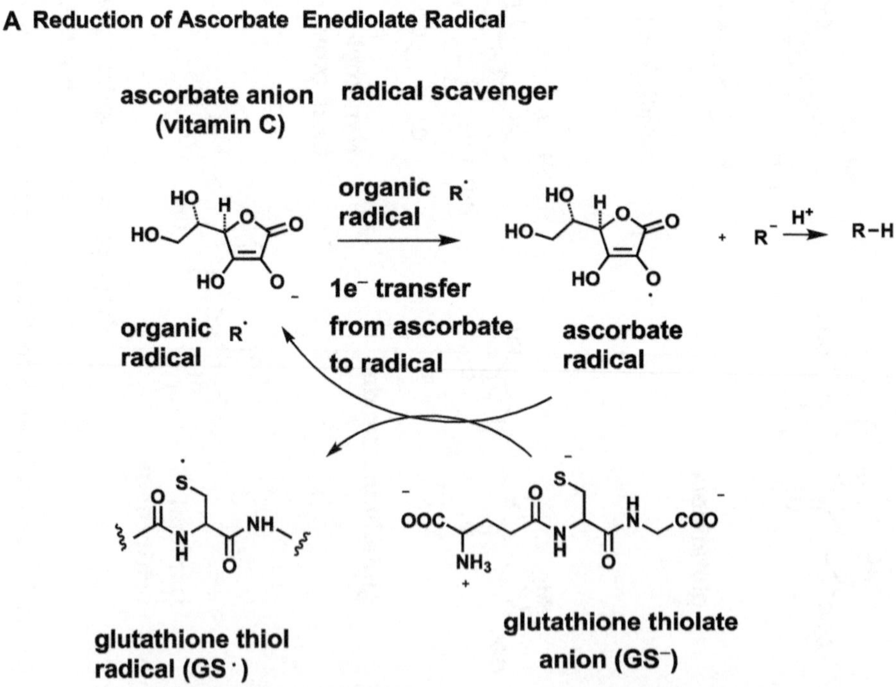

Figure 5.9 Glutathione, normally thought of as a two-electron thiol reagent, can function *via* sulfur radicals. Reduction of both (A) oxidized vitamin C in aqueous compartments, and (B) oxidized vitamin E at aqueous–membrane interfaces, is mediated by reduced glutathione.

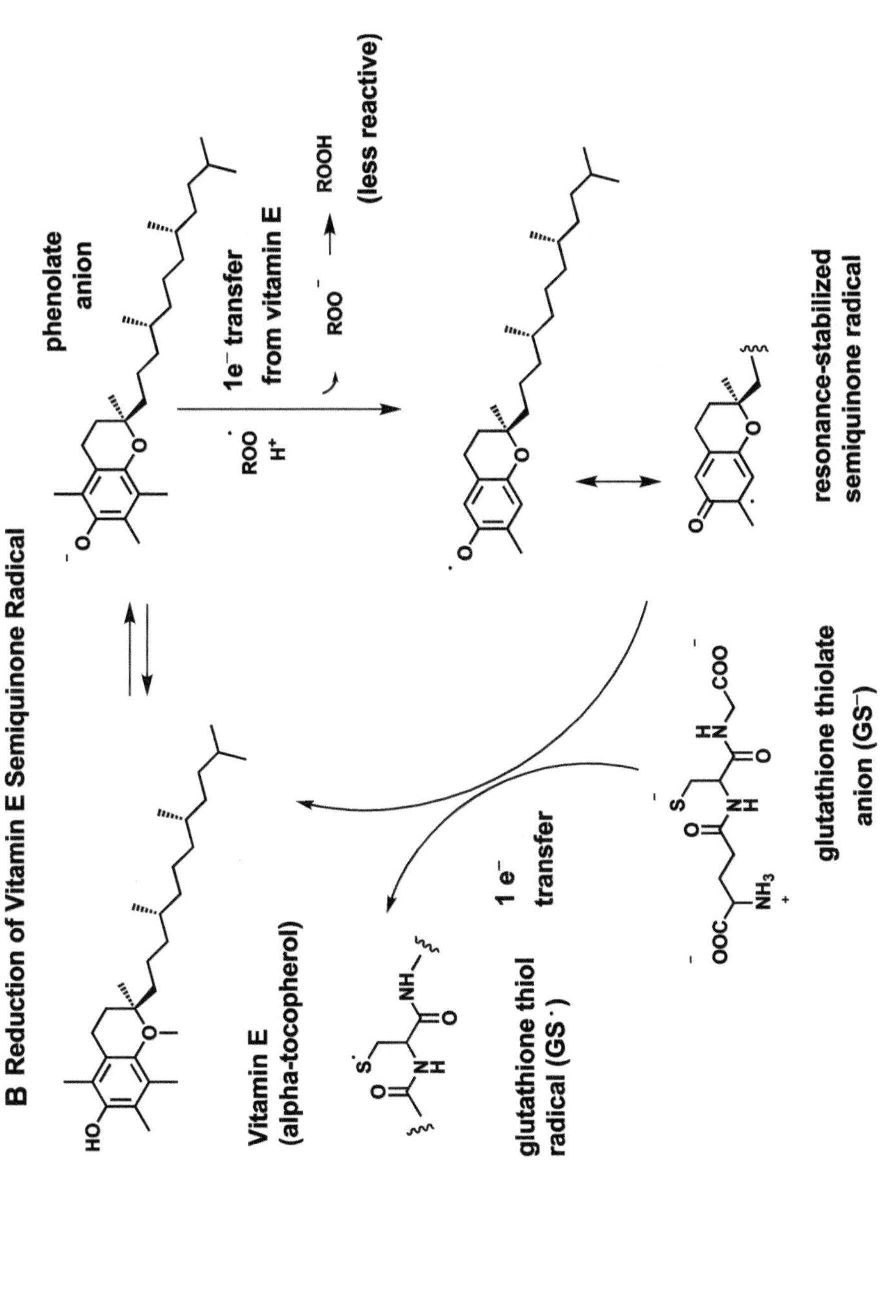

Figure 5.9 Continued.

($O_2^{-\bullet}$). This path can lead to glutathione sulfenic acid (GS-OH) discussed in Chapter 7.[25]

Two examples of glutathione utilization as a cosubstrate for capture, deactivation, and/or rerouting of electrophilic metabolites[26] are depicted in Figure 5.10. The first involves scavenging of methyl glyoxal, also known as pyruvaldehdye, which is a minor but toxic metabolite that arises by unwanted dehydrative reaction of the normal glycolytic intermediate dihydroxyacetone-phosphate (Figure 5.10A).

The aldehyde group of methylglyoxal, as a 1,2 dicarbonyl scaffold, is problematically reactive with cellular molecules. One path for its detoxification is enzymatic addition of the thiolate of glutathione into the C1 aldehyde by the enzyme glyoxalase I,[27] as shown. That tetrahedral adduct can deprotonate at C1 to give an enolate. However, inspection shows it is actually an *enediolate* that can equilibrate between two monoanionic enediolate isomers. The second enediolate isomer can collapse to the acyl thioester D-lactylglutathione.

This is an internal dismutation. C1 has been oxidized by two electrons, from aldehyde to thioester. C2 has been reduced by two electrons, from the ketone to the hydroxyl, with stereochemical control to generate the D-isomer of the lactyl scaffold. Detoxification has occurred and the next two enzymes act to retrieve the glutathione as GSH and to return D-lactate to the usable metabolic pool. Enzymatic hydrolysis of the thermodynamically activated thioester reclaims GSH and gives free D-lactate. Cells have a D-lactate dehydrogenase, generating reduced nicotinamide adenine dinucleotide (NADH) and pyruvate, a primary achiral metabolite (Figure 5.10A).

The orientation of the glutathione and methylglyoxal in the glutathione *S*-transferase active site sets up attack by the thiolate on the aldehyde carbonyl. The resultant tetrahedral adduct is a reversible hemithioacetal. Proton abstraction and creation of the dienolates, then leads to the thermodynamically activated acyl thioester.

This logic presages the enzymatic oxidation of other aldehydes. Prior addition of a thiolate to form the thiohemiacetal that is then oxidized yields an energy rich acylthioester product. (This is the case in the sole redox step in glycolysis where 3-P-glyceraldehyde is converted to 1,3-phosphoglycerate.) The thermodynamically activated D-lactoyl-*S*-glutathione acyl thioester is hydrolyzed to the favored equilibrium products of free acid D-lactate and GSH.

The second example in mammalian metabolism again utilizes reduced glutathione as a nucleophilic cosubstrate. The second substrate is leukotriene A4, with an electrophilic 5,6 epoxide group. Leukotriene A4 is generated by enzymatic epoxidation of free arachidonic acid. It is an intermediate on the way to cysteinyl-containing leukotrienes C4, D4 and E4.[28] The three leukotrienes C, D and E all arise from glutathione addition to the epoxide ring in LTA4 followed by peptidase trimming of the glutathionyl tripeptide chain to the dipeptide Cys–Gly (in LTD4) and the cysteinyl adduct

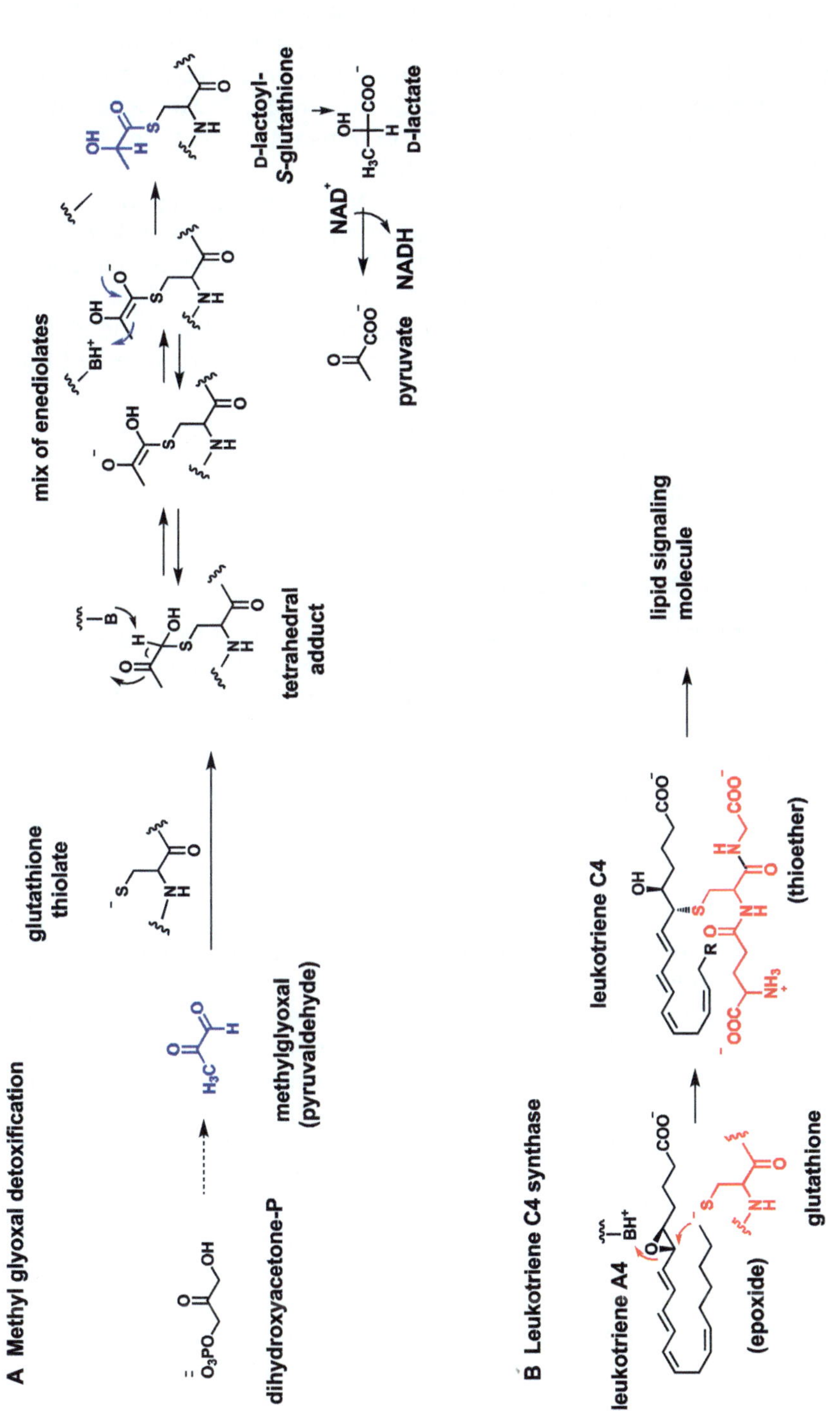

Figure 5.10 Glutathione as a nucleophilic cosubstrate in cellular reactions. (A) Glutathione adds in to the aldehyde carbonyl of methylglyoxal in the glyoxylase catalytic cycle. (B) Glutathione is the thiolate nucleophile opening the epoxide of leukotriene A4 on the way to leukotriene C4.

in LTE4. LTB4 is not a peptidyl leukotriene and arises instead from a doubly vinylogous hydrolysis of the epoxide by water attack at C10 (not shown). The cysteinyl leukotriene scaffolds are lipid mediators in allergy, asthma and some presentations of rheumatoid arthritis. Each leukotriene binds to G-protein-coupled membrane receptor proteins on the surfaces of target cells to initiate intracellular signaling pathways.

The three-ring epoxide electrophile of LTA4 opened by leukotriene C4 synthase as shown in Figure 5.10B creates a stable thioether linkage. The biologically active leukotriene product LTC4 has the glutathionyl tripeptide moiety attached to C6 in a metabolically stable thioether linkage. The epoxide opening has left the –OH at C5. The glutamyl moiety and the glycine moiety can be trimmed away by stepwise action of peptidases to create LTD4 and LTE4 (not shown).

This is a classic example of a covalent addition of the glutathione thiolate (GS^-) to a cellular electrophile, in this case an endogenous metabolite, with an increase in downstream bioactivity.

5.3.2 Geranylgeranyl Glutathione

The thiolate of the Cys residue in glutathione can also act as a nucleophile towards the endogenous metabolite geranyl-geranyl diphosphate, an intermediate on the way to C_{30} squalene and ultimately cholesterol (Figure 5.11). This third example highlights a third kind of metabolic carbon electrophile that is capturable by the thiolate in glutathione. This C_{20} tetraenoyl-diphosphate can readily dissociate at the C1–OPP bond to an allyl cationic transition state or intermediate. That delocalized and kinetically accessible cation is capturable by GS^- to again yield a stable thioether link.

In lymphoid tissue germinal centers, one of the glutathionyl *S*-transferase family of isozymes catalyzes the formation of that thioether linkage in geranyl-geranyl-*S*-glutathione (GGG)[29] (Figure 5.11). (This prefigures similar geranylgeranylation of cysteinyl side chains in ras proteins, noted below in Figure 13.3 in Part III.) This lipidated tripeptide–prenyl glutathione adduct, GGG, is a nanomolar ligand for the receptor purine–pyrimidine receptor 2Y family member 8 (P_2RY_8) to activate retention and/or confinement of B cells in the follicle germinal centers where antibody gene rearrangements and maturation of antibody affinities, essential for effective adaptive immunity, occur. Deactivation of GGG occurs by action of a peptidase that cleaves the amide bond between the γ-glutamyl and the Cys–Gly moieties of the glutathione backbone. This recent discovery may open a search for broader roles of GGG in organization and/or growth control in other human organs and tissues. It also reveals that GSH can be utilized for purposes other than deactivation of endogenous electrophiles, here to form a signaling molecule.

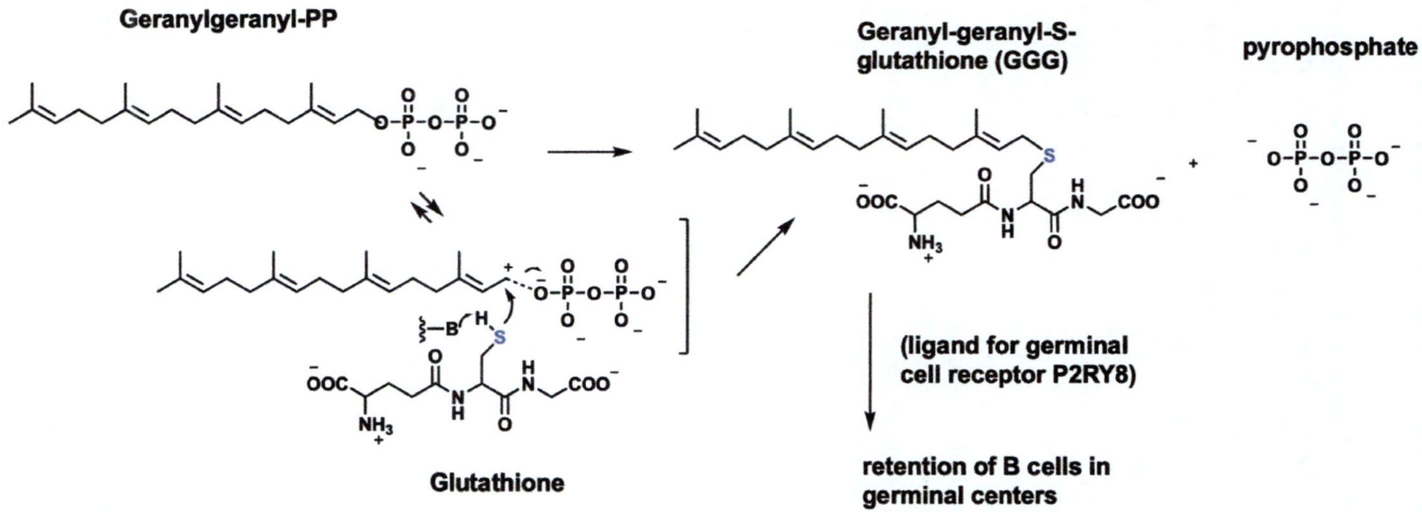

Figure 5.11 Glutathione attacks C1 of geranyl-geranyl–PP to give the peptidolipid geranyl-geranyl glutathione (GGG) as a ligand for B cell retention and maturation in germinal centers. The GGG product acts as a potent ligand for receptor P_2RY_8 that signals to keep B cells from exiting the follicles until they have undergone antibody chain rearrangements for adaptive immune diversification.

5.3.3 Reaction of Reduced Glutathione with Xenobiotic Electrophiles

One can find a long list of xenobiotic electrophilic molecules that get to the liver in first pass of blood from the digestive tract and are covalently captured by GS^-. This detoxification strategy is used by hepatic glutathionyl *S*-transferase isozymes during metabolism of the exogenous drug molecule acetaminophen (tylenol). In this case the electrophile is not an epoxide group. Instead it is an iminoquinone that arises from the parent acetaminophen by two prior enzymatic steps (Figure 5.12). The first is *N*-hydroxylation by one or more hepatic cytochrome P450 monooxygenases. The second enzyme-catalyzed step is sulfuryl transfer from phosphoadenosine phosphosulfate (PAPS). With its mixed sulfuric–phosphoric anhydride side chain, PAPS is reminiscent of APS as a donor of activated $-SO_3^-$ group in Chapter 2 in inorganic sulfate reduction, and it prefigures Chapter 10 dealing with other metabolic sulfuryl transfers.

For now, we note that the sulfur in the itinerant SO_3^- group is electrophilic and it is added enzymatically to the newly introduced N–OH group. This is in service of adding the water-solubilizing sulfate group to increase

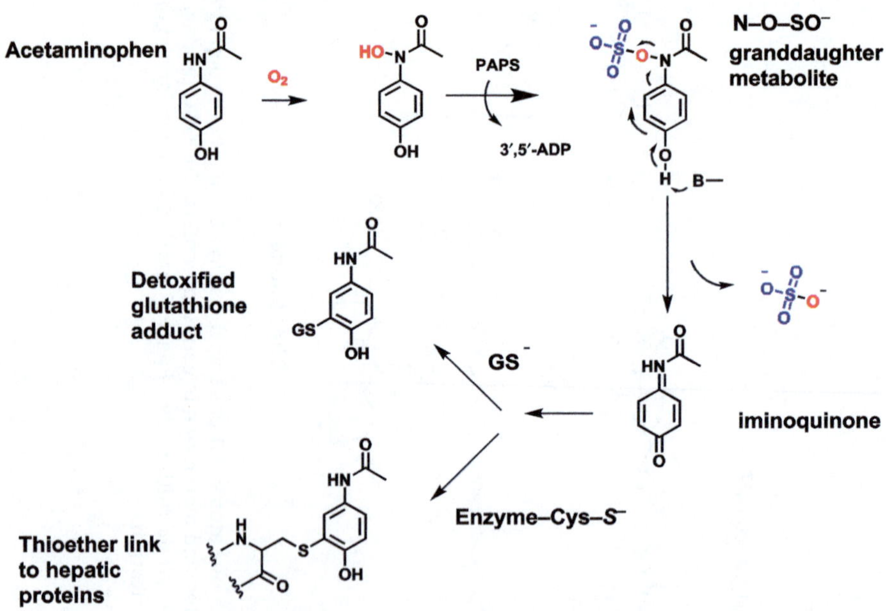

Figure 5.12 Two-step metabolic activation of acetaminophen: *N*-hydroxylation in hepatocytes is followed by sulfuryl transfer from PAPS to give the N–O–sulfate granddaughter metabolite. Intramolecular assistance to N–O bond cleavage by the *para*-OH group gives the toxic electrophilic iminoquinone. Capture by GSH in a conjugate addition reaction reflects detoxification, until stores of GSH are consumed by excess iminquinone.

the aqueous solubility of the drug metabolite so a larger fraction can be excreted by the kidneys.

However, this two-step drug metabolism strategy goes awry because now the *para*-OH group, dissociated to the phenoxide anion, uses its electron pair to expel the SO_3 group and the oxygen that had been initially added to the drug nitrogen. The products are the sulfate dianion and the *para*-iminoquinone. The quinone is a reactive electrophile.

As long as there is sufficient GSH in hepatocytes, that thiolate anion attacks in a conjugate addition as shown, to trap and detoxify the reactive iminoquinone species.[30] Unfortunately, at high doses of acetaminophen, if hepatic cellular glutathione has been consumed by high levels of ingested tylenol, there is insufficient GSH or GS^- left to capture all the iminoquinone granddaughter metabolites. Instead, the iminiquinone starts to be captured by protein side chain thiols and amines. The result is protein modification, dysfunction and degradation. Enough iminiquinone metabolic flux can cause fulminant hepatic necrosis that can be acutely fatal.[31]

Genomic analyses indicate that the encoded human proteomes contain up to 16 glutathione *S*-transferases.[32–34] Some are likely to be specific as in the methylglyoxal and leukotriene A4 metabolic conversions noted in Figure 5.10. Others are likely to be involved in detoxification of xenobiotic electrophiles such as the iminoquinone metabolite from tylenol processing in the liver as well as capture of acrolein and aflatoxin, noted in Figure 5.13. It is probable that there are overlapping specificities for xenobiotic electrophilic functional groups (*e.g.* epoxides, *vs.* conjugated aldehydes, *vs.* iminoquinones) and a spectrum of catalytic efficiencies of various GST isozymes in different tissues.

5.3.4 Cysteine *vs.* Glutathione as a Sulfur Nucleophile

All the amino acids have at least one potential nucleophile, the free base form of the amino group. Cysteine has a second potent one in the side chain sulfur and so is classifiable as a bidentate nucleophile.[1,2] In the "hard–soft" terminology of nucleophiles the nitrogen is hard and the thiol sulfur is soft. Reaction with saturated aldehydes reveals both nucleophiles: an initial imine adduct from nitrogen attack can be followed by sulfur addition to the imine to produce the cyclic thiazolidine ring framework (Figure 5.13). When the aldehyde is acetaldehyde, the 2-acetylthiazolidine is a known meat-flavoring molecule,[35] formed after loss of CO_2. Analogously, in reaction with conjugated carbonyl molecules, αβ-unsaturated aldehydes, ketones, and esters, the thiolate is the primary nucleophile followed by attack of nitrogen in cyclic imine formation.

Glutathione, bearing the thiol group as a nucleophile but with the amine tied up in a peptide linkage is the highest abundance thiol molecule in cells, ranging up to levels of 1–15 mM.[13] Thus, thiol capture of electrophiles, both physiologic and xenobiotic αβ-unsaturated compounds, yields the deactivated covalent adducts[36] (Figure 5.13). Enzymatic cleavage

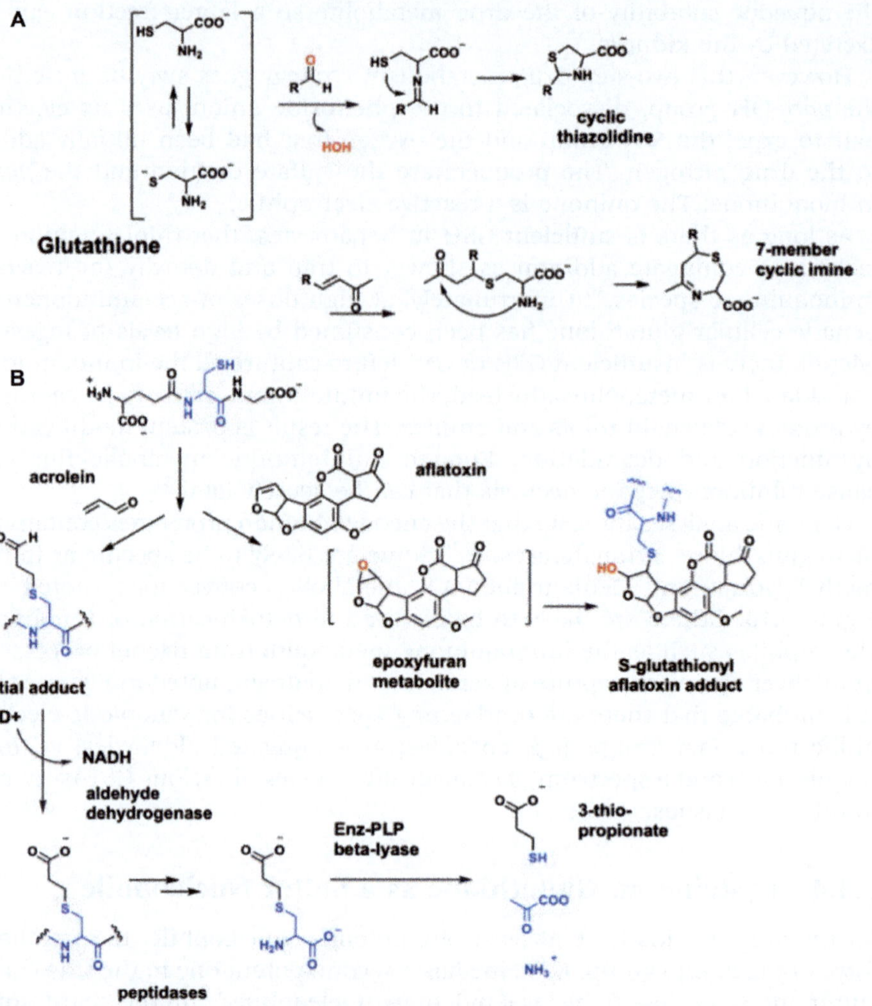

Figure 5.13 *S*-Alkylations and acylations by cysteine and glutathione (A) Cysteine reacts with linear aldehydes to form the imines that are then captured by the thiolate side chain to form the cyclic thiazolidines However, reaction with a conjugated carbonyl involves sulfur as the initial nucleophile, followed by cyclic imine formation. (B) Glutathione has the cysteine amine tied up and so reacts only *via* the thiol side chain as shown for capture of acrolein (conjugate addition) or in ring-opening of an aflatoxin epoxide. The acrolein adduct is further processed to 3-thipropionate as shown.

of the tripeptide backbone of the conjugated glutathione releases the conjugated cysteine moiety that is then substrate for pyridoxal-P-dependent lyases that cleave the C_3–S bond of the cysteinyl portion to generate pyruvate, ammonia and the *S*-alkylated product fragment. These are often volatile, flavor components of plant and bacteria metabolism (*e.g.* 3-thiopropionate).[35]

A comment about the possible rationale of using tripeptide glutathione rather than free cysteine as the thiol scavenger at high abundance in cells. While the thiol group in each is the kinetically significant nucleophile, when acyl groups are captured *S*-acyl-cysteines rearrange to the thermodynamically more stable *N*-acyl-cysteines and are no longer transferable. In contrast, when cysteine is in peptide linkage, as in glutathione, the amine is tied up in amide linkage and the *S*- to *N*-acyl shifts do not occur.

As we will note in Part III, the *S*-acylation of cysteinyl residues (with NH tied up in a peptide bond) is subsequently reversible. This is relevant as noted in Chapter 13 for *S*-palmitoylation of cysteine residues to be reversible and thereby dynamically control membrane association of protein targets. Conversely, the *S*- to *N*-acyl shift is the final and irreversible step in reforming the peptide bond between newly joined exteins in biologic protein splicing mechanisms (Chapter 14).

5.4 *S*-Nitrosylation of Glutathione

Nitric oxide acts as a biological signaling agent largely by binding to the regulatory heme site in the enzyme guanylate cyclase.[37] The ligation of NO to the iron in the heme equatorial plane sets off conformational changes that result in the conversion of guanosine triphosphate (GTP) to 3′,5′-cyclic guanosine monophosphate (GMP), which, in turn, serves as a diffusible activator of smooth muscle contraction. Nitric oxide, a gaseous free radical, is generated from arginine by one of three isoforms of nitric oxide synthase. (Figure 5.14).

Among the downstream consequences of NO• production is the accumulation of nitrosylated thiols on reduced glutathione and then, when local GSH has been depleted, certain cysteinyl residues across many proteins in different tissue and cellular proteomes.[38] While it may be one-electron reactions that generate GS• for direct capture of the NO•, it is also possible that the glutathione capture of nitric oxide is indirect and a two-electron rather than a one-electron reaction process.[39] For example, the odd electron NO• may be converted to the nitrosyl cation NO^+ by electron transfer to O_2 or other reactive oxygen species. The electrophilic NO^+ could then be captured by glutathione thiolate anion side chains.

The resultant adduct is *S*-nitrosyl glutathione (abbreviated GSNO)[40] (Figure 5.14). It is proposed that GSNO is in turn a reservoir for release of NO•, presumably by homolysis of the S–N bond and transient formation of the glutathione radical, which could in turn be captured by reactive oxygen species (ROS) to give glutathionyl sulfenate.[41] Alternatively, glutathione can be retrieved from the NO adduct by one of the isozymes of aldehyde dehydrogenase (ADH5) which can deliver a hydride from NADH to the N=O group, converting it to *N*-hydroxy-*S*-glutathione. This S–N–OH species can

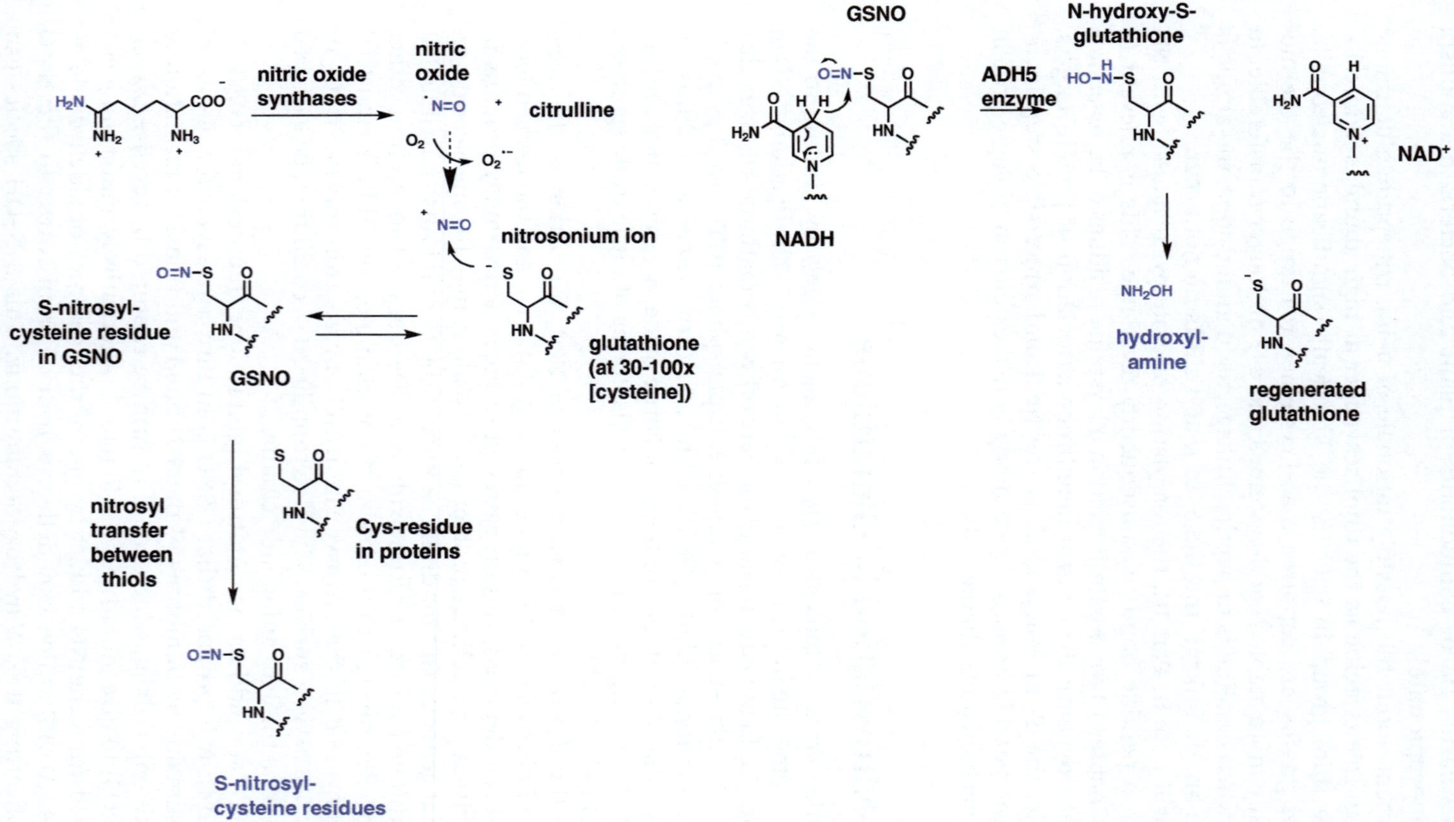

Figure 5.14 Reaction of reduced glutathione with nitric oxide. Possible conversion of nitric oxide to nitrosonium ion before a two-electron reaction manifold involving attack of glutathione anion on the cationic NO species. *S*-Nitrosyl glutathione (GSNO) can act as a nitrosylating agent for protein cysteinyl thiolate side chains. GSNO can also be reduced enzymatically by hydride transfer to the tetrahedral *N*-hydroxy-*S*-glutathione that unravels spontaneously to hydroxylamine and GSH, destroying NO and restoring GSH levels.

unravel nonenzymatically to hydroxylamine and the regenerated GSH[40] (Figure 5.14).

S-Nitrosyl-glutathione can also function as a mobile transfer agent of the nitrosyl group to other thiolates, most notably a subset of cysteinyl side chains in a large variety of proteins[42] noted later in Part III on the cysteinyl proteome. That flux is largely if not totally, presumed to be nonenzymatic, and a function of the differential accessibility, pK_a and nucleophilicity of cysteinyl residues in proteins dependent on their local microenvironments in their protein scaffolds.[1,2,43]

Cells place a high premium on keeping reduced glutathione levels high and we shall examine how they return the oxidized, disulfide form of glutathione back to the nucleophilic, protective surveillance molecule GSH. To set the context in which that enzyme, glutathione reductase, operates we note the major routes to oxidized glutathione in the first place.

5.5 Glutathione: Sulfenates and Disulfides: Thiol Oxidations

In addition to its scavenging role for organic electrophiles, both endogenous and xenobiotic, the second major surveillance role for reduced glutathione (GSH) in cells is capture of reactive oxygen species. This is predominantly hydrogen peroxide (H_2O_2), which can arise metabolically either from oxidases that reduce O_2 to hydrogen peroxide or by Fenton-type iron radical chemistry that generates one-electron-reduced superoxide anion (Figure 5.15). Dismutation of two superoxide ions by the enzyme superoxide dismutase yields back O_2 and also H_2O_2.

In recent years evidence has accrued that hydrogen peroxide is not just a reactive byproduct of a combined action of superoxide dismutase and

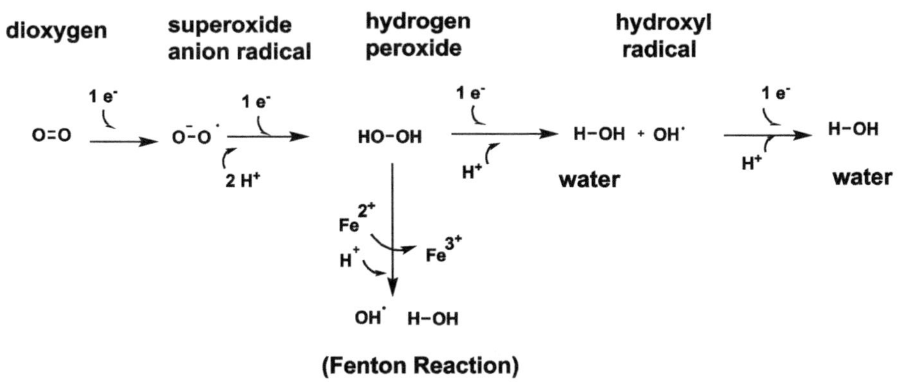

Figure 5.15 The reduction of O_2 to two H_2O molecules in one-electron steps with superoxide anion, hydrogen peroxide and hydroxyl radical as partially reduced intermediates.

A

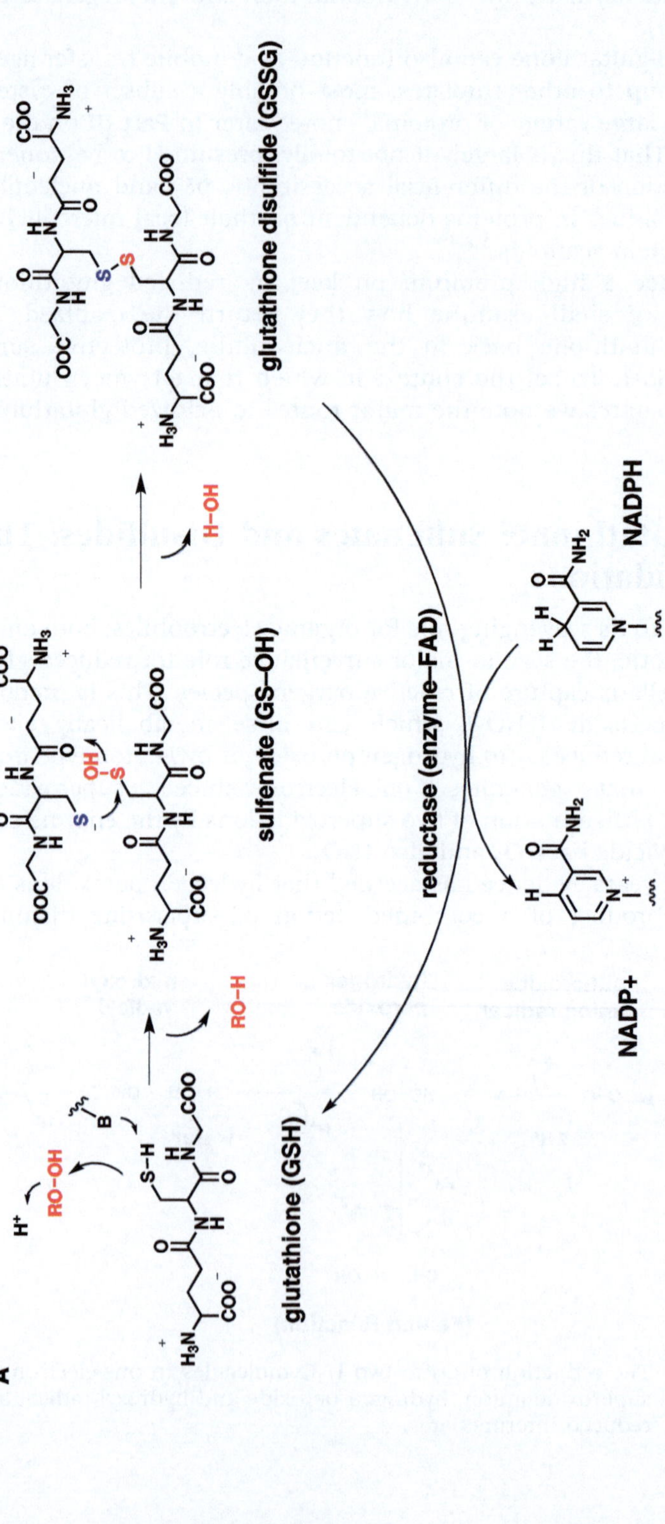
H+
RO–OH
RO–H
H–OH
glutathione (GSH)
sulfenate (GS–OH)
glutathione disulfide (GSSG)
reductase (enzyme–FAD)
NADP+
NADPH

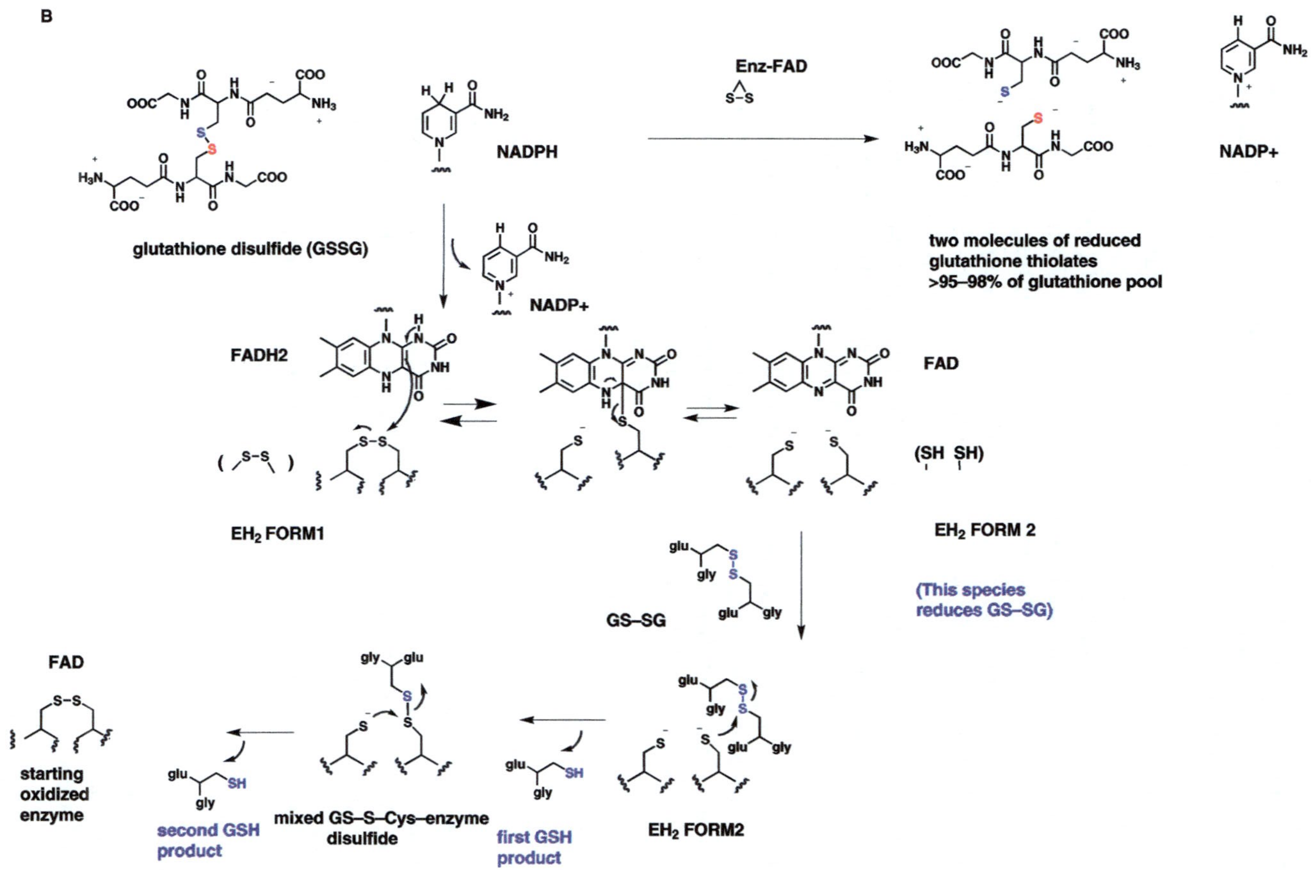

Figure 5.16 Tandem action of glutathione peroxidase and glutathione reductase. (A) Glutathione peroxidase involves capture of RO–OH by a GS^- thiolate anion to give the sulfenic acid. This is captured as sulfur electrophile by a second GS^-. The peroxide has been reduced by two electrons and the weak O–O bond cleaved. The two GSH molecules have been oxidized to the intermolecular disulfide with a new S–S bond. (B). Glutathione reductase spends an NADPH to reduce the enzyme-bound FAD to $FADH_2$. The $FADH_2$ C4a can then attack the active site disulfide and reduce it to the dithiol as the flavin reoxidizes. This form of EH_2 can proceed to reduce GS–SG back to two GSH molecules by disulfide–dithiol exchange reactions.

the suite of enzymatic oxidases but is actually a signaling molecule. Its spatio-temporal distribution is substantially controlled by its capture by GSH accelerated by multiple forms of glutathione peroxidases.

This detoxification catalyst family is composed of different forms of glutathione peroxidases.[44,45] Catalyzed attack by the glutathione thiolate on the weak O–O bond in HO–OH forms a glutathionyl–S–OH (GS–OH) and H_2O as coproducts (Figure 5.16). GS–OH is a sulfenic acid (RS–OH). The sulfur is still at a formal oxidation state of −2. However, the sulfur atom is now electrophilic in sulfenate functional groups.

Its major fate is to be captured by a second molecule of GSH, with that sulfur thiolate acting as a nucleophile. As shown in Figure 5.16, the result is another water molecule product and now the intermolecular disulfide form of glutathione (GS–SG). Thus, glutathione peroxidase action involves two additional forms of the key sulfur atom, the *S*-oxygenated sulfenate where the sulfur is now electrophilic, and the more stable disulfide (where both sulfurs are electrophilic). Hydrogen peroxide has been reduced by two electrons and split to two water molecules while two thiols (2 GSH) have undergone net two electron oxidation to the disulfide (Figure 5.16).

In the GS–SG, oxidized, disulfide form of glutathione, the dimer no longer has a nucleophilic thiol group and so cannot perform any surveillance or protective functions. Cells therefore spend nicotinamide adenine dinucleotide phosphate (NADPH) electron-reducing currency by action of the flavoenzyme glutathione reductase to regenerate the two molecules of reduced glutathione to function as chemoprotectant metabolites (Figure 5.16B).

This reductase[46,47] is a member of a class of FAD-enzymes that also have an active site Cys–Cys disulfide, in the same family as the hydrogen sulfide quinone oxidoreductase discussed in Part I. In general, proteins in cell cytoplasms have their cysteine residues as the reduced dithiols, not in the oxidized disulfide form, given the high concentrations of GSH. However, in the small family of FAD-containing disulfide reductases the two active site cysteine residues start and end each catalytic cycle as disulfides (Figure 5.16B).

Here, NADPH reduces the bound FAD by hydride transfer to N5 of FAD. The resultant $FADH_2$ can act at C4a as nucleophile to reduce the Cys–Cys active-site disulfide *via* a flavin thiol intermediate. This is a two-electron reduced form of the enzyme (EH_2) with both electrons in the $FADH_2$ and the Cys–Cys oxidized disulfide still intact. This form-1 EH_2 has a ready pathway to convert to EH_2 form 2 *via* the indicated flavin sulfur chemistry, involving attack of the C4a nucleophilic carbon of $FADH_2$ on the active site disulfide. The intermediate 4a-thiol flavin adduct can decompose intramolecularly. As the flavin–sulfur bond cleaves, the electrons go with the sulfur. This generates an EH_2 form where the flavin is oxidized and the pair of active-site cysteinyl residues as the reduced dithiolate forms.

This is the form of EH_2 that can reduce the oxidized disulfide form of GS–SG by straightforward dithiol–disulfide interchange (Figure 5.16B). One of the Cys–S^- side chains in the dithiol form of the reductase can now attack the disulfide form of bound GS–SG, engaging in a disulfide interchange. The result is first a mixed enzyme–Cys–SG plus the first released GSH monomer. The second GSH is next released as the remaining Cys–S^- attacks that mixed enzyme–Cys–S–SG (Figure 5.16B). Glutathione (GSH) is so important in its reduced redox state to cellular economies that cells spend as much NADPH as necessary to keep the 2GSH : GSSG ratio above 95 : 5. Indeed, this is a major flux of NADPH utilization in many cell types.

We emphasize again that glutathione reductase is unusual in maintaining the active site Cys pair as a disulfide. In most proteins in the cytoplasm, reduced glutathione is at sufficient concentration to keep Cys–S–S–Cys protein disulfides in reduced bis-thiol forms. In eukaryotic cells as proteins get extruded into the endoplasmic reticulum and proceed through the Golgi organelles to the cell surface and on to secretion to extracellular milieu, the GSH level gradually falls (without regenerating reductase and NADPH equivalents), nascent disulfide bonds accumulate. Indeed, most proteins in cell surface membranes or extracellular spaces are in crosslinked disulfide forms. We will examine the thiol *vs.* disulfide oxidation state of cysteine proteomes more generally in Part III, Chapter 13 when we examine the role of disulfide-containing protein disulfide isomerases in the secretory compartments that oxidize nascent proteins extruded into the endoplasmic reticular lumen.

In bacteria lacking glutathione, the bacillothiol, mycothiol and glutathionyl amide low molecular weight thiols are thought to play equivalent protective roles against organic electrophiles and any ROS in the microenvirons. In turn there will be comparable disulfide reductases to return the inactive oxidized disulfide forms to the active surveillance dithiol forms.

In kinetoplastid-containing single-cell eukaryotes including *Crithidia fasciculata* and both the South American and African types of tryanosomatid parasites, glutathione is converted biosynthetically to the bis-spermidine derivative known as trypanothione. A comparable NADPH-oxidizing trypanothione reductase keeps trypanothione predominantly in the dithiol form (Figure 5.17). Whereas reduced glutathione is a bis-carboxylate, trypanothione has a neutral internal bis-amide to spermidine and carries two net positive charges. Moreover, oxidized trypanothione has a macrocyclic amide structure, so the molecules can be distinguished by the distinct host reductases.

The facile oxidation of cysteine-based thiol side chains to sulfenates as transient intermediates on the way to the more stable disulfides is a chemical property of high biological significance. Isolated sulfenates have sufficient stability to exist under physiological conditions.[48] In the presence of neighboring thiols, the bimolecular conversion to disulfide is moderate

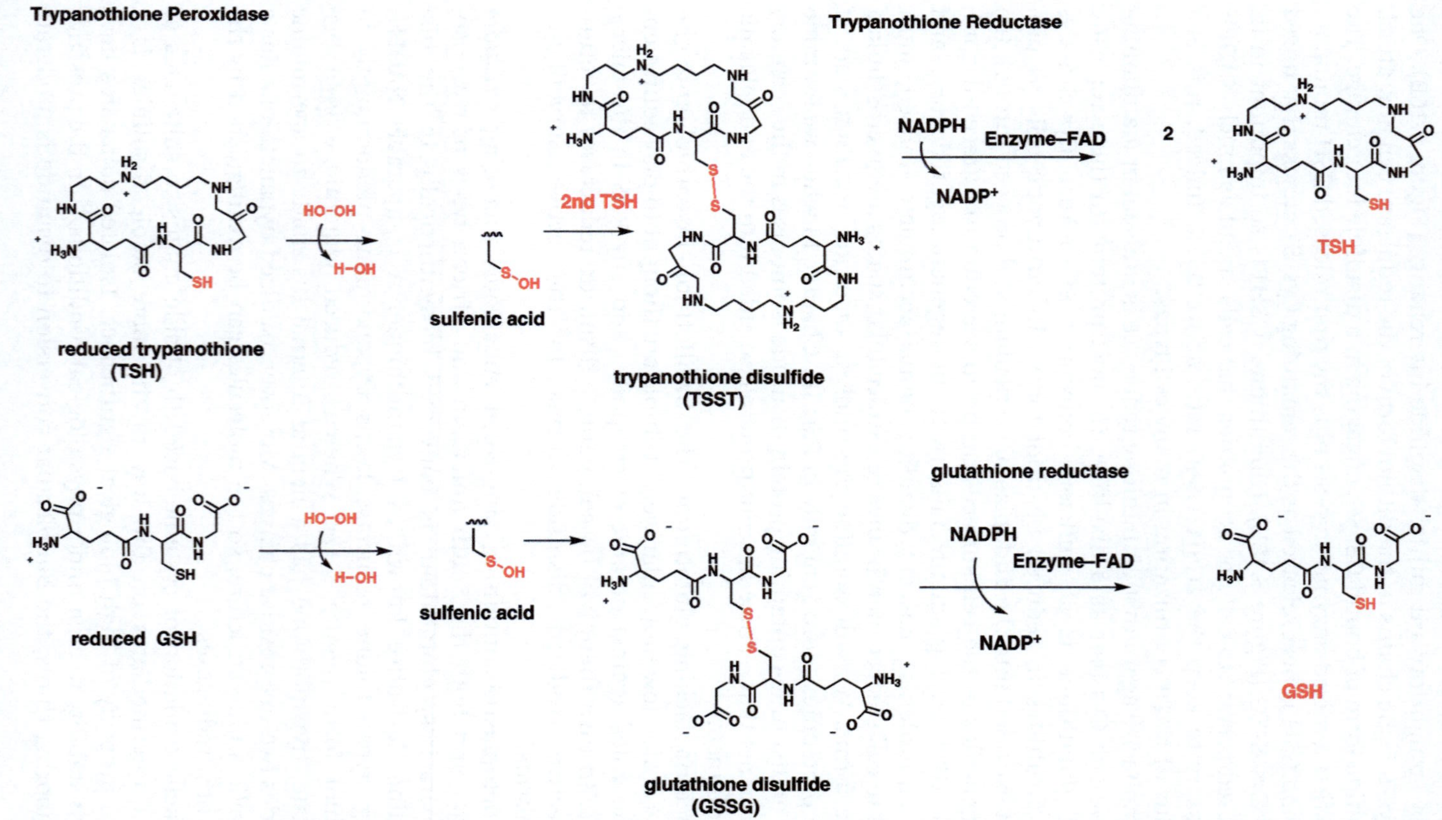

Figure 5.17 Comparison of glutathione and its trypanosomal homolog trypanothione in redox biology. The reduced trypanothione has the two tripeptidyl moieties connected by a spermidine bridge in one scaffold. The oxidized intermolecular disulfide form (TSST) has two macrocyclic peptide rings tethered by a disulfide link.

(5–10 $M^{-1} s^{-1}$) but can be accelerated by many orders of magnitude by certain enzyme classes. The oxidative dimerization of two cysteines *via* the side chain sulfurs to the cystine disulfide is a unique property among the 20 proteinogenic amino acid metabolites. Cysteine is the sole member that has a side chain thiol group and it is the "on pathway" intermediate for many but not all Cys–S–S–Cys protein disulfide formations as noted in Chapter 13. We compare sulfenates and disulfides with persulfides in the next chapter.

References

1. L. B. Poole, The basics of thiols and cysteines in redox biology and chemistry, *Free Radical Biol. Med.*, 2015, 148–157.
2. D. W. Bak, T. Bechtel, J. A. Falco and E. Weerapana, Cysteine reactivity across the subcellular universe, *Curr. Opin. Chem. Biol.*, 2019, **48**, 96–105.
3. M. H. Stipanuk, J. J. Dominy, J.-I. Lee and R. M. Coloso, Mammalian cysteine metabolism: new insights into the regulation of cysteine metabolism, *J. Nutr.*, 2006, **136**, 1652S–1659S.
4. A. J. Crisjoe and M. Maroney, Cysteine dioxygenase structure and mechanism, *Chem. Commun.*, 2007, 3338–3349.
5. C. T. Walsh and Y. Tang, *The Chemical Biology of Human Vitamins*, Royal Society of Chemistry Press, London, 2018.
6. E. Strauss and T. Begley, Mechanistic studies on phosphopantetheinylcysteine decarboxylase, *J. Am. Chem. Soc.*, 2001, **123**, 6449–6450.
7. E. Strauss, H. Zhai, L. A. Brand, F. W. McLafferty and T. P. Begley, Mechanistic studies on phosphopantothenoylcysteine decarboxylase; trappping of an enethiolate intermediate with a mechanism-based inactivating agent, *Biochemistry*, 2004, **43**, 15520–15533.
8. W. P. Jencks, *Catalysis in Chemistry and Enzymlogy*, McGraw-Hill, New York, 1969.
9. C. T. Walsh, B. Tu and Y. Tang, Eight kinetically stable but thermodynamicallay activated molecules that power cell metabolism, *Chem. Rev.*, 2018, **118**, 1460–1494.
10. M. Fischbach and C. Walsh, Assembly line enzymology for polyketide and nonribosomal peptide antibiotics, *Chem. Rev.*, 2006, **106**, 3468–3496.
11. R. H. Lambalot, A. Gehring, R. S. Fugel, P. Zuber, M. Lacelle, M. A. Marahiel, R. Reid, C. Khosla and C. T. Walsh, A new enzyme superfamily: the phosphopantetheinyl transferases, *Cell Chem. Biol.*, 1996, **3**, 923–936.
12. T. Brautaset, N. Sekurova, SlettaH, T. E. Ellingsen, A. R. Strom, S. Valla and S. B. Zotchev, Biosynthesis of the polyene antifungal antibiotic nystatin in *Streptomyces noursei* ATCC 11455: analysis of the gene cluster

and deduction of the biosynthetic pathway, *Cell Chem. Biol.*, 2000, **7**, 395–403.
13. A. Meister, Glutathione metabolism and its selective modification, *J. Biol. Chem.*, 1988, **263**, 17205–17208.
14. H. J. Forman, H. Zhang and A. Rinna, Glutathione: overview of its protective roles, measurement and bisynthesis, *Mol. Aspects Med.*, 2009, **30**, 1–12.
15. O. W. Griffith and R. Mulcahy, The enzymes of glutathione synthesis: gamma glutamylcysteine synthetase, *Adv. Enzymol. Relat. Areas Mol. Biol.*, 1999, **73**, 209–267.
16. M. V. Fawaz, M. Topper and M. Firestone, The ATP grasp enzymes, *Bioorg. Chem.*, 2011, **39**, 185–191.
17. M. Y. Galperin and E. Koonin, A diverse superfamily of enzymes with ATP-dependent carboxylate/amine thiol ligase activity, *Protein Sci.*, 1997, **6**, 2639–2643.
18. J. P. Pizzorno, Glutathione!, *J. Integr. Med.*, 2014, **13**, 8–12.
19. K. Van Laer, C. Hamilton and J. Messens, Low molecular weight thiols in thiol-disulfide exchange, *Antioxid. Redox Signaling*, 2013, **18**, 1642–1653.
20. R. G. Bartsch, G. Newton, C. Sherrill and R. C. Fahey, Glutathione amide and its perthiol in anaerobic sulfur bacteria, *J. Bacteriol.*, 1996, **178**, 4752–4756.
21. G. L. Newton and M. Rawat, N-methyl bacillothiol: a novel thiol from anaerobic bacteria, *mBio*, 2019, **10**, e02634–02618.
22. G. L. Newton, M. Rawat, J. J. La Clair, V. K. Jothivasan, T. Budiarto, C. J. Hamilton, A. Clairborne, J. Helmann and R. C. Farhey, Bacillothiol is an antioxidant thiol produced in Bacilli, *Nat. Chem. Biol.*, 2009, **5**, 625–627.
23. R. C. Fahey, Novel thiols of prokaryotes, *Annu. Rev. Microbiol.*, 2001, **55**, 333–356.
24. S. B. Nimse and D. Pal, Free radicals, natural antioxidants and their reaction mechanisms, *RSC Adv.*, 2015, **5**, 27986.
25. H. Wefers and H. Sies, Oxidation of gluathione by the superoxide radical to the disulfide and the sulfonate yielding singlet oxygen, *Eur. J. Biochem.*, 1983, **137**, 29–36.
26. M. Deponte, Glutathione catalysis and the mechanism of glutathione-dependent enzymes, *Biochim. Biophys. Acta*, 2013, **1830**, 3217–3266.
27. P. J. Thornalley, Glyoxalase I- structure, function, and a critical role in the enzymatic defence against glycation, *Biochem. Soc. Trans.*, 2003, **31**, 1343–1348.
28. D. Niegowski, T. Kleinschmidt, U. Olsson, S. Ahmad, A. Rinaldo-Matthis and J. Z. Haeggstrom, Crystal structures of leuktriene C4 synthase in complex with product analogs: implications for the enzyme mechanism, *J. Biol. Chem.*, 2014, **289**, 5199–5207.
29. E. Liu, F. Wolfreys, J. R. Muppidi, Y. Xu and J. G. Cyster, S-Geranylgeranyl-L-glutathione is a ligand for human B cell confinement receptor, *Nature*, 2019, **568**, 245–248.

30. L. L. Mazaleuskaya, K. Sangkhui, C. F. Thrn, G. A. Fitzgerald, R. B. Atman and T. E. Klein, PharmGKB summary: pathways of acetaminophen metabolism at the therapeutic vs toxic doses, *Pharmaocogenet, Genomics*, 2015, **25**, 416–426.
31. J. A. Hinson, D. Roberts and L. P. James, Mechanisms of acetaminophen-induced liver necrosis, *Handb. Exp. Pharamcol.*, 2011, **196**, 369–405.
32. A. Oakley, Glutathione transferases: a structural perspective, *Drug Metab. Rev.*, 2011, **43**, 138–151.
33. D. W. Nebert and V. Vasiliou, Analysis of the glutathione S-transferase (GST) gene family, *Hum. Genomics*, 2004, **1**, 460–464.
34. N. Allocati, M. Masulli, C. D. Ilio and L. Frederici, Glutathione transferases: substrates, inhibitors and pro-drugs in cancer and neurodegenerative diseases, *Oncogenesis*, 2018, **7**, 1–15.
35. C. Starkenmann, M. Troccaz and K. Howell, The role of cysteine and cysteine-S-conjugates as odour precursors in the flavour and fragrance industry, *Flavour Fragrance J.*, 2008, **23**, 369–381.
36. T. Baumann, S. Bergmann, T. Schmidt-Rose, H. Max, A. Martin, B. Enthaler, L. Terstegen, D. Schweiger, H. Kahlbacher, H. Wenck, G. Jeditschky and Z. Jovanovic, Glutathione-conjugated sulfanylalkanols are substrates for ABCC11 and glutamyl transferase1: a potential new pathway for the formation of odorant precursors in the apocrine sweat gland, *Exp. Dermatol.*, 2014, **23**, 247–252.
37. M. Russwurm and D. Koesling, NO activation of guanylyl cyclase, *EMBO J.*, 2004, **23**, 44435–44450.
38. J. S. Stamler, S. Lamas and F. C. Fang, Nitrosylation. The prototypic redox-based signaling mechanism, *Cell*, 2001, **106**, 675–683.
39. A. Keszler, Y. Zhang and N. Hogg, The reaction between nitric oxide, glutathione and oxygen in the presence and absence of protein: how are S-nitrosothiols formed?, *Free Radical Biol. Med.*, 2010, **48**, 55–64.
40. S. D. Barnett and I. Buxton, The role of S-nitrosyl-gluathione reductase in human disease therapy, *Crit. Rev. Bochem. Mol. Biol.*, 2017, **52**, 340–354.
41. M. W. Foster and J. Stammler, New Insights into Protein Nitrosylation, *J. Biol. Chem.*, 2004, **279**, 25891–25897.
42. S. P. Singh, J. Wishnok, M. Keshive, W. M. Deen and S. R. Tannenbaum, The chemistry of the S-nitrosylglutathione/glutathione system, *Proc. Natl. Acad. Sci. U. S. A.*, 1996, **93**, 14428–14433.
43. N. J. Fowler, C. Blanford, S. P. de Visser and J. Warwicker, Features of reactive cysteines discovered through computation: from kinase inhibition to enrichment around protein degrons, *Sci. Rep.*, 2017, **7**, 16338.
44. O. Epp, R. Ladenstein and A. Wendel, The refined structure of the selenoenzyme glutathione peroxidase at 0.2 nM resolution, *Eur. J. Biochem.*, 1983, **133**, 51–69.
45. R. Brigelius-Flohe and M. Maiorino, Glutathione peroxidases, *Biochim. Biophys. Acta*, 2013, **1805**, 3289–3303.

46. D. S. Berkholz, H. Faber, S. N. Savvides and P. A. Karplus, Cataytic cycle of human glutathione reductase near 1 Agstrom resolution, *J. Mol. Biol.*, 2008, **382**, 371–384.
47. E. Pai and G. Schulz, The catalytic mechanism of glutathione reductase as derived from x-ray diffraction analyses of reaction intermediates, *J. Biol. Chem.*, 1983, **258**, 1752–1757.
48. V. Gupta and K. Carroll, Sulfenic acid chemistry detection and cellular lifetime, *Biochim. Biophys. Acta*, 2014, **1840**, 847–875.

CHAPTER 6

Protein Cysteine Persulfides and Fe–S Clusters: Intersection of Organic and Inorganic Sulfur Biochemistry

One of the strands of cysteine chemical biology that has become clearer over the past two decades is the relationship between cysteine persulfide (Cys S–SH) groups and the delivery of inorganic sulfide ions for assembly of the ubiquitous 2Fe–2S and 4Fe–4S iron–sulfur clusters that serve in many protein contexts as one-electron conduits in redox reactions (Figure 6.1).

This is a well-defined intersection of organic thiols with inorganic sulfide biochemistry. In a strategic sense the delivery of inorganic sulfide atoms for Fe–S cluster assembly *via* cysteine as a carrier is the reverse of the organification reaction carried out by *O*-acetylserine sulfhydrylase, as described in Chapter 4. One might have thought that the sulfide ions from assimilatory sulfite reductases (Chapters 2 and 3) could be channeled directly into Fe–S cluster assembly. Instead those sulfur atoms are processed through free cysteine and then protein-bound cysteine residues.[1–4] Perhaps, this strategy evolved to keep H_2S and HS^- at low concentrations in cells to avoid toxicity (Chapter 4). By delivery of single sulfur atoms to target proteins where the Fe–S clusters get assembled, freely diffusible sulfide and sulfane forms of sulfur may be avoided.

Because the single sulfur atoms that build the Fe–S clusters are carried by cysteine residues in proteins rather than free cysteine molecules, this topic could be as easily placed in Part III as its present location. However, the process starts with free cysteine molecules. Furthermore, it represents the intersection with the inorganic sulfur biochemistry featured in Part I, so we address the topic in this chapter.

Chemical Biology No. 12
The Chemical Biology of Sulfur
By Christopher T. Walsh

Published by the Royal Society of Chemistry, www.rsc.org

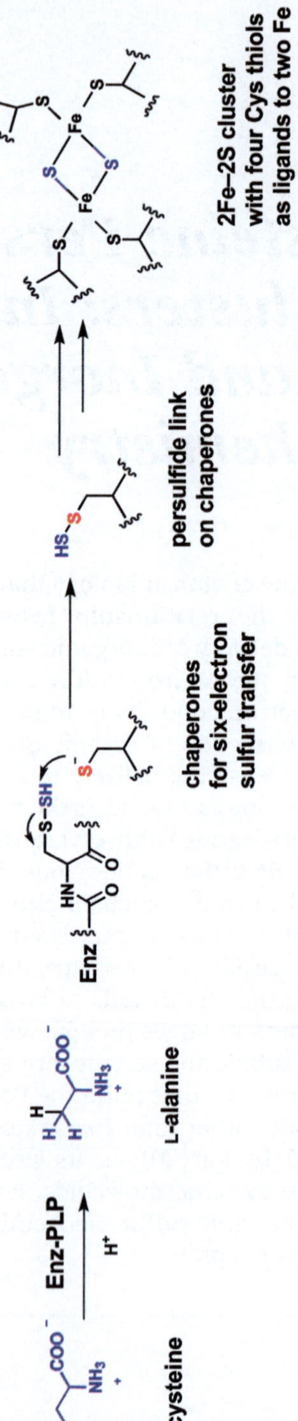

Figure 6.1 The inorganic sulfide atoms in Fe–S clusters come from the SH group of cysteine *via* persulfide protein intermediates.

6.1 Persulfide-forming Cysteine Lyases

Persulfide functional groups comprise the category R–S–SH. The S–S bond makes one think of disulfides (RS–SR) by analogy, but in the R–S–SH group the terminal sulfur is termed a sulfane atom.[5] Cysteine S–SH, as the free amino acid persulfide and as protein bound –Cys–S–SH residues are the most common forms of cellular persulfides, although glutathione persulfide is also a known metabolic entity.[6]

We have already met classes of enzymes that generate enzyme bound –Cys–SSH residues as nascent products, in the sulfide quinone oxidoreductase (Chapters 3 and 4) and the mercaptopyruvate sulfotransferase example of Chapter 4. Also, we commented briefly at the end of Chapter 4 on the ability of persulfide groups to act either as nucleophiles or electrophiles. We pursue this topic in more depth here.

A small family of pyridoxal-phosphate-dependent cysteine lyase enzymes can reductively cleave L-cysteine to L-alanine while transferring the substrate sulfur to an active site cysteine residue side chain thiolate[1,7,8] (Figure 6.2A). This is *not* release of HS^- but instead a formal release of a sulfur atom with six, not eight, electrons.

The result of catalytic turnover is a covalent adduct of the enzyme's active site cysteinyl-S^- with the itinerant sulfur atom, to generate an enzyme-persulfide as coproduct This is a formal two-electron reduction of C3 (–CH_2SH) of cysteine to C3 of alanine (–CH_3). To balance the redox stoichiometry, the active site Cys–SH has undergone a *two-electron oxidation* in formation of the persulfide (R–S–S–H) functional group. This observation emphasizes that the Cys–SSH and the Cys–SH side chain are at distinct oxidation states.

To proceed to subsequent catalytic cycles, the enzyme must be returned to its original Cys–SH state, requiring transfer of the sulfane sulfur atom to a partner protein as we shall note below. The issue of how the enzyme–S–SH transfers the terminal thiol to ready itself for a subsequent catalytic cycle impinges on the role of these lyases as sources of the inorganic sulfide in Fe–S clusters.

6.1.1 β-Replacement *vs.* β-Elimination Cysteine Lyases: Six-electron *vs.* Eight-electron Sulfur Atom Products

The cysteine to alanine plus enzyme–SSH conversions are examples of β-replacement enzymes (Figure 6.2).[9,10] The replacement *vs.* elimination categorization category focuses on the oxidation state of C2 of amino acid substrate: that carbon, with an amino group, stays at a constant oxidation state between substrate and product. At C3, the SH group has been lost and formally replaced with a hydrogen, converting the CH_2SH to CH_3 (formally an H^+ replacing an $[SH]^+$). The actual chemical steps are more complex. The product distribution (cysteine to alanine) differs from the enzymatic conversion of cysteine to pyruvate, ammonia, and H_2S or HS^- noted in Chapter 4 when cystathionine β-synthase and cystathionine-γ-lyase use cysteine as alternate substrate. In those reaction manifolds the enzymes act as

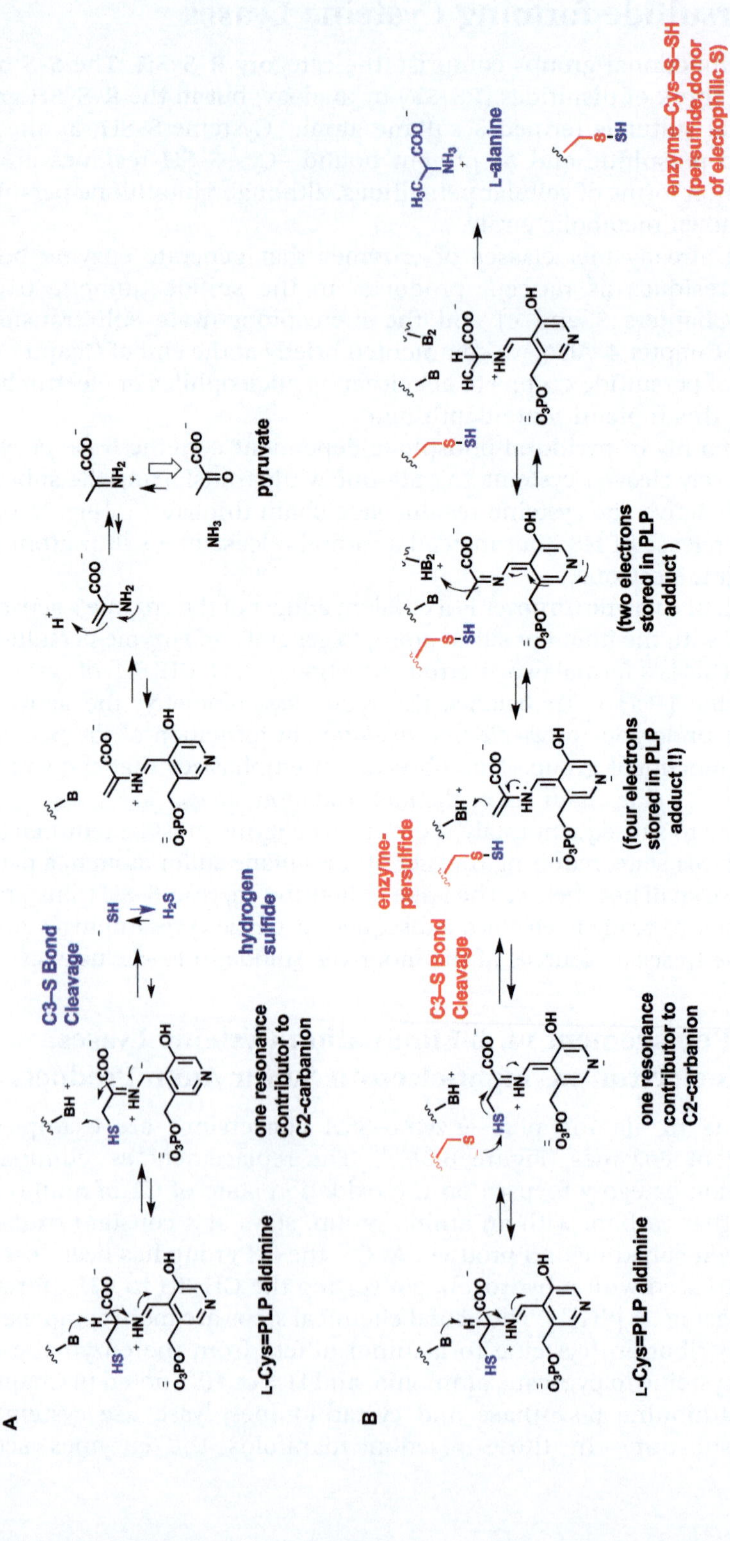

Figure 6.2 Two modes of C3–S bond cleavage in cysteine by PLP–enzymes. (A) Beta elimination *vs.* (B) beta replacement mechanism and outcomes.

β-elimination catalysts (Figure 6.2). The C3–S bond of substrate–pyridoxal phosphate (PLP) adduct is cleaved with the electron pair departing with sulfur (eight-electron sulfur as SH^-).

In contrast, for the β-replacement Cys to Ala lyases, the C3–S bond cleaves with the electron pair *staying* in the three-carbon-PLP scaffold. The itinerant S is moving as a six-electron, not an eight-electron sulfur atom. This remarkably different route for cleaving the C3–S bond of cysteine is a tribute both to the versatility of the PLP coenzyme and to the ability of sulfur to move readily at two different oxidation states: as a six-electron or eight-electron inorganic atom.

Examination of the mechanism shows the anticipated α-carbanion–PLP intermediate. However, its reactivity is controlled so that the electron pair *does not* participate in β-elimination of an SH^- equivalent. Instead the –SH group of the cysteinyl–PLP adduct is captured by the active site Cys–thiolate without the pair of electrons that had been in the HS–C3 bond. As shown, that electron pair is stored transiently in the PLP electron sink. This is the second example of PLP acting as a four-electron sink (cystathionine-γ-lyase in Chapter 4 was the other case) and is the crucial feature of this reaction pathway. The four electrons come out two at a time, one pair at C3 to pick up a proton, the other pair at C2 for protonation. The net result is L-alanine as the product. This is a formal β-replacement of a "SH^+" equivalent at C3 by H^+ as electrophile.

Thus, remarkably, pyridoxal-dependent cysteine-processing lyases can release either nucleophilic SH^- (eight electrons) or electrophilic sulfane sulfur (six electrons). The reversal of sulfur polarity ("*ompolung*") is a tribute to the chemical versatility of the PLP coenzyme to store two or four electrons as an electron sink, the timing and accessibility of one *vs.* two protonations in the enzymes active sites, and the ability to break a C3–S bond with the electron pair in the bond either departing with sulfur or remaining on C-3 (and picking up a proton).

6.1.2 The Differential Reaction of Persulfides *vs.* Persulfide Anions

Persulfides are an interesting complement to the universe of protein posttranslational modifications, explored in Part III. We can compare the Cys–S–SH to the Cys–S–OH sulfenic acid noted briefly at the start of Chapter 4 and examined more fully in the next two chapters.[5,11] The sulfenic acid sulfur, and also the terminal sulfur in the persulfide functional group, behave as electrophiles when protonated.

Also, as it happens, the pK_a of the terminal S–H in Cys–S–S–H is 6.2, so at the physiological intracellular pH of 7.2 the Cys persulfide side chains are 90% in the persulfide anion form, about 1.5 to 2 logs more basic than Cys–SH itself[12] (Figure 6.3).

So, the terminal thiolate in Cys–S–S^- is a strong nucleophile. However, consideration of how the –Cys–S–SH group forms during lyase action shows that the terminal –SH was transferred as an electrophile. This unusual property makes the persulfide group a uniquely valuable addition to the sulfur biochemistry inventory and we will see it played out below (Figure 6.3).

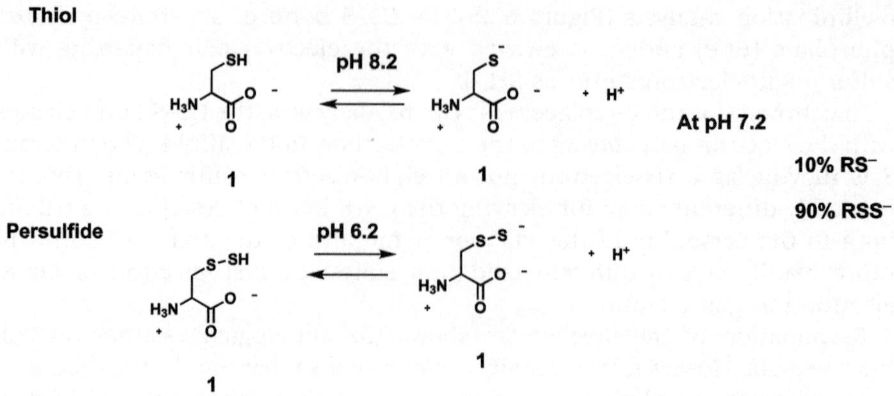

R–SH	sulfur not elecrophilic	RS⁻	nucleophlic thiolate anion
R–SSH	terminal sulfur electrophilic	RSS⁻	nucleophilic persulfide anion
R–S–OH	sulfur electrophilic		

Figure 6.3 Comparison of cysteine thiol and cysteine persulfide pK_a values. Estimate of nucleophilicity and electrophilicity of anionic *vs.* protonated species.

6.2 Iron–Sulfur Cluster Assembly

For purposes of this discussion, we note that more than 200 proteins are known with some variant of 2Fe–2S or 4Fe–4S (Figure 6.4) clusters as one-electron redox transfer species.[1,3,13] The Fe–S clusters are found in all forms of life (except viruses) and thought to be ancient machinery for redox chemical biology in many cellular contexts. They represent the assembly of iron 2^+ or 3^+ cations with the soft anionic ligands of inorganic sulfide. They are crucial electron conduits in the enzyme nitrogenase, in hydrogenases and in electron transfer chains in chloroplasts. The midpoint reduction potentials for both 2Fe–2S and 4Fe–4S clusters can be varied by the nature of the specific protein environment, the geometry of the cluster, location of H bonds, charged side chains, distortion by coordination to the four cysteine side chain thiolates that anchor the clusters to the proteins, the presence of other nearby redox groups, and so on.[14]

Of the 15 known Fe–S clusters attached to human mitochondrial proteins, there are eight Fe–S clusters as part of the single enzyme NADH oxido-reductase that feeds electrons in two at a time, steps down to one at a time through the Fe–S clusters, down a series of potential drops into the mitochondrial respiratory electron transport chains[15] (Figure 6.4). As befits these universal Fe–S redox clusters that have been studied for over half a century,[1,13] the primary and even the review literature is immense and continuing. The exposition here does not do justice to the many varieties and roles of particular Fe–S clusters. Nor do we take up more complex Fe–S cluster organizations, such as those found in hydrogenases and nitrogenases

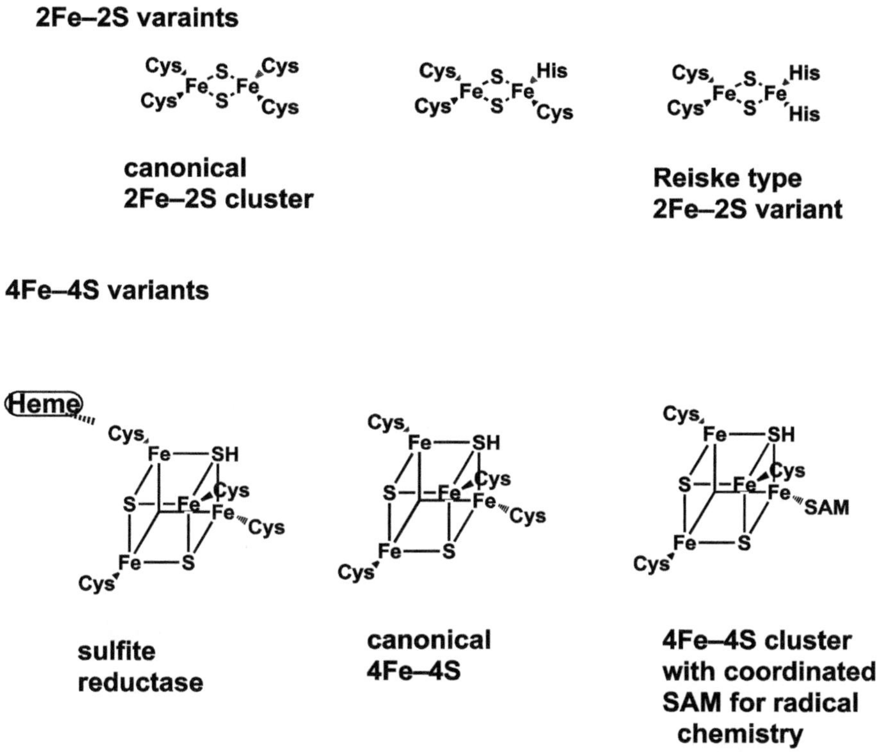

Figure 6.4 Schematic of some of the common types of 2Fe–2S and 4Fe–4S iron sulfur clusters in proteins. While four cystenyl thiolate ligands are found in canonical 2Fe–2S clusters there are examples of one-His and two-His side chain imidazole replacements. These have altered redox properties and sit in distinct biologic niches. The 4Fe–4S line shows a version lacking one of the four cysteinyl protein thiolate ligands, as well as a canonical 4Fe–4S cluster. The third 4Fe–4S structure is a schematic of the 4Fe–4S siroheme redox center in sulfite reductase, discussed in Chapter 2. Not shown are more complex Fe–S clusters, for instance of the type found in nitrogenases.

(such as Fe_8S_9C centers).[16,17] Readers are directed to the appropriate review literature.[1] Here, we emphasize that Fe–S cluster biology represents the intersection of organic and inorganic sulfur chemistry in living organisms.

In the 2Fe–2S clusters, typically one sulfur at the formal oxidation state of sulfide dianion acts as a bridge to two Fe cations. (We note that the second pK_a for H_2S, dissociation of HS^- to S^{2-} and H^+ is 19, implying 50% dissociation at pH 19.) Thus, organisms must manage the Fe–S interactions to avoid high energy free sulfide dianions during the assembly process where protons are stripped away from inorganic sulfides. Usually, two additional cysteine sulfur atoms from the host protein fill out the coordination shell to each of the two iron atoms, anchor the Fe–S cluster in a specific region of the protein and create the stable coordination shell for each iron atom. This produces two inorganic sulfides and two organic thiolates as ligands, a total of four sulfur ligands to each iron in a 2Fe–2S cluster.

Figure 6.5 shows a panel of Fe–S clusters rendered as pimol diagrams taken from protein databases.[18] Panel A shows the classic 2Fe–2S in a ferredoxin protein. Panels B and C show protein microenvironments where either one or two of the canonical four cysteine ligands have been replaced by the imidazole ring of histidine residues in these variant 2Fe–2S clusters. Panel D is a 3Fe–4S cluster in a cyanobacterial glutamate dehydrogenase. Panel E is a classic 4Fe–4S cluster with four protein thiolate ligands. Panel F harks back to sulfite reductase but is actually the corresponding structure for a cyanobacterial nitrate reductase. It shows the communication between the 4Fe–4S cluster and the siroheme group that coordinates nitrate (or sulfite in the sulfite reductase case) as the top axial ligand. The various ligand substitutions change redox potential and electron transfer properties, placing those clusters in different metabolic intersections.

As depicted in Figure 6.5, panel E, the 4Fe–4S clusters typically have distorted cubane architecture.[18] In these arrays, each iron atom now has three inorganic sulfides as ligands and one single organic thiolate as the

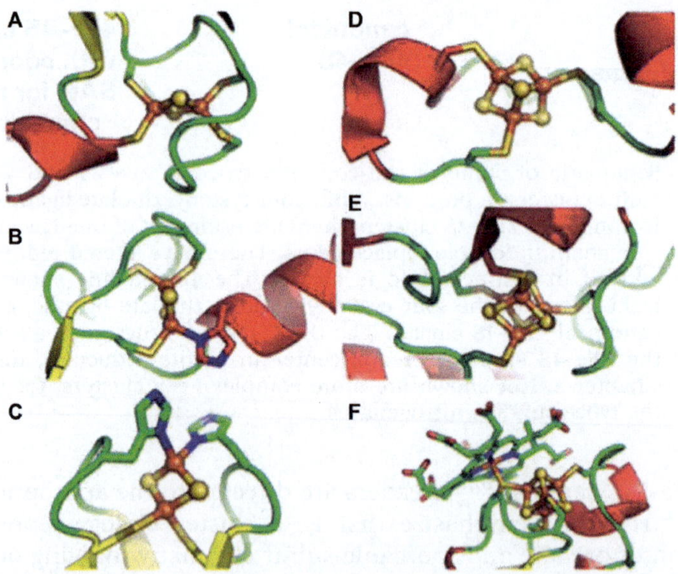

Figure 6.5 Schematic of Fe–S clusters in pimol representations from the indicated proteins; **A**, 2Fe–2S cluster in ferredoxin; **B**, 2Fe–2S cluster with 3 Cys and 1His; *Arabidopsis thaliana* NEET [protein name derived from its Asn–Glu–Glu–Thr (NEET) c-terminal sequence]; **C**, 2Fe–2S Reiske cluster b6/f complex *Spinchia oleracea*; **D**, 3Fe–4S glutamate dehydrogenase, *Synechocystis*; **E**, 4Fe–4S cluster in *Synechocystis* ferredoxin thioredoxin reductase; **F**, 4Fe–4S cluster linked to siroheme by intervening Cys–S in nitrate reductase. (B) Reproduced from ref. 18, https://doi.org/10.3389/fpls.2013.00259, under the terms of a CC BY 3.0 license, https://creativecommons.org/licenses/by/3.0/.

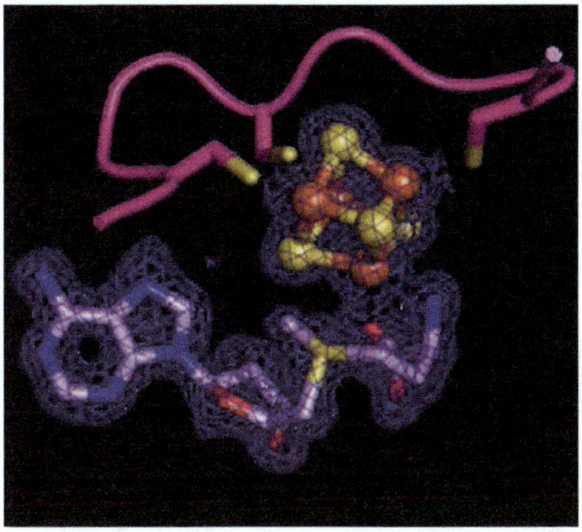

Figure 6.6 Family of 4Fe–4S clusters with a coordinated *S*-adenosylmethionine as ligand. Reaction center for subsequent one-electron chemistry *via* 5′-deoxyadenosyl radical.
Reproduced from ref. 41, http://www.jbc.org/content/293/45/17349, under the terms of a CC BY 4.0 license, https://creativecommons.org/licenses/by/4.0/.

anchoring cysteinyl residue, quite distinct iron microenvironments from the 2Fe–2S clusters. There are also other architectural variants with distinct redox or non-redox functions. For example a 3Fe–4S variant is converted to the full 4Fe–4S cluster and anchors the tricarboxylate scaffolds during the catalytic cycle of the enzyme aconitase in the citric acid cycle as citrate and isocitrate are converted *via cis*-aconitate.[19] There is a large class of 4Fe–4S variants where one of the cysteine thiolate ligands is absent and a substrate molecule is coordinated instead.[20] That is the case of a very large predicted family of 4Fe–4S proteins with the amine and carboxylate of *S*-adenosylmethionine (SAM) coordinated to that fourth Fe atom (Figure 6.6). We will examine this class of SAM-proteins as radical SAM catalysts in Chapter 9.

At rest, the 2Fe–2S clusters are formally oxidized, in the Fe^{3+}/Fe^{3+} oxidation state. On one-electron reduction the formal charge on iron is a mixed Fe^{3+}/Fe^{2+}. The 4Fe–4S clusters typically exist in the $3Fe^{3+}/Fe^{2+}$ oxidized state and in the one-electron-reduced $2Fe^{3+}/2Fe^{2+}$ state, with precise reduction potentials tuned by the architecture of the protein cysteines thiol ligands, among other inputs. In the reduced state, both the 2Fe–2S and 4Fe–4S clusters are obligate one-electron donors.

Let's return now to the mechanism of the cysteine lyase β-replacement enzyme family as the catalysts initiating conversion of cysteine organic thiol atoms to enzyme-bound persulfides and ultimately to the inorganic sulfides that are shunted into Fe–S cluster assemblies (Figure 6.7A).

Consistent with the two-electron reduction of cysteine to alanine during each catalytic turnover, the proposed route for C–S bond cleavage occurs at

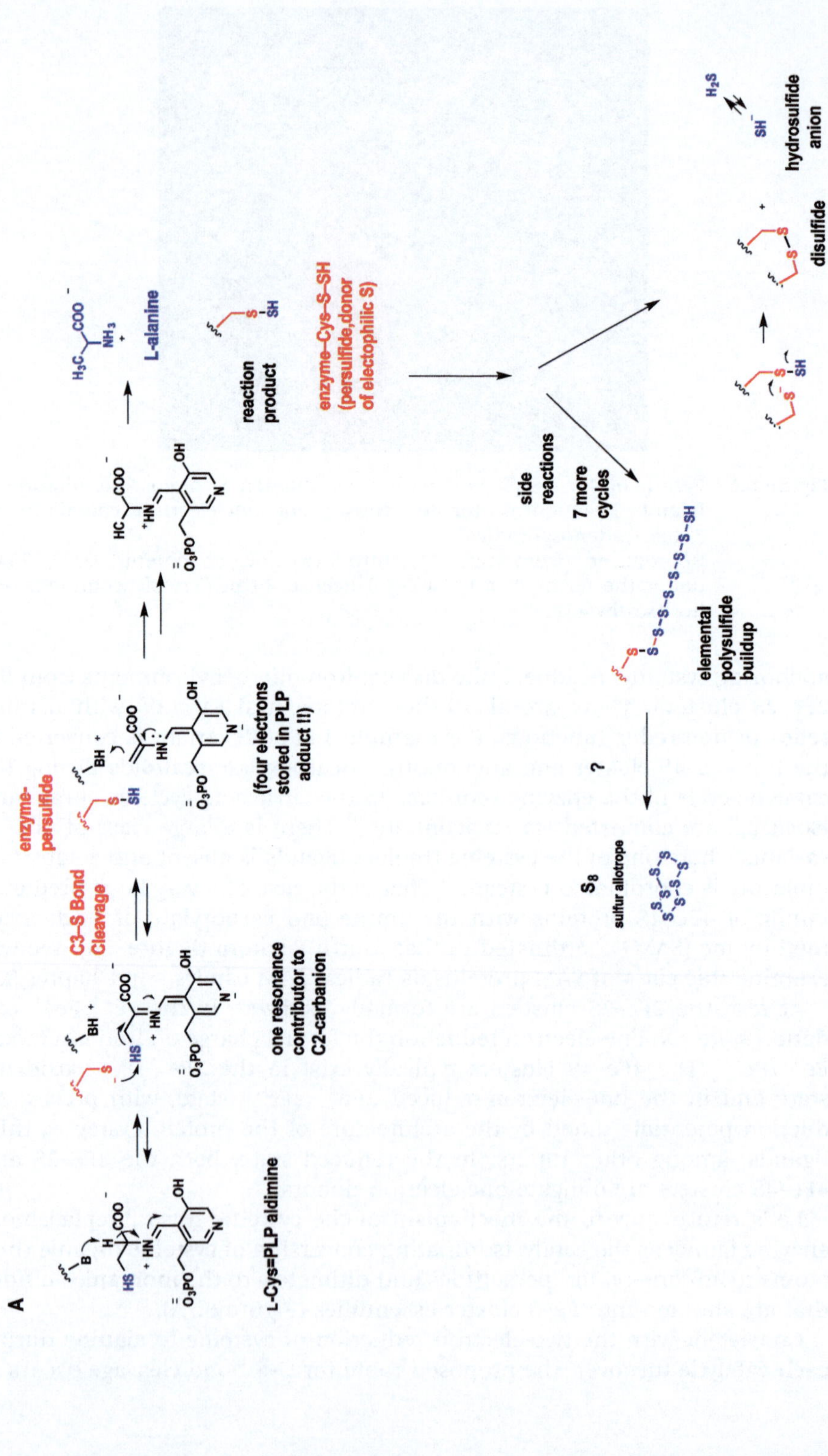
A
L-Cys=PLP aldimine
one resonance contributor to C2-carbanion
C3–S Bond Cleavage
enzyme-persulfide
(four electrons stored in PLP adduct !!)
L-alanine
reaction product
enzyme–Cys–S–SH (persulfide,donor of electophilic S)
side reactions
7 more cycles
elemental polysulfide buildup
?
S_8
sulfur allotrope
disulfide
hydrosulfide anion
H_2S

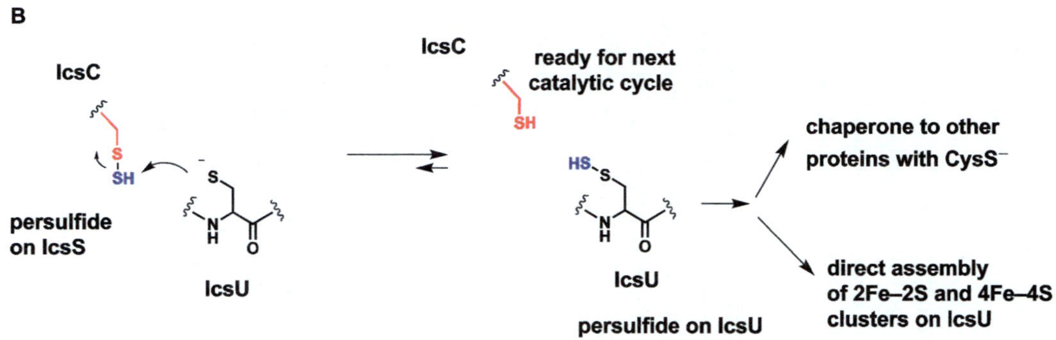

Figure 6.7 (A) Proposed mechanism for cysteine lyase involves S–S bond formation with electrophilic transfer of substrate-derived sulfur atom to enzyme cysteine thiolate to create a persulfide linkage (B) The enzyme–Cys–SSH nascent product is a donor of electrophilic sulfur to other proteins with cysteine residues that serve as chaperones in the Fe–S cluster biosynthetic pathways. The active cysteine lyase is actually a heterodimer of IcsS–IcsU and the nascent persulfide sulfur on IcsS is transferred to IcsU to enable subsequent IcsS turnover cycles.

the level of the stabilized C-2-carbnion of the Cys–PLP adduct in the active site of the lyases. Attack by the thiolate form of an active site cysteinyl residue, is shown to capture the itinerant S atom from the substrate–PLP adduct. The pair of electrons in the C3–S bond are left behind and flow into the PLP–electron sink. As shown above in Figure 6.2 this C–S bond-breaking polarity is remarkable and relies on the PLP cofactor to act transiently as a *four-electron sink.*

If the lyase remains as the enzyme–Cys–S–SH covalent nascent product, it cannot turn over again, unless it can be disgorged of the persulfido sulfur. That is the role of iron–sulfur cluster assembly scaffold protein (IcsU) as an acceptor and scaffold protein for Fe–S cluster formation. In the absence of the partner scaffold, IscS from some sources can slowly produce small amounts of a poly-persulfide chain (*e.g.* think of the cyclic S_8 allotrope of elemental sulfur), or release the SH as H_2S. Both of these products can be seen as mixtures from some lyases (Figure 6.7).

However, the major flux is to deliver the sulfur atom to cysteine thiolates of partner proteins that then act as persulfide chaperones for eventual release of sulfide dianion equivalents in Fe–S assembly protein complexes. Indeed, the potential toxicity of those itinerant sulfur atoms (H_2S equivalents if they get loose) and/or the difficulty in their efficient recapture has led organisms to implement a series of proteins acting as chains of persulfide chaperones.

Fe–S clusters can be built in eukaryotic mitochondria (presumably using ancestral machinery from bacterial ancestors), where 17 proteins are required for the assembly, or in the cytoplasm and nucleus, where 13 proteins have been identified.[21] (The protein machinery is complex but robust, as some 70 Fe–S clusters are assembled on human proteins across mitochondria, cytoplasm and nuclei.) Some of the 17 assembly proteins help the cysteine lyases transfer the mobile sulfur atom to form Cys–SSH persulfides on IcsU and onto other chaperone proteins. Some are involved in 2Fe–2S cluster assembly. A third subgroup appears to be involved in dimerization of the 2Fe–2S preformed clusters to the cubane 4Fe–4S architectures. Others chaperone iron cations to the target proteins. Of the 70 members of the human Fe–S proteome, about 40 are 4Fe–4S clusters and the remaining 30 are mostly 2Fe–2S clusters.[22] Yet other chaperone proteins, such as TusA, deliver a sulfur atom for thiouridine residue formation in tRNA molecules, discussed in Section 6.4.

The IcsS–IcsU dimer is the active species for the cysteine lyase activity. While the first turnover of IcsS lodges the itinerant sulfane sulfur on Cys328 (close to the PLP–coenzyme), it is IcsU that removes that sulfide *via* attack of its own Cys63 thiolate[23] (Figure 6.7B). Intriguingly, IcsU is not only an acceptor of the sulfane sulfur.[24] It is also the scaffold where one 2Fe–2S complex, then a second 2Fe–2S complex is built up between each of the IscS–IscU heterodimer pairs[25] (Figure 6.8). There are additional similar lyase–chaperone–scaffolding pairs in humans,[14,26] and in plant plastids for Fe–S cluster assemblies.[14]

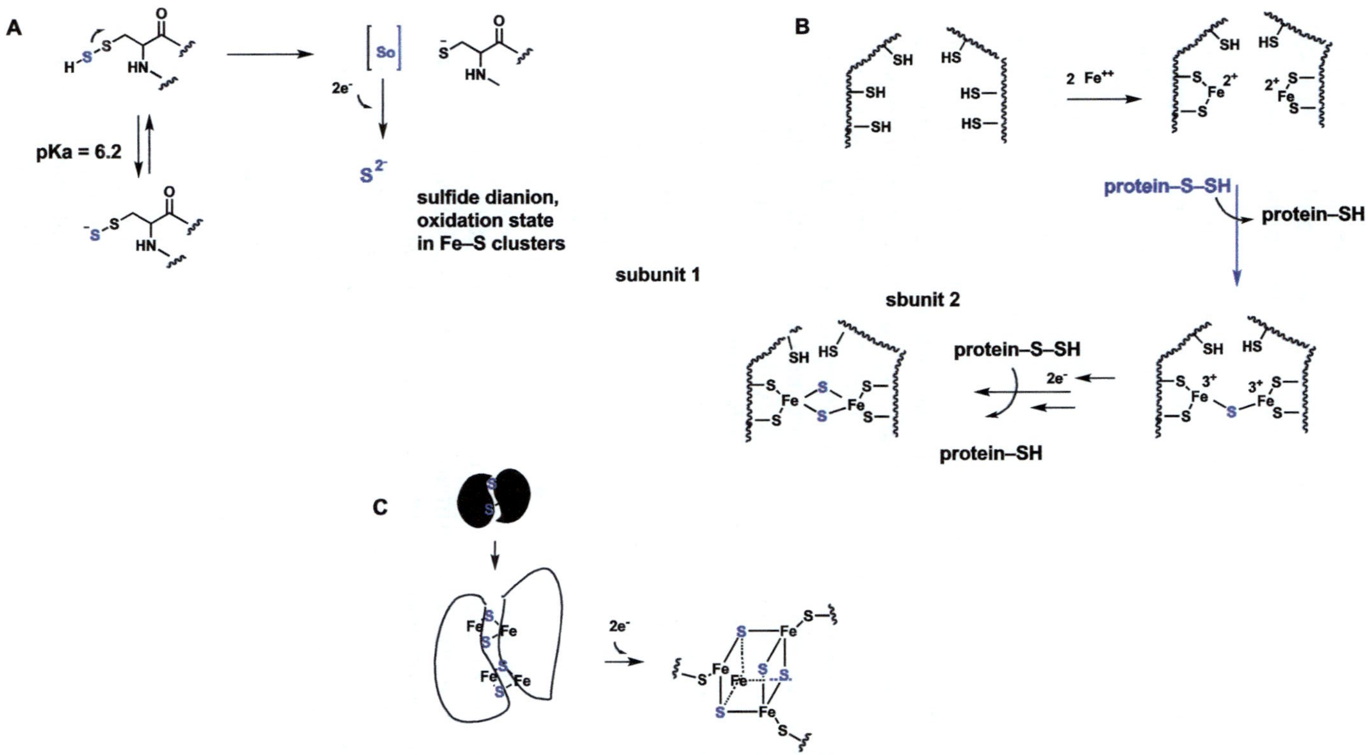

Figure 6.8 Fe–S cluster formation from protein persulfides. (A) The formal oxidation state of the terminal sulfur in a protein–Cys–SSH linkage is divalent. If one looks at S–SH bond cleavage as leaving the two electrons in the S–S bond with the protein Cys residue, that yields a S^0 atom and a protein Cys-S^-. The inorganic sulfur atoms in Fe–S clusters are at −2 oxidation states so will need two-electron input for each sulfide incorporated. (B) A putative path to 2Fe–2S cluster assembly requires an apoprotein conformation with four available cysteine thiol side chains, binding of two Fe^{2+} ions (presumably delivered from iron binding protein chaperones), and then a net two-electron reduction step ($2Fe^{3+}$ to $2Fe^{2+}$). (C) Proposal for formation of 2Fe–2S clusters across the subunit interface of IcsU dimers as way stations for reductive dimerization to 4Fe–4S cubane geometry.

These 2Fe–2S complexes can be shipped out to other apoprotein chaperones or on to target proteins for the Fe–S clusters. Even more economically, the pair of 2Fe–2S complexes in the IcsS–IcsU tetramer can mature to a 4Fe–4S cluster, with the reductive input of two electrons, and that 4Fe–4S cluster can be exported to other chaperones and client proteins as schematized in Figure 6.9.

Figure 6.9A illustrates that the cysteine to alanine lyases provide persulfide sulfur atoms, not only for Fe–S cluster assembly but also for molybdopterin sulfur insertion (*e.g.* for sulfite oxidase, Chapter 3) and for the thiouridine and thiocytidine modifications discussed in a subsequent section of this chapter. Figure 6.9B serves as a reminder that the IcsS lyase makes protein–protein interactions with a variety of client proteins in addition to the IcsU scaffold protein for sulfane atom transfer without free diffusion of the wandering six-electron sulfur atom.[27]

Finally, Figure 6.9C acts as a reminder that sulfide atoms are only half of the Fe–S clusters. The other half is iron in the form of iron cations, Fe^{3+} or Fe^{2+}. Just as free cysteine and free H_2S are kept at low concentrations to minimize promiscuous reactivity, ferrous ions are also problematic, engaging in ready one-electron transfer chemistry to generate a cascade of reactive oxygen species. Figure 6.9C therefore indicates that there are carrier proteins that chaperone iron in cells and provide it to the scaffold proteins and presumably avoid having it diffuse freely away.

The basic schematic for 2Fe–2S cluster formation in a cysteine-rich site in a target protein is sketched out in Figure 6.8B. Two ferrous irons (Fe^{2+}) are proposed to coordinate with a pair of protein-cysteines as preamble for receipt of one sulfur from a Cys–S–SH-containing chaperone protein. The net transfer is formally a sulfide dianion, ending up bridging the two iron atoms. The scaffolding could arise within a subunit or between subunit interfaces as suggested in cartoon style in Figure 6.8B for each IcsS–IcsU nascent pair of 2Fe–2S clusters. One can assume that Reiske type Fe–S clusters are transferred to their ultimate client proteins with two histidine ligands replacing two of the four cysteine anchor residues.

Presumably, the proximal protein donor of the migrating sulfur atom ends up as the reduced Cys–SH. Note that in a formal sense the two iron atoms are each oxidized from Fe^{2+} to Fe^{3+} to accommodate the formal reduction of the itinerant sulfane sulfur equivalent to S^{2-} with its high affinity for iron coordination. The mechanism of the reductive insertion for the sulfide dianion between the two iron atoms in the platform proteins' binding sites is not yet deciphered (17 proteins in the machinery!) and could in principle involve one-electron, stepwise pathways rather than a concerted two-electron process. The process must be iterated to transfer a second sulfur at the oxidation state of sulfide ion to complete the 2Fe–2S cluster formation (Figure 6.8B). For purposes of stoichiometry only, not mechanism, we note that the net four-electron reduction of the two inserted sulfide atoms (from -Cys–S–SH donors) is represented as generation of two Fe^{3+} ions and conversion of host protein dithiol to disulfide. Further electron transfers could

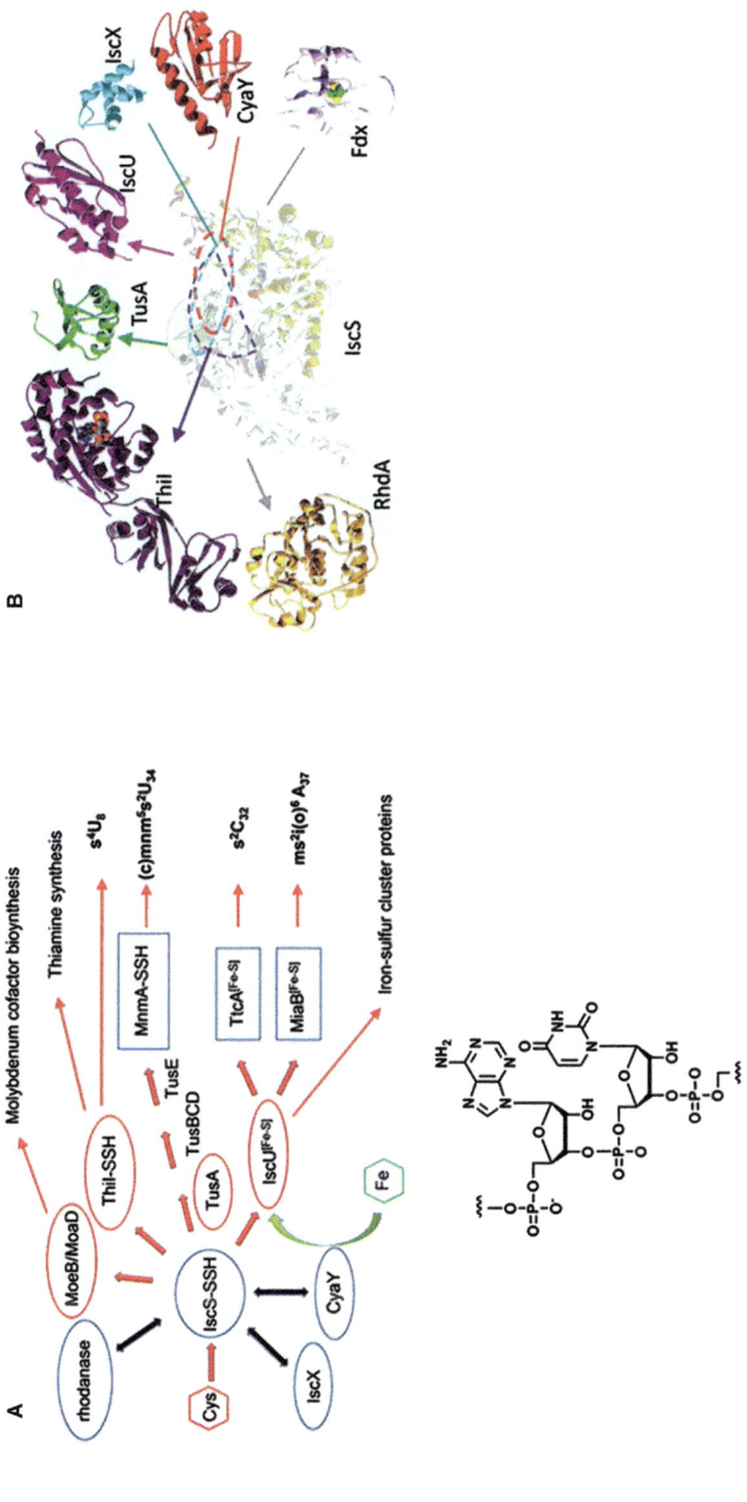

Figure 6.9 Continued on next page.

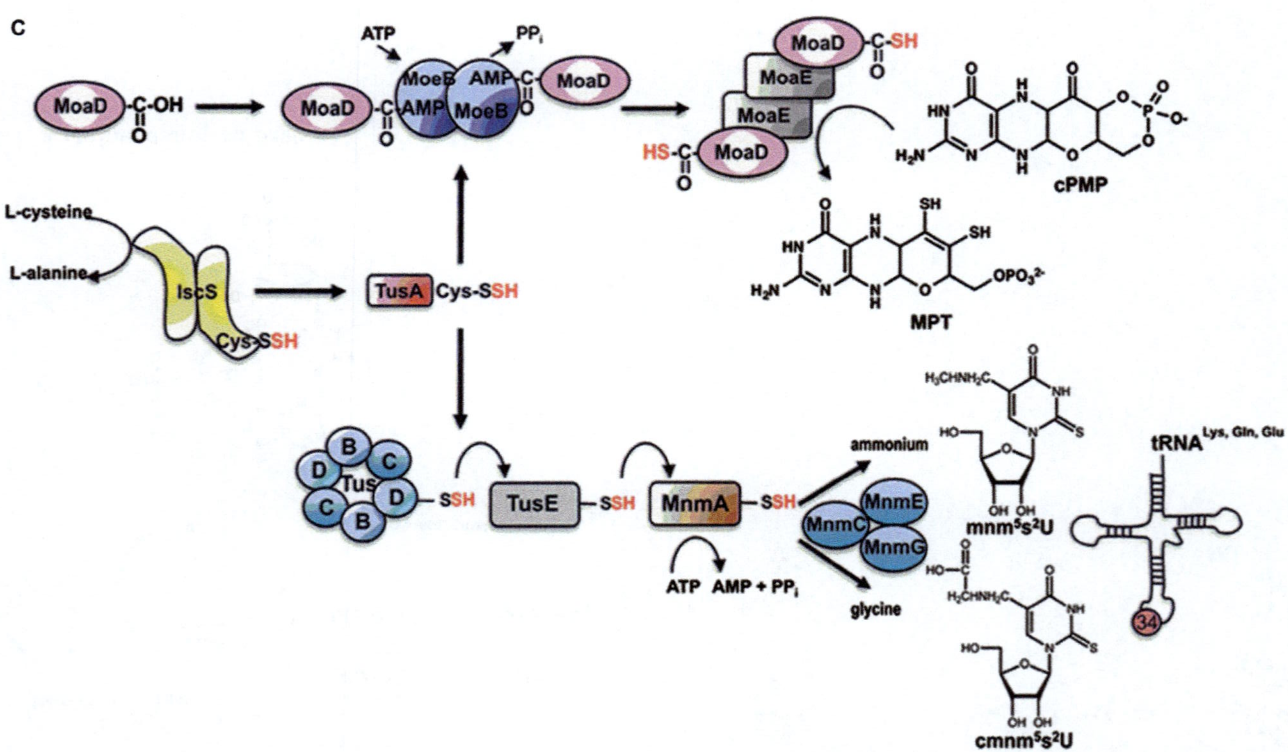

Figure 6.9 Schematic for persulfide sulfur transfer from IcsS to the Fe–S cluster network. (A) Schematic for IcsS (cysteine lyase) nascent persulfide product as electrophilic sulfur donor to many proteins involved both in Fe–S cluster assembly, molybdopterin assembly and pyrimidine thiocarbonyl groups in tRNAs. (B) Schematic of protein architectures for sulfur transfer from IcsS to partner and client proteins. (C) Schematic for role of IcsS as sulfur donor and partner protein Fe donors to scaffold proteins where Fe–S clusters get assembled and then transferred to dozens of apo forms of client proteins with the appropriate cysteine thiolate side chain distributions and orientations.

mature a 2Fe–2S cluster to a dithiol, $2Fe^{2+}$ oxidation state or a mixed Fe^{3+}–Fe^{2+} cluster.

The details of mechanism for dimerization of preexisting 2Fe–2S clusters to 4Fe–4S clusters is also murky, yet of deep biologic consequence. The kinetics of cluster assembly on the *Escherichia coli* iscU apo-protein show by spectroscopic measurements, the formation of one 2Fe–2S cluster, then a second one. At later times the signal of a 4Fe–4S cluster replaces the two 2Fe–2S cluster signals. On the other hand, the details of how Fe-sulfide bonds rearrange and go from a 2Fe planar to a 4Fe distorted cubic array are not well understood.

Like the 2Fe–2S clusters, the 4Fe–4S clusters appear to be one-electron redox boxes (formal charge of Fe^{3+} to or from Fe^{2+}, at only one of the four Fe centers), but the electron density is spread over and felt by all the irons in the 4Fe and 2Fe cluster versions.

A closer look at the series of eight Fe–S clusters that act *ad seriatim* over almost 100 Ångstroms distance in the mitochondrial membrane NADH oxidoreductase reveal that two of the eight Fe–S clusters are 2Fe–2S and the other six are the 4Fe–4S cubanes (Figure 6.10). The 2Fe–2S versions are the first, proximal to and directly receiving one electron at a time from the active site $FMNH_2$, and then the third cluster in the series of eight. In terms of directionality of electron flow, the redox potentials indicate a net gradient in potential drop from -0.37 to -0.17 volts, reflecting an equilibrium constant of almost 10^6 in favor of electron flow down to the external coenzyme Q one-electron acceptor[15] (Figure 6.10).

6.3 Variant Fe–S Clusters

Two variants of the 4Fe–4S clusters are particularly relevant to the sulfite reductase topic we described in Part I, Chapter 2, and also to the radical SAM paradigm in Chapter 9. The sulfite reductase electron transfer route takes advantage of both the 4Fe–4S cubane and the bridging cysteine thiol to the heme group (Figure 6.5, panel F) ultimately to feed in six electrons, presumably one at a time, as sulfite is reduced all the way to axially co-ordinated sulfide dianion (Figure 2.1). This is a clear intersection of in-organic sulfide, cysteine thiolate and inorganic sulfite in one catalytic machine.

The second case is exemplified in Figure 6.11. The radical SAM enzyme family has evolved active site catalytic architecture such that one of the four external cys ligands to the 4Fe–4S cube is missing, leaving room for the amino and carboxylate of the SAM substrate to coordinate to that access-ible iron atom on that face of the cube. This ensures positioning of the SAM sulfonium sulfur in electron transfer distance to the Fe–S cube. Electron transfer from an exogenous reductant (such as a reduced flavodoxin protein)[28] into that 4Fe–4S cube, as a catalytic cycle begins, leads to one-electron transfer to the sulfonium sulfur, breaking the S–C5′-adenosyl bond (Figure 6.11). That C–S bond cleavage yields the 5′-deoxyadenosyl

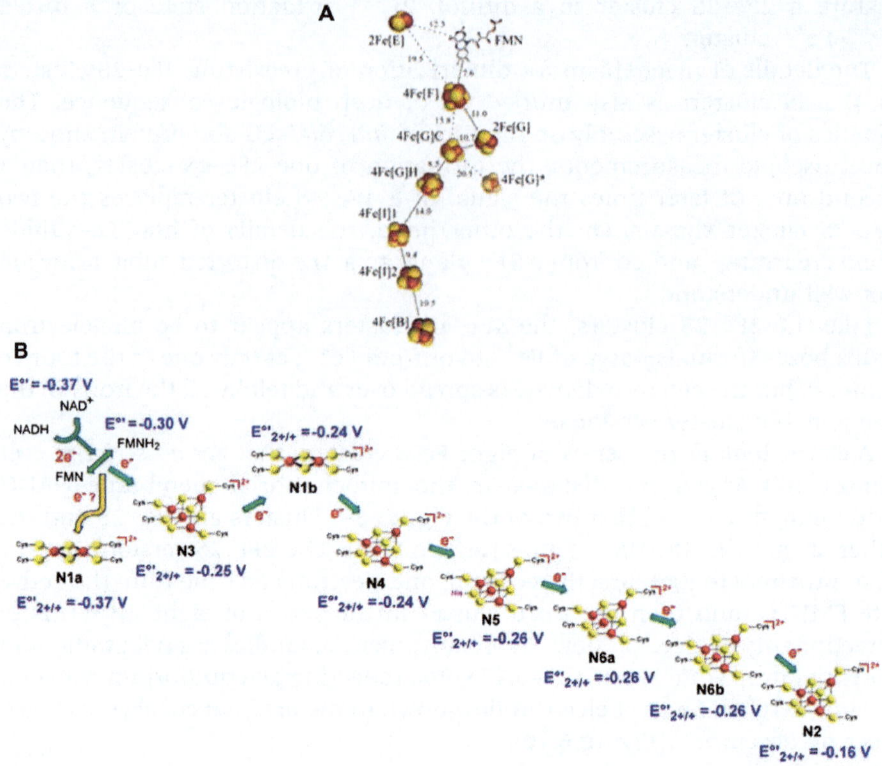

Figure 6.10 (A) The NADH oxidase involved in electron transport chain function, transfers a hydride from NADH to FMN two electrons at a time. Then the $FMNH_2$ reoxidizes in one-electron steps, passing electrons down a series of eight Fe–S clusters before reaching the mobile quinone acceptor in the membrane. The Fe–S cluster arrangements in the *Thermus thermophilus* enzyme. (B) The redox potentials are successively more positive, mediated by protein microenvironment, such that the potential drop favors electron flow from NADH to the lipid-soluble quinone.
(A) Reproduced from ref. 40, https://doi.org/10.1073/pnas.0705593104, under the terms of a CC BY 4.0 license, https://creativecommons.org/licenses/by/4.0/. (B) Reproduced from ref. 15 with permission from Elsevier, Copyright 2017.

radical and jump starts one-electron chemistry reaction manifolds in many radical SAM catalytic microenvironments. We examine those consequences in Chapter 9.

6.3.1 Sacrifice of Fe–S Clusters in Biotin and Lipoic Acid Sulfur Insertions

Two additional Fe–S clusters, a 2Fe–2S cluster and a 4Fe–4S cluster in biotin synthase and lipoyl-acyl carrier protein synthase, respectively, undergo

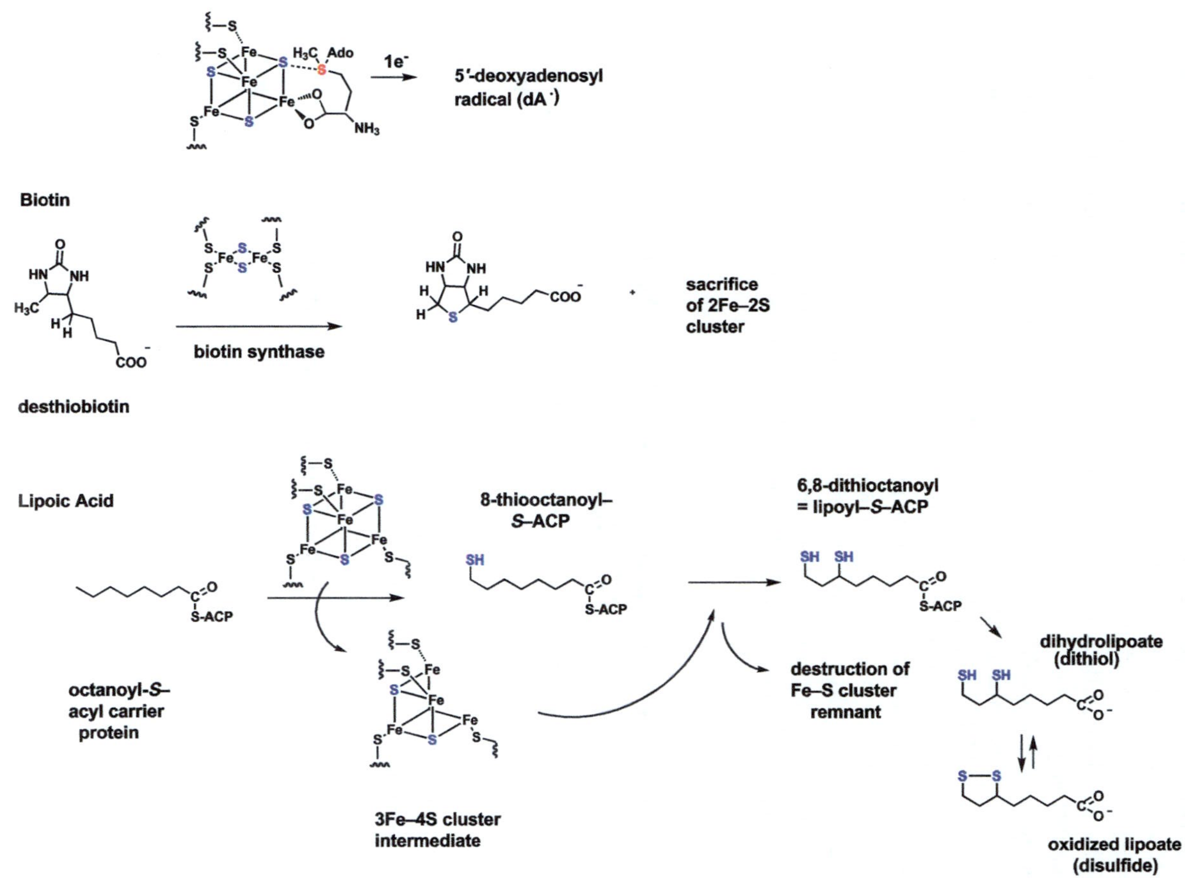

Figure 6.11 Biotin and lipoamide C–S bond enzymatic constructions involve early generation of 5′-deoxyadenosyl radicals (dA•) from SAM homolysis to drive one-electron chemistry. In particular, each scaffold is built by sacrificial destruction of one or more Fe–S clusters: 2Fe–2S clusters for biotin and a 4Fe–4S cluster for lipoamide.

sacrificial destruction during insertion of one or two sulfurs into the scaffolds of biotin and lipoamide, respectively (Figure 6.11). In each of those cases one-electron chemistry is in play from a 5′-deoxyadenosyl radical. We defer the mechanistic discussion to Chapter 9, Section 9.4, but note here that one or two sulfur atoms (presumably as the $S^{\bullet}$ radical) get inserted into C–H bonds of the biotin and lipoate groups as they form. These processes lead to destruction of adjacent 2Fe–2S cluster (biotin) and a 4Fe–4S cluster (lipoate) as the inorganic sulfide radicals are converted to thioether and thiols, respectively in the organic frameworks of the products.

The early, ancient evolution of Fe–S clusters as widely dispersed one-electron transfer machinery in cells illustrates the remarkably useful pairing of $Fe^{2+/3+}$ and sulfide anions for fundamental electron transfer reactions of life. Their assembly also highlights the three-way intersection of organic thiol, organic persulfide and inorganic sulfide chemical biology that makes sulfur a uniquely central constituent at so many metabolic intersections. The consecutive protein chaperoning of six-electron sulfur equivalents and then the assembly of Fe–S clusters in protected protein microenvironments emphasizes the care and complexity that cells have evolved to avoid free inorganic sulfide equivalents.[29,30]

6.4 Sulfur as Electrophile in Sulfenic Acids, Persulfides and Disulfides

Much of the organic sulfur chemistry in metabolic pathways described thus far have been by two-electron reaction pathways and emphasized the nucleophilic properties of thiolate anions accessible under physiologic conditions. We will take up one-electron reaction manifolds separately in Chapters 8 and 9 while revisiting one more time cysteine-based sulfur atoms as electrophilic equivalents here.

6.4.1 Persulfides as Sulfur Donors to Pyrimidines

That β-replacement family of PLP-dependent cysteine lyases, that generate -Cys–S–SH residues as covalent adducts in partner–acceptor proteins, are also involved (Figure 6.10) in sulfur transfer to C4 of uridine, C2 of uridine and C2 of cytidine residues at positions 8, 34 and 32, respectively of some bacterial tRNAs (Figure 6.12A).[27,31] The exact mechanism of replacement of an sp2 C=O bond by a C=S double bond in those thio-nucleosides is not clear, but involves prior activation of the C=O groups as enamide-*O*-AMP derivatives[32] (Figure 6.12B).

The replacement of oxygen by sulfur could occur from attack of the terminal S^- of the enzyme–Cys–S–S^- (distal sulfur as nucleophile). Cleavage might then occur by formation of an active site disulfide to break the persulfido S–S bond. That would leave the catalyst with a Cys–S–S–Cys disulfide that would need an external dithiol exchange protein to regenerate it

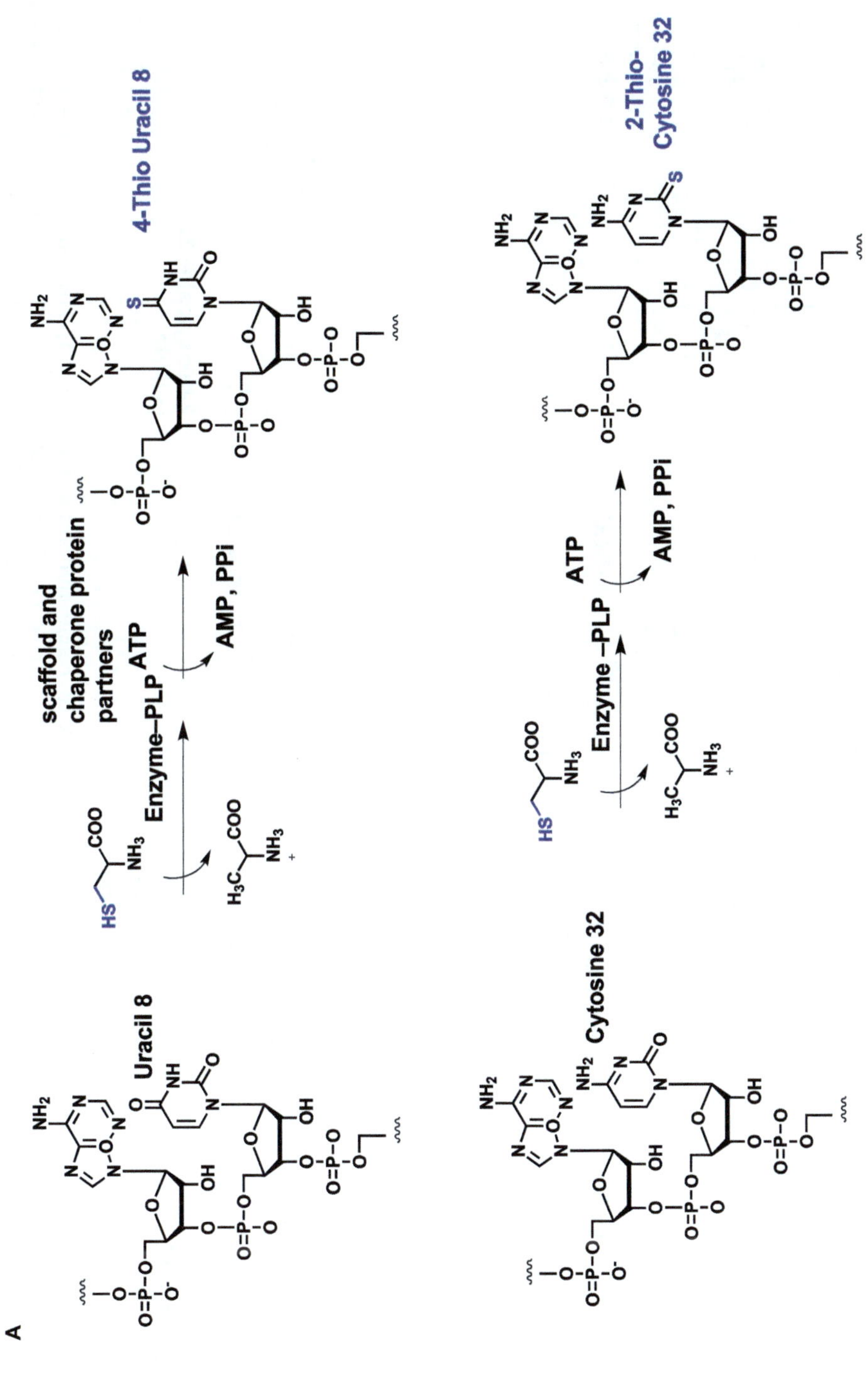

Figure 6.12 Continued on next page.

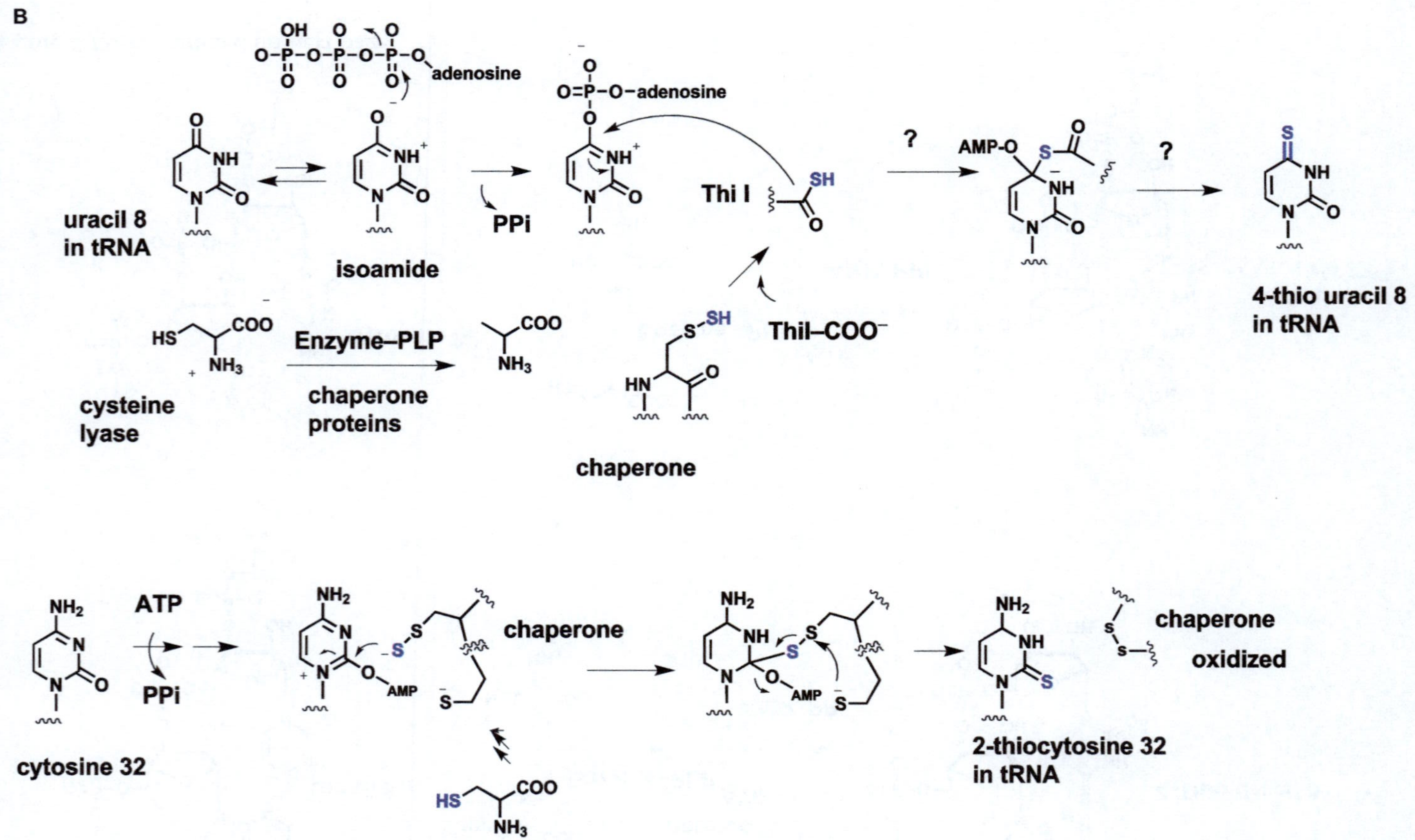

Figure 6.12 Sulfur transfer to pyrimidine residues in tRNA to create thiocarbonyl groups. (A) Formation of 4-thiouracil-8 and 2-thiocytosine-32 in tRNAs; (B) Mechanistic proposal involving ATP cleavage by isoamide oxyanions to form phosphate esters that are displaceable in addition–elimination paths by sulfur derived from cysteine lyase action.

for the next catalytic cycle. However, it appears that the persulfide sulfur gets transferred first to a small carrier protein partner as a C-terminal thiocarboxylate. This is examined in the following subsection.

6.4.2 Intersection of Protein Persulfides, Carrier Protein Acyl Persulfides and Protein Thiocarboxylates as Sulfur Donors

An unusual metabolic logic obtains for a limited set of protein persulfide intermediates involved in the biosynthesis of the sulfur-containing thiazole ring of vitamin B1 thiamin, of the ene dithiol moiety of molybdopterin (see sulfite oxidase in Chapter 3), of the thiocarboxylic acid of the iron-chelating bacterial siderophore thioquinolobactin and in formation of the thiocarbonyl groups in 2-thiocytidine and 4-thiouridine and 2-thiocytidine residues in some tRNAs (Figure 6.13). The terminal thiane sulfur of Cys-persulfide residues in proteins such as IscS (in cysteine lyase action for Fe–S cluster construction) is the progenitor of C-terminal thiocarboxylate residues in a set of small proteins that act as intermediate sulfur carriers.[33–35]

These carrier proteins (ThiS[36] in thiamin assembly, TtuB in thiouridyl tRNA maturation, MoaD in molybdopterin formation, QbsE in quinlobactin sulfuration) all end in the C-terminal dipeptide Gly–Gly–COOH (analogous to the flexible, small C-terminal diglycine moiety of ubiquitin (see Chapter 12). That C-terminal glycine of the specific carrier protein is activated to attack the α-P of ATP to form the C-terminal mixed glycyl-AMP anhydride (Figure 6.13A). This electrophilic C-terminus is now the partner for transfer of a sulfur atom from a protein persulfide donor like IscS–Cys–S–SH, presumably as the nucleophilic persulfide anion shown in Figure 6.13A. Collapse of the resultant tetrahedral adduct in the forward direction would yield an *acyl persulfide*, a novel species in our ongoing catalog of organosulfur metabolites.

The presumption is that such acyl persulfides would be reducible in cells, akin to the reduction of disulfides, but these presumably unstable species have not yet been reported to be substrates for enzymatic reduction. Reduction would liberate the IscS and related persulfide carriers as the simple cysteinyl residue partner products while creating the unusual C-terminal glycine thiocarboxylic acid anion in the carrier protein (Figure 6.13A). The thiocarboxylic acid intermediate at the C-terminus of ThiS has been isolated and characterized. It was the first example of a biological *thiocarboxylic acid* group.[33]

There are parallels to the logic of acyl carrier proteins in fatty acid synthase, polyketide synthase and non-ribosomal peptide synthase assembly lines, briefly noted in Chapter 5. In those cases, the terminal cysteamine of the P-pantetheinyl prosthetic group carries elongating acyl chains as covalent acyl thioesters. In this alternate carrier protein family bearing C-terminal thiocarboxylates, the sulfur atom itself is the transferable nucleophile. Note that Figure 6.13A poses two novel protein-bound sulfur intermediates: acyl disulfides and thiocarboxylic acids.

Four examples of C-terminal thiocarboxylic acid residues as (nucleophilic) sulfur donors are shown in Figure 6.13B. Example (A) features formation of the quinolobactin thiocarboxylic acid as the terminal step in assembly of that siderophore. The prior ATP-dependent conversion of the quinolobactin carboxylate to the acyl-AMP species lowers the energy for C–O bond cleavage on attack of the QbsE thiocarboxylate as cosubstrate. This example is a simple exchange of sulfur between two carboxylates, achieved by spending

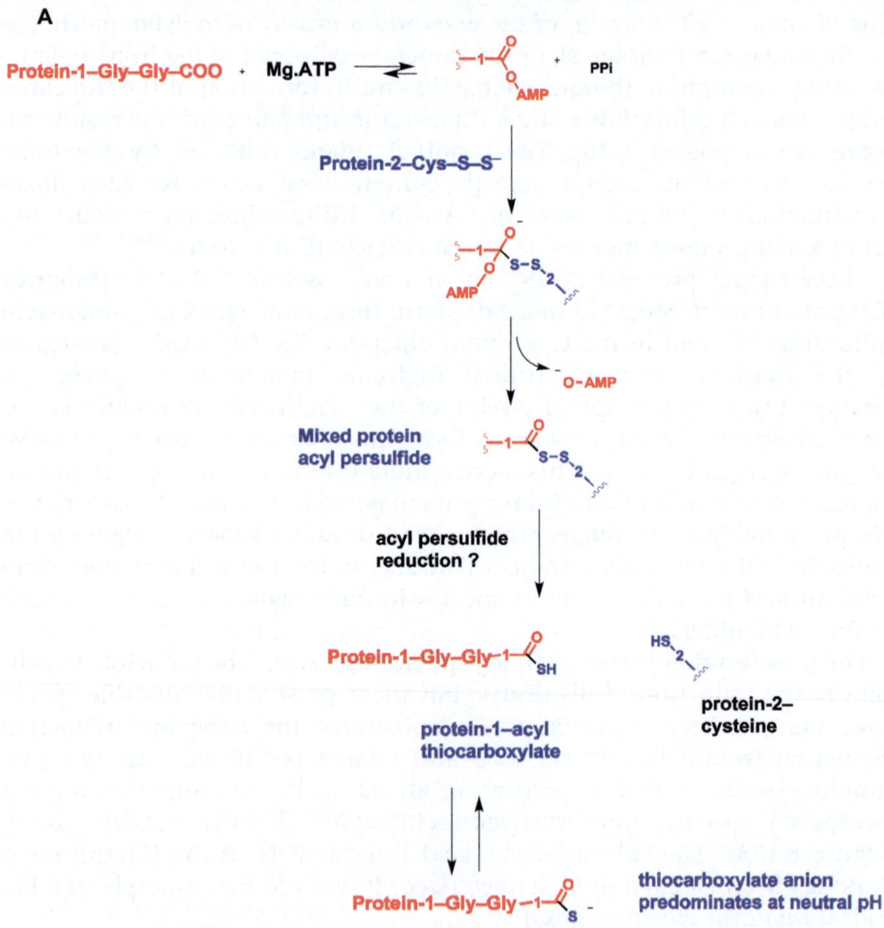

Figure 6.13 A Small family of carrier proteins in the form of protein-C-terminal thiocarboxylates: (A) Formation of protein acyl persulfides and then thiocarboxylates at the C-termini of a family of sulfur carrier proteins. (B) Four examples of transfer of the nucleophilic sulfur atom from a carrier protein thiocarboxylate: path A: quinolobactin; path B thiazole ring of thiamin; path C the bis-thiolation of the molybdopterin cofactor; path D thiocarboxylate as sulfur donor in 4-thio U tRNA.

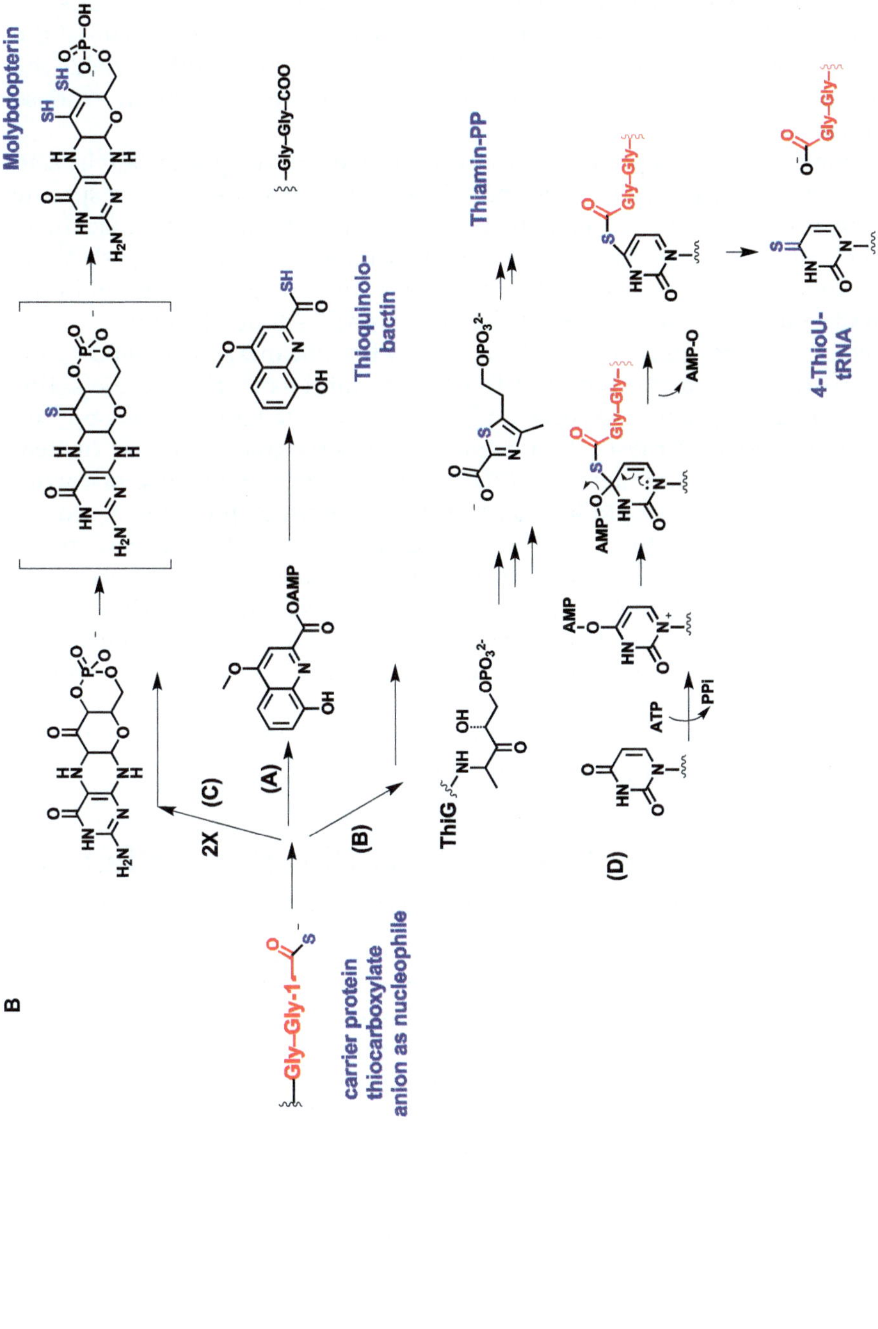

Figure 6.13 Continued.

an ATP. Path (B) shows attack of the ThiS C-terminal thiocarboxylate on the C_2 carbonyl of an aminopentose cosubstrate bound to a separate protein ThiG.[37] The resultant tetrahedral adduct isomerizes to a thiocarbonyl on the way to the dehydrated aromatized thiazole ring of thiamin-PP.

Example (C) illustrates the tandem addition of *two* sulfur atoms from a dimeric form of (thiocarboxy)–MoaD. The first sulfur becomes a transient thioketone. The second sulfur transfer occurs as the cyclic phosphate of the tetracyclic framework is opened hydrolytically. Isomerization then yields the enedithiol moiety that is the coordination site for Mo^{VI} in sulfite oxidase (Chapter 3).

Example (D) shows a possible pathway for transfer of the thiocarboxylic acid sulfur at the C-terminus of carrier protein TtuB to the 4-position of a specific uridine residue in tRNAs. Presumably, an analogous logic and machinery would be involved in 2-thiocytidine tRNA maturation. Again, as in path A one needs to activate the C–O bond (here a ketone but a hydroxylate anion in the iminium tautomer) to set up C–O cleavage as the C=S bond forms.

One of the questions one might raise is what advantage is gained by organisms that use carrier protein Gly–Gly–C-termini as carriers of the sulfur atom to be incorporated in these and other metabolic pathways as opposed to a direct attack of the IscS (and homologs) persulfide anions as nucleophiles. One possibility may be added layers of specificity in use of carrier proteins to ferry nucleophilic thiane sulfurs to partner proteins. A second rationale may be that the enzymatic machinery for activation of the C-terminal diglycine moiety of the small protein ubiquitin (and congeners) was conscripted for the corresponding activation of the ThiS type of proteins as the AMP mixed anhydrides (albeit ubiquitin is not a widespread signaling pathway in prokaryotes).[36] The relative nucleophilicities of protein persulfide anions *vs.* protein thiocarboxylates have not been reported to evaluate the balance of reactivity *vs.* hydrolytic stability of the two activated forms of transferable organic sulfur atoms.

6.5 One More Look at the Versatility of Persulfides as Electrophilic or Nucleophilic Forms of Sulfur

A reprise of the strategy that organisms use to build the ancient and ubiquitous Fe–S redox clusters is to "wash" inorganic sulfide through cysteine as an organic carrier of divalent sulfur (Figure 6.14A). In turn free cysteine is coerced by cysteine lyases into fragmentation at the C3–SH bond to transfer the itinerant sulfur as a six-electron sulfur atom to protein-bound cysteine thiolates to create protein persulfides. These are, on the one hand, carriers of electrophilic sulfur while the persulfide is protonated, *e.g.* to form new persulfide linkages in partner chaperones That chain terminates when the persulfide distal sulfur is liberated with its pair of electrons (and a proton) as a hydrosulfide anion for Fe–S cluster assembly.

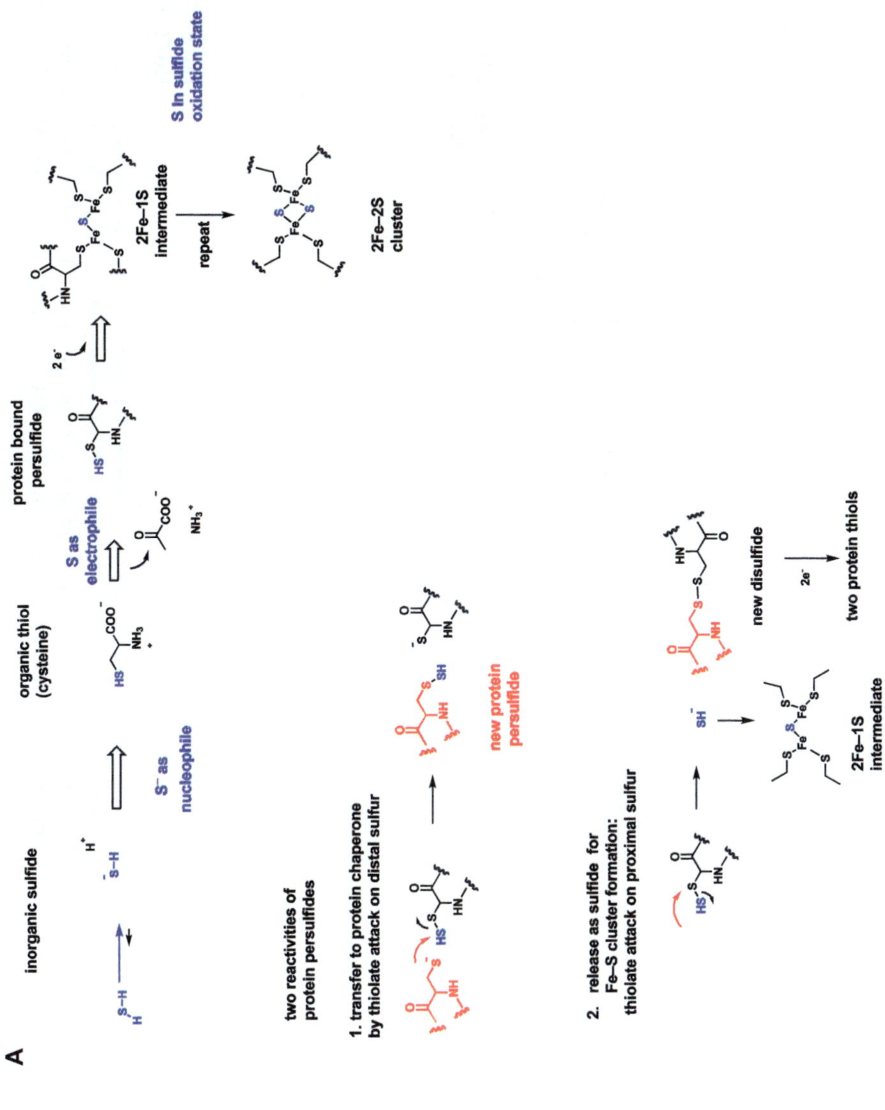

Figure 6.14 Continued on next page.

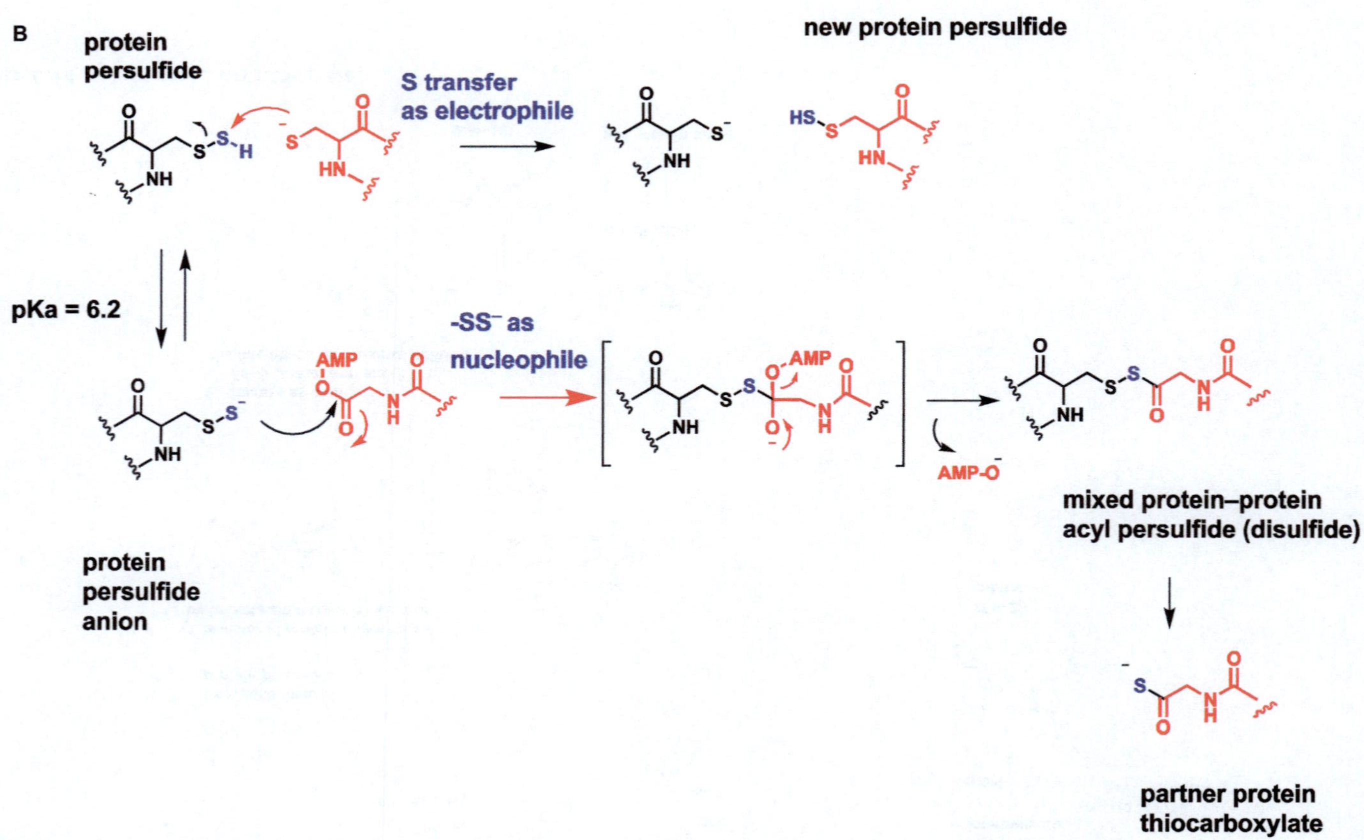

Figure 6.14 (A) The inorganic sulfide ions that end up in Fe–S redox clusters ultimately derive from inorganic sulfide (from assimilatory sulfate reduction) but pass through free cysteine, then protein cysteinyl persulfide oxidation states as intermediates. (B) Attack on distal *vs.* proximal sulfurs of the persulfides creates distinct products. Distal attack by a protein cysteinyl thiolate leads to electrophilic sulfane transfer while proximal attack generates a protein disulfide and liberates the electron-rich hydrosulfide anion.

While there are several possible routes to a net two-electron reductive cleavage of the sulfur–sulfur bond in protein–Cys–S–SH groups, this could be achieved by a simple change in regioselectivity of which sulfur atom in the persulfide group gets attacked. Route 2, in Figure 6.14A shows that attack by a protein thiolate (probably *in cis*) on the *proximal* sulfur rather than the *distal* sulfur of the S–SH linkage would lead to a protein disulfide rather than a new protein persulfide. The coproduct from proximal attack on the –S–SH bond is a protein disulfide. The requisite two-electron input could then be a conventional thioredoxin reductase type of disulfide–dithiol interchange. It is known that both sulfurs in persulfide linkages are electrophilic (in analogy to both sulfurs in disulfide linkages).

Control of regiospecificity of attack on protein persulfides could be a simple approach to achieve diversity of outcomes. An overview of the ultimate provision of sulfido sulfurs in iron–sulfur clusters from inorganic sulfide (from assimilatory sulfate reduction, Chapter 2) shows that inorganic hydrosulfide is "washed" through the three-carbon thiol cysteine framework of cysteine, the thiol then mobilized as a persulfide and transferred until it reaches a protein Fe–S site and is then regiospecifically cleaved to release HS^- only in the desired protein microenvironment, embraced and surrounded by cysteines and iron.

Figure 6.14B meanwhile serves as a reminder that the persulfide group is more basic than a thiol group, with a pK_a of 6.2, such that the persulfide anion, a strong nucleophile, will be an accessible species. Instead of behaving as an electrophile when protonated, the persulfide anion is utilized to capture the C-terminal Gly–Gly dipeptide of partner proteins when they have been activated as the mixed acyl-AMP anhydrides. That action, after decomposition of the tetrahedral intermediates with low energy expulsion of AMP-O^-, yields the covalently connected two proteins *via* an acyl persulfide or acyl disulfide linkage.

To the extent that this S–S link is electrophilic, in analogy to a simple disulfide link, its *in vivo* reduction would result in transfer of the distal sulfur of the initial persulfide from the first protein to the second protein. However, the transfer has involved nucleophilic sulfur chemistry by the itinerant sulfur atom, not electrophlic chemistry. Those second proteins, with the now C-terminal thiocarboxylate anion group are members of sulfur carrier proteins as discussed in the previous section.

The versatility of protein cysteine–persulfide groups for both electrophilic and nucleophilic sulfur transfer offers expanded chemical biology roles that are probably yet to be explored in metabolic logic.

In Chapter 14 we will discuss the biosynthetic conversion of peptide carbonyl oxygens to peptide thiocarbonyls. An equivalent strategy of converting the carbonyl group to an *O*-phosphoryl ester is the preamble to its elimination and net replacement by sulfur.

References

1. C. Ayala-Castro, A. Sain and F. W. Outten, Fe-S cluster assembly pathways in bacteria, *Microbiol. Mol. Biol. Rev.*, 2018, **72**, 110–125.
2. S. Bandyopadhyay, K. Chandram and M. K. Johnson, Iron-sulfur cluster biosynthesis, *Biochem. Soc. Trans.*, 2008, **36**, 1112–1119.
3. E. G. Moore, Trafficking in persulfides: delivering sulfur in biosynthetic pathways, *Nat. Chem. Biol.*, 2006, **2**, 185–194.
4. M. R. Filipovic, J. Zivanovic, B. Alvarez and R. Bannerjee, Chemical Bbology of H2S signaling through persulfidation, *Chem. Rev.*, 2018, **118**, 1253–1337.
5. J. I. Toohey and A. Cooper, Thiosulfoxide (sulfane) sulfur: new chemistry and new regulatory roles in biology, *Molecules*, 2014, **19**, 12789–12813.
6. P. Rose, P. K. Moore and Y. Z. Zhu, H2S biosynthesis and catabolism: new insights from molecular studies, *Cell. Mol. Life Sci.*, 2016, **74**, 1391–1412.
7. P. Rose, P. K. Moore and Y. Z. Zhu, Hydrogen sulfide biosynthesis and catabolism: new insights from molecular studies, *Cell. Mol. Life Sci.*, 2017, **74**, 1391–1412.
8. E. Beshad, S. Parkin and J. M. Bollinger, Mechanism of cysteine desulfurase Slr0387 from Synechocystis 6803; kinetic analysis of cleavage of the persulfide intermediate by chemical reductants, *Biochemistry*, 2004, **43**, 12220–12226.
9. M. D. Toney, Controlling reaction specificity in pyridoxal phosphate enzymes, *Biochim. Biophys. Aca*, 2011, 1407–1418.
10. C. T. Walsh and Y. Tang, *The Chemical Biology of Human Vitamins*, Royal Society of Chemistry Press, London, 2018.
11. L. B. Poole, The basics of thiols and cysteines in redox biology and chemsitry, *Free Radical Biol. Med.*, 2015, 148–157.
12. C.-M. Park, L. Weerashinge, J. J. Day, J. M. Fukuto and M. Xian, Persulfides: current knowledge and challenges in chemistry and chemicalbioloy, *Mol. BioSyst.*, 2015, **11**, 1775–1785.
13. H. Beinert, R. H. Holm and E. Munck, Iron sulfur clusters: nature's modular, multipurpose structures, *Science*, 1997, **277**, 653–659.
14. Y. Lu, Assembly and transfer of iron-sulfur clsuetrs in the plasmid, *Front. Plant Sci.*, 2014, 14.
15. P. Zanello, Structure and electrochemistry of proteins harboring iron-sulfur clusters of different nuclearities. Part I. [4Fe-4S] and [2Fe-2S] iron-sulfur proteins, *J. Struct. Biol.*, 2017, **200**, 1–19.
16. L. Zheng and D. Dean, Catalytic formation of a nitrogenase iron-sulfur cluster, *J. Biol. Chem.*, 1994, **269**, 18723–18726.
17. L. A. Rettberg, J. Wilcoxen, C. C. Lee, M. T. Stiebritz, K. Tanifuji, R. D. Britt and Y. Hu, Probing the coordination and function of Fe4S4 modules in nitrogenase assembly protein NifB, *Nat. Commun.*, 2018, **9**, 2824.
18. J. Couturier, B. Touraine, J.-F. Briat, F. Gaymard and N. Rouhier, The iron-sulfur cluster assemly machinery in plants: current knowledge and open questions, *Front. Plant Sci.*, 2013, **4**, 259.

19. A. H. Robbins and C. Stout, Structure of activated aconitase: formation of the [4Fe-4S] cluster in the crystal, *Proc. Natl. Acad. Sci. U. S. A.*, 1989, **86**, 3639–3643.
20. J. B. Broderick, B. Duffus, K. S. Duschene and E. M. Shepard, Radical S-ademosylmethionine enzymes, *Chem. Rev.*, 2014, **114**, 4229–4317.
21. S. Ciofi-Baffoni, V. Nasta and L. Banci, Protein networks in the maturation of human iron-sulfur proteins, *Metallomic*, 2018, **1**, 1–55.
22. N. Maio and T. Roualt, Mammalian Fe-S proteins: definition of a consensus motif recognized by the co-chaperone HSC20, *Metallomics*, 2016, **8**, 1032–1046.
23. A.-i. Kato, H. Mihara, T. Kurihara, Y. Takahashi, U. Tokumoto, T. Yoshimura and N. Esaki, Cys-328 of IscS and cys-63 of IscU are the sites of disulfide bridge formation in a covalently bound IscS/IscU complex: implications for the mechanism of iron-sulfur cluster assembly, *Proc. Natl. Acad. Sci. U. S. A.*, 2002, **99**, 5948–5952.
24. H. D. Urbina, J. Silberg, K. G. Hoff and L. E. Vickery, Transfer of sulfur from IscS to IscU during Fe/S cluster assembly, *J. Biol. Chem.*, 2001, 44521–44526.
25. E. N. Marinori, J. de Olivera, Y. Nicolet, E. C. Raulfs, P. Amara, D. R. Dean and J. C. Fontecilla-Campls, IscS-IscU)2 complex structures provideinsghts into Fe2S2 biogenesis and transfer, *ACHIE*, 2012, **51**, 5439–5442.
26. Y. Liu and J. Cowan, Iron sulfur cluster biosynthesis. Human NFU mediates sulfide delivery to ISU in the final step of [2Fe-2S] cluster assembly, *Chem. Commun.*, 2007, 3192–3194.
27. K. A. Black and P. dos Santos, Shared intermediates in the biosynthesis of thio-cofactors: mechanism and function of cystene desulfurases and sulfur acceptors, *Biochim. Biophys. Acta*, 2015, **1853**, 1470–1480.
28. J. Sancho, Flavodoxins: sequence, folding, binding, function, and beyond, *Cell. Mol. Life Sci.*, 2006, **63**, 855–864.
29. H. Ye and T. A. Roualt, Human iron-sulfur cluster assembly, cellular iron homeostasis and disease, *Biochemistry*, 2010, **49**, 4945–4956.
30. R. K. Yadav, M. Martinov, V. Vitvitsky, J. Seravalli, R. Wedmann, M. R. Filiopvic and R. Banerjee, Biosynthesis and reactivity of cysteine persulfides in signaling, *J. Am. Chem. Soc.*, 2016, **138**, 289–299.
31. S. Leinkuhler, M. Buhning and L. Beilschmidt, Shared sulfur mobilization routes for tRNA thiolation and molybdenum cofactor biozynthesis in prokaryotes and eukaryotes, *Biomolecules*, 2017, **7**, 5.
32. R. Kambampati and C. Luahin, IcsS is a sulfotransferase for in vitro biosynthesis of 4-thiouridine in Eschericia coli, *Biochemistry*, 1999, **38**, 16561–16568.
33. S. V. Taylor, N. Kelleher, C. Kinsland, H. J. Chiu, C. A. Cstello, A. D. Backstrom, F. W. McLafferty and T. P. Begley, Thiamin biosynthesis in Eschricia coli. Identification of ThiS thiocarboxylate as the immediate sulfur donor in the thiazole formation, *J. Biol. Chem.*, 1998, **273**, 16555–16660.

34. G. Gutzke, B. Fischer, R. R. Mendel and G. Schwartz, Thiocarboxylation of molybdopterin synthase provides evidence for the mechanism of dithiolene formatin in metal-binding pterins, *J. Biol. Chem.*, 2001, **276**, 36268–36274.
35. S. Leimkuhler and K. V. Rajagopalan, A sulfotransferase is required in the transfer of cysteine sulfur in the in vitro synthesis of molybdopterin from precursor Z in Eschericia cli, *J. Biol. Chem.*, 2001, **276**, 22024–22031.
36. C. Lehmann, T. P. Begley and S. E. Ealick, Structure of E.coli ThiS-ThiF complex, a key component of the sulfur transfer system in thiamin biosynthesis, *Biochemistry*, 2006, **45**, 11–19.
37. P. C. Dorrestien, H. Zhai, F. W. McLafferty and T. P. Begley, The biosynthesis of the thiazole phosphate moiety of thiamin: the sulfur transfer mediated by the sulfur carrier protein This S, *Chem. Biol.*, 2004, **11**, 1373–1381.
38. R. Shi, *et al.*, Structural basis for Fe-S cluster assembly and tRNA thiolation mediated by IscS protein-protein interactions, *PLoS Biol.*, 2010, DOI: 10.1371/journal.pbio.1000354.
39. C. Zheng and P. Dos Santos, Metallocluster transactions: dynamic protein interactions guide the biosynthesis of Fe-S clusters in bacteria, *Biochem. Soc. Trans.*, 2018, **46**, 1593–1603.
40. G. Yakoviev, T. Reda and J. Hirst, Reevaluating the relationship between EPR spectra and enzyme structure for the iron-sulfur lusters in NADH:quinne oxidoreductase, *Proc. Natl. Acad. Sci. U. S. A.*, 2007, **104**, 12720–12725.
41. T. A. Grell, W. M. Kincannon, N. A. Bruender, E. J. Blaesi, C. Krebs, V. Bandarian and C. L. Drennan, *J. Biol. Chem.*, 2018, **293**, 17349–17361.

CHAPTER 7

Cysteine to Taurine: Thiol to Sulfonic Acid

We have noted earlier, for example in Chapter 6, that the nucleophilic side chain of free cysteine,[1] of cysteine residues in proteins[2] (also see Chapter 9), and of glutathione (GSH) is prone to oxygenation to sulfenic acids. We noted that those are usually transient species in the presence of another molecule of thiol. The subsequent RS–SR disulfides are the stable products. In this chapter we note that the *S*-oxygenation of cysteine can be purposeful, enzyme-catalyzed, and lead to the decarboxylated *C*-sulfonic acid metabolite taurine.

There are five major metabolic routes leading from cysteine (Figure 7.1). One involves flux to methionine biosynthesis, as noted in Chapter 4. A second is as a precursor of the terminal cysteamine moiety of coenzyme A and the coenzyme A-derived phosphopantetheine prosthetic group found covalently linked to a serine side chain in fatty acid synthases, polyketide synthases and non-ribosomal peptide synthases (also Chapter 4). A third route is the ribosome-independent assembly of the tripeptide glutathione (γ-glutamyl-cysteinyl-glycine) (Chapter 5). The fourth route is in protein biosynthesis at the ribosome where abundance among the 20 proteinogenic amino acid residues of cysteine is in the 2–3% range. The role of cysteine side chains in proteome chemistry and posttranslational modification is the subject matter of Part III.

7.1 Cysteine *S*-Oxygenation

A fifth route is oxygenative, controlling the levels of free cysteine, and thereby the potential toxicity of unrestrained nucleophilic behavior of the side chain thiolate anion. The strategy is to disable the nucleophilicity of the sulfur atom by enzymatic dioxygenation in a single enzyme active site, *bypassing a free sulfenic acid intermediate.* The enzyme catalyst is cysteine dioxygenase, transferring both atoms from cosubstrate O_2 to the sulfur of cysteine, converting it,

Chemical Biology No. 12
The Chemical Biology of Sulfur
By Christopher T. Walsh

Published by the Royal Society of Chemistry, www.rsc.org

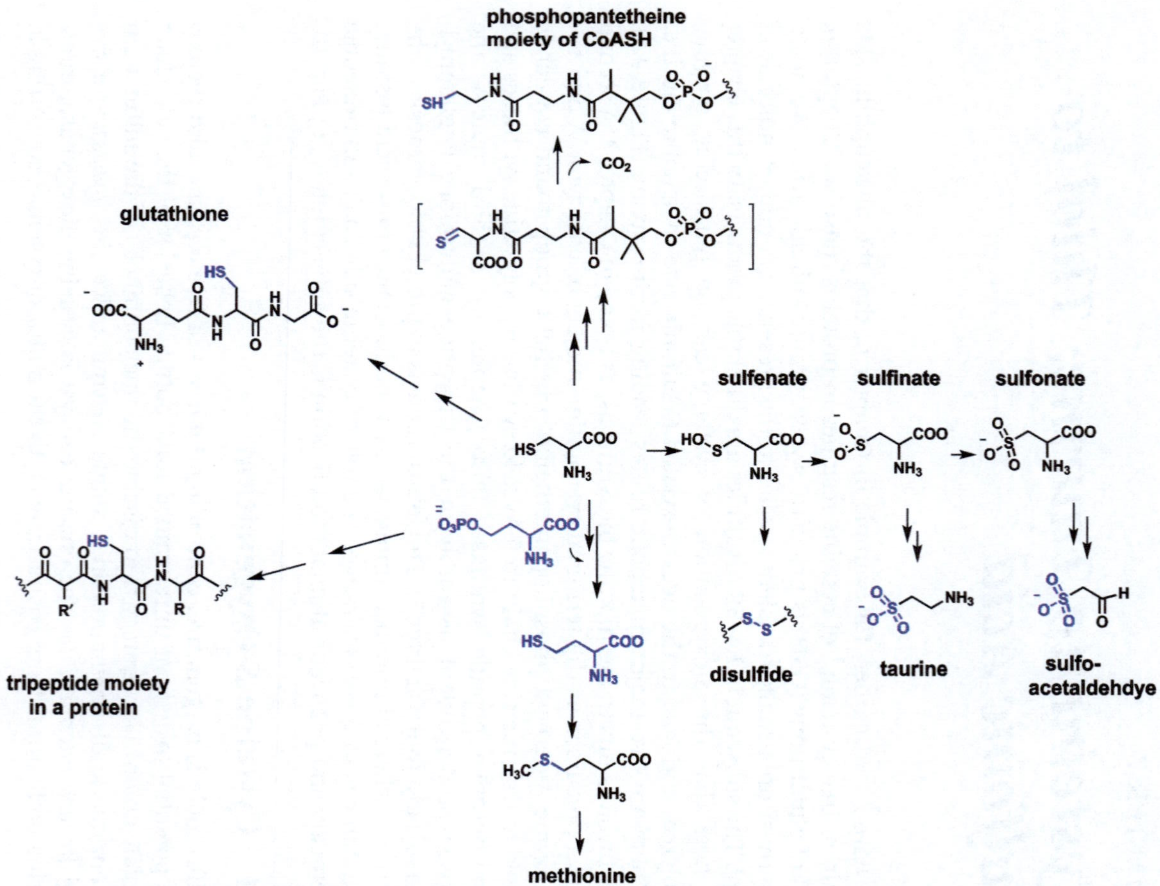

Figure 7.1 Five metabolites from the free amino acid cysteine include the *S*-oxygenation pathway.

presumably *via* a bound sulfenate, to the released cysteine sulfinate (Figure 7.2A). There is evidence that accumulation of cysteine in cysteine dioxygenase deficiency is correlated with neurodegenerative diseases.[3]

In the sulfenate and sulfinate oxidation state the sulfur is non-nucleophilic. We have noted that oxidation of thiols to sulfenate is reversible *via* two subsequent steps (Figure 7.3): conversion to a disulfide and then two-electron reduction (*e.g.* the action of glutathione reductase, Chapter 5). In contrast, double oxygenation to the free sulfinic acid is irreversible in biological contexts. This may be why the cysteine oxygenase is a dioxygenase rather than the typical monooxygenase. That outcome assures unidirectional flux of free cysteine.

(In Chapter 13, we will examine the role of sulfiredoxin as an ATP-dependent enzyme that can rescue protein cysteine sulfinic acid side chains by net reduction, with over 50 proteins now recognized as substrates.[4])

The dioxygenase is a member of the mononuclear nonheme iron family of catalysts that usually act as monooxygenases not dioxygenases. There are two unusual wrinkles that may tilt the enzyme to transfer of both atoms of the dioxygen cosubstrate. The first is that the enzyme's three side chain ligands to Fe are histidine imidazole side chains (rather than two His and one Asp or Glu side chain carboxylate) that populate the monooxygenase ligand sphere.[5] A second unusual feature is the absence of the typical cosubstrate α-ketoglutarate that would undergo decarboxylation in the course of generating the high-valency oxoiron oxygen transfer species in the active site.

Both oxygen atoms from a single O_2 cosubstrate molecule are incorporated into the cysteine substrate, consistent with a true dioxygenase catalytic role. Thus, the cysteine dioxygenase catalyst must avoid the O–O bond fragmentation step that leads to an $Fe^{IV}{=}O$ that then acts as monooxygenation catalyst. The enzyme must instead transfer both oxygen atoms from O_2. A possible reaction pathway[6–8] is shown in Figure 7.2B.

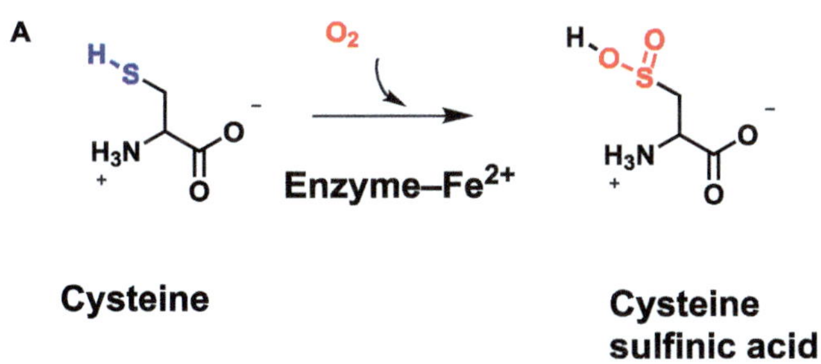

Figure 7.2 Cysteine sulfinate dioxygenase: (A) The enzymatic conversion of cysteine to cysteine sulfinic acid involves a four-electron oxidation of the sulfur atom, with transfer of both oxygen atoms from a single molecule of cosubstrate O_2. (B) The dioxygenase contains an active site nonheme mononuclear iron with, unusually, three histidine ligands from the enzyme as anchors. Whereas most catalysts in this enzyme family use an external 2-ketoglutarate as reductant and generate a high valent $Fe^{IV}{=}O$, the cysteine dioxygenase reaction manifold must transfer both oxygen atoms, perhaps *via* an initial Fe^{III}–superoxide anion species.

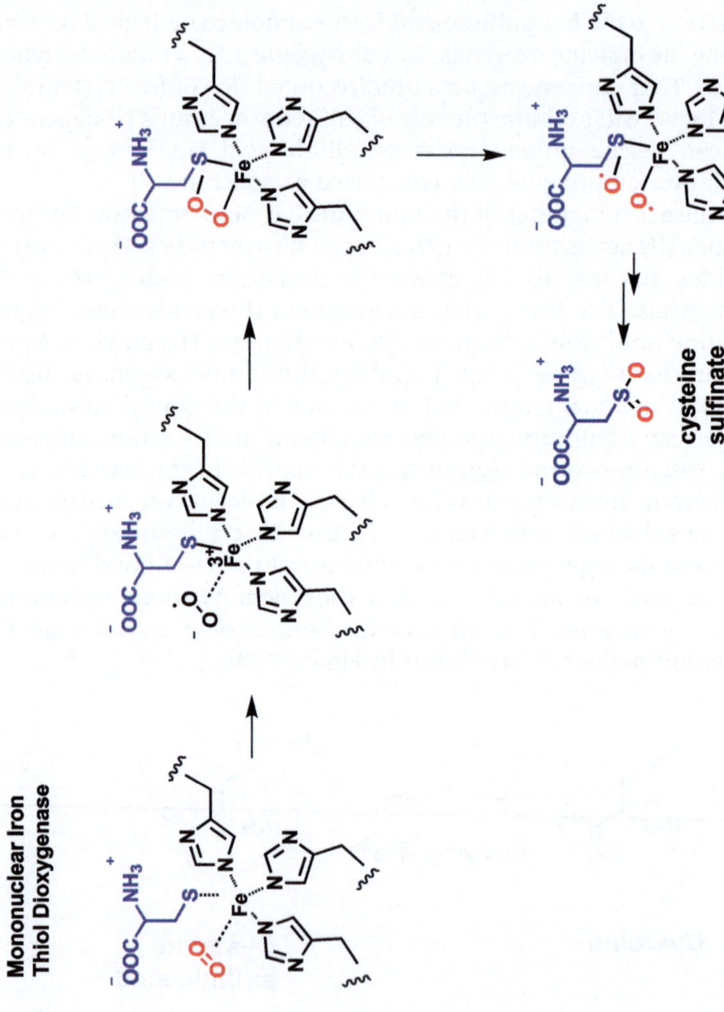

Figure 7.2 Continued.

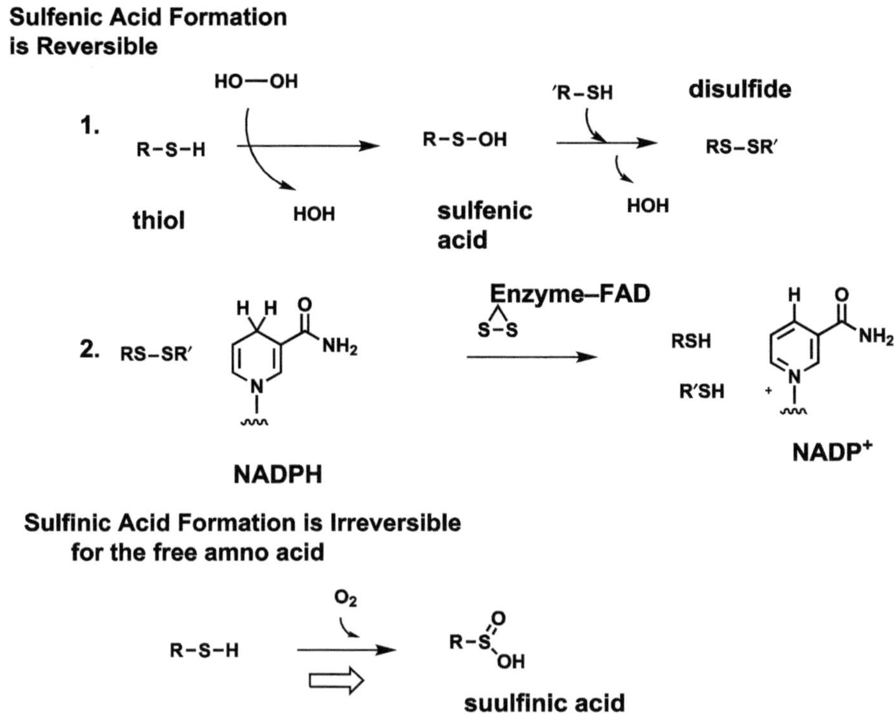

Figure 7.3 Different fates for dioxygenated cysteine in free *vs.* protein-bound contexts.

Given the sluggishness of ground state triplet O_2 to two-electron chemistry, it is likely that the oxygenations proceed by *one-electron transfer*, involving thiol and then sulfenate radicals. Indeed, the reactivity of thiols with O_2 by one-electron manifolds is one of the reasons that high levels of cysteine carry toxic liabilities.

Cysteine sulfinate is not an endpoint metabolite. Rather, it is the substrate for the pyridoxal phosphate (PLP)-dependent cysteine sulfinate decarboxylase.[9] Thus, while cysteine turns out not to be readily decarboxylated in cells (which would *not* have deactivated the nucleophilic thiolate group), its dioxo derivative is. The mechanism is *via* a classical amino-acid–PLP adduct, with the COO^- presumably oriented orthogonal to the plane of the delocalized carbanionic PLP adduct that will develop as the C–COO bond breaks.[10] The products are CO_2 and the aminosulfinate which is known as hypotaurine (Figure 7.4A).

Cysteine sulfinate is also a substrate for transaminases to yield β-sulfinylpyruvate (Figure 7.4A). In back reactions with transaminases this scaffold can undergo a β-elimination to sulfite anion and pyruvate. The sulfite anion is then metabolizable by the sulfite oxidase noted in Chapter 3. This is a route to convert organic sulfur back to inorganic sulfite and then sulfate.

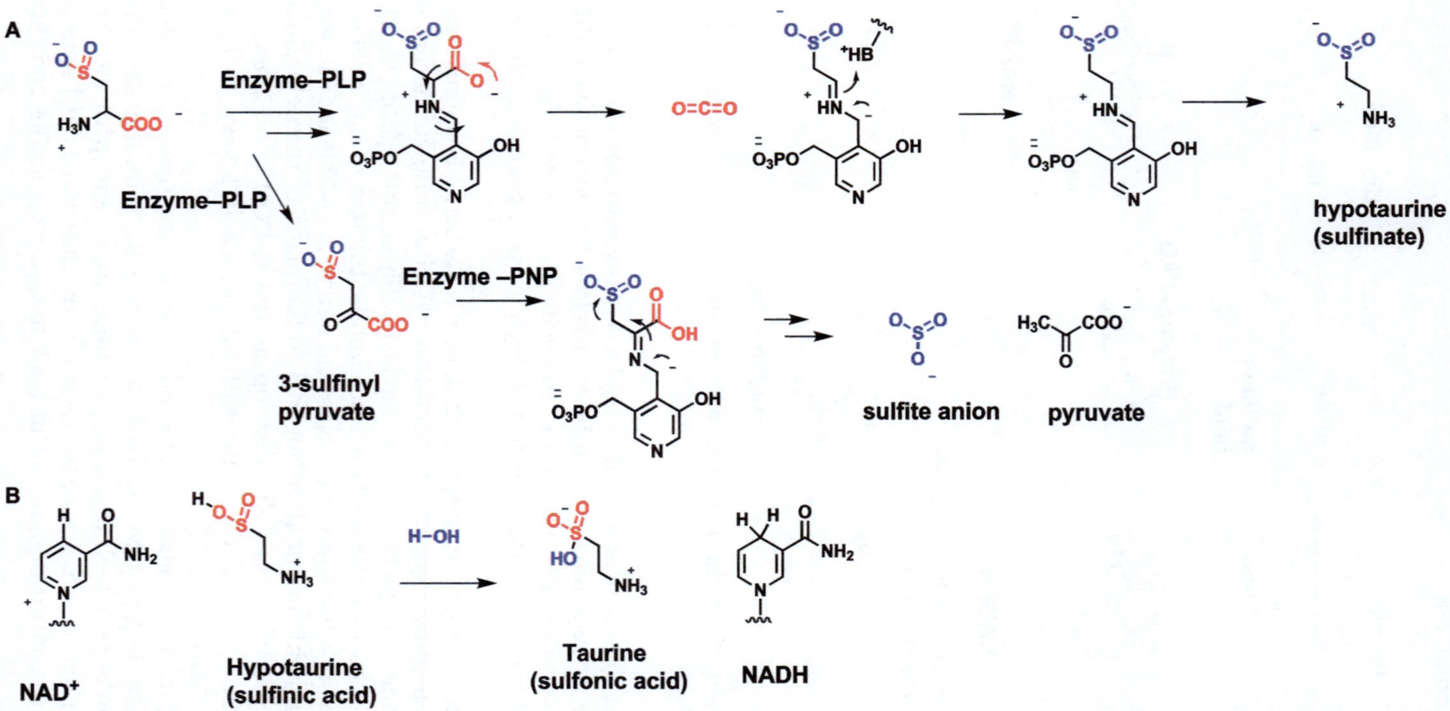

Figure 7.4 Cysteine sulfinate to taurine. (A) Cysteine sulfinate decarboxylase acts as a typical PLP-dependent amino acid decarboxylase *via* the amino acid–PLP aldimine complex to give hypotaurine. The carbanion formed on loss of CO_2 is lowered in energy by delocalization into the PLP electron sink scaffold. Also shown is a transaminase-mediated oxidative deamination of cysteine sulfinate to the sulfinopyruvate which can undergo intramolecular C–S bond fragmentation to sulfite and pyruvate. (B) Oxidation of hypotaurine to taurine is actually an NAD^+-dependent dehydrogenation leading to the (trioxo)sulfonate metabolite taurine, the end product of the pathway.

In Chapter 13 on cysteine proteome posttranslational modifications, we will return briefly to this cysteine dioxygenase (CDO) family of sulfinate-forming nonheme mononuclear iron oxygenases. There, we will note homologous enzymes working, not on free cysteine but on N-terminal cysteinyl residues, to create dioxygenated sulfinate side chains as initiators of enhanced proteasomal degradation. That machinery is termed Cys-dependent *N*-rule degron recognition. It involves the rapid degradation of client proteins with N-terminal cysteinyl residues, and serves as rapid sensing of $[O_2]$ levels in cells as one arm of hypoxia responses.

7.2 Hypotaurine to Taurine

Hypotaurine, in turn, is progenitor to taurine by one further enzymatic oxidation at sulfur, producing the stable $-SO_3^-$ (sulfonate) group of taurine[11] (Figure 7.4B). This third oxygen atom derives from water, not O_2, and the two electrons removed from sulfur are transferred to NAD^+, although mechanistic details are not known. It may be that a formal sufinate sulfur cation is the species captured by water.

The sulfonate moiety imparts aqueous solubility to the product taurine, enabling it to function as a water-soluble, conjugating group for hydrophobic metabolites. Taurine levels are higher than those of many free amino acids in brain, muscle, retina (where it is a cytoprotectant) and other tissues.[12]

Taurine levels are high in meat diets but low in vegetables. Energy drinks can contain in a serving as much as 1.6 micromoles of taurine.[13] Excess taurine can be secreted in urine or conjugated in amide linkage to sterol bile acids. Secretion of bile acids into the intestine is a major route for elimination of the cholesterol scaffold in humans. The two major bile acids in humans are cholic acid and chenodeoxycholic acid (Figure 7.5).

Enzymatic conjugation of the taurine amine to the carboxylate of cholic acid yields taurocholate which is amphipathic, joining hydrophobic and hydrophilic moieties. To activate the carboxyl group for amide bond formation hepatocytes first synthesize the cholyl CoA thioester.[14] This is now activated as an electrophile for amide bond formation by the bile acid aminotransferase that can use either taurine or glycine as the amine co-substrate. Taurine has a fivefold lower Michaelis constant (K_m = substrate concentration that yields half maximal velocity of the enzyme). Thus, the abundance of taurine *vs.* glycine determines the product outcome and is often diet-dependent.

The conjugated bile acid pool in an adult human is about 2–3 g and is estimated to circulate approximately six times a day from liver to gall bladder to intestines and back to liver (the enterohepatic circulation). During the passage through the intestine, resident microbiome bacteria can remove the 7-OH group reductively to form deoxycholate conjugates. Taurocholate is prone to form micelles, with hydrophilic portions oriented to solvent and hydrophobic ones in the interior. Such micelles act as sequestering and solubilizing agents in the gastrointestinal (GI) tract for absorption of dietary lipids.

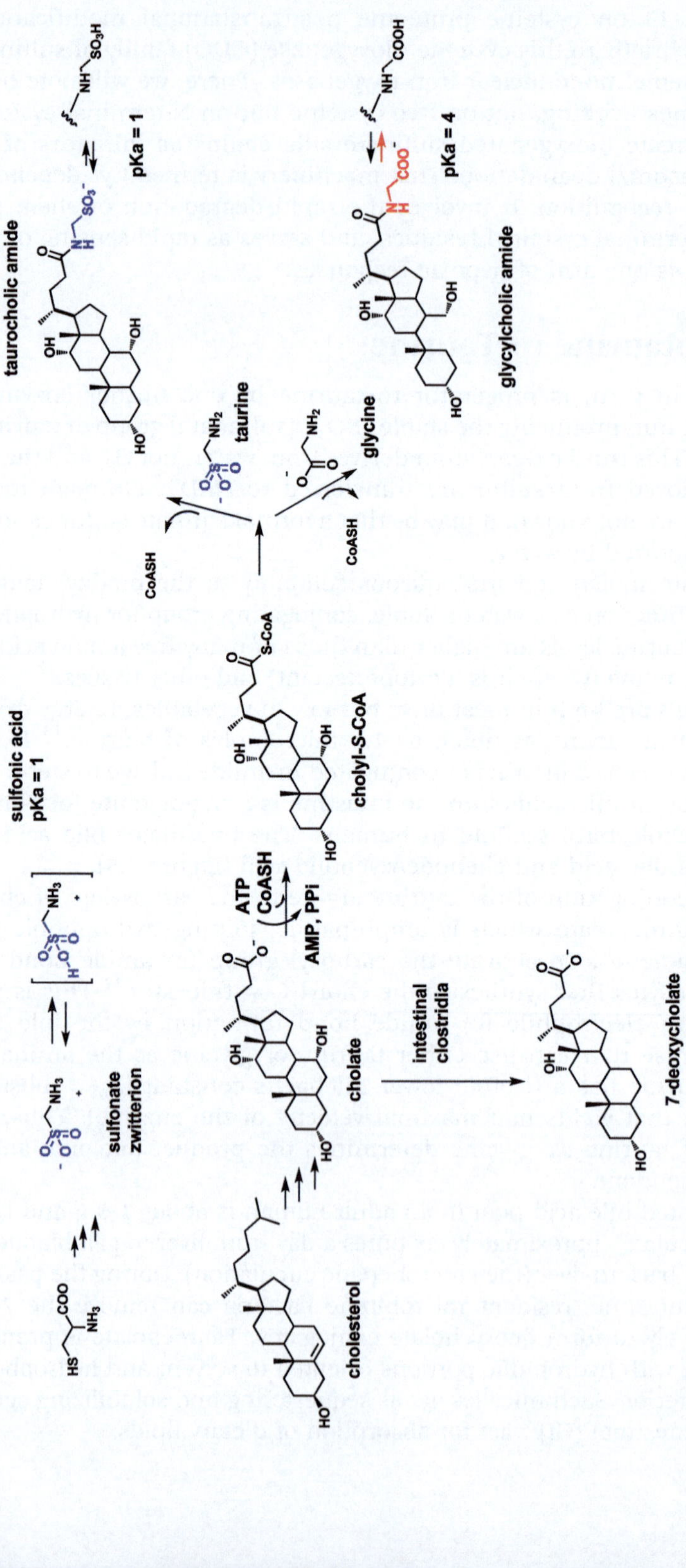

Figure 7.5 Taurine and cholic acid metabolism. Cholate as a major bile acid is activated as the cholyl-*S*-CoA thioester and serves as the electrophilic partner in the amide-bond forming reaction with the amine of taurine to yield taurocholic acid. The amide bond is stable, the strong acid sulfonic acid group remains anionic at all physiological pH values, creating an amphiphilic bile acid amide that recirculates and enables lipid uptake from the gastrointestinal tract.

The sulfonic acid has a pK_a of 1 so even at low pH values in the GI tract the taurocholates are negatively charged. This is in contrast to many carboxylic acids with pK_a values around 4–5. The sulfonate group anion's persistent negative charge at all pH ranges disrupts the ability of the otherwise hydrophobic bile acids from crossing cell membranes and is crucial for micelle formation of the taurocholates and consequently for bile recirculation and lipid uptake.

Figure 7.6 shows three distinct metabolic outcomes for mono-, di- and trioxygenated sulfur derivatives of cysteine. The sequential enzymatic addition of three oxygen atoms to the thiol group of free cysteine has deactivated it as a promiscuous nucleophile and radical generator but leads to different scaffold fates. We have noted that net mono-oxygen transfer to the cysteinyl sulfur to the sulfenic acid oxidation state is common, especially for the high-abundance, freely diffusing glutathione–SH. Sulfenates are in general precursors to disulfide.

As noted in Chapter 5 and in Figure 6.3, the traverse from thiol to sulfonate to disulfide and then back to a pair of thiols indicates that thiol mono-oxygenation is in principle a reversible process, assuming enough dihydronicotinamide as reductant and the availability of disulfide reductases. In Part III on protein disulfide formation we will note that protein linked Cys–S–S–Cys covalent adducts persist when GSH concentrations are low, as in secretory compartments.

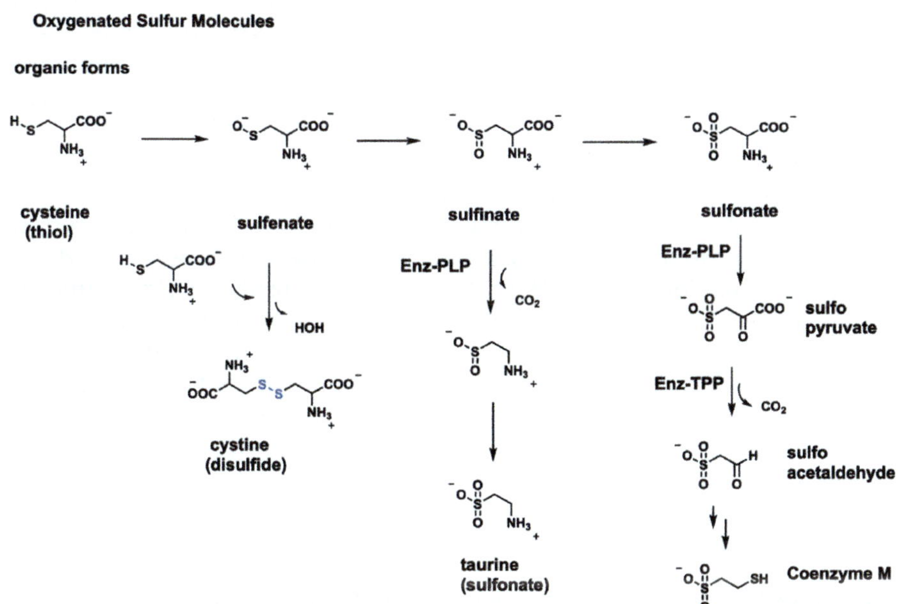

Figure 7.6 Distinct metabolic fates for cysteine sulfenate, cysteine sulfinate and cysteine sulfonate. The sulfonate can yield sulfopyruvate as a precursor of sulfoacetaldehyde and eventually the thiol and sulfonic acid scaffold of coenzyme M in methanogenic bacteria.

Addition of two oxygens to the cysteine sulfur converts the thiol group to the sulfonate. This is irreversible in cells on the free amino acid, (but reversible in some protein contexts, and see Section 7.4 below). It is the major route for control of levels of free cysteine. To avoid having the sulfonate accumulate as an otherwise useless metabolite, cells use a selective PLP-containing decarboxylase that removes CO_2 from the sulfonate framework, but not from free cysteine. The resultant two-carbon sulfonate hypotaurine, is then susceptible to oxidative hydration to the trioxygenated sulfonic acid taurine. Two of the oxygens come from O_2, the third from H_2O, reflecting different underlying logic for getting to the trioxygenated sulfur atom in the sulfonate group of taurine.

7.3 Taurine Amide Conjugation

Taurine is unique as an amino sulfonic acid primary metabolite, a close structural analog of glycine, the corresponding two-carbon aminocarboxylate scaffold. Indeed, glycine and taurine compete as substrates for cholic acid conjugation. Taurine levels can rise to relatively high abundance and are cytoprotective in the retina.[12] The sulfonic acid grouping in taurine is about 1000-fold stronger on the acid scale than the carboxylic acid of glycine (pK_a of 1 *vs.* pK_a of 4). The strength of the sulfonic acid in bile acid conjugates may have a specialized value in the multiple niches of enterohepatic circulation of bile acid-conjugate lipid micelles for fat absorption.

Figure 7.6 also summarizes that some organisms can convert the sulfonate cysteine sulfonate to the trioxygenated sulfonate. With the amino acid moiety still intact, this sulfinic, carboxylate metabolite is a substrate for a PLP-dependent transaminase to yield sulfopyruvate. Now the α-keto acid functionality enables it to serve as substrate for a nonoxidative thiamin-PP-containing decarboxylase. The carboxylate is shed enzymatically as CO_2 and the two-carbon remainder is sulfoacetaldehyde.

This metabolite is an intermediate in methanogenic bacteria to produce the interesting two-carbon thiosulfonic acid known as coenzyme M, an unusual juxtaposition of two sulfurs at such different oxidation states (thiol *vs.* sulfonic acid). M in this context stands for methanogenesis.[15] Coenzyme M is a carrier for the one-carbon methyl group, as methyl-*S*-CoM, that acts as donor in the last reductive step to produce methane gas by methanogenic bacteria. In contrast to the O–SO_3 linkage in the sulfuryl transfer metabolite 3′-phosphoadenosine-5′-phosphosulfate (PAPS) noted in Chapter 10, the C–S bond in C–SO_3^- metabolites is stable. Taurine, for example, is not an –SO_3^- donor because the C–SO_3^- is not chemically activated.

7.4 Cysteine to Sulfoxide to Sulfenate to Thiocarbonyl Group

In Chapter 5 in the section on low-molecular-weight thiol alternative structures to glutathione we noted the fungal metabolite ergothioneine

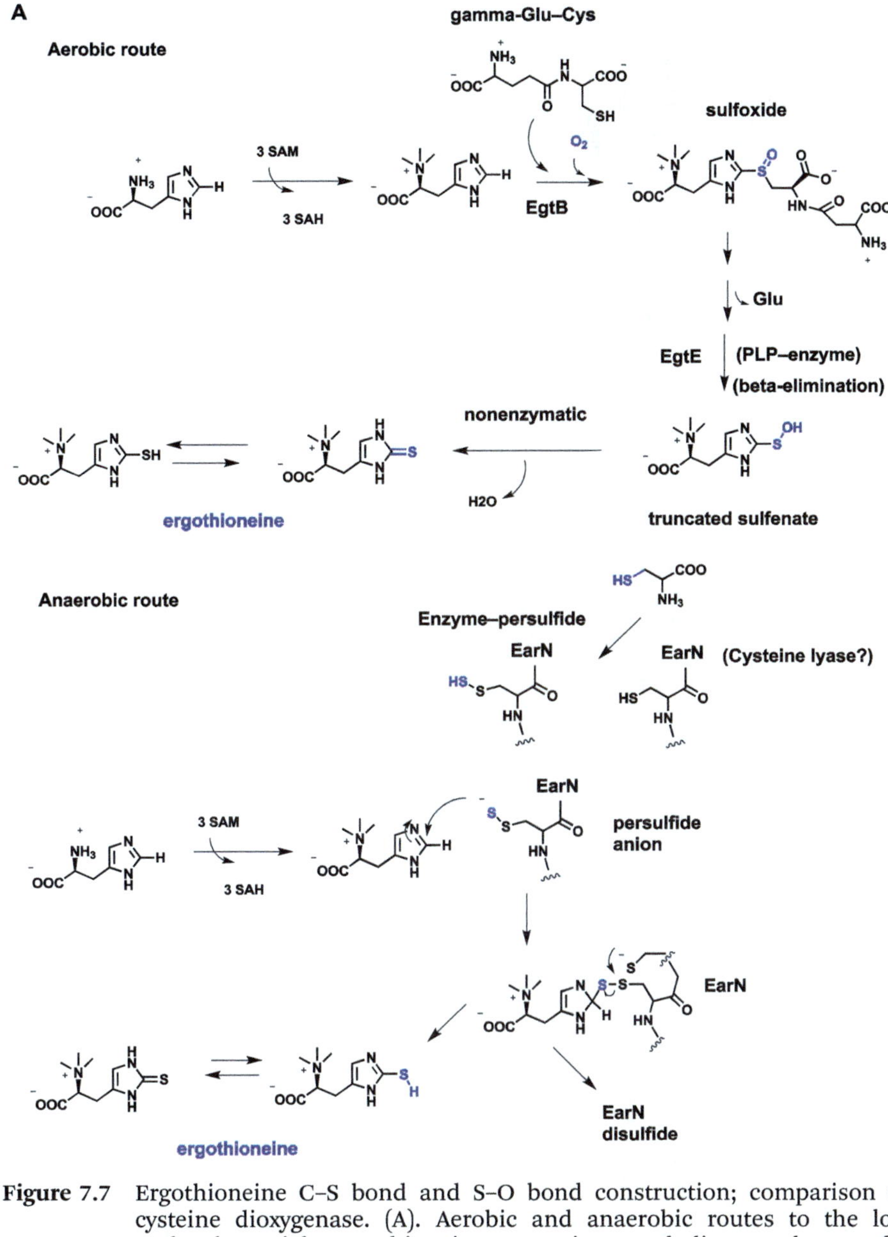

Figure 7.7 Ergothioneine C–S bond and S–O bond construction; comparison to cysteine dioxygenase. (A). Aerobic and anaerobic routes to the low molecular weight ergothioneine protective metabolite are shown. The aerobic route goes *via* an *S*-oxygenated intermediate while the anaerobic route goes *via* an oxidized persulfide intermediate. (B). Both CDO and EgtB are nonheme iron oxygenases with a triad of histidine residues in the iron-coordination shell. EgtB catalyzes C–S bond formation and one *S*-oxygenation while CDO catalyzes two S–O oxygenations. The distinction may occur at the stage of the initial high valency oxo-iron intermediates.

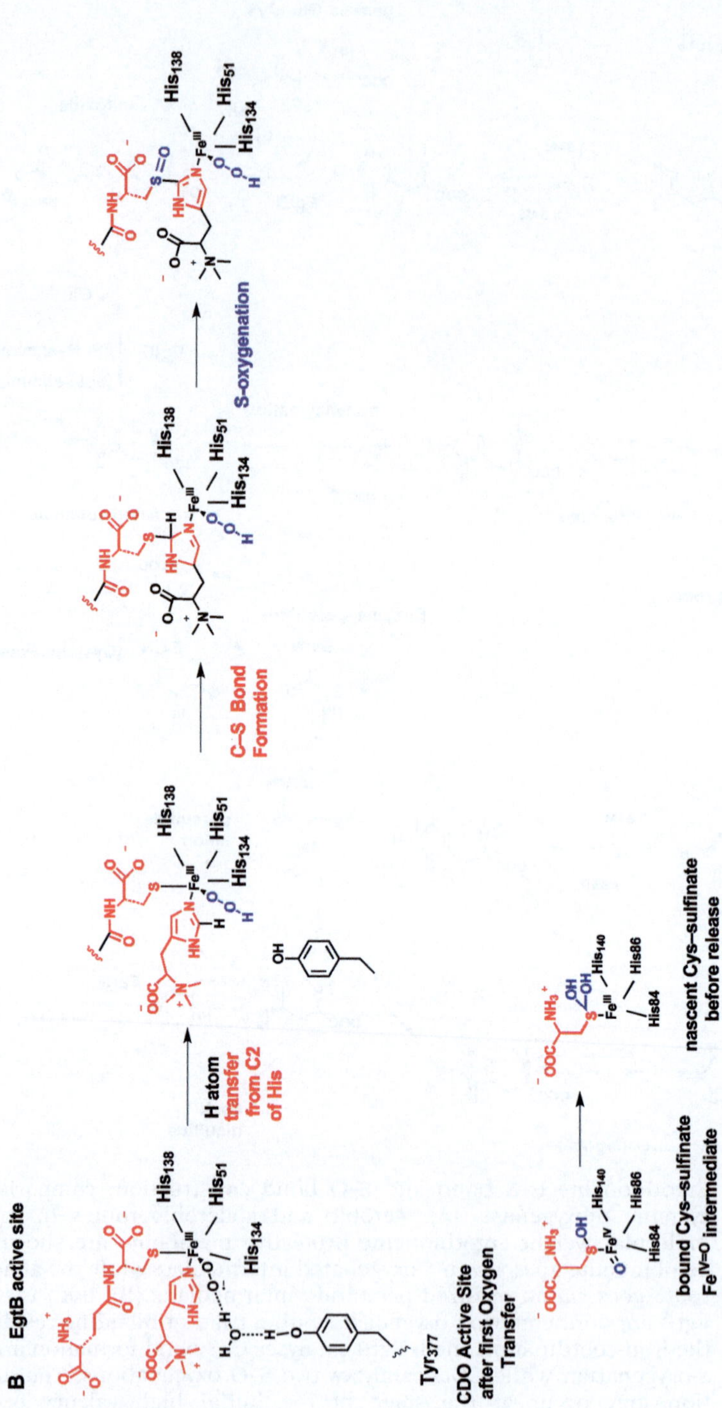

Figure 7.7 Continued.

(2-mercaptohistidine trimethylbetaine) (Figure 5.8) is a protectant against oxidants,[16] without comment on how the thiocarbonyl gets installed on the imidazole ring of histidine. Figure 7.7 shows that one can write interconverting structures with a thio-imidazole ring or a thiono-dihydroimidazole, emphasizing the tautomeric relationship. Recent studies have uncovered two different routes, one aerobic, the other anaerobic and presumably earlier in evolution, to biosynthetic installation of the sulfur atom, albeit both probably arise from cysteine sulfur.[17,18]

The aerobic route in fungi and mycobacteria is known to proceed by trimethylation of the histidine amino nitrogen and then an O_2-dependent introduction of the cysteinyl scaffold at the C2 carbon of the imidazole ring as the sulfoxide. The actual substrate is the dipeptide γ-glutamylcysteine as shown.

Ergothioneine biosynthesis protein EgtB catalyzes two sequential reactions in ergothioneine biosynthesis (Figure 7.7). First is the C–S bond-forming step, followed by *S*-oxygenation to the sulfoxide. Intriguingly, both EgtB and cysteine dioxygenase (CDO) share underlying catalytic logic, as nonheme iron oxygenases with the unusual set of three His ligands in the first coordination shell of the catalytic iron atom.[19] A tyrosine phenol side chain in EgtB is thought to route the flux of iron oxo intermediates away from the canonical $Fe^{IV}{=}O$ intermediate instead to an Fe^{III}–OOH species where the S–C bond has formed (Figure 7.7B). The two enzymes are thought to each carry out subsequent oxygen transfer to the substrate sulfur. For cysteine dioxygenase this is a second oxygen transfer to sulfur converting the sulfenic acid intermediate to the nascent sulfinate product. For EgtB this first *S*-oxygenation step converts the thioether to the γ-Glu–Cys sulfoxide. Two distinct product fates, C–S and an S–O bond formations in the EgtB active site *vs.* two S–O bond formations in the CDO active site, show the subtle ways that active site architectures can alter reaction flux. The C–S bond formation by EgtB action is both rare and the mechanism unusual and may occur in one-electron steps (not shown).

In the ergothioneine pathway the three-carbon cysteine scaffold is removed by C3–S bond cleavage in a net β-elimination reaction by a PLP enzyme EgtE (Figure 7.7A). This action has shortened the side chain to an S–OH sulfenate. This ergothioneine precursor can undergo nonenzymatic elimination of water to the end product ergothioneine. The interconversion between dihydroimidazole–thiol and imidazolothione focuses attention on the oxidation state of the cysteine-derived sulfur.

Intriguingly, studies of anaerobic green algae reveal that they too can synthesize ergothioneine, but in an O_2-independent pathway.[18] Although details are still to be worked out, it is reasonable that the donor of the sulfur atom is still cysteine but now by action of a cysteine lyase enzyme catalyst of the type that liberates six-electron sulfur for Fe–S cluster assembly in anaerobes. The presumed enzyme persulfide, electrophilic when protonated, is nucleophilic when deprotonated (pK_a approximately 6) and could attack C2 of the trimethylhistidine ring as shown in Figure 7.7B.

To get from there to ergothioneine, requires S–S bond cleavage, with the terminal sulfur departing with the electron pair in that S–S bond. We have seen that enzymes can obtain this polarity of cleavage by disulfide formation. This proposed, more ancient, route to ergothioneine shows the convergence of anaerobic persulfide biochemistry *vs.* aerobic sulfoxide-sulfenate chemistry to form C–S bonds.

References

1. V. Gupta and K. Carroll, Sulfenic acid chemistry detection and cellular lifetime, *Biochim. Biophys. Acta*, 2014, **1840**, 847–875.
2. M. L. Conte and K. Carroll, The redox biochemistry of protein sufenylation and sulfinylation, *J. Biol. Chem.*, 2013, **288**, 26480–26488.
3. M. H. Stipanuk, J. J. Dominy, J.-I. Lee and R. M. Coloso, Mammalian cysteine metabolism: new insights into the regulation of cysteine metabolism, *J. Nutr.*, 2006, **136**, 1652S–1659S.
4. J. Yang, K. Carroll and D. C. Liebler, The expanding landscape of the thiol redox proteome, *Mol. Cell. Proteomics*, 2016, **15**, 1–11.
5. C. A. Joseph and M. Maroney, Cysteine dioxygenase: structure and mechnism, *Chem. Commun.*, 2007, 3338–3349.
6. S. Ye, X. Wu, L. Wei, D. Tang, P. Sun, M. Bartlam and Z. Rao, An insight into the mechanism of human cysteine dioxygenase: key roles of the thioether-bonded tyrosine-cysteine cofactor, *J. Biol. Chem.*, 2007, **282**, 3391–3402.
7. A. J. Crisjoe and M. Maroney, Cysteine dioxygenase structure and mechansim, *Chem. Commun.*, 2007, 3338–3349.
8. L. Wenning, N. Stoveken, J. H. Wubbler and A. Steinbuchel, Substrate and cofactor range differences of two cysteine dioxygenases from *Ralstonia eutropha* H16, *Appl. Environ. Microbiol.*, 2016, **82**, 910–921.
9. M. Tappaz, K. L. F. Alrmarghini and A. Remy, Taurine biosynthesis enzyme cysteine sulfinate decarboxylase (CSD) from brain: the long and tricky trail to identification, *Neurochem. Res.*, 1992, **17**, 849–859.
10. H. C. Dunathan, Conformation and rection specificity in pyridoxal phosphate enzymes, *Proc. Natl. Acad. Sci. U. S. A.*, 1966, **55**, 712–716.
11. V. Vitvitsky, S. Garg and R. Banerjee, Taurine biosynthesis by neurons and astrocytes, *J. Biol. Chem.*, 2011, **286**, 32002–32010.
12. H. Ripps and W. Shen, Taurine: a very essential amino acid, *Mol. Vision*, 2012, **18**, 2673–2686.
13. J. M. Ridlon, P. Wolf and H. R. Gaskins, Taurocholic acid by gut microbes and colon cancer, *Gut Microbes*, 2016, **7**, 201–215.
14. J. B. Wheeler, D. Shaw and S. Barnes, Purification and characterization of rat liver bile acid CoA ligase, *Arch. Biochem. Biophys.*, 1997, **348**, 15–24.
15. D. E. Graham, S. Taylor, R. S. Wolf and S. C. Namboori, Convergent evolution of coenzyme M biosynthesis in Methanosarcinales: cysteate synthase evolved from threonine synthase, *Biochem. J.*, 2009, **424**, 467–478.

16. K. J. Sheridan, B. Lechner, G. O'Keefe, M. A. Keller, E. R. Werner, H. Lindner, G. W. Jones, H. Haas and S. Doyle, Ergothioneine biosynthesis and functionality in the opportunistic fungal pathogen *Aspergillus fumigatus*, *Sci. Rep.*, 2016, **6**, 35306.
17. F. P. Seebeck, *In vitro* reconstitution of mycobacterial ergothioneine biosynthesis, *J. Am. Chem. Soc.*, 2010, **132**, 6632–6633.
18. R. Burb, L. Misson, M. Meury and F. P. Seebeck, Anaerobic Origin of Ergothioneine, *Angew. Chem., Int. Ed.*, 2017, **56**, 12508–12511.
19. A. S. Fapponle, F. P. Seebeck and S. P. de Visser, Sulfoxide synthase vs cysteine dioxygenase reactivity in a nonheme iron enzyme, *J. Am. Chem. Soc.*, 2017, **139**, 9259–9270.

CHAPTER 8

Methionine: Thioether Biological Chemistry

8.1 Introduction

In comparison to the reactive thiolate side chain of cysteine, methionine with its sulfur in *thioether* linkage is much less reactive and does not pose the same kind of promiscuity and toxicity as a nucleophile. In protein biosynthesis it is specified by the single mRNA codon AUG and that is most often the start codon for proteins so methionine is most frequently the *N*-terminal amino acid in nascent proteins as they roll off the ribosome. There are methionyl aminopeptidases that cleave the *N*-terminal methionine residue cotranslationally in a significant subset of proteins to liberate the second amino acid in the primary sequence as the new *N*-terminus. For internal methionyl residues in proteins there is almost no catalytic action or posttranslational modifications. *S*-oxygenation at the thioether sulfur is discussed later in this chapter and in Chapter 13.

The most remarkable reaction of the thioether sulfur in the methionine side chain is its function as a nucleophile towards carbon 5′ of the ribose moiety of ATP. This generates *S*-adenosylmethionine and carries the methionine scaffold to a distinct chemical biology universe from that of cysteine as will be discussed in Section 8.3.

Animals lack the full biosynthetic pathway to make methionine: for humans it is one of the nine essential amino acids that must be provided in the diet. Humans do have the last enzyme in the pathway, methionine synthase, as noted in the next section.

Chemical Biology No. 12
The Chemical Biology of Sulfur
By Christopher T. Walsh

Published by the Royal Society of Chemistry, www.rsc.org

8.2 Methionine Biosynthesis in Plants and Microbes

The full biosynthetic logic used in plants and microbes is to start with L-aspartate [available from the citrate cycle central metabolite oxaloacetate by pyridoxal phosphate (PLP)-dependent transaminase action].[1] The C4 carboxylate of aspartate is activated as the mixed acyl–phosphoric anhydride by an ATP–dependent aspartate kinase. In this pathway, that acyl phosphate undergoes consecutive hydride-mediated reductions from two equivalents of NADH (passing through the aldehyde) to generate homoserine (Figure 8.1A).

The homoserine to homocysteine conversion is logic we have seen previously in the transsulfuration pathway to get to cysteine. In this route, succinyl CoA is used as the active acylating cosubstrate to form *O*-succinyl homoserine, setting the C4–O bond for a low energy cleavage route. The catalyst is a PLP-dependent cystathionine-γ-synthase, a parallel to the cystathionine β-synthase in the cysteine-forming pathway (Figure 8.1B). Cystathionine is then resolved by a companion PLP-dependent lyase enzyme. Again, this is now a β-lyase rather than the γ-lyase in the cysteine transsulfuration pathway. The result is homocysteine, ready for methylation by the methionine synthases noted in the next section. Note that in the cysteine biosynthetic pathway homocysteine is the sulfur donor to the C3 scaffold. In the methionine biosynthetic pathway cysteine is the sulfur donor to the C4 scaffold. Thus, cells have devised tandem pairs of PLP–enzymes that make and break the seven-carbon cystathionine framework with exquisite regioselectivity to build up either cysteine or homocysteine.

8.3 Methionine Biosynthesis from Homocysteine: The Last Step in the Pathway

Conversion of homocysteine to methionine involves transfer of a one-carbon methyl group from a cosubstrate that can act as a net methyl group donor. That process can be written as attack of the homocysteine thiolate anion on a methyl cation equivalent in the transition state. There are two types of methionine synthases known.

The simplest version involves direct transfer from the methyl donor substrate *N*-5-methyl tetrahydrofolate (Figure 8.2A). This is the enzyme form used by plants.[1] The tetrahydro oxidation state in the right hand pterin ring is the biologically active form of folate, vitamin B_9. In that oxidation state N5 of the pterin ring is basic and capable of capturing nascent formaldehyde released in the enzymatic fragmentation of L-serine to the two-carbon product glycine and a one-carbon aldehyde, formaldehyde, derived from C3 of serine. The adventitously reactive formaldehyde is sponged up, by N5 and the side chain N10 of the *para* aminobenzoyl moiety, as the N5,10-methylene adduct. The formaldehyde is latent, reversibly stored as the strain-free five membered ring-an unreactive adduct.

Enzymatic reduction *via* hydride transfer from NADPH yields the N-5–CH_3–tetrahydrofolate (THF) metabolite. Direct attack of the homocysteine thiolate

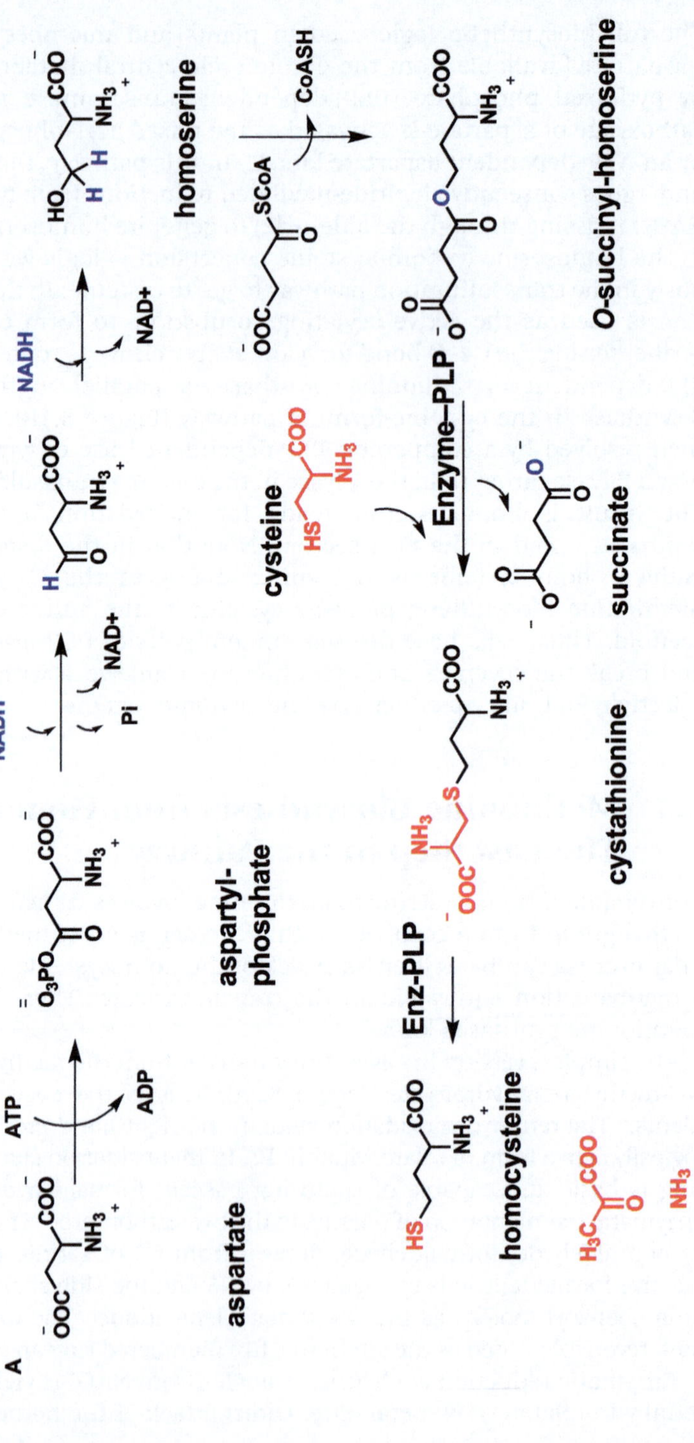
A
aspartate
ATP
ADP
aspartyl-phosphate
NADH
Pi
NAD+
NADH
NAD+
homoserine
SCoA
CoASH
O-succinyl-homoserine
cysteine
Enzyme–PLP
succinate
cystathionine
Enz-PLP
homocysteine

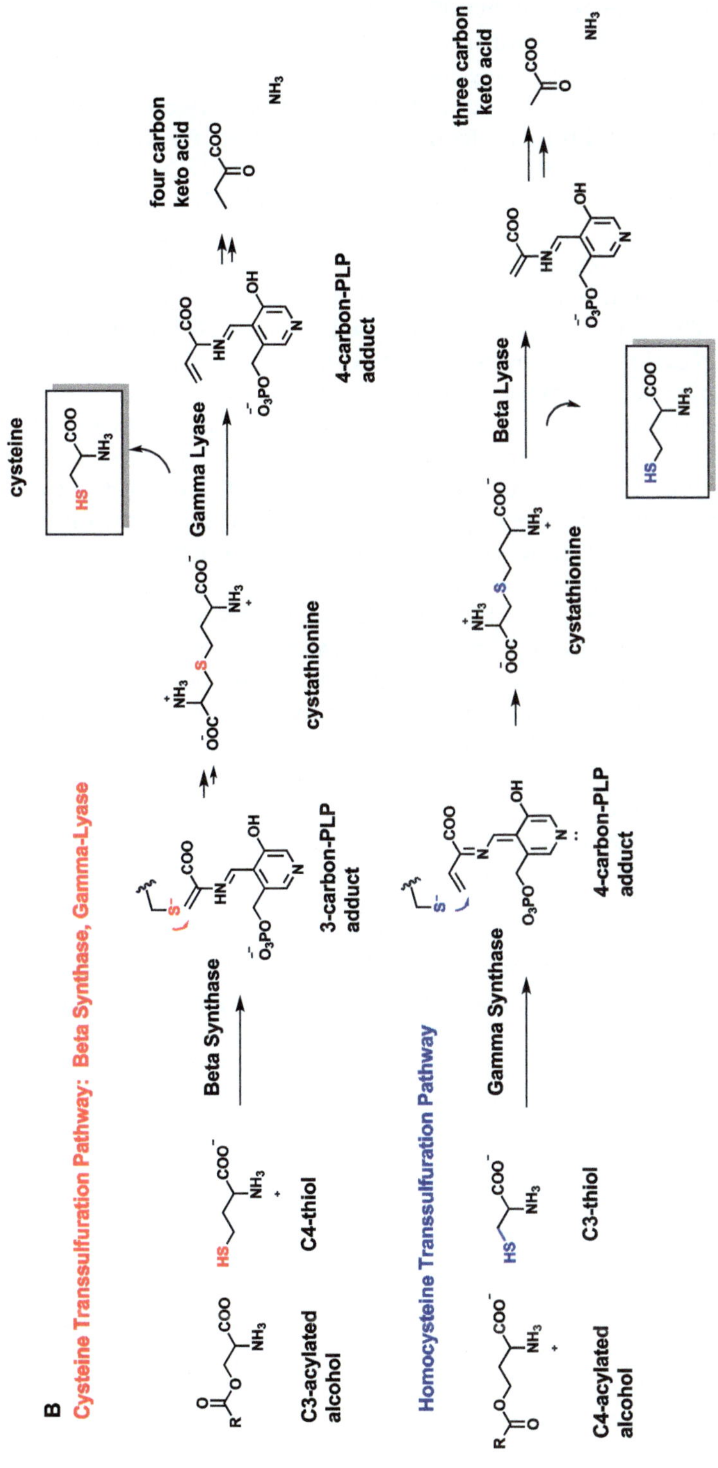

Figure 8.1 Methionine biosynthetic route in plants and microbes. (A) Generation of homoserine. (B) Cysteine-forming pathway.

anion on the methyl of the N5–methyl–THF cosubstrate produces L-methionine and the THF coproduct. This form of methionine synthase is termed the coenzyme B_{12}-independent version (Figure 8.2).

The B_{12}-dependent version carries out the same reaction stoichiometry but involves intermediate methyl transfer from N-5–CH_3–THF to the cobalt atom in the equatorial plane of the B_{12} corrin ring system, tightly held in non-dissociating form in the synthase active site[2] (Figure 8.2B). The resting stage Co^{III} form of the B_{12} coenzyme is primed by a reductase to yield the super-nucleophilic Co^{I} oxidation state. This attacks the N5–CH_3–THF to transfer the CH_3 as a methyl cation equivalent, yielding the CH_3–Co–B_{12} as an isolable intermediate. The CH_3 is then transferred, again as a [CH_3^{+}] equivalent, in the second half reaction to the homocysteine thiolate. The Co^{I} form of the B12-synthase complex is competent to perform subsequent methyl transfer cycles, unless adventitiously autoxidized. Adventitious oxidation to the Co^{II} intermediate oxidation state occurs about once every 2000 turnovers and requires a rescue redox system to reduce the central cobalt back to the Co^{I} oxidation state. This repair is necessary since tens of billions of methionine molecules are needed every cell cycle, both for protein biosynthesis and for *S*-adenosyl methionine reactions.[3]

Methionine synthase is one of two B_{12}-dependent reactions in humans. The other is methylmalonyl CoA mutase,[4] rearranging methylmalonyl CoA to succinyl CoA by radical chemistry that we will touch on in the next chapter on *S*-adenosylmethionine radical chemical biology. B_{12} is the only cobalt-containing molecule in primary metabolism. Its biosynthetic route diverges from heme at the level of the common tetrapyrrolic precursor uroporphyrinogen III. Compared to siroheme, in sulfite reductases and to heme itself, the corrinoid frameworks are neutral, with the acetic and propionic acid chains converted to amides. Most notable has been a contraction of one of the methylene bridges between two of the pyrrole rings, creating a macrocyclic cavity that optimally fits Co as the central equatorial ligand to the four pyrrolic nitrogens.

Cobalt can access three oxidation states in the B_{12} microenvironment. In air the Co^{III} form is stable and B_{12} is isolated with the top axial ligand as the aquo complex. It can also trap adventitious amounts of cyanide ion and be isolated as cyanocobalamin (Figure 8.2B). The bottom axial ligand is the indicated dimethylbenzimidazole nitrogen tethered at the end of a long substituent chain. In some B_{12} enzyme active sites, including methionine synthase, that bottom axial ligand can be replaced by the imidazole group of a histidine residue with changes in redox potential of the cobalt atom.

The air-stable Co^{III} is not a relevant starting redox state for B_{12} biology. The one-electron-reduced Co^{II} state is involved in B_{12}-dependent enzyme-catalyzed scaffold rearrangements as in the methylmalonyl CoA mutase noted above. It is the two-electron-reduced Co^{I} oxidation state that is active as a nucleophile in methionine synthase. While Co^{I} has been characterized as a super nucleophile, it is susceptible to rapid one-electron oxidation by O_2 such that in contemporary aerobic organisms the methionine synthase active site is the only locus where it is known to function in metabolism.[4,5]

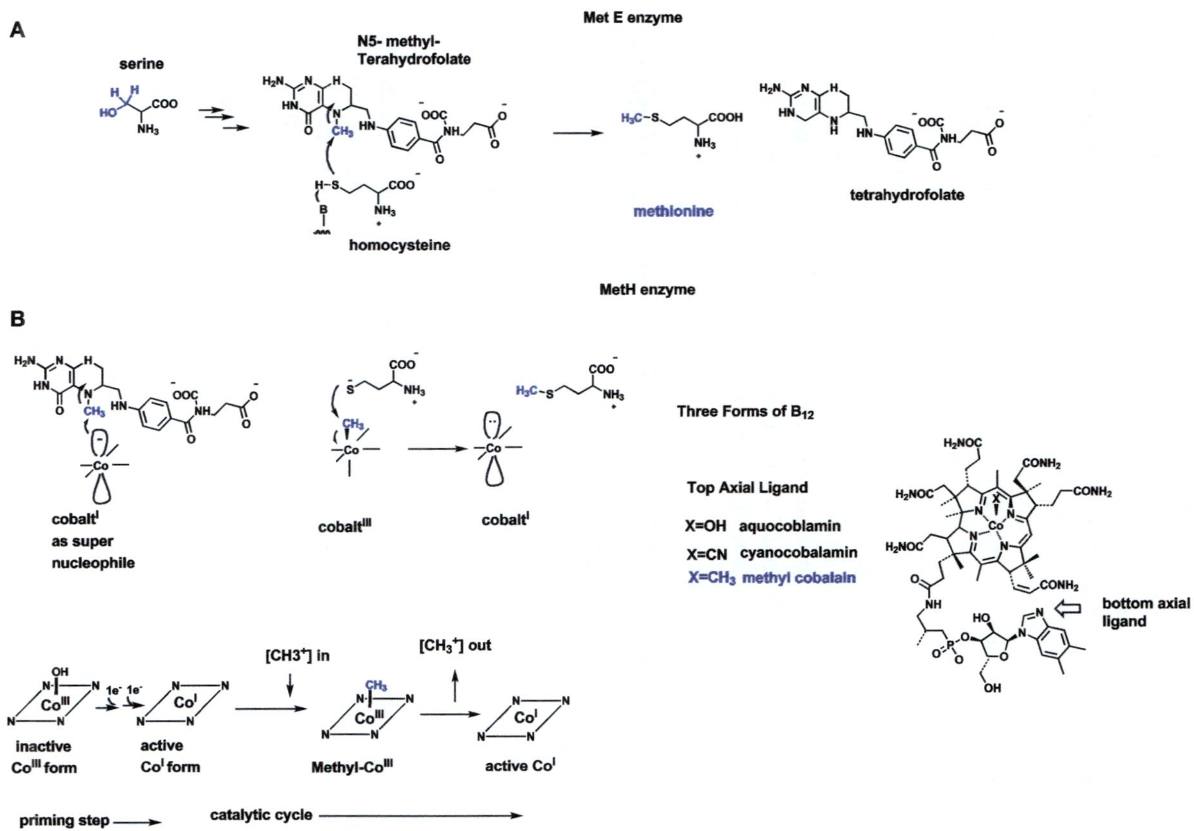

Figure 8.2 Two methionine synthases. (A) Direct transfer *vs.* (B) Methyl-B$_{12}$ intermediate.

The dichotomy between the MetE and MetH enzymes may reflect a holdover from anaerobic eons. In anaerobes, the control of reduced Co^{I} may have been easier to manage. In turn, the super nucleophilicity of $cobalt^{I}$ may have facilitated the transfer of a $[CH_3^+]$ equivalent from N5 of N5–CH_3–tetrahydrofolate. The second half of both the MetE and MetH reactions involves transfer of the methyl group, again as a $[CH_3^+]$ equivalent in both cases, to the thiolate anion of homocysteine. The evolution of an active site in MetE that dispensed with the binding of B_{12}, the need to pump it down from Co^{III} to Co^{I} and the intermediacy of the organometallic CH_3–Co methyl B_{12} intermediate, may have simplified the catalytic machinery to a point where the direct attack of the homocysteine thiolate on the relatively stable N5–CH_3 of tetrahydrofolate gave overall advantage to the direct methyl transfer mechanism.

Because of its dual role as the chain initiation monomer in protein biosynthesis and a source of billions of methyl groups *via* *S*-adenosylmethionine, methionine availability can be rate-limiting in some rapidly dividing cell populations. That appears to be the case for T cells undergoing rapid expansion as the immune system responds to infections or tumors. Such activated T cells avidly take up methionine to enable proliferation to carry out their immune functions.[6]

8.4 *S*-Adenosylmethionine: Thioether to Trivalent Sulfonium Cation

Beyond its role in protein biosynthesis, the dominant role of methionine in metabolism is its conversion to the premier biologic methylating reagent *S*-adenosylmethionine (SAM) (Figure 8.3). Hundreds of millions to billions of methyl group transfers in every mammalian cell cycle originate with SAM as methyl donor.

The responsible enzyme is *S*-adenosylmethionine synthase.[7,8] It catalyzes an almost unique reaction in the long list of ATP enzymatic fragmentations: net adenosyl transfer to the incoming nucleophile. The other known example is where the super nucleophilic cobalt (Co^{I}) oxidation state in vitamin B_{12} attacks the same C5′ of the ribose moiety of ATP to create adenosyl-B_{12} (see next chapter). Even though the C5′–O bond in ATP is not notably electrophilic, it is reasonable that supernucleophilic Co^{I} can break that bond (adenosyl–B_{12}). It is still surprising that the thioether sulfur of the methionine side chains can be coaxed into such powerful nucleophilic behavior (*S*-adenosylmethionine). It may be that there is an early cleavage of the C5′–O bond of bound ATP that is then captured by the thioether sulfur of methionine.[7] This is clearly a particular feature of the SAM synthetase active site, as the methionine thioether sulfide is minimally promiscuous elsewhere in cells.

As methionine is adenosylated on its sulfur atom by the enzyme SAM synthetase, the divalent thioether sulfur atom becomes trigonalized and attains a positive charge as a sulfonium cation (Figure 8.3). The synthetase is unusual in multiple respects.

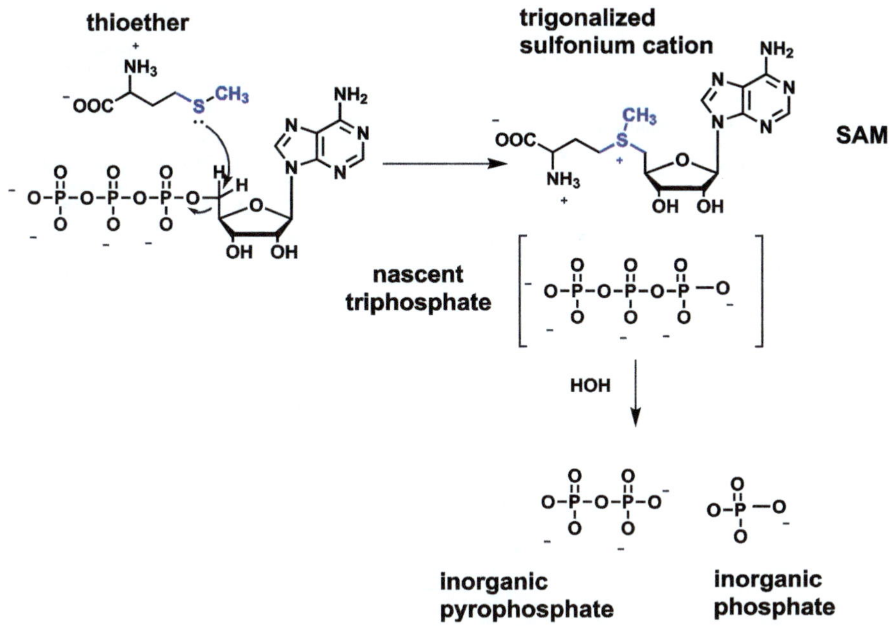

Figure 8.3 *S*-Adenosylmethionine synthase: unusual regioselectivity of C5′–S bond formation.

First is the directed attack on the ribose C5′ carbon. In addition, to draw off the equilibrium in the forward direction, the nascent inorganic triphosphate (PPPi) is not released but instead is cleaved regiospecifically to inorganic pyrophosphate (PPi) and Pi, adding a 10^5 kick in the equilibrium constant for SAM formation.

8.5 SAM as an Electrophilic Group Transfer Substrate

The consequences of converting the neutral thioether sulfur to the permanently trigonalized sulfonium cation are monumental. Positive charge on the trivalent sulfur in SAM provides thermodynamic activation for each of the three groups attached to that sulfur[7] (Figure 8.4).

For example, the synthase back reaction establishes that the adenosyl group can be transferred as an electrophilic fragment to one of the oxygen anions of inorganic triphosphate (Figure 8.5) Likewise, the queuosine base in certain tRNAs derives from the adenosyl moiety of SAM.

The biosynthesis of the seven-carbon triamine spermidine occurs after separate enzymatic decarboxylation of two amino acids, ornithine and SAM. Ornithine decarboxylase yields the four-carbon 1,4-diaminobutane. SAM decarboxylase yields the aminopropyl side chain of descarboxy-SAM. Catalyzed attack of one of the primary amines of the diaminobutane on the C3 carbon of the aminopropyl chain of descarboxy-SAM yields spermidine and thiomethyladenosine (thioether) as the other product.

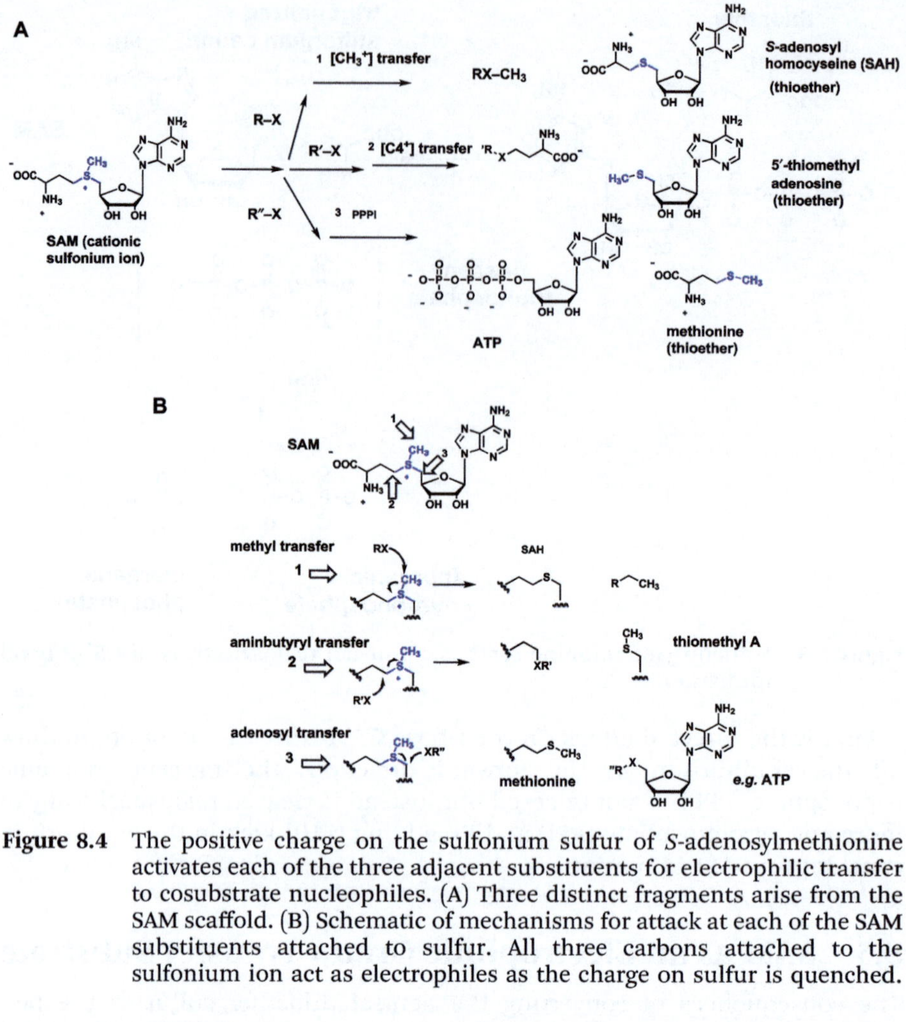

Figure 8.4 The positive charge on the sulfonium sulfur of *S*-adenosylmethionine activates each of the three adjacent substituents for electrophilic transfer to cosubstrate nucleophiles. (A) Three distinct fragments arise from the SAM scaffold. (B) Schematic of mechanisms for attack at each of the SAM substituents attached to sulfur. All three carbons attached to the sulfonium ion act as electrophiles as the charge on sulfur is quenched.

In the biosynthesis of the fruit ripening hormone ethylene by plants, an intriguing conversion of SAM to 1-amino-1-carboxy-propane is the first step and involves cyclizing disconnection of the aminobutyryl side chain from SAM.[8] This requires generation of a C2 carbanion in a PLP adduct to set up the internal capture of C4 of the methionyl side chain (Figure 8.5).

8.5.1 Transfer of Electrophilic One-carbon $[CH_3]^+$ Fragment from SAM to Cosubstrate Nucleophiles

It is the third substituent, the one-carbon methyl group on the sulfonium sulfur of SAM that is the main group transferred in dozens of methyl transferase reactions (Figure 8.6). A wide range of enzymes catalyze electrophilic

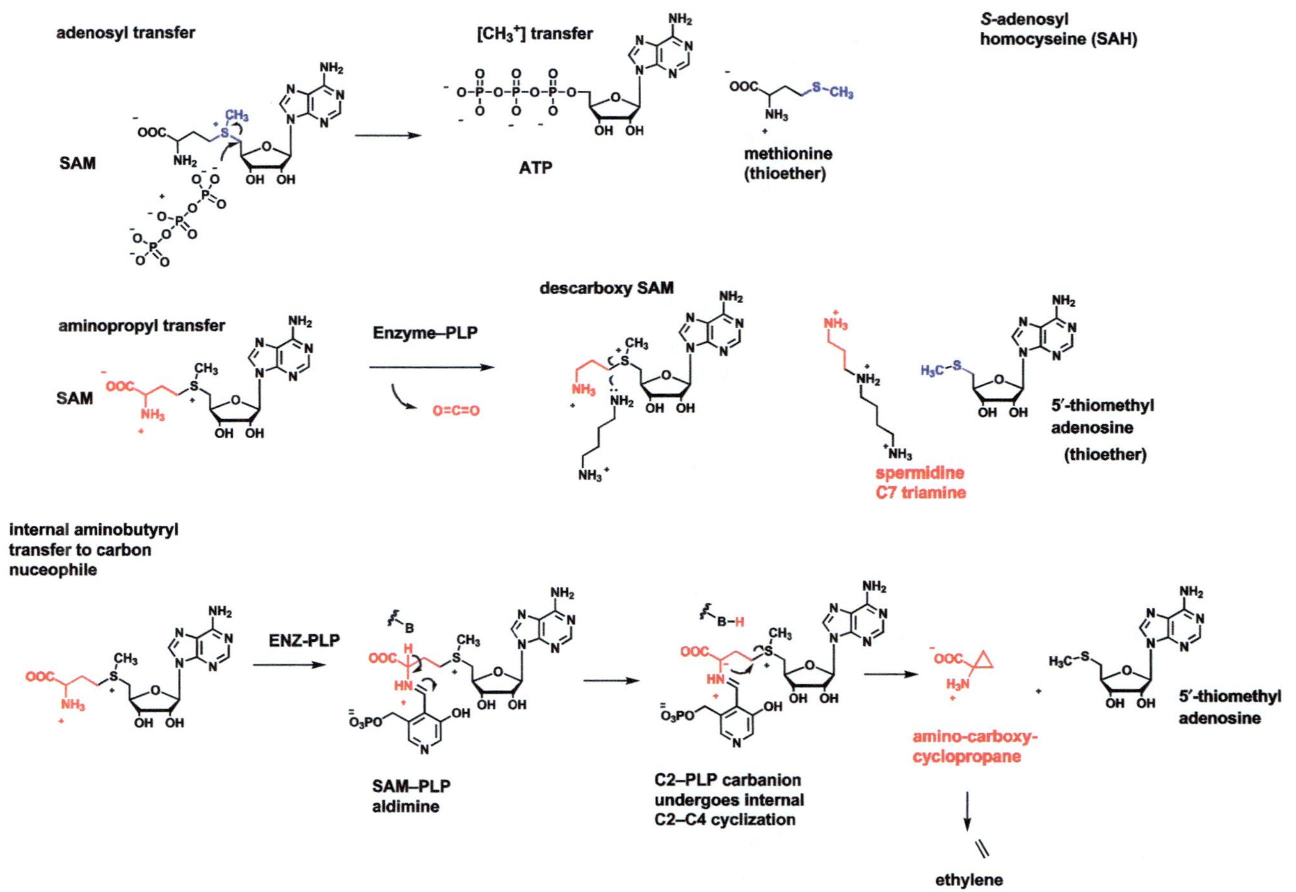

Figure 8.5 Three specific examples of SAM as electrophilic fragment donor: ATP synthesis; spermidine synthesis; aminocarboxycyclopropane biosynthesis by intramolecular cyclization of the aminobutyryl side chain of SAM.

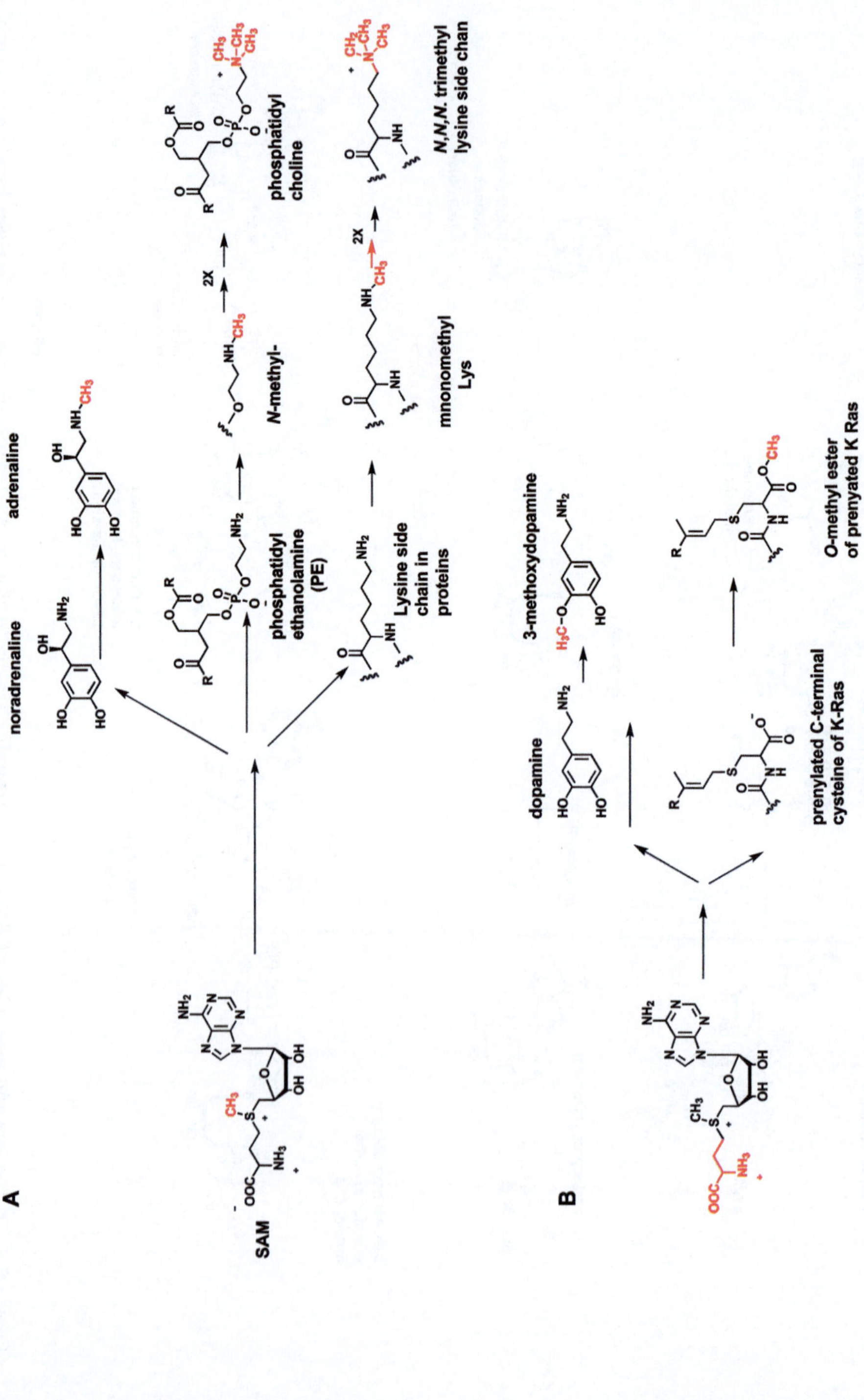
A
SAM
noradrenaline
adrenaline
phosphatidyl ethanolamine (PE)
N-methyl-
2X
phosphatidyl choline
Lysine side chain in proteins
mnonomethyl Lys
N,N,N. trimethyl lysine side chan
B
dopamine
3-methoxydopamine
prenylated C-terminal cysteine of K-Ras
O-methyl ester of prenyated K Ras

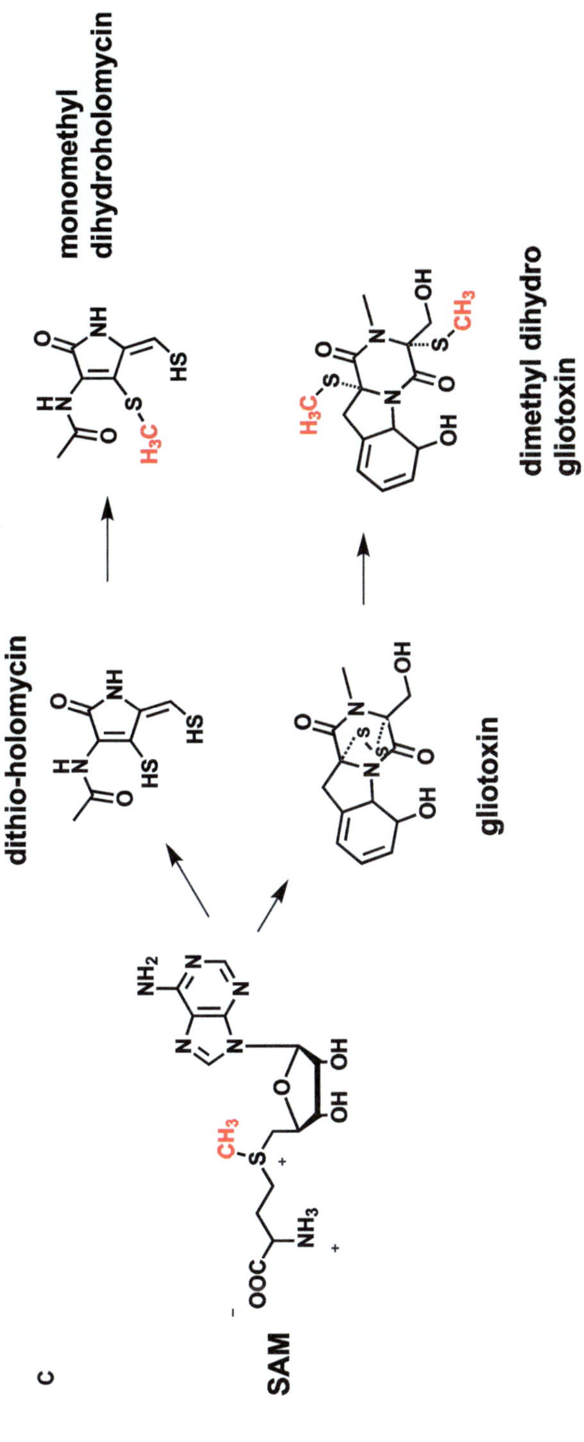

Figure 8.6 Continued on next page.

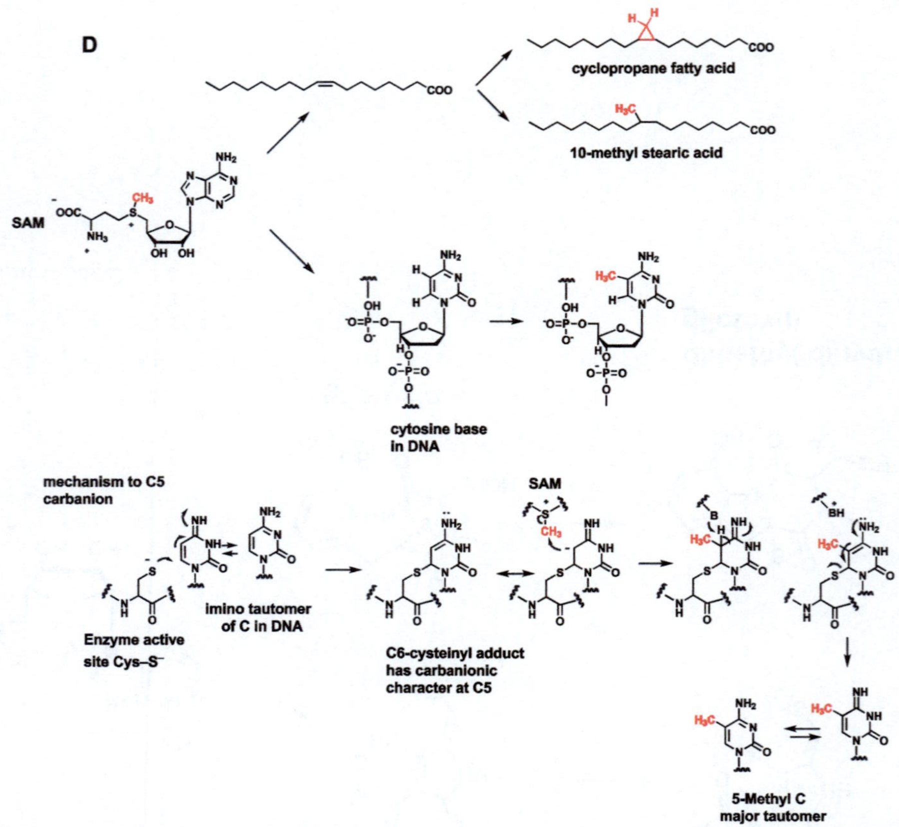

Figure 8.6 Examples of enzymatic $[CH_3^+]$ transfer to different classes of cosubstrate nucleophilic atoms. (A) N, (B) O, (C) S, (4) C atoms as nucleophiles.

$[CH_3^+]$ equivalent transfer to nitrogen, oxygen, sulfur and the occasional carbon nucleophile. The substrates encompass the full range of low- and high-molecular-weight substrates in cells.

Figure 8.6A shows methyl group transfer to three nitrogen nucleophiles in cosubstrates. The terminal step in biogenesis of the hormone adrenaline is *N*-methylation of noradrenaline. The second example occurs in 50% of the phospholipids in cell membranes. Phosphatidylethanolamine is *N*-methylated sequentially three times to the trimethylammonium choline headgroup. Phosphatidylcholine is a net neutral phospholipid. The third example is *N*-methylation of lysine side chains in the proteome. Among the most celebrated in recent years are the amino terminal tail of histones H3 and H4. Up to three methyl groups are transferred per residue and the *N*,*N*,*N*-trimethyl lysine side chains bind specific cotranscriptional partner proteins.

Two cases of oxygen nucleophiles in cosubstrates are also shown in Figure 8.6B. One is an alcoholic oxygen, the other a carboxylate oxygen anion. In the dopamine–adrenaline class of neurotransmitters *O*-methylation of the catechol 3-OH is a deactivating event. The second case is the final step in the C-terminal modification of Ras family proteins (discussed in Chapter 13). The thiolate on the fourth residue from the C-terminus of Ras family protein members is the initial site of posttranslational modification. Prenylation of the side chain cysteine four residues from the C-terminus of Kras protein (–CAAX) activates a protease that trims off the final three residues to liberate the prenylcysteine C-terminal carboxylate. Then, a SAM-dependent *O*-methyl transferase converts the anionic carboxylate to the neutral methyl ester (Figure 8.6). This aids in partition of the posttranslationally modified Ras proteins to membrane subcellular locations.

S-Methylation is a less frequent metabolic transformation. Figure 8.6C shows two examples of self-protection by natural product-generating bacteria (holomycin) and fungi (gliotoxin). The penultimate intracellular forms of holomycin and gliotoxin, respectively, are the dithiols.[7,8] The producing bacteria (streptomyces) and fungi (aspergilli) express *S*-methyltransferases to capture any dithiols that may accumulate in the cytoplasm, before oxidative export as the disulfide natural products.

Finally, three examples of carbon nucleophiles are denoted in Figures 8.6D and 8.7. In general, there are few free carbanions in metabolism. The π electrons of the double bond in the C16 monounsaturated fatty acid oleic acid constitute one of the carbon nucleophiles. Mycobacteria generate the branched 10 methyl stearic acid product while other bacteria instead install the one-carbon unit as a cyclopropane ring. A second carbon nucleophile is the stabilized C2 carbanion of the SAM–PLP adduct in Figure 8.6D. That carbanion is directed to an internal cyclization to create a different cyclopropane functionality in 1-amino-1-carboxy-cyclopropane.

The third case occurs in the many C5-methylations of cytosine bases in DNA. At first glance it is not obvious how to generate the requisite carbon nucleophile at the sp^2 hybridized carbon five of cytidine units in DNA double helices.

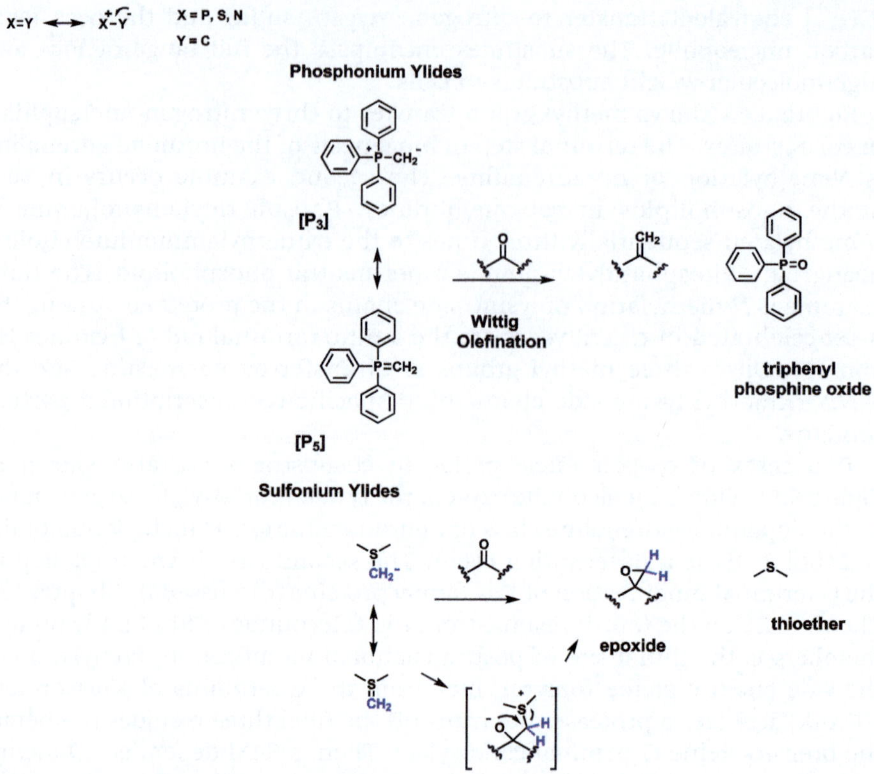

Figure 8.7 Abiotic chemistry of phosphonium and sulfonium ylides. Carbanion chemistry in C–C bond formations. Wittig olefination from phosphonium ylides, epoxidation from sulfonium ylides.

The catalytic strategy employed by these cytosine methyltransferases is first to add an active site cysteine thiolate across the 5,6 double bond of a bound cytosine unit *via* regiospecific attack at C6.[9] This most likely occurs on the minor cytosine C4 imino tautomer as shown in Figure 8.6D. The resultant enzyme–DNA thioether adduct has sp^3 hybridization at C5 in the enamine group. This is a stabilized carbanion equivalent that can attack the activated CH_3 group in SAM. Formation of that C5–CH_3 bond generates the product *S*-adenosylhomocysteine and the 5-MeC-containing DNA–enzyme adduct. Expulsion of the initially added active site Cys–thiolate completes a catalytic cycle.

These few examples out of many in small molecule and macromolecule metabolism call attention to the nature and strategy for generation of cosubstrate nucleophiles and the promiscuity of SAM to deliver the $[CH_3^+]$ equivalent to a wide range of electron rich cosubstrates. The trivalent sulfonium center in *S*-adenosylmethionine is thus responsible for perhaps billions of methyl transfers in every cell cycle and reaches every corner of metabolism and class of biomolecules as substrates. The invention/evolution of SAM as a kinetically stable, thermodynamically activated alkyl group transfer agent, because of its cationic trivalent sulfur, seems a necessary condition for life in ancient and contemporary organisms.[10]

We turn in the next chapter to SAM's complementary role as an initiator not of electrophilic [CH_3^+] transfers, but instead of a wide range of *radical* reactions. We do so in the context of both thiols and sulfonium cations undergoing one-electron transfer routes for metabolite conversions.

8.6 SAM-ylide Formation, Carboxy-SAM Formation and Carboxymethyl Transfers

One additional mode of reactivity of the one-carbon methyl group in *S*-adenosylmethionine would be deprotonation, *e.g.* by an enzyme active site base, to create a SAM–[CH_2^-] carbanion equivalent (Figure 8.7). The placement of a carbanion adjacent to a positively charged atom creates a charged pair, X^+–Y^-(both atoms with a stable octet shell) known as an *ylide.* The X^+ species in abiotic organic chemical reactions is typically phosphorus, sulfur or nitrogen (Figure 8.7). The Y^- atom is often a carbanion. Once formed, the carbanion can be stabilized by resonance forms that involve a double bonded X=Y form as shown.

The most extensively studied ylides are probably phosphorus ylides. Figure 8.7 shows the transfer of the methylene group to ketone coreactant in the widely used Wittig reaction for organic syntheses. There are no known biological correlates of the Wittig reaction yet uncovered.

The next most investigated ylide category in organic synthesis is probably the sulfonium ylide. As shown in Figure 8.7 these are useful in epoxide creation from carbonyl groups by the indicated mechanism. In both the phosphonium and sulfonium ylides the ability of d–p orbital hybridization allows phosphorus and sulfur to expand their shells beyond eight electrons. In the triphenylphosphonium ylide hybrid contributors to structure the formal oxidation state of phosphorus can go between $P=+3$ and $P=+5$. In sulfonium ylides, it can be $S=[+2]$ or $S=[+4]$ in the comparable resonance hybrids.

In contradistinction to phosphonium ylides, a sulfonium ylide in the context of SAM turns out to be an accessible intermediate in sulfur chemical biology. The enzymatic fate is not epoxidation of cosubstrate, but rather carboxylation of the SAM ylide carbon (Figure 8.8). The first, and at the time singular, example was deciphered in 2010 in examination of the formation of the modified base 5-acetoxoyuridine in the wobble position of several bacterial tRNAs (Figure 8.8).[11,12] The conversion of the uracil ring to 5-acetoxyuracil occurs in two enzymatic steps. The first enzyme is a monooxygenase, generating the 5-hydroxyuracil residue in the tRNA molecule. The second enzyme transfers a two-carbon carboxymethyl group onto the 5-hydroxyl oxygen to yield the *O*-carboxymethylated uridine that functions as the wobble base in mRNA decoding.

The source of the two carbon carboxymethyl group is a modified SAM molecule known as carboxy-SAM (Figure 8.8).[11] *Inter alia*, this is the first report of SAM being used to transfer a two-carbon electrophilic fragment [$^+CH_2COO^-$], rather than the a one-carbon [CH_3^+] equivalent. In turn, enzymatic formation of carboxy-SAM is shown in Figure 8.8B, invoking the ylide form of SAM as the carbanion equivalent to attack a nascent CO_2 molecule. That CO_2 molecule arises from decarboxylation of the cosubstrate prephenate in the active site of the carboxy-*S*-adenosyl-L-methionine synthase (CmoA) enzyme. It is not yet clear if prephenate decarboxylative aromatization is concerted with CO_2 transfer to the SAM–ylide carbanion or

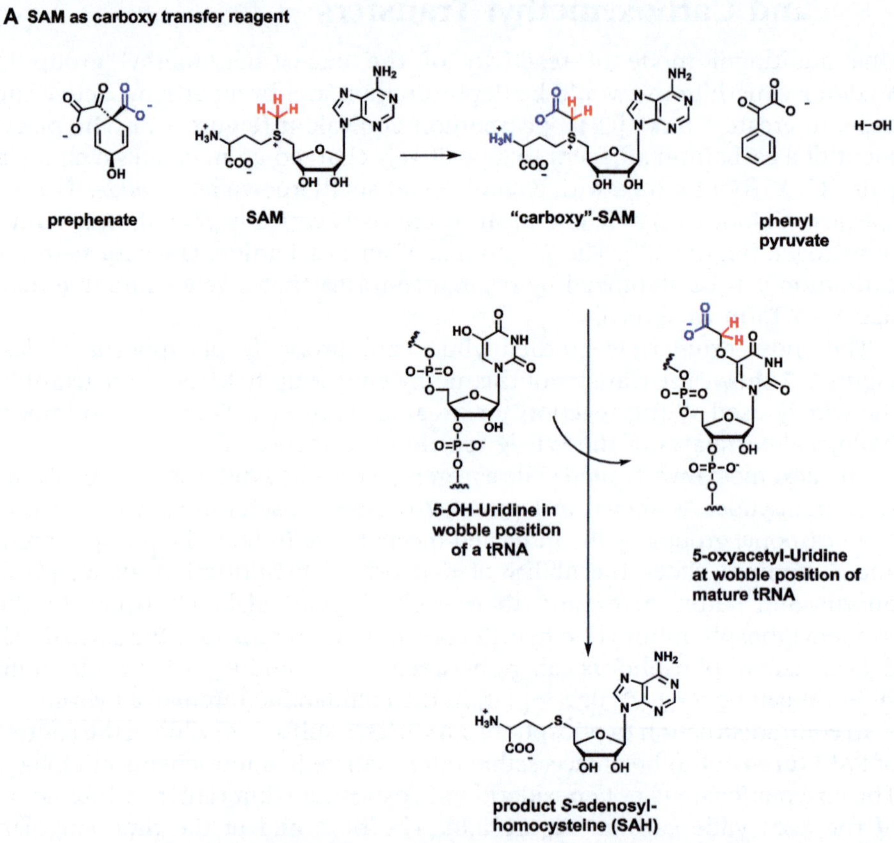

Figure 8.8 Carboxy-SAM formation and enzymatic carboxymethyl transfers. (A) Conversion of a 5-hydroxyuracil residue at the wobble base in tRNA to the 5-*O*-carboxymethyl-uracil ring by carboxymethyltransfer from carboxy-SAM as it undergoes conversion to *S*-adenosyl homocysteine (SAH). Carboxy-SAM is formed during aromatizing decarboxylation of prephenate. (B) SAM ylide formation in the active site of CmoA and capture of nascent CO_2 being released on prephenate decarboxylation. (C) Two distinct polarities for the CH_3 group of SAM: contrast between electrophilic [CH_3^+] fragment transfer and nucleophilic behavior of the SAM ylide [CH_2^-] equivalent. (D) Carboxymethyl transfer from carboxy-SAM in posttranslational 3-thiaGlu residue formation.

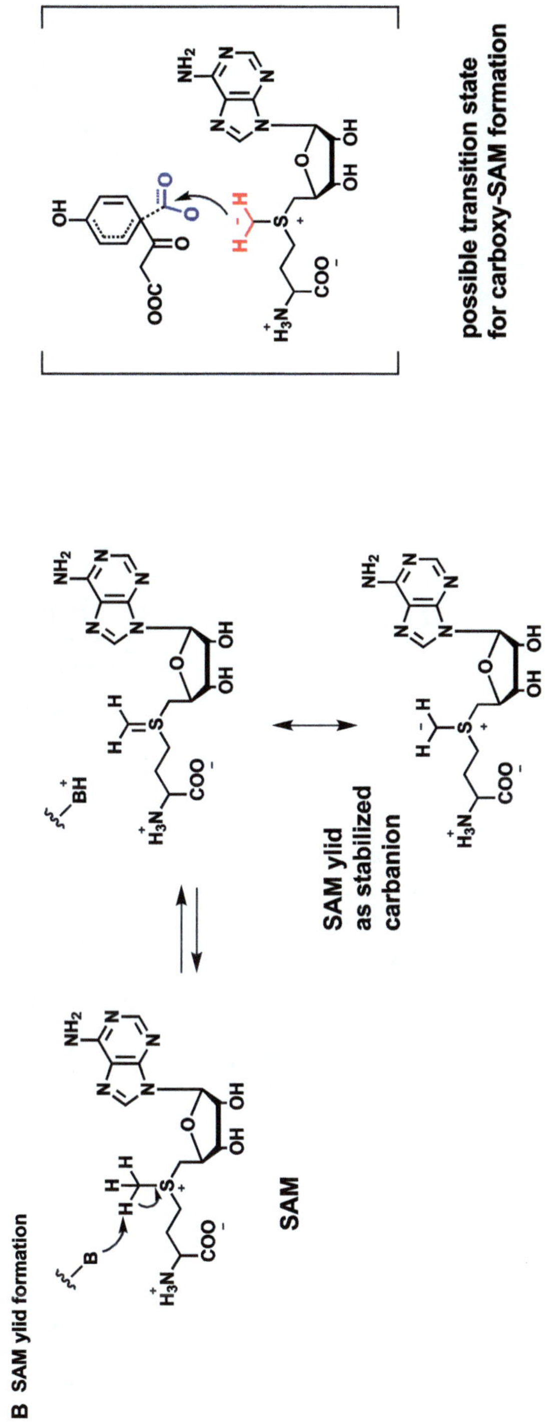

Figure 8.8 Continued on next page.

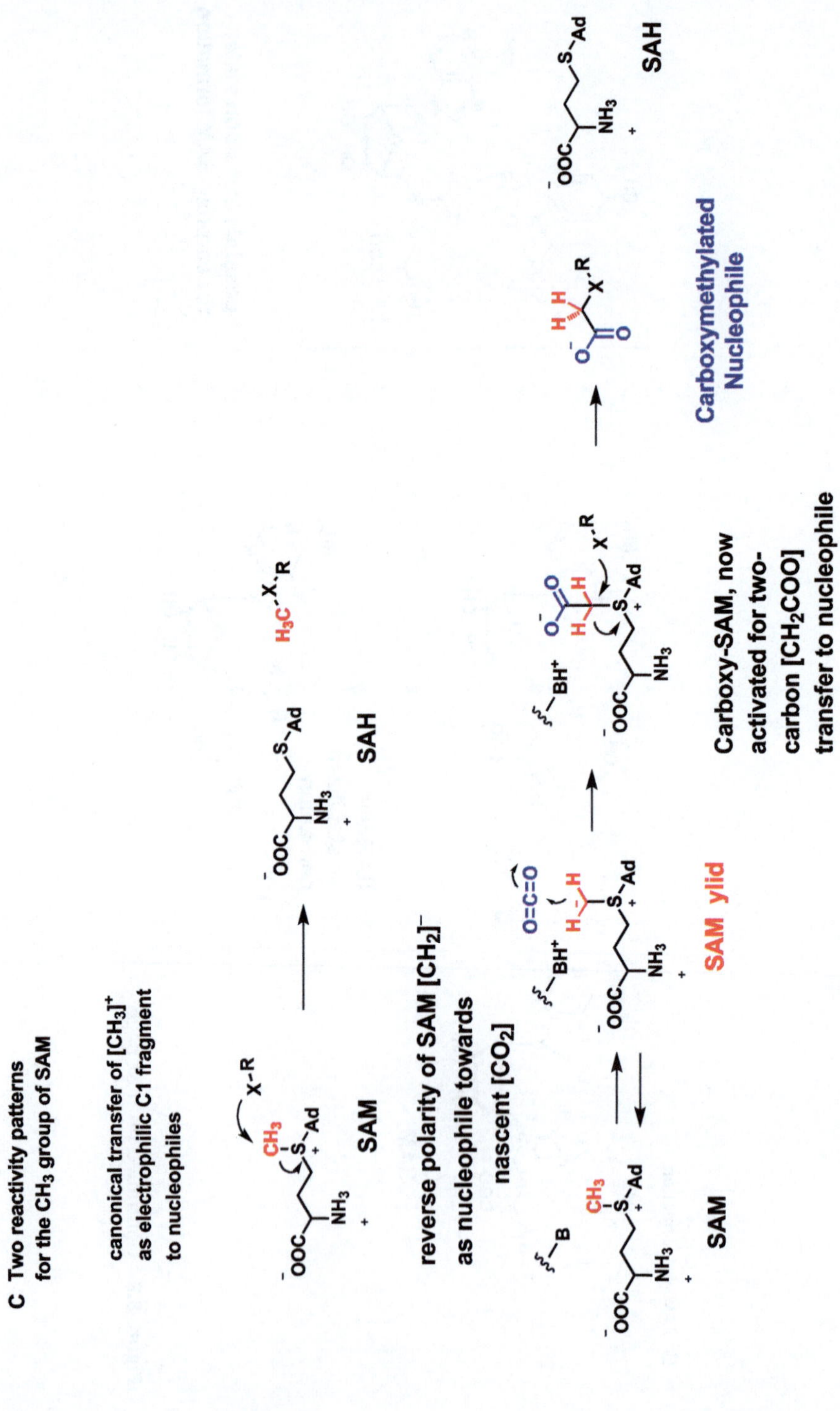
C Two reactivity patterns for the CH3 group of SAM
canonical transfer of [CH3]+ as electrophilic C1 fragment to nucleophiles
SAM
SAH
reverse polarity of SAM [CH2]– as nucleophile towards nascent [CO2]
SAM
SAM ylid
Carboxy-SAM, now activated for two-carbon [CH2COO] transfer to nucleophile
Carboxymethylated Nucleophile
SAH

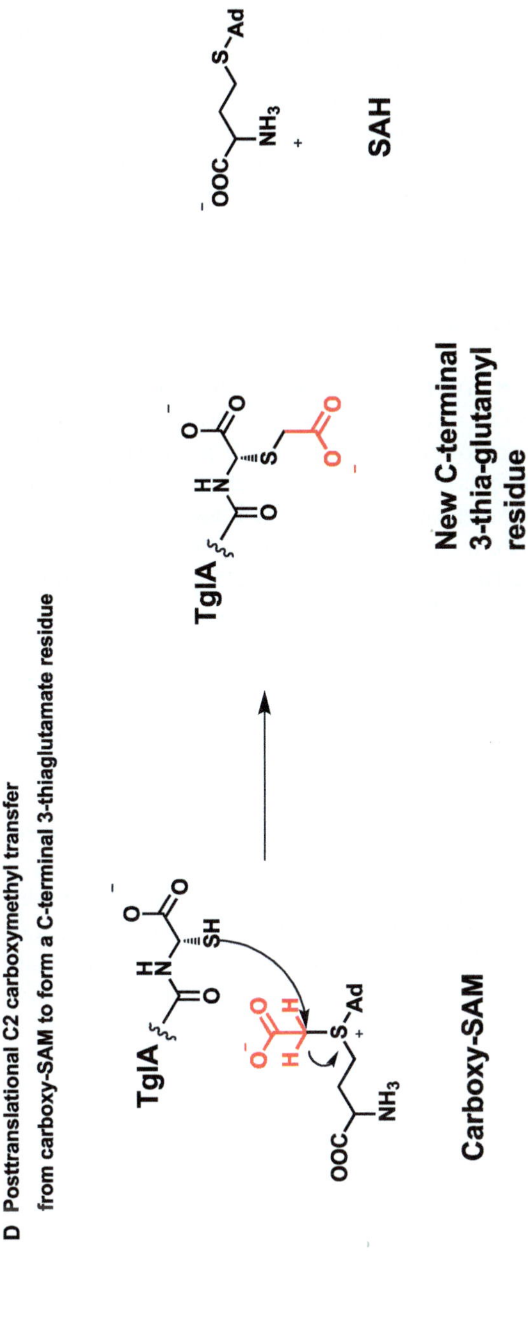

Figure 8.8 Continued.

whether nascent CO_2 is a discrete intermediate. For sake of illustration only, the concerted possibility is depicted in Figure 8.8B. Carboxy-SAM was discovered and characterized by solving a cocrystal of this metabolite bound in the active site of the CmoA synthase.[11]

The discovery of the CmoA enzyme in the tRNA maturation pathway reveals that two quite distinct two-electron reaction manifolds exist for the CH_3 substituent at the sulfonium center of *S*-adenosylmethionine (Figure 8.8C). The canonical pathway in hundreds of cellular methyltransferase active sites is electrophilic one-carbon transfer of [CH_3^+] to nucleophilic cosubstrates. In contrast, the sulfonium ylide route reverses the polarity of that CH_3–S^+-center from functional electrophile to functional nucleophile as the ylide captures nascent CO_2 from prephenate decomposition.

The resultant carboxy-SAM now has a two-carbon (carboxymethyl) substituent rather than the conventional one-carbon [CH_3] substituent. That two-carbon substituent on the cationic sulfur is next used as an electrophilic fragment by the subsequent enzyme CmoB as the 5-OH uracil residue acts as a nucleophile (Figure 8.8C).[12] The 5-OH of the uracil ring is carboxymethylated (two carbons) rather than conventionally methylated (one carbon). The hydrophilic carboxymethyl group carries the negative charge of the carboxylate, quite different from a hydrophobic methyl substituent.

A second example of carboxy-SAM as a donor of the two-carbon –CH_2COO as an electrophilic fragment, this time to a thiolate nucleophile, has been reported in the posttranslational modification of a small, 50-residue lanthipeptide-like protein (TglB) produced by the plant pathogen *Pseudomonas syringae*.[13] The mature protein has a C-terminal 3-thiaglutamate residue. It arises first by posttransational addition of a cysteinyl residue from a Cys–tRNA, followed by a mysterious oxygenative conversion of the cysteine to a thioglycine. That C-terminal thioglycine thiolate is the nucleophile attacking carboxy-SAM and transferring the CH_2COO^- group (Figure 8.8D). The result is a new S–CH_2COO^- side chain in the TglB modified protein. This constitutes a C-terminal 3-thiaglutamate residue.

The Cmo pathway is a novel intersection of cysteine and SAM chemical biology. One could as readily have placed this example in Chapter 14 as an exotic modification of a cysteine residue in cysteine proteomes. There may be more carboxy-SAM enzymes yet to be unearthed, indicating a broader metabolic role for the ylide form of SAM and its attached methyl group as a carbon nucleophile rather than a carbon electrophile. The next chapter explores yet a third reaction manifold for SAM in biology; a one-electron route rather than the orthogonal two-electron routes examined in this chapter.

References

1. S. Ravanel, B. Gakiere, D. Job and R. Douce, The specific features of methionine biosynthesis and metabolism in plants, *Proc. Natl. Acad. Sci. U. S. A.*, 1998, **95**, 7805–7812.

2. R. V. Banerjee and R. G. Matthews, Cobalamin dependent methionine synthase, *FASEB J.*, 1990, **4**, 1450–1459.
3. M. Koutmos, S. Datta, K. A. Partridge, J. L. Smith and R. G. Matthews, Insights into the reactivation of cobalamin-dependent methionine synthase, *Proc. Natl. Acad. Sci. U. S. A.*, 2009, **106**, 18527–18532.
4. M. Ludwig and R. G. Matthews, Structure-based perspectives on B12 enzymes, *Annu. Rev. Biochem.*, 1997, **66**, 299–313.
5. C. T. Walsh and Y. Tang, *The Chemical Biology of Human Vitamins*, Royal Society of Chemistry Press, London, 2018.
6. R. I. K. Geltink and E. Pearce, T cell activation: the importance of methionine, *E Life*, 2019, e47221.
7. J. Komoto, T. Yamada, Y. Takata, G. F. Mrkham and F. Takusegawa, Crystal structure of S-adenosylmethionine synthetase ternary complex: a novel catalytic mechanism of S-adenosymethionine synthesis from ATP and methionine, *Biochemistry*, 2004, **43**, 1821–1831.
8. L. A. Wessjohann and W. Brandt, Biosynthesis and metabolism of cyclopropane rings in natural compounds, *Chem. Rev.*, 2003, **103**, 1625–1627.
9. G. Vilkaitis, E. Merkiene, S. Serva, E. Weinhold and S. Klimasauskas, The mechanism of DNA cytosine-5-methylation, *J. Biol. Chem.*, 2001, **276**, 20924–20934.
10. M. Fontcave, M. Attta and E. Mulliez, S-adenosyl methionine: nothing goes to waste, *Trends Biochem. Sci.*, 2004, **29**, 243–249.
11. J. Kim, H. Xiao, J. B. Bonano, C. Kalyanaraman, S. Brown, X. Tang, N. F. Al-Obaudu, Y. Patkovsky, P. C. Babbitt, M. P. Jacobson, Y.-S. Lee and S. C. Almo, Structure-guided discovery of the metablite carboxy-SAM that modulates tRNA function, *Nature*, 2013, **498**, 123–126.
12. J. Kim, H. Xiao, J. Koh, Y. Wang, J. B. Bonanno, K. Thomas, P. C. Babbitt, S. Brwn, Y.-S. Lee and S. C. Almo, Determinants of the CmoB carboxymethyltransferase utilized for selective tRNA wobble modification, *Nucleic Acids Res.*, 2015, **43**, 4602–4613.
13. C. P. Ting, M. Funk, S. L. Halaby, Z. Zhang, T. Gonen and W. A. van der Donk, Use of a scaffold peptide in the biosynthesis of amino acid-derived natural products, *Science*, 2019, **365**, 280–284.

CHAPTER 9

One-electron vs. Two-electron Reaction Manifolds in Sulfur Enzymology

Thus far, in Chapters 6–8, we have focused on two-electron reaction manifolds for organic sulfur scaffold reactivity. This has included the protonated and deprotonated forms of cysteine thiol groups as prime reactant for S–H bond cleavage in Chapters 6 and 7. In Chapter 8 we have examined two-electron reaction mechanisms for attack on each of the three carbon substituents adjacent to the permanent positive charge on the sulfonium sulfur of *S*-adenosylmethionine. Those are all categorizable as C–S bond cleavages.[1]

In this chapter we turn our attention from heterolytic cleavages of the S–H bond (*e.g.* cysteine) and the C–S bond (in SAM) to *homolytic reaction modes*, as key aspects of cysteine and *S*-adenosylmethionine chemical biology roles (Figure 9.1). We examine thiol radicals first.[2]

9.1 Thiol Radicals: One-electron Enzyme Chemistry

One of the parameters that accounts for the versatility of sulfur in organic frameworks in cell metabolism is its ability to engage in both one-electron and two-electron reaction manifolds. The two major organic scaffold sulfur functional groups, thiols and the sulfonium center in *S*-adenosylmethionine, can react in either mode. For the two-electron reaction pathways, we have noted above the behavior of thiols as nucleophiles, disulfides as electrophiles, and persulfides as both electrophiles and nucleophiles ($R–S–S^-$). Likewise, we have denoted the two-electron modality of *S*-adenosylmethionine, most prominently in S_N2-type attack on its methyl group by several classes of cellular nucleophiles.

Here, we discuss some aspects of one-electron transfer chemistry both from S–H bond cleavage in thiols to generate thiol radicals ($RS^•$)

Chemical Biology No. 12
The Chemical Biology of Sulfur
By Christopher T. Walsh

Published by the Royal Society of Chemistry, www.rsc.org

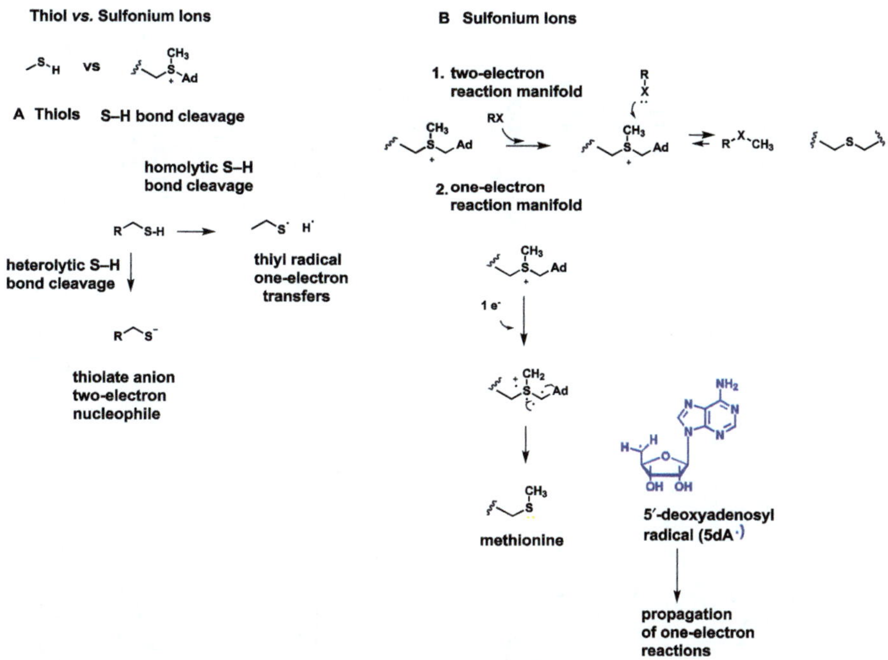

Figure 9.1 (A) Thiols *vs.* (B) sulfonium ions: one- *vs.* two-electron reaction manifolds.

(Figure 9.1A) and from C5′–S bond cleavage on SAM. That radical path generates a carbon-centered radical (5′-deoxyadenosyl radical = dA•) and a transient sulfur radical that is quenched by one-electron transfer from the 4Fe–4S cluster (Figure 9.1B).

Actually, four of the twenty amino acid side chains in proteins can participate in one-electron transfer steps in enzymatic catalyses: the side chain sulfur radical of cysteinyl residues, the phenolate side chain radical of tyrosyl residues, the heteroaromatic radical cation of tryptophan, and the C2-carbon-centered radical of glycyl residues (Figure 9.2A). The phenolate and indolyl radicals are stabilized by delocalization through the aromatic or heteroaromatic ring systems. In a formal sense these four residues offer radical intermediates with unpaired electron density at sulfur, oxygen, nitrogen and carbon atoms, respectively.

9.1.1 Ribonucleotide Reductases: A Bevy of Thiol Radicals in the Active Site

The first three of these radicals participate in electron transfer across the two subunits of ribonucleotide reductases, the key enzymes converting the RNA precursor ribonucleoside diphosphates to the DNA precursors 2′-deoxyribonucleoside diphosphates[3] (Figure 9.2B).

The tyrosyl radical is generated from an adjacent iron center, then the odd electron travels *via* intervening tryptophanyl residues some 35 Ångstroms to

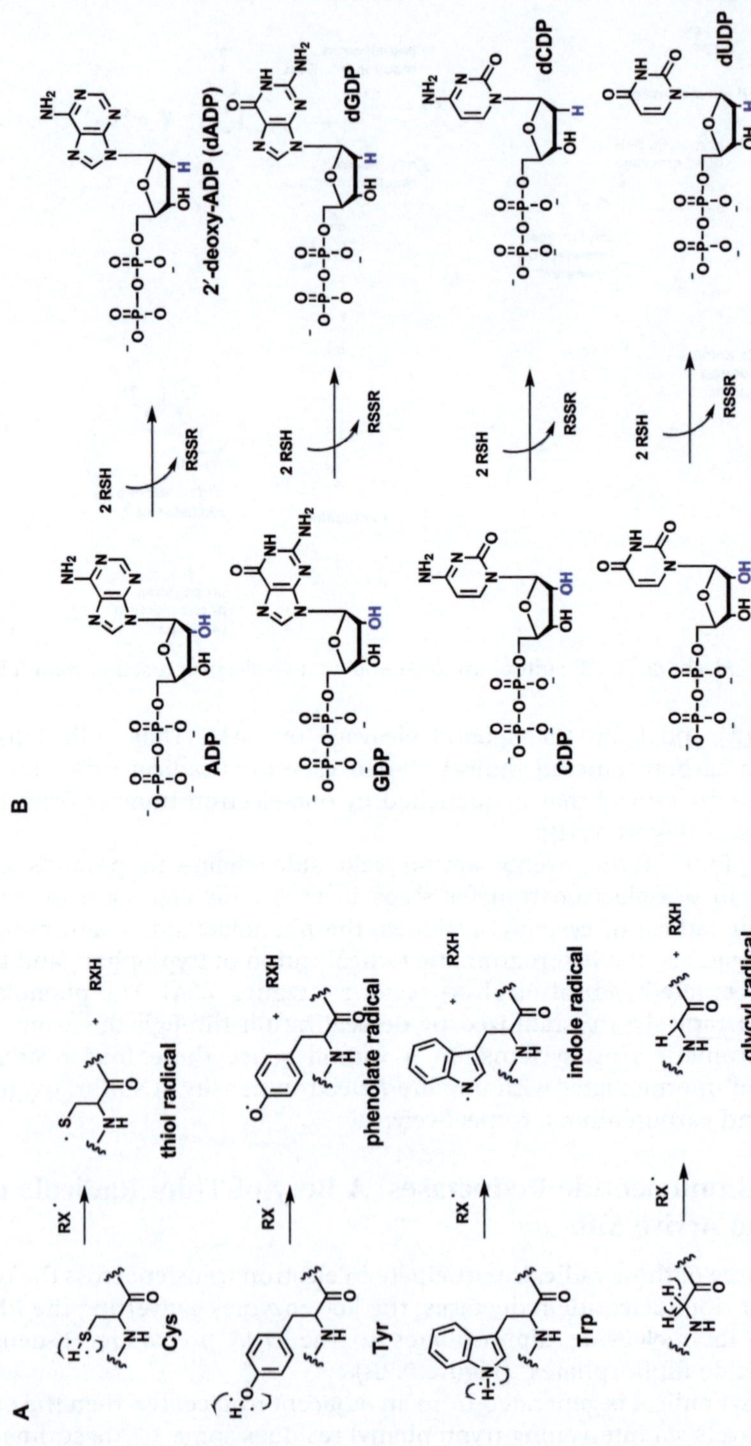

Figure 9.2 Enzymatic cycles involving radicals on oxygen, nitrogen, sulfur and carbon atoms. (A) Cysteinyl, tyrosyl, tryptophanyl and glycyl radical enzyme intermediates. (B) Nucleoside diphosphate substrates for ribonucleotide reductase.

the active site in the other subunit where there are three cysteine side chains that participate in the catalytic cycle. That electron arrives at Cys439 and generates the thiol radical as the S–H bond is cleaved homolytically. That is where Figure 9.3 begins. The one-electron traverse over some 35 Ångstroms distance and across the subunit interface is an astonishing architectural array. The electron passes through Trp45 and then Tyr365 in the originating β-subunit on the way to Ty731 and Tyr730 in the α subunit before arriving at the active site Cys439.

The reaction flux involves a set of one-electron transfers, starting from thiol radical Cys439 $S^{\bullet}$, involving sequential radicals at C3′ and then C2′ of the ribonucleoside diphosphate substrate (here ADP is shown but GDP, CDP and UDP are also all substrates for reduction) Two more sulfur-based cysteinyl side chain residue radicals also are formed and quenched as well, at Cys224 and Cys462. Figure 9.3 shows three discrete hydrogen atom ($H^{\bullet}$) transfer steps involving these thiol side chains in the reductase active site.

Thus, all three active site cysteine side chains participate as sulfur radicals at some stage in each catalytic cycle. In every cell division period, to replicate the three billion base pairs of human genomic DNA, there must be at least 3×10^9 catalytic turnovers per cell of ribonucleotide reductase with its cascade of four distinct ribonucleoside diphosphate substrates and its suite of three thiol radicals. In the presumed evolution of an RNA world to a DNA world in organisms living under anaerobic conditions, protein–cysteine $S^{\bullet}$ intermediates may have been more abundant than today.

Note that in Figure 9.3 the ribonucleotide reductase is left in an oxidized disulfide form at the end of a catalytic cycle. Reduction by two electrons back to the dithiol form occurs by thiol–disulfide interchange with thioredoxin. In turn the disulfide form of thioredoxin is reduced by oxidation of a molecule of NADH by thioredoxin reductase. The two-enzyme couple emphasizes that the conversion of the ribose moiety to a 2′deoxyribose moiety is a two-electron reduction and the enzyme undergoes a two-electron oxidation. Only in the presence of thioredoxin, thioredoxin reductase and NADPH can ribonucleotide reductase turn over.

The reductive conversion of ribonucleotides to 2′-deoxyribonucleotides is a difficult transformation to achieve chemically and with regioselectivity. The various forms of ribonucleotide reductases that have evolved over eons are literally the gatekeepers from an RNA world to a DNA world of biological information storage and retrieval. The participation of sulfur radicals even today in otherwise aerobic organisms is in some sense a testament to how a difficult piece of enzyme chemistry has been retained down the ages.

9.1.2 Tandem Glycyl and Thiol Radical Intermediates in Enzyme Catalysis

An example of an enzyme that uses, in tandem fashion, a glycyl backbone radical and then a cysteine thiol radical to carry out decarboxylation of pyruvate is the enzyme pyruvate formate lyase[4] (Figure 9.4). This enzyme is of interest for multiple reasons.

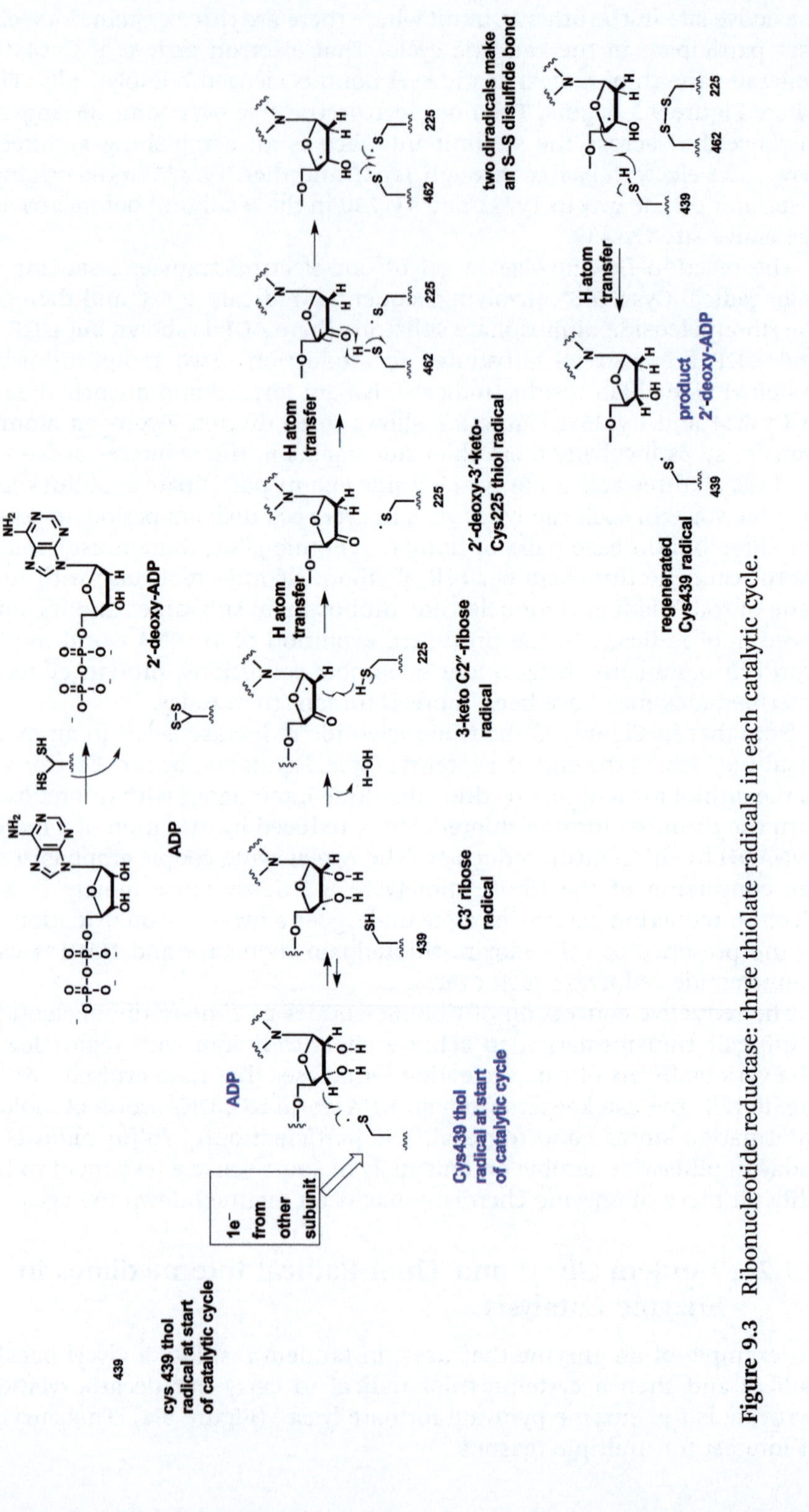

Figure 9.3 Ribonucleotide reductase: three thiolate radicals in each catalytic cycle.

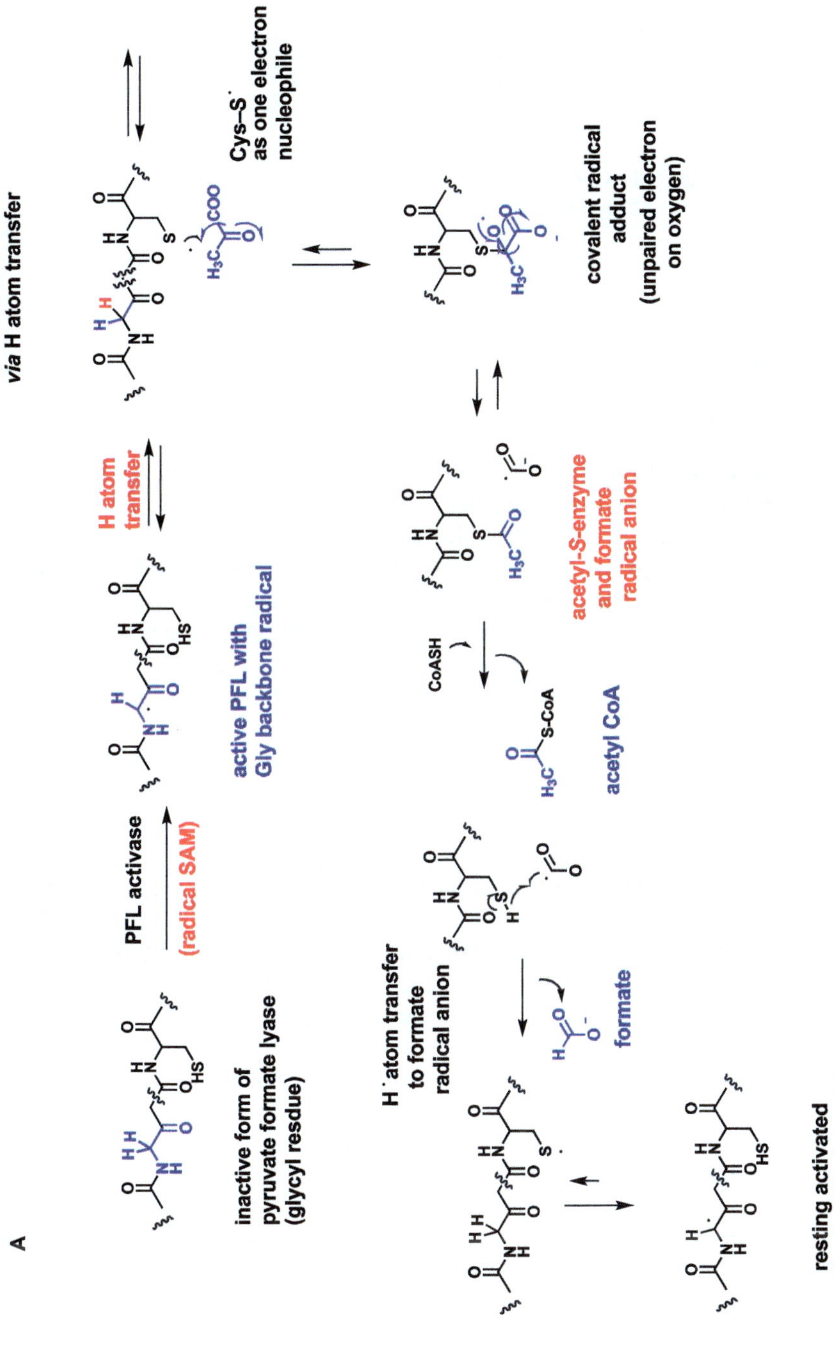

Figure 9.4 Continued on next page.

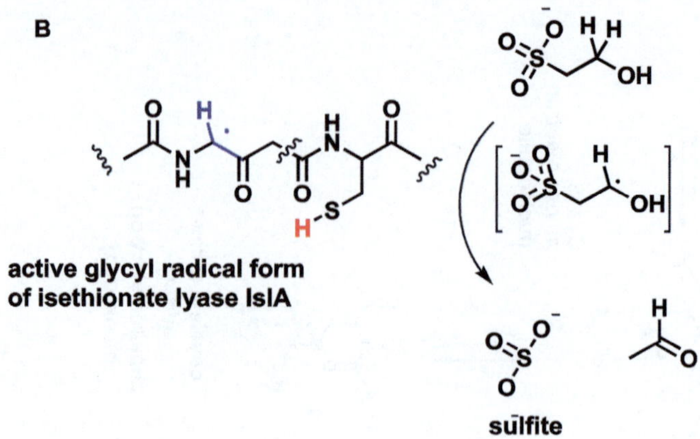

Figure 9.4 (A) Pyruvate formate lyase: tandem glycyl backbone radical and cysteine side chain radical in the catalytic cycle of activated enzyme. (B) Isethionate C–S lyase *via* a one-electron pathway.

First, it functions in anaerobic microenvirons. If it gets oxidized at some point in a catalytic cycle it will require a reductive activase ancillary enzyme to get it back to the active state. The action of the activase to convert a backbone methylene group of a glycyl residue in the active site of the lyase to a glycyl radical form is prequel to getting the active lyase catalyst. That helper enzyme, pyruvate formate lyase activase, is a radical SAM enzyme. The logic of such an activase is discussed in Section 9.2, where we will note the activase creates a 5′-deoxyadenosyl radical to generate the backbone glycyl radical *in trans* on the inactive lyase (Figure 9.4).

A second notable feature of pyruvate formate lyase, once activated as the glycyl radical species, is that on decarboxylation of substrate keto acid pyruvate, the one-carbon fragment is not CO_2 but rather formate. This differs from the more common thiamin-pyrophosphate-dependent pyruvate decarboxylases that generate CO_2 and either a two-carbon acetaldehyde or acetyl CoA coproduct.

The glycyl radical in the now activated pyruvate formate lyase is itself not the direct actor in one-electron transfers to and from bound pyruvate substrate. Instead, as schematized in Figure 9.4, the glycyl radical engages in hydrogen atom ($H^{\bullet}$) transfer from an active site Cys–S–H side chain to generate a thiol radical form of an active site cysteine residue. Now, the enzyme has arrived at a species, the thiol radical, that initiates a one-electron pathway involving the bound pyruvate substrate carbonyl group.

This is a one-electron carbonyl addition mechanism by the thiol radical and generates a radical alkoxy form of the indicated enzyme–*S*-pyruvyl adduct. Decomposition of the tetrahedral radical adduct can occur by *homolytic* cleavage between C2 and C1 of the substrate. This C–C bond breakage yields an acetyl-*S*–Cys–enzyme along with the formate anion radical. Note that this is an intersection of one-electron thiol radical and two-electron acyl thioester logic.

Hydrogen atom (H•) transfer from the Cys–SH side chain, creates the one-carbon formate anion product and the thiol radical form of active enzyme. Presumably this goes on to H• transfer from the glycine methylene backbone, resetting the cysteine–glycine radical equilibrium for the next catalytic cycle. There is facile, temporally and directionally controlled H• atom movement between the Cys–SH and the glycine methylene carbon at different stages of each catalytic cycle. The glycyl radical and the cysteine thiol radical must be close in energy for this logic to operate.

Note that thiolytic release by CoASH of the acetyl-thioester enzyme generates thermodynamically activated acetyl CoA as product of the C–C lyase reaction: a one-electron reaction coordinate rather than the standard two-electron reaction coordinate of pyruvate dehydrogenase complexes that likewise generate acetyl CoA. The activase is needed only every few thousand catalytic cycles to reductively resuscitate the starting glycyl residue radical. This is reminiscent of the reductive resuscitation of the B_{12}-dependent methionine synthase which occasionally gets left in the Co^{III} oxidation state at the end of a catalytic cycle (Chapter 8). Presumably, these resuscitation companion enzymes are an adaptation of oxygen-labile intermediates to the gradual oxygenation of the planet.

A recent article has also described a glycyl radical intermediate in a bacterial anaerobe from the human gastrointestinal tract that processes taurine (see Chapter 7) ultimately to hydrogen sulfide.[5] The immediate substrate for C–S bond one-electron cleavage is 2-hydroxyethane sulfonate, yielding sulfite and acetaldehyde by action of the C–S lyase.[5] The liberated sulfite is then a substrate for bacterial sulfite reductases (Chapter 2). The initial steps in the pathway are transamination of taurine to sulfoacetaldehyde then reduction to the hydroyxyethane sulfonate.

9.1.3 β-Lactam Formation *via* Substrate Thiol Radicals

There are several examples of thiol radicals formed on substrates, not just on enzyme cysteinyl residue side chains during enzyme action, especially in the realm of natural product biosynthetic pathways. Two of the most intensively studied are the tandemly acting enzymes in biosynthesis of the penicillin and then the cephalosporin class of antibiotics, historically the best-selling antibiotic class globally over the past 85 years.[6]

Isopenicillin *N* synthase (IPNS) takes the acyclic non-ribosomal tripeptide aminoadipyl-cysteinyl-D-valine (ACV) and O_2 and generates the 4,5-fused β-lactam-thiane ring systems characteristic of the β-lactam antibiotic class[7] (Figure 9.5). This bacterial and fungal synthase is a nonheme mononuclear iron enzyme that reduces cosubstrate O_2 by four electrons to two product molecules of H_2O (see Section 12.2.1 for non-heme iron enzyme ligands). Inspection of ACV tripeptide substrate reveals a corresponding four-electron oxidation as the lactam (two electrons) and then the thiane ring (two electrons) is formed. The designation of an isopenicillin framework *vs.* a penicillin framework is in the stereochemistry of the aminoadipyl side chain. Figure 9.5B summarizes that an epimerase converts isopenicillin N to penicillin N before action of the tandem enzyme, desoxyacetyl cephalosporin synthase (DAOCS), trivial name "expandase".

The minimal mechanistic scheme in Figure 9.5A shows two steps in which one-electron chemistry at the cysteinyl sulfur is crucial. The first one

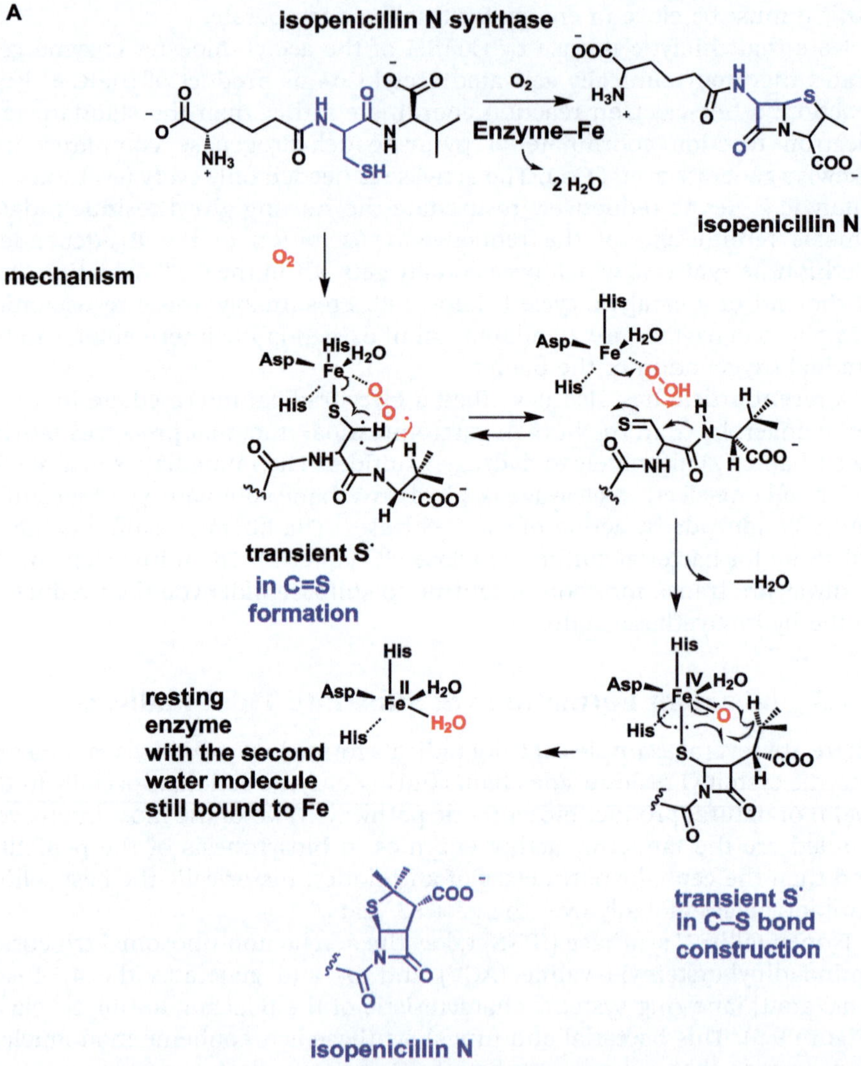

Figure 9.5 Penicillins and cephalosporins. (A) Conversion of the non-ribosomal peptide aminoadipyl–cysteinyl–D-valine (ACV) to isopenicillin N by the nonheme iron isopenicillin N synthase. (B) O_2-Based ring expansion of the thiane ring in penicillin N to the cephem ring in desoxyacetyl cephalosporin C by the "expandase" enzyme. Substrate-derived thiyl radicals are proposed in each of the two enzyme active sites.

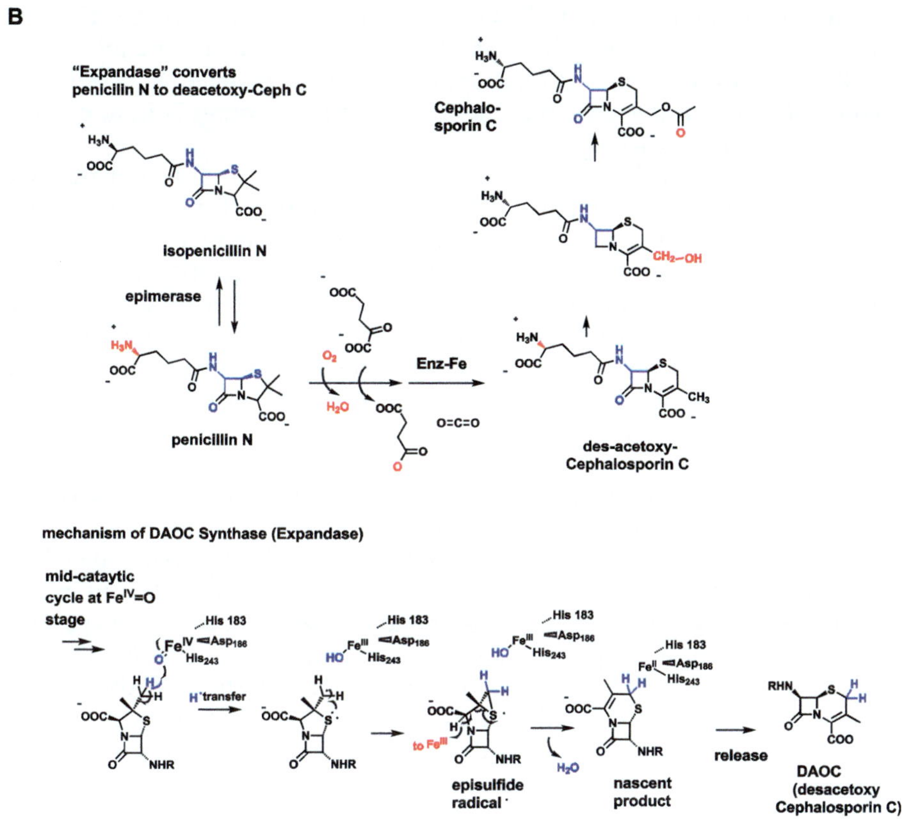

Figure 9.5 Continued.

is during formation of the four-membered lactam, featuring a radical rearrangement to a thioaldehyde. The second occurs in cyclization that yields the fused five-membered thiane ring with C–S bond construction. This is a remarkable refashioning of the thiol side chain of substrate to the cyclic thioether linkage of product. However, it is the strained β-lactam, the thermodynamically activated warhead as an acylating agent for targeted bacterial cell wall transpeptidases, that gets most of the attention. Even so, the thiol–iron radical intermediates are crucial to building that warhead.

An equally remarkable transformation in the chemical biology of sulfur occurs in the next biosynthetic step practiced by some fungi. The 4,5-fused penicillin nucleus is expanded to a fused 4,6-bicyclic system by a cephalosporin synthase known colloquially as expandase (Figure 9.5B). This is another nonheme mononuclear iron enzyme, consuming O_2 and reducing it by four electrons.[8] The expandase is a more typical member of the nonheme oxygenase class in that the typical third substrate, α-ketoglutarate, is concurrently oxidatively decarboxylated to succinate and CO_2, in service of generating the high valent $Fe^{IV}{=}O$ (iron-oxo) species, a robust active site oxidant. On the other hand, the expandase, like the IPNS enzyme just before it, fails to incorporate an oxygen atom into product. Instead, as noted, H_2O accounts for one of the oxygen atoms in O_2, while the other is in one of the succinate product's carboxylates.[9]

Figure 9.5B shows the expandase (DAOCS) reaction in mid-catalytic cycle with the $Fe^{IV}{=}O$ oxidant in place. $H^{\bullet}$ transfer from one of the penicillin *gem* dimethyl groups to the iron atom, reduces the $Fe^{IV}{=}O$ by one electron to Fe^{III}–OH. Rather than undergoing a typical subsequent [$OH^{\bullet}$] transfer to a substrate radical to complete an oxygenation event, the –OH is stranded and eventually released as HOH by protonation. Instead, the initial substrate radical ($CH_2^{\bullet}$) rearranges internally, by one-electron transfer from the thioether sulfur atom. This creates a new three-member ring by homolytic formation of the new $CH_2{=}S^{\bullet}$ bond in the episulfide radical structure shown.

That strained three-membered ring can unravel, again by one-electron paths, one of which is shown in Figure 9.5B. The tricyclic 5-4-3 sulfide radical undergoes net ring expansion to the 4-6 bicyclic scaffold. One of the two C–S bonds in the episulfide ring collapses. As shown, the unpaired electron density is quenched by double bond formation in the now expanded six-membered sulfur ring. This is termed a cephem ring system. The full scaffold is a cephalosporin. It is two enzymatic steps away, by hydroxylation of the remaining exocyclic methyl and then acetyl transfer from acetyl *S*-CoA, to yield cephalosporin C. The 4,6-fused bicyclic framework of cephalosporin C has served as the core for hundreds of semisynthetic cephalosporins including some that have become mainstays of hospital antibiotics for treatment of life-threatening bacterial infections.[6]

It is perhaps worth comparing the fates of two tripeptide metabolites with a central cysteinyl residue. We summarized several of the reactions of glutathione (γ-Glu–Cys–Gly) in Chapter 5. Much of its chemical biology goes through the nucleophilicity of the cysteinyl side chain in two-electron transfer steps, whether capturing HOOH on the way to the glutathionyl sulfenate and then the GS–SG disulfide, or by addition to electrophilic functional groups in organic cosubstrate frameworks. (This emphasis is not meant to deny the biological relevance of involvement of $GS^{\bullet}$ radicals in cellular scavenging, as in the reductions of vitamin C and vitamin E one-electron-oxidized forms.)

The related δ-aminoadipoyl–Cys–D-Val tripeptide is put to a quite distinct set of biological uses through one-electron reaction manifolds, rather than two-electron reaction manifolds. In the tandem reactions of the iron-containing penicillin and cephalosporin synthases, sulfur radicals feature prominently in setting up the four-member β-lactam warhead, the five-member fused thane ring and the expanded 4,6-bicyclic scaffold of cyclosporins.

The different niches occupied by these two non-ribosomally generated cysteinyl tripeptides are measures of the versatility provided by accessible two-electron or one-electron chemistry for sulfur in thiol oxidation states.

9.2 Radical SAM Enzymatic Logic

9.2.1 Two Convergent Routes to the 5′-Deoxyadenosyl Radical

The most ubiquitous organic scaffold initiator of enzyme-mediated one-electron reaction paths, presumably originally in (anaerobic) microbial metabolism also reaches into eukaryotic radical metabolite transformations. The *in situ* generation of the 5′-deoxyadenosyl radical (dA•) in enzyme active sites appears to have evolved convergently from two distinct starting molecules[10–12] (Figure 9.6).

One dA• is the metabolite of the cobalt-containing tetrapyrrole vitamin B_{12} described in Chapter 8 (Figure 8.2) as a cofactor for methionine synthase. The intermediacy of methyl B_{12}, a rare organocobalt metabolite in the MetH catalytic cycle, was noted. A second, stable organometallic form of B_{12} is found in adenosyl B_{12} (Figure 9.7) which occupies a distinct set of metabolic niches from the methyl B_{12} of methionine synthase (MetH). That adenosyl–cobalt organometallic bond arises when the reduced Co^{I} form of vitamin B_{12} attacks ATP at the active site of the B_{12} adenosyl transferase enzyme[13] (Figure 9.7). The only

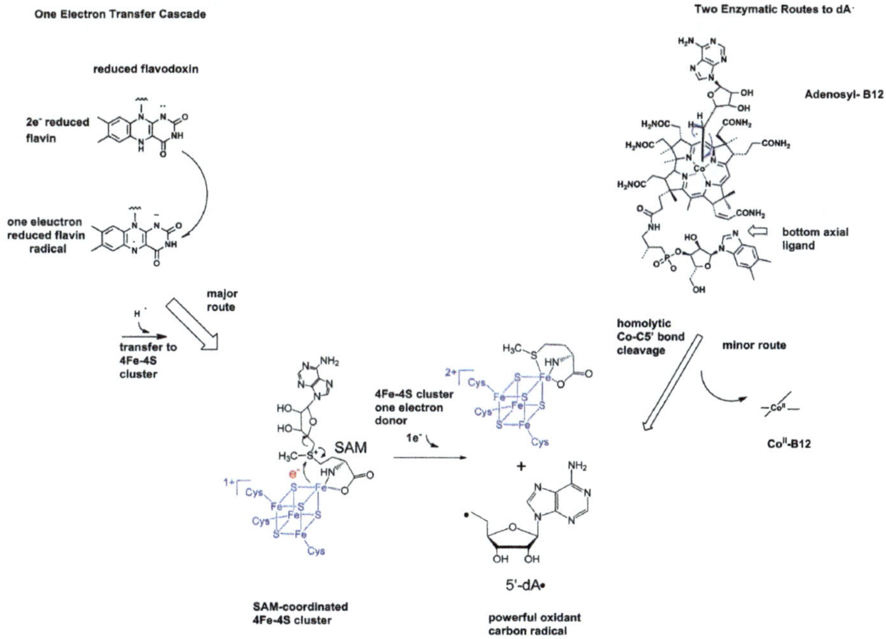

Figure 9.6 Two enzymatic one-electron transfer routes to the 5′-deoxyadenosyl radical (dA•). The left side shows a one-electron transfer cascade in a radical SAM enzyme active site from flavin semiquinone radical to 4Fe–4S cluster to cleave the C5′–S bond in the coordinated SAM molecule homolytically. The right side shows a convergent route to dA• from homolysis of the Co–C5′ bond in adenosyl B_{12} coenzyme. In terms of abundance, the radical SAM enzymes are the major routes to dA• while the B_{12} enzymes provide the minor route.

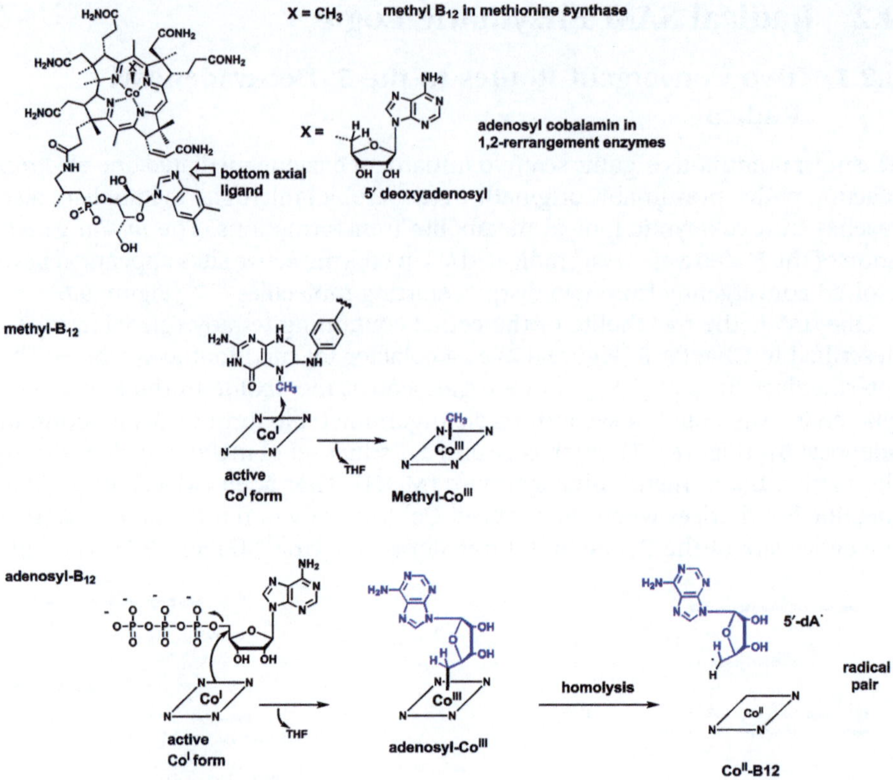

Figure 9.7 Two organometallic forms of vitamin B_{12} in metabolism. Methyl B_{12} is involved as the transfer agent in the MetH version of methionine synthase. Adenosyl B_{12} is the coenzyme form in 1,2-radical rearrangement reactions.

other case of enzymatic adenosyl transfer is in the biosynthesis of *S*-adenosylmethionine (Figure 8.1) discussed in the preceding chapter.

Currently there are a dozen or so known adenosyl B_{12}-dependent enzymatic transformations involving conversion of specific substrate to product. They all proceed by radical pathways (Figure 9.8). The specific example shown is for a bacterial 1,2-propanediol dehydrase. The net reaction is a 1,2 migration: an –OH substituent goes from C2 to C1 while an H from C1 goes to C2. This yields the *gem* diol hydrate of propionaldehyde. Loss of water to the aldehyde happens before release and is stereospecific. The mechanism involves $H^{\bullet}$ and $C^{\bullet}$ radicals as shown in Figure 9.8.

In each case the homolytic cleavage of the Co^{III}–C5′-adenosyl bond to produce Co^{II}–B_{12} and 5′-deoxyadenosyl radical ($dA^{\bullet}$) is the initial step. The $dA^{\bullet}$ abstracts a hydrogen atom ($H^{\bullet}$) from a carbon atom of bound cosubstrate and rearrangements begin. At the end of a catalytic cycle, the product radical $P^{\bullet}$ re-abstracts the $H^{\bullet}$ from dA–H, to give back $dA^{\bullet}$. The final step is recombination of the single electron on $dA^{\bullet}$ with one electron from the Co^{II}–B_{12} to regenerate the Co^{III}–adenosyl organometallic bond in adenosyl-B_{12}. Comparative analysis of the MetH methionine synthase, cycling between Co^{I} and Co^{III} states, and the adenosyl-B_{12} rearrangement enzymes that cycle between Co^{I} and Co^{II}, show the remarkable redox versatility of the cobalt ion embedded in the tetrapyrrolic corrinoid macrocycle. The ease of homolysis of the Co–C5′ bond in adenosyl B_{12} is a ready route to the deoxyadenosyl radical (Figure 9.8).

The second route to reversible generation of the same robust dA• one electron oxidant is from *S*-adenosylmethionine in enzymes where it is co-ordinated to one of the iron atoms in a 4Fe–4S cluster (Figures 6.6 and 9.6). In comparison to the approximately one dozen adenosyl B_{12} radical-generating enzymes, there are now over 100 000 bioinformatically predicted "radical SAM" enzymes. Only a few dozen have been purified and studied thus far, but there is no reason to doubt the high predictive value of the algorithms. It appears that *S*-adenosylmethionine has won out over adenosyl B_{12} in the evolutionary process as a dA• radical generator.

Both the adenosyl B_{12} enzymes and the few dozen radical SAM enzymes that have been purified and characterized to date are oxygen-sensitive, consistent with evolution of both adenosyl B_{12} and *S*-adenosylmethionine in the eons before substantial global oxygenation of the atmosphere. The radical SAM enzyme family oxygen lability is typically due to autoxidative decomposition of the Fe–S clusters. The adenosyl-B_{12} bond homolyzes both photolytically and by O_2 electron transfer.

9.3 The Scope of Radical SAM Reactions

Radical SAM catalytic cycles are typically initiated by one-electron transfer from an exogenous reductant (*e.g.* a flavodoxin protein)[14] into the oxidized 4Fe–4S cluster (Figure 9.6). In turn one electron gets transferred out from the reduced 4Fe–4S cluster to the coordinated SAM molecule, perhaps directly to the sulfonium center. As that itinerant electron ends up at the cationic sulfonium center of the bound SAM, homolytic cleavage of the S–C5′-adenosyl bond proceeds. One electron from the bond goes to the sulfur, a net two-electron reduction of the trigonal sulfonium to the thioether of methionine that stays coordinated to the cluster *via* its carboxylate, The other electron in the S–C5′ bond, as it breaks, goes to the 5′ carbon to yield *in situ* dA• (Figure 9.6). This would be the temporal sequence if the electron from the 4Fe–4S cluster arrives on the SAM sulfur before the S–C5′ bond breaks. If the process is concerted then a discrete SAM sulfur cation radical would be avoided but the transition state would still involve single electrons coming from two sources to give the net reduction of the trigonal sulfonium cation of SAM to the neutral thioether of Fe–S-coordinated methionine.

With the generation of the dA• radical, the substrate half of each radical SAM catalytic cycles begins. As we noted in the adenosyl B_{12} cases, the dA• is a sufficiently robust oxidant that it can abstract H• atoms from unactivated C–H bonds of cosubstrates bound in the active sites. Those substrate carbon-centered radicals set off a host of different outcomes, dependent on specific substrate structure. Figure 9.9 shows a partial list of the kinds of

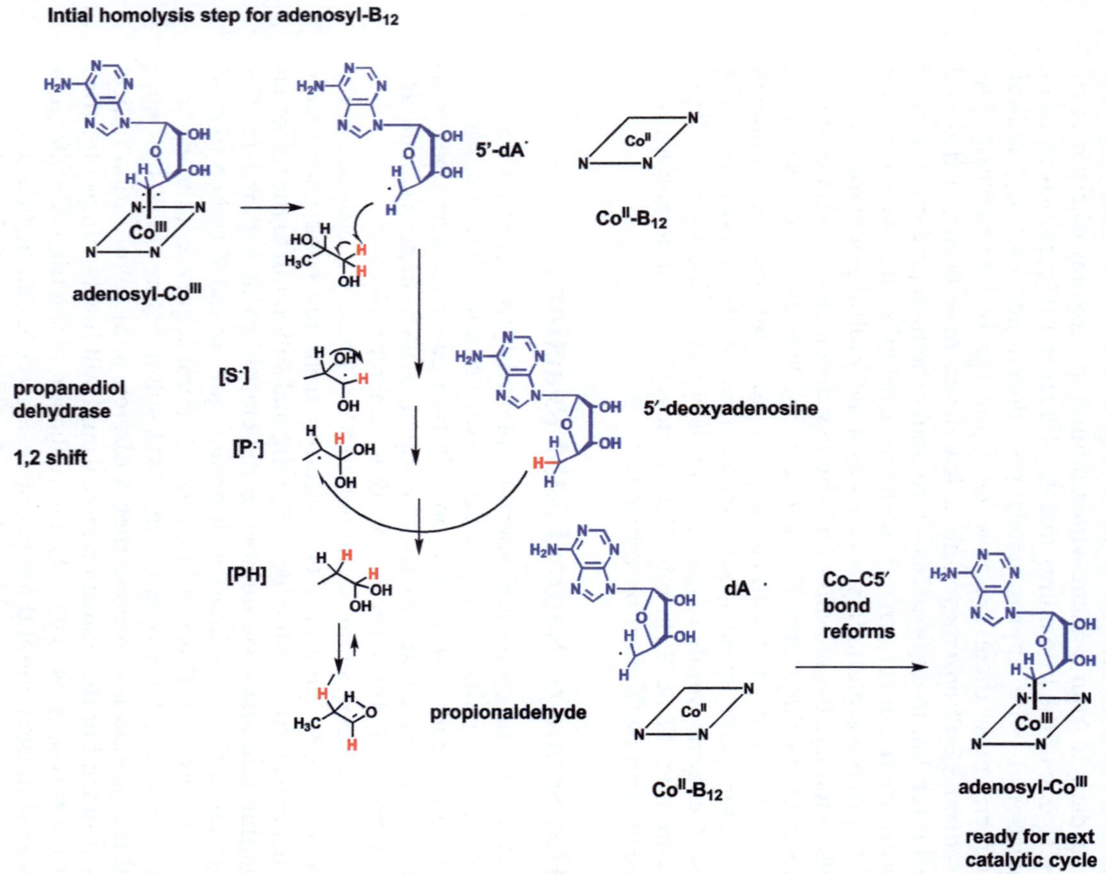

Figure 9.8 Adenosyl-B_{12}-dependent propanediol dehydrase catalyzes homolysis of the B_{12} coenzyme Co–C5′-adenosyl bond to yield $dA^{\bullet}$ as the species that initiates radical chemistry on bound diol substrate. $H^{\bullet}$ abstraction from substrate generates dA–H and the substrate radical transiently. Once the product radical has formed, it is quenched by $H^{\bullet}$ transfer back from dA and the starting adenosyl B_{12} bond is reformed from Co^{II} B_{12} and $dA^{\bullet}$ for the next catalytic cycle.

radical-based transformations seen. The generation of dA• in catalytic microenvironments opened up a universe of chemical biology.

From the point of view of the starting SAM substrate molecule, three kinds of fates have been catalogued (Figure 9.10).

One fate is where the SAM molecule is regenerated at the end of each catalytic cycle. That is, when the substrate to product dust settles, the dA• is regenerated

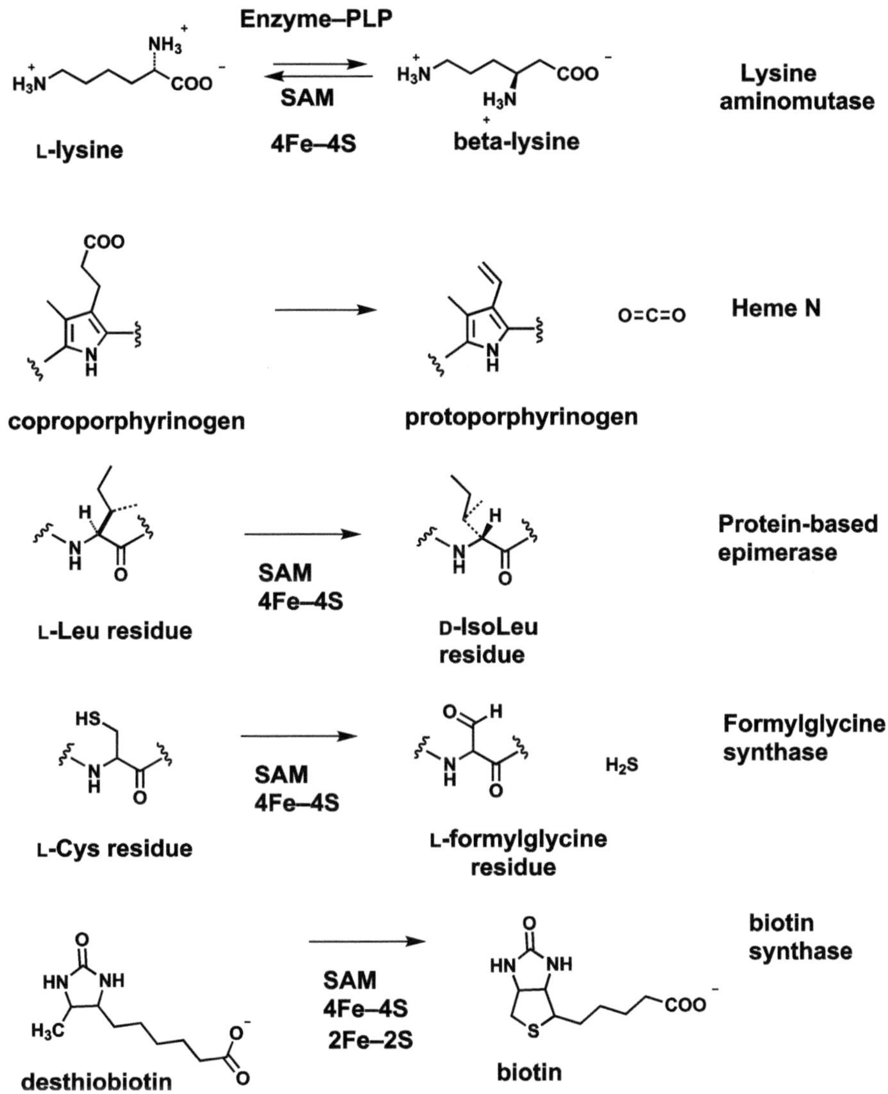

Figure 9.9 Five examples of radical SAM enzyme-dependent conversions: lysine to β-lysine; decarboxylation of two propionyl side chains in coproporphyrinogen III to protoporphryinogen IX; epimerization of a protein bound L-leucine residue to a D-isoleucine residue; oxidation of a cysteinyl side chain to a formylglycyl residue; desthiobiotin to biotin.

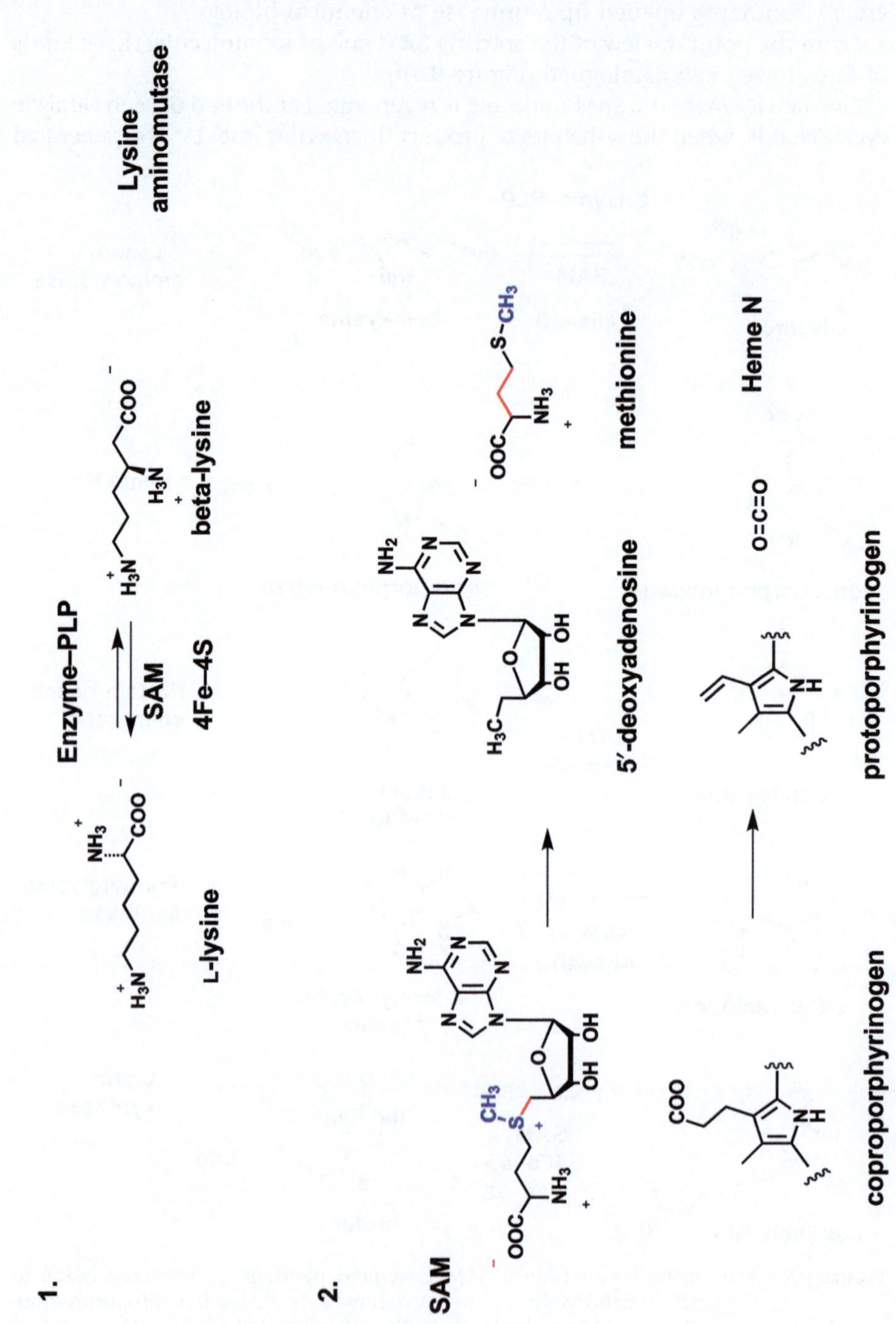
1.
L-lysine
Enzyme–PLP
SAM
4Fe–4S
beta-lysine
Lysine aminomutase
2.
SAM
5′-deoxyadenosine
methionine
coproporphyrinogen
protoporphyrinogen
O=C=O
Heme N

3.

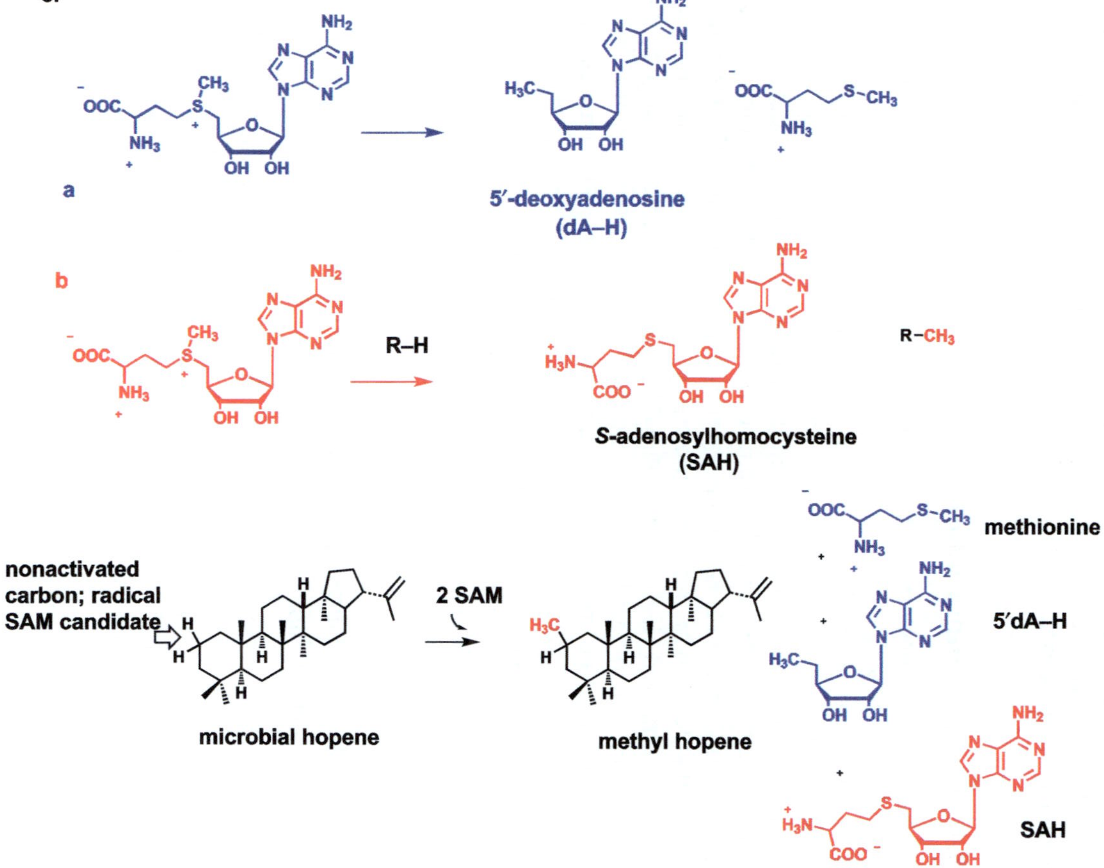

Figure 9.10 Three variants of radical SAM enzymes based on *S*-adenosylmethionine consumption and product identity. (1) SAM acts catalytically, not consumed. (2) One equivalent of SAM is cleaved to dAH and methionine. (3) Two equivalents of SAM consumed; one to dAH and methionine, one to *S*-adenosylhomocysteine (SAH) and methylated cosubstrate.

(from a bound product radical) and can recouple with the methionine still coordinated to the Fe–S center (This is strategically akin to the reformation of the intact adenosyl B12, Co^{III}–C5′ carbon bond at the end of each 1,2-rearrangement catalytic cycle above). The second electron needed for regeneration of the S–C5′ bond comes from the thioether sulfur, while its second electron goes back to the Fe–S cluster to reduce it back to starting state. In this mode SAM acts catalytically, is not consumed, and so does not appear in reaction stoichiometries.

9.3.1 Lysine Aminomutase

Such a catalyst is bacterial lysine isomerase, interconverting L-lysine and β-lysine[12] (Figure 9.11).

The flux starts with cleavage of Fe–S-coordinated SAM to the 5′-deoxyadenoxyl radical. The cosubstrate L-lysine is presented as the PLP aldimine adduct. This is a case where the amino acid–PLP adduct engages in one-electron, radical chemistry rather than two-electron carbanion chemistry, as suggested in Figure 9.11. Hydrogen atom transfer from one of the C3 hydrogens of the lysine–PLP adduct temporarily creates the dA–H reservoir of hydrogen atoms and yields the Lys–PLP C3 radical. This can rearrange to the indicated three-membered aziridine radical. That can reverse to the initial C3–PLP radical or go in the forward direction to the C2–PLP radical. That partition is the essence of the mutase reaction. Now, H• transfer back from dA–H creates the product β-Lys–PLP product adduct and the dA•. The last step in the catalytic cycle is radical recombination to reform the S–C5′ bond in SAM. One electron comes from the methionine thioether to complete the S–C5′ bond while the other electron from sulfur goes back into the Fe–S cluster.

9.3.2 SAM to 5′-Deoxyadenosine and Methionine: HemN Decarboxylase

A second mode of radical SAM enzyme catalysis is where the product radical gets an electron or hydrogen atom from a distinct source, perhaps a metal center or exogenous redox agent.[11] Then, the starting SAM will have been cleaved irreversibly and stoichiometrically to the products methionine and 5′-deoxyadenosine. This is perhaps the most common reactivity mode observed to date in purified radical SAM enzymes. This product pair is telltale evidence for a radical SAM mechanism. The conversion of propionyl to vinyl side chains on two of the four tetrapyrrole rings during heme biogenesis by the bacterial HemN decarboxylase is such an example in Figure 9.12.

The HemN bacterial enzyme has a 4Fe–4S cluster, works in the absence of O_2, and catalyzes the conversion of coproporphyrinogen III to protoporphyrinogen IX as penultimate step in the heme biosynthetic pathway in microbes. The presence of 5′-deoxyadenosine and methionine as products signals the intermediacy of the dA• radical and one-electron pathways.

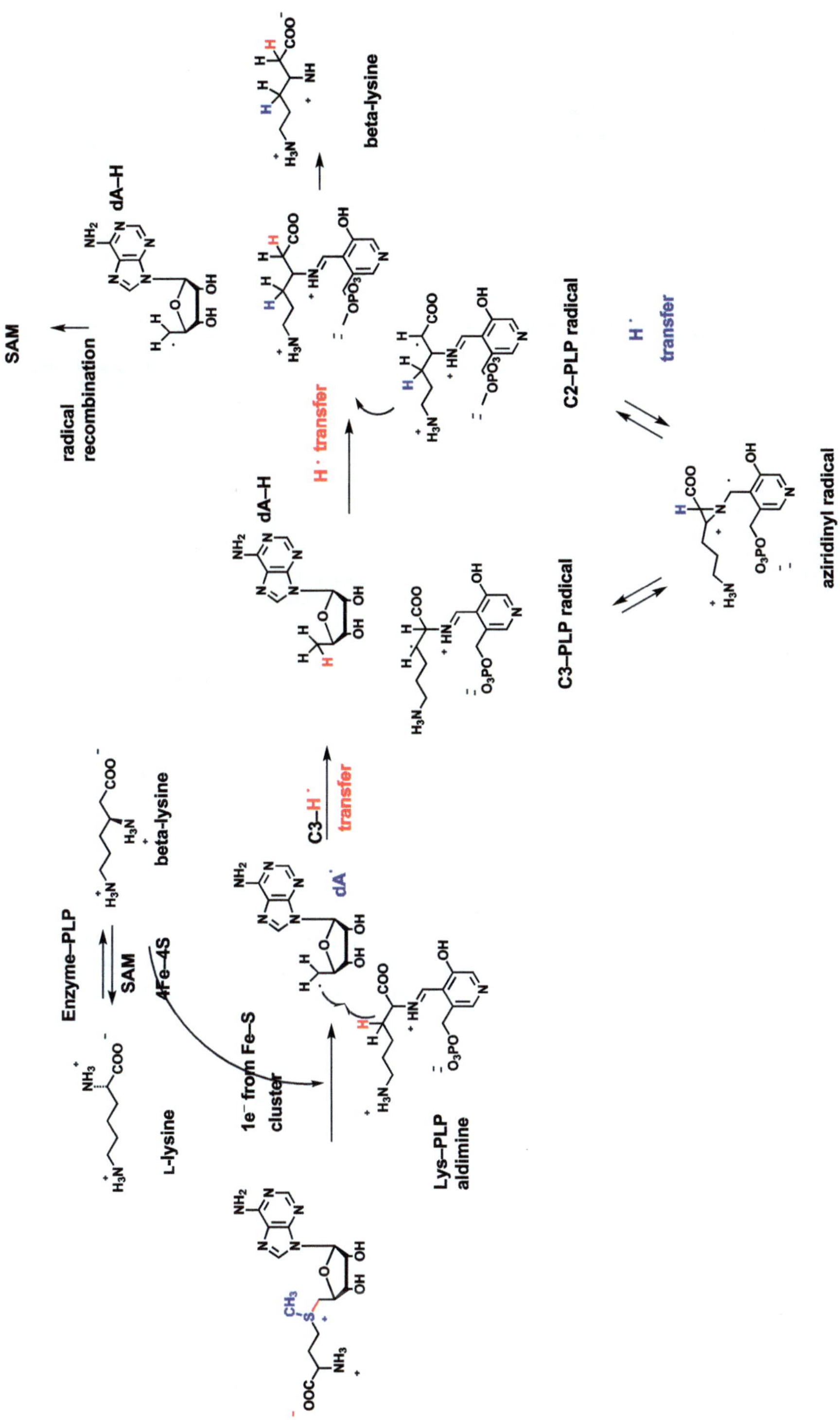

Figure 9.11 Lysine aminomutase epimerizes (α) lysine to β-lysine and *vice versa* as the amino acid–PLP aldimine. Radical chemistry is involved as shown and at the end of the catalytic cycle, SAM is reformed, so it does not show up in reaction stoichiometry.

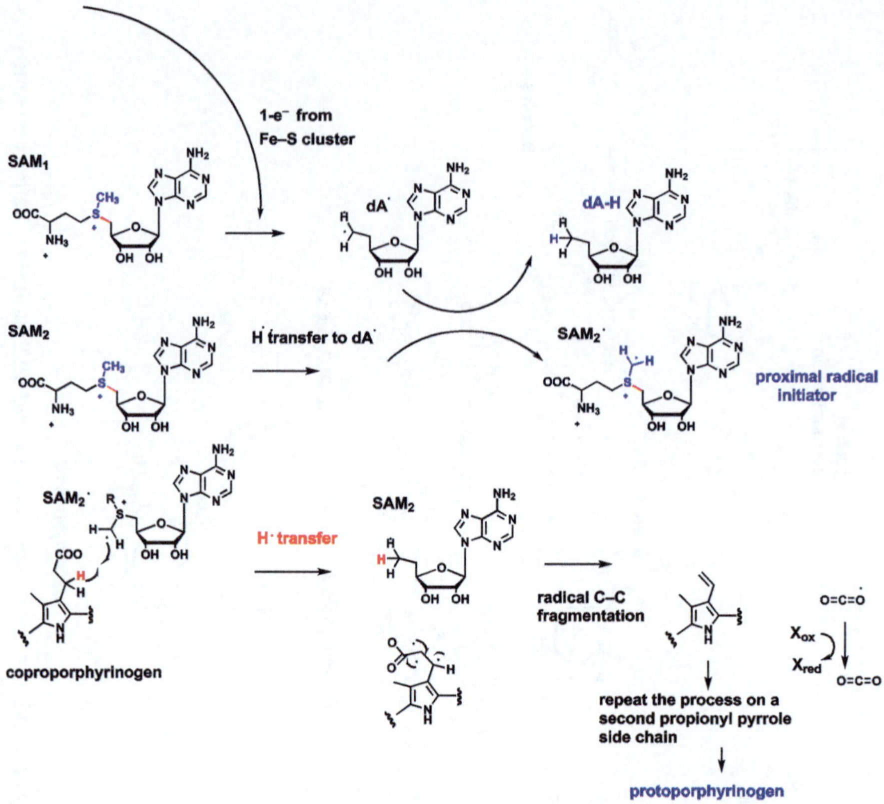

Figure 9.12 The anaerobic bacterial enzyme HemN in the pathway of heme biosynthesis carries out late-stage decarboxylation of two of the four propionyl carboxylate side chains to the two vinyl groups in protoporphyrinogen IX. A radical pathway is proposed: HemN has two SAMs in the active site, the first SAM is cleaved to methionine and dA, the second SAM donates H to dA· to give dAH and SAM_2–CH_2, SAM_2 methylene radical initiates substrate radical decarboxylation but is regenerated and so acts catalytically, the whole process is repeated a second time.

Surprisingly, though, it was observed that two SAM molecules, SAM_1 and SAM_2,were bound in the HemN active site and that both are required in each catalytic cycle.

Furthermore, there is a twist in the fate of the 5′-deoxyadenosyl radical. Rather than directly abstracting a hydrogen atom from one of the two propionyl side chains of the coproporphyrinogen substrate, the dA$^•$ abstracts a hydrogen atom from the methyl group of SAM_2, forming the dAH product before any coproporphyrinogen-based radical chemistry occurs (Figure 9.12).[15] It is instead the $CH_2^•$ radical form of SAM_2 that initiates H$^•$ transfer from Cβ of a propionyl side chain. The pyrrolyl side chain radical can decompose by various one-electron routes. One is shown in Figure 9.12 and indicates that one-electron transfer out from a nascent $CO_2^•$ to an as yet unidentified electron acceptor would complete the reaction. Analysis of the fate of the two SAM molecules shows that the first, SAM_1, is cleaved as expected to methionine and dAH. The second SAM acts catalytically and does not show up in the reaction stoichiometry. However, it is that SAM_2 radical, not the initial 5′-deoxyadenosyl radical, that engages in one-electron chemistry with the substrate.

This overall process must occur tandemly on two pyrrole rings (two separate catalytic turnover events) to get to protoporphyrinogen IX.

9.3.3 Consumption of Two SAM Molecules to Two Sets of Products

A third reaction stoichiometry occurs when a radical SAM enzyme actually functions as a methyltransferase. In that case, the first SAM is fragmented to methionine and 5′-deoxyadenosine products. A second SAM substrate is the methyl donor and exhibits the typical cleavage pattern to *S*-adenosylhomocysteine and methylated coproduct. Thus, the same enzyme catalyst can conduct both modes of SAM reactivity, with two distinct fragmentation patterns. In the biosynthetic maturation of the antibiotic gentamicin C1 to the compound G418, precisely that stoichiometry is observed.

Based on the sets of products from usage of the two SAM substrates, one would propose that the first molecule of SAM would function as a radical generator *via* fragmentation to the 5′deoxyadenosyl radical. This suggests formation of the indicated radical on the target CH_2–OH group of gentamicin C1 by hydrogen atom transfer as shown in Figure 9.13. The second SAM substrate molecule is processed to SAH as a methyl group ends up in the G418 product. The conventional wisdom would be a carbanion equivalent attacks the CH_3 substituent on the sulfonium cation of SAM#2, transferring the one carbon moiety as a $[CH_3^+]$ equivalent.

The question in the methyl hopene case in Figure 9.12 and the gentamicin case in Figure 9.13 with the stoichiometry of cleavage of two SAM molecules to the two distinct sets of fragmentation products raises the question of how the methyl transfer is effected. Whether the presumptive radicals gain another electron and become transient carbanion equivalents is unknown. Whether the second SAM substrates actually function instead as transient $[CH_3^\bullet]$ donors is another possibility[16] (not shown). This enzyme family clearly deserves more mechanistic scrutiny.

9.3.4 Likely Ancient Coenzymes from Radical SAM Pathways

In addition to heme, several other cofactor and coenzyme biosynthetic pathways (to thiamin, 5-deazariboflavin, molybdenum cofactor and menaquinone, the vitamin K core framework) involve one or more radical SAM-mediated catalytic steps, indicating that both the one-electron routes and the cofactors themselves may be ancient molecules (Figure 9.14). Without

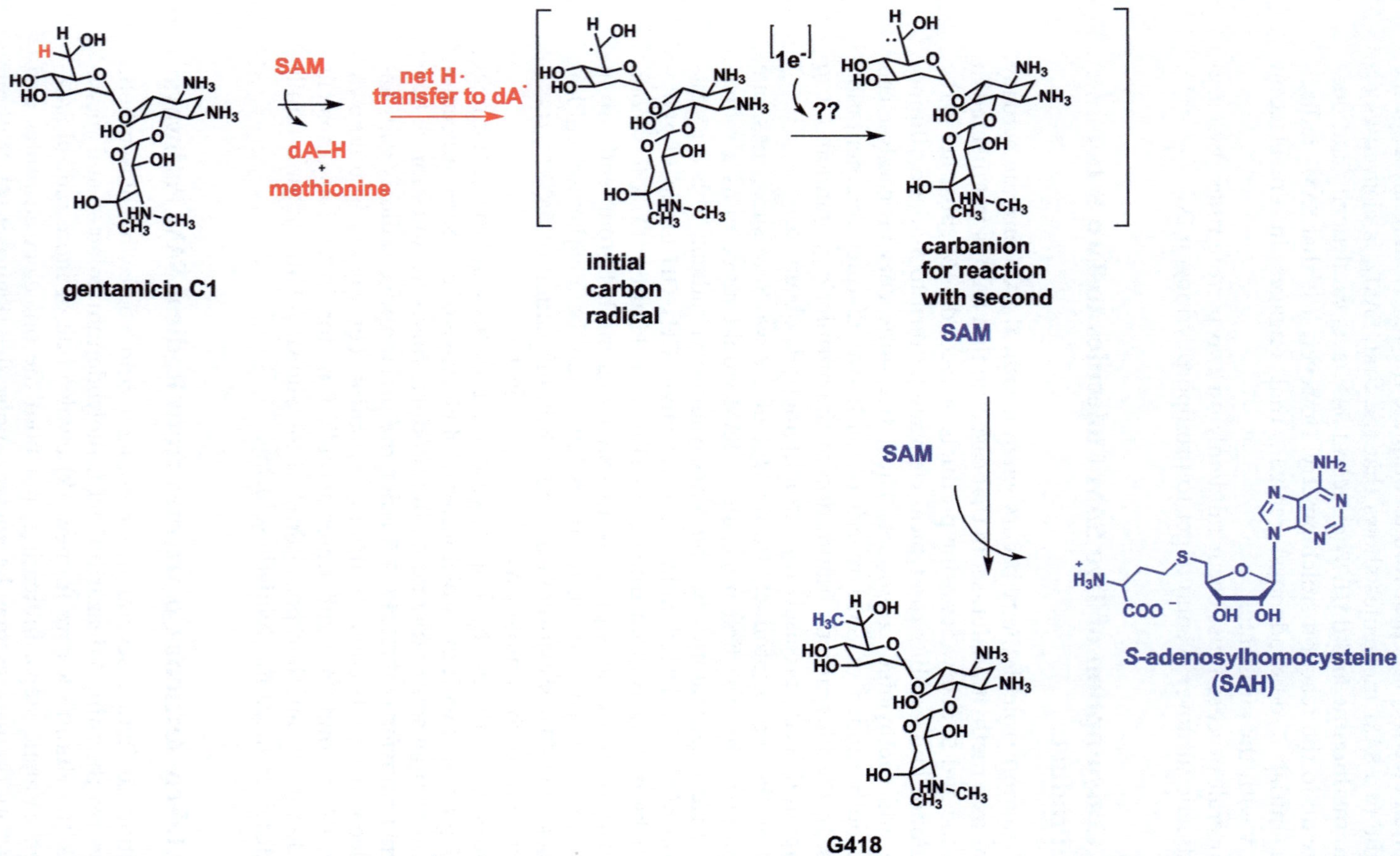

Figure 9.13 Biosynthetic conversion of gentamicin C1 to the metabolite G418 consumes two molecules of cosubstrate SAM. The first SAM is cleaved to dAH and methionine, consistent with the proposed radical intermediate. The second SAM donates a methyl group and yields SAH as coproduct. The mechanism of the CH_3 transfer is not clear but would typically involve $[CH_3^+]$ transfer.

going into any biosynthetic detail, these examples give additional scope into the distinct chemical scaffolds that are likely to be constructed by one-electron chemistries.

9.4 Cannibalization of 2Fe–2S or 4Fe–4S Clusters in Biotin and Lipoic Acid Biosynthesis

Two remarkable examples of cannibalization of either one or two of the sulfide atoms in Fe–S clusters occur in the biosynthesis of the thioether moiety in biotin and in the disulfide ring in the coenzyme lipoic acid[17] (Figures 9.15 and 9.16). Biotin, vitamin B7, has two functional groups in its bicyclic scaffold, the ureido group in the top five-membered ring and the thioether sulfur in the bottom five-membered ring. The N1 amide of the ureido group is the reactive site for CO_2 fixation and chaperoning, but the

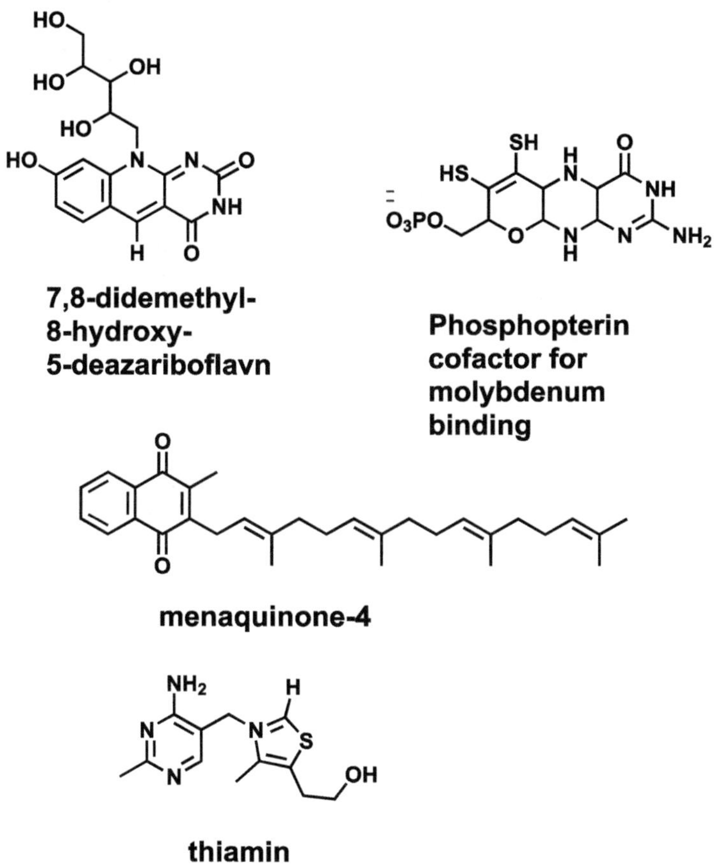

Figure 9.14 Coenzymes and cofactors presumed to be assembled *via* one or more radical SAM enzyme actions: 8-hydroxy-5-deazariboflavn, molybdopterin, menaquinone, thiamin.

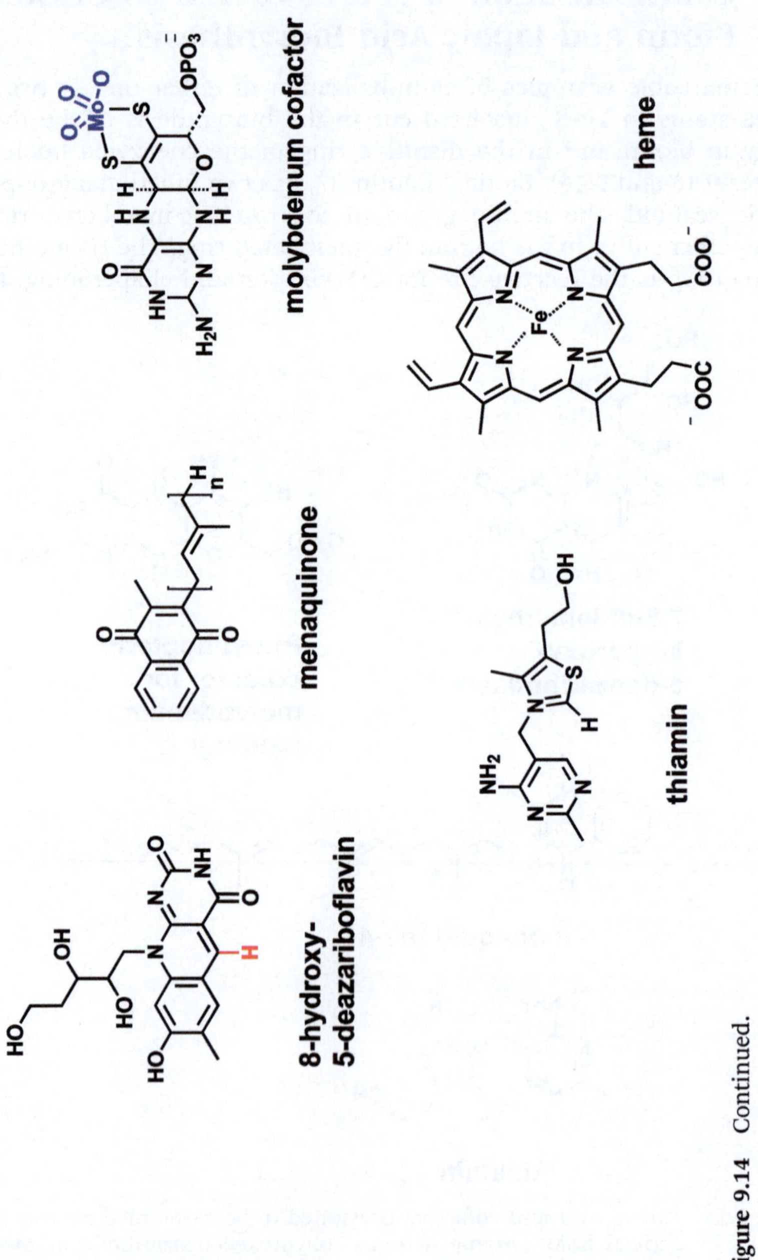

Figure 9.14 Continued.

cis-fused thiane ring, which imposes a right-angle geometry to the bicyclic scaffold, is essential. Lipoic acid with its 6,8-dithio group when reduced and a cyclic disulfide when oxidized, is the key redox active cofactor in oxidative decarboxylation of α-keto acid metabolites.

The last step in biotin biosynthesis is conversion of desthiobiotin to biotin, a formal insertion of one sulfur atom in between the pendant CH_3 group and one of the CH_2 groups of the valeric acid side chain of desthiobiotin (Figure 9.15).

This is a radical SAM enzyme that commits suicide in each turnover and has yielded its mechanistic secrets only after some decades of intense scrutiny. The 5′dA$^\bullet$ intermediate from the first SAM substrate molecule initiates radical chemistry on the bound desthiobiotin cosubstrate, removing a hydrogen atom (H$^\bullet$) from the indicated CH_3 group. The resultant $CH_2^\bullet$ is then able to extract a sulfur atom (S$^\bullet$) from an adjacent 2Fe–2S cluster. This 2Fe–1S cluster decomposes (the suicidal outcome) and the biotin scaffold intermediate has a pendant CH_2SH. In a formal sense, a sulfur atom has been inserted into one of the C–H bonds of the CH_3 group of the substrate.

Now this happens a second time, presumably with a second molecule of SAM fragmenting to dA$^\bullet$. This (second) dA$^\bullet$ is proposed to also abstract a hydrogen atom (H$^\bullet$) from the monothiol biotin intermediate, converting the C6 methylene of the side chain to CH$^\bullet$. This is proposed to act intramolecularly with the sulfur of the pendant thiol group, in a one-electron reaction manifold to create the second CH_2–S bond in mature biotin. The residual unpaired electron is presumably transferred back to the 2Fe–2S cluster in its oxidized state as shown in Figure 9.15.

The net outcome is use of two molecules of SAM, presumably in one enzyme active site, to create the two CH_2–S bonds in the biotin product. Alternatively, the second step could come from a second enzyme molecule that does not cannibalize its own 2Fe–2S destruction as the second of the two C–S bonds forms in the last step of biotin thiane ring closure. This is remarkable radical chemistry, resulting in a net insertion of a sulfide ion between two carbon atoms, producing the final cyclic thioether by way of a thiol intermediate.

The lipoic acid synthase was equally difficult to purify and study in active form to deconvolute its molecular logic. It now appears that a similar mechanism is used, this time on an enzyme-bound 4Fe–4S cluster to cannibalize two of its four sulfide atoms (Figure 9.16). As in the biotin example, initial cleavage of SAM into dA$^\bullet$ goes *via* one-electron input into, and then out from the 4Fe–4S cluster. The dA$^\bullet$ radical then converts the C8 methyl of the precursor octanoyl-*S*-acyl carrier protein (diverted from bacterial fatty acid synthase machinery) to the $CH_2^\bullet$ substrate radical. Extraction of one of the sulfides from the 4Fe–4S cluster yields the 8-CH_2–SH thiol linkage as in the first step of the biotin synthase scheme above.

Then, the C-6–CH_2 group of that 8-thiooctanoyl-*S*–ACP nascent product is subjected to another round of SAM cleavage, dA$^\bullet$ formation, and abstraction

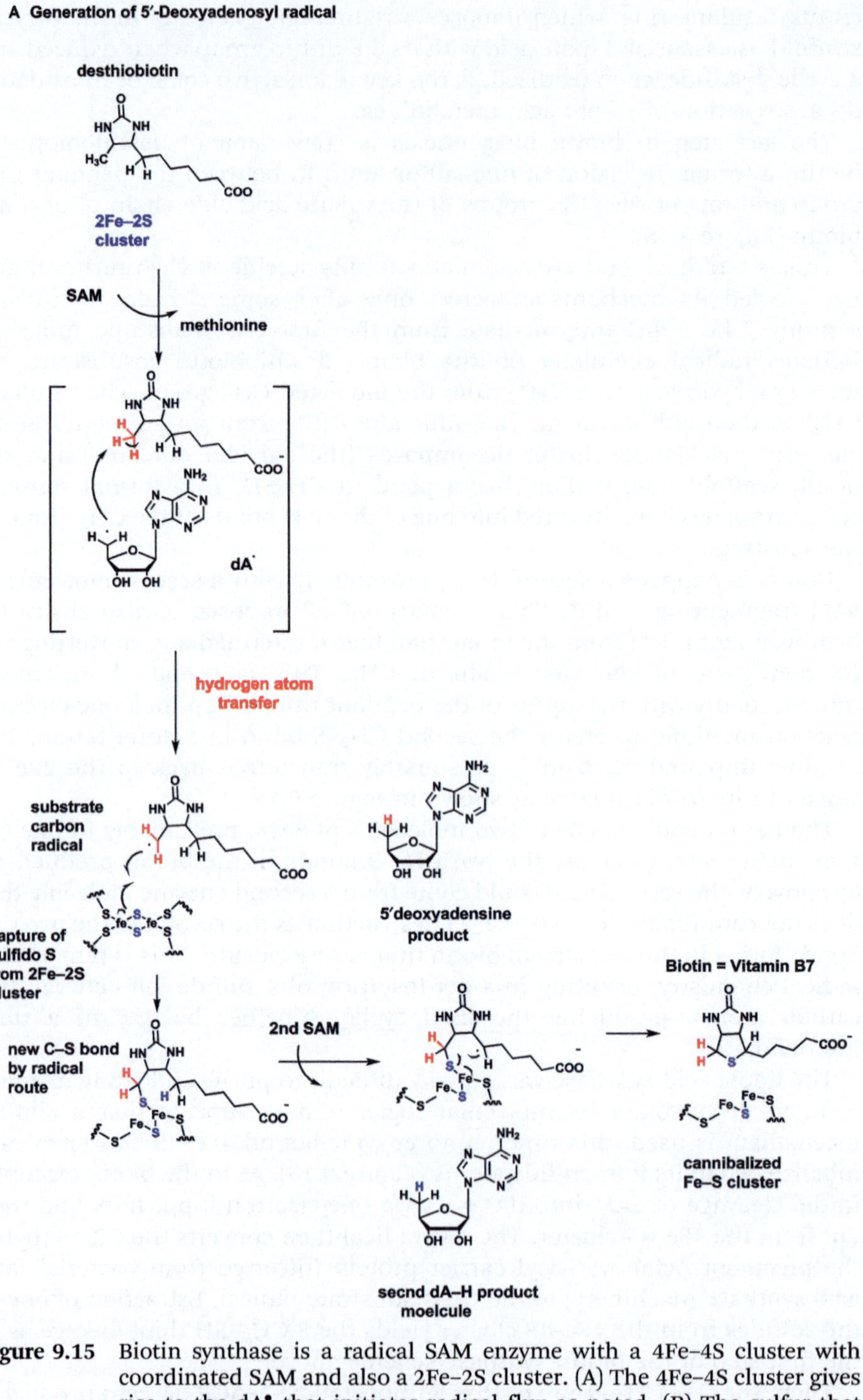

Figure 9.15 Biotin synthase is a radical SAM enzyme with a 4Fe–4S cluster with coordinated SAM and also a 2Fe–2S cluster. (A) The 4Fe–4S cluster gives rise to the dA• that initiates radical flux as noted. (B) The sulfur that becomes inserted in the newly formed biotin thiane ring comes from suicidal destruction of an adjacent 2Fe–2S cluster.

B Biotin Synthase: Radical SAM 2X and Cannibalism of the 2Fe–2S cluster

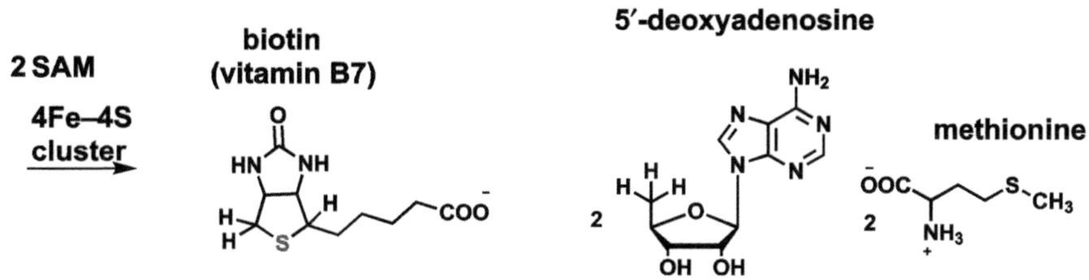

Figure 9.15 Continued.

of one of the two prochiral C6-CH_2 hydrogens as a hydrogen atom ($H^•$). This generates dA–H (5′-deoxyadenosine) from the second molecule of SAM. The C6 methylene radical can scavenge one more of the three remaining sulfide ions in the wounded 4Fe–3S cluster by radical recombination to yield the 6,8-dithiol octanoyl-*S*–ACP.[18] This is the scaffold of the dithiol form of lipoate. Hydrolysis of the thioester to free dihydrolipoate is followed ultimately by covalent attachment of the lipoic acid to a small set of partner proteins *via* amide linkage to a side chain lysine residue.

In the resting form of the enzyme complexes where lipoamide functions it has undergone oxidative conversion to the intramolecular disulfide form, akin to the cystine disulfide in the glutathione reductase enzymes.[19] For example, in the three-enzyme pyruvate dehydrogenase complex, the disulfide form of tethered lipoamide is captured by the hydroxyethyl C-2-carbanion, itself tethered to the thiazolium ring of thiamin-PP in the neighboring enzyme. The two-carbon transfer to lipoamide constitutes a net oxidation of the migrating two-carbon unit from aldehyde to acid oxidation state.

The mechanistic strategy yields the thermodynamically activated acetyl thioester of lipoamide (Figure 9.17). Thiol exchange with free CoASH then produces the mobile activated cellular acetyl transfer currency acetyl-*S*–CoA.

The lipoamide example is a reminder that disulfide sulfurs are electrophilic, whether in cysteine, glutathione or lipoamide and also that acyl thioester chemistry is pervasive in metabolic logic. Finally, the use of a disulfide instead of a thiol for acyl *S*–enzyme formation, allows for polarity reversal in acyl thioester formations: from the thiolate as nucleophile on an acyl group *vs.* the disulfide as electrophile as it undergoes oxidation to the acyl thioester (Figure 9.17).

Overall, the sacrificial scavenging of sulfide ions (one or two) during biotin and lipoic acid assembly show a novel role for Fe–S clusters. In the approximately 100 000 or more as yet unexplored radical SAM open reading frames predicted in databases, one may see more such cannibalization for sulfur insertions.[16] The large set of predicted, but as yet unexplored, open reading frames augur for new types of SAM-mediated radical chemistry reaction manifolds.

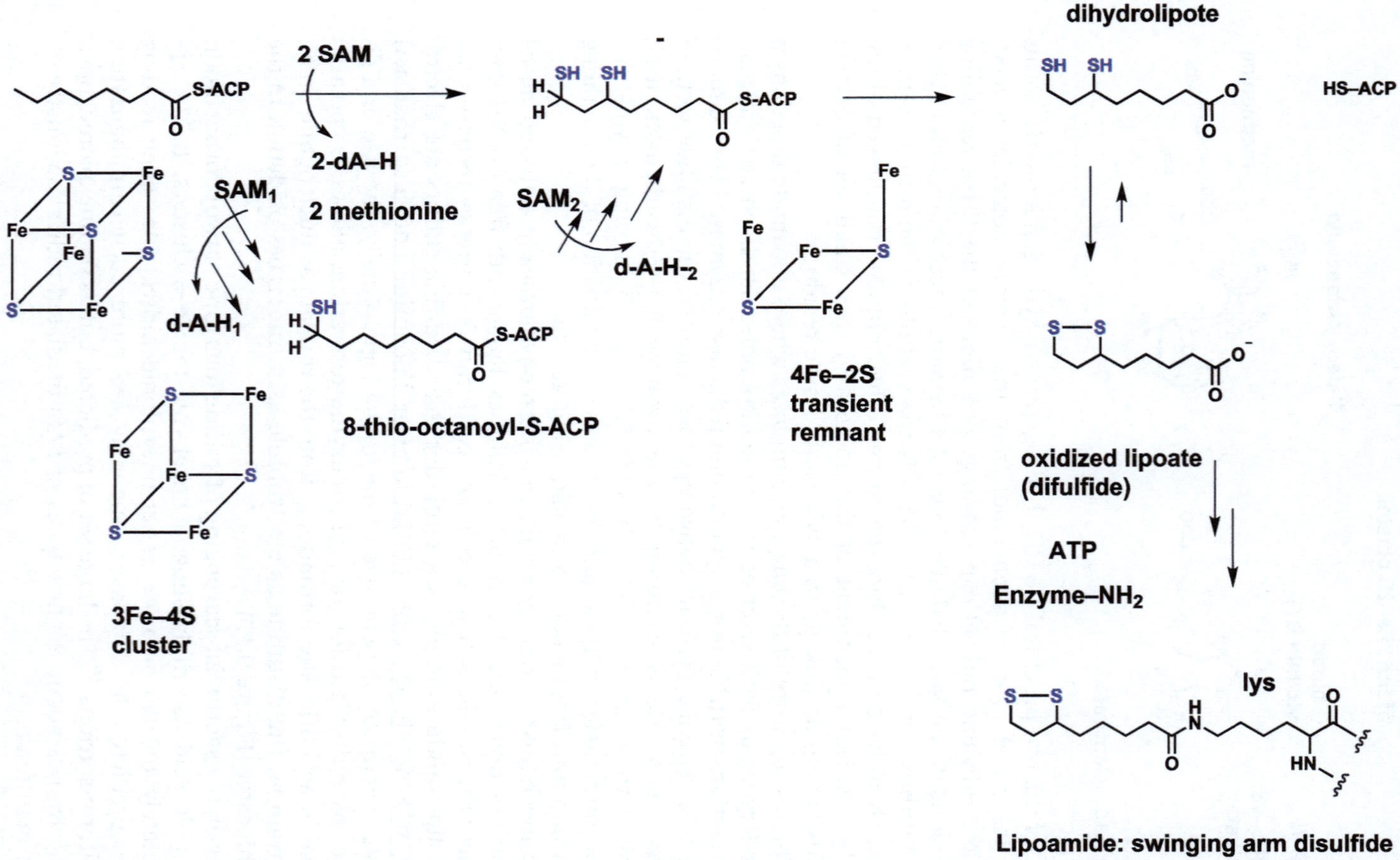

Figure 9.16 Lipoyl-*S*–ACP synthase converts octanoyl-*S*-acyl carrier protein to the 6,8-dithiol product. It uses two SAMs in tandem to do radical chemistry, first at C8 and then at C6 of the lipoyl moiety. Two of the four sulfide ions from a 4Fe–4S cluster are extracted as sulfur radicals for the two S–H bond-forming steps.

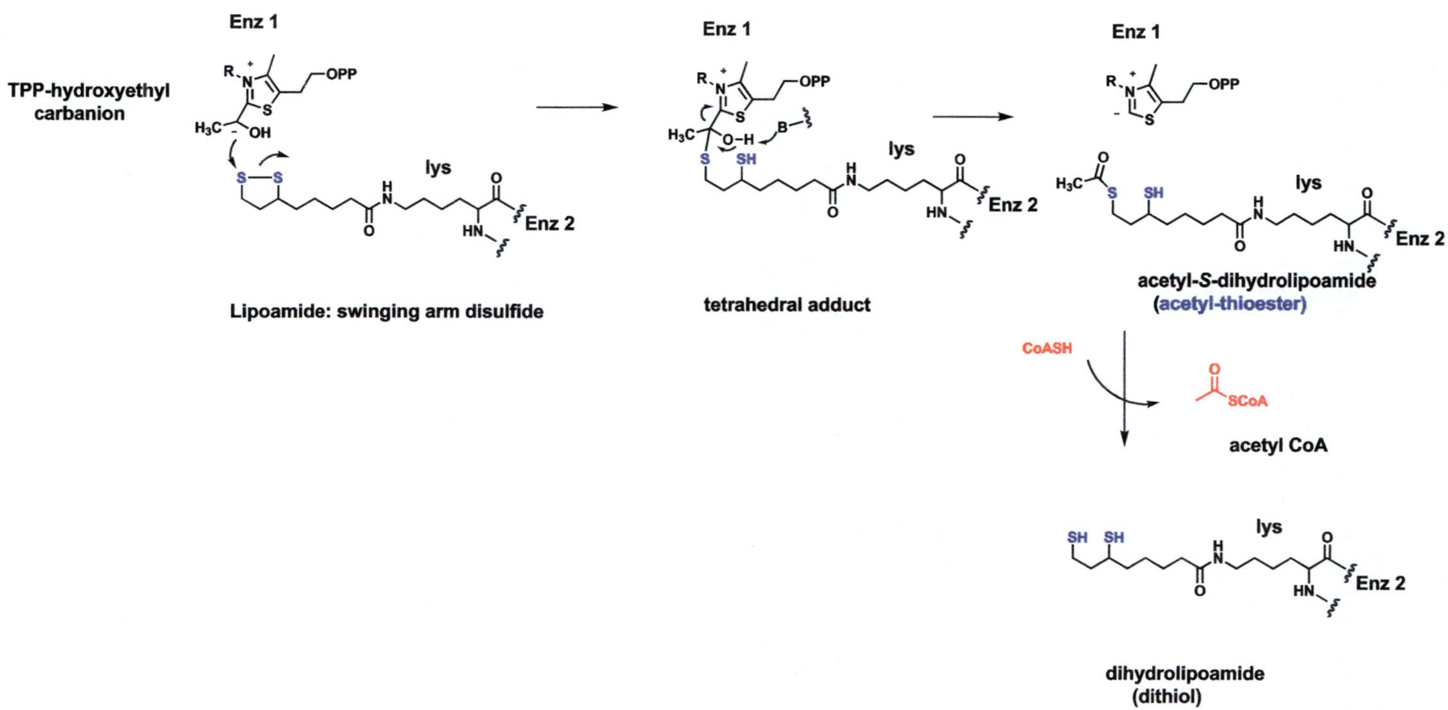

Figure 9.17 The resting state of the lipoyl group in its host protein is the cyclic disulfide oxidation state. That oxidation state serves as the electrophile for transfer of a hydroxyethyl carbanion equivalent from covalent connection to the thiamin–PP coenzyme in enzyme 1 to the lipoyl disulfide in enzyme 2. In the course of that transfer, the hydroxyethyl moiety is oxidized to an acetyl fragment and the lipoyl disulfide is reduced to an acetyl thioester and a thiol side chain.

The biotin synthase and lipoyl-ACP synthases illuminate two distinct roles for one-electron enzyme chemistry to and from sulfur: (1) the reaction-initiating cleavage of the SAM sulfonium S–C5′ bond and (2) the radical-mediated destructive extraction of sulfur atoms from Fe–S clusters to build the C–S bonds in biotin and liopate frameworks.

They emphasize the propensity of the arrays of 4Fe–4S clusters with coordinated SAM molecules to carry out one-electron chemistry at the sulfonium oxidation state. Coupled to the orthogonal reactivity of SAM for billions of $[CH_3^+]$ transfers in every cell cycle to a plethora of high and low molecular weight cosubstrate nucleophiles, it is likely that the sulfonium atom in SAM enables the largest number of reaction modes among all the molecules of life.

9.5 Combined Methylthiolation *via* Two SAMs and a 4Fe–4S Sacrifice

An additional twist on the radical SAM, Fe–S cluster sacrificial donation of one of its sulfur atoms to cosubstrate occurs in the MiaB and RimO subfamily of radical SAM enzymes.[20] MiaB encodes the second enzyme in a two-enzyme pathway that leads to formation of a modified adenosine residue in the wobble position of tRNAs that decode mRNA triplets beginning with a U.[21] The modified adenine is 2-methylthio-*N*-6-isopentenyladenine. MiaA transfers the isopentenyl group from isopentenyl diphosphate first. Then, MiaB, acting in two steps, transfers in a methylthio group to the unactivated C2-position of adenine, replacing the C2–H with a methylthio group (Figure 9.18A). In some analogy the RimO radical SAM enzyme converts the side chain of Asp88 in the small ribosomal subunit protein S12 into a 3-methylthio-Asp residue, a universally conserved posttranslational modification.[22]

The stoichiometry of both RimO and MiaB reactions involves utilization of two molecules of SAM in each catalytic cycle. The first SAM is the source of the 5′-deoxyadenosyl radical (dA•), while the second SAM is the methyl donor in the net methylthiolation. The quandary was the source of the thiol sulfur atom. This was resolved when structural analyses of both MiaB and RimO indicated a second 4Fe–4S cluster, over and above the 4Fe–4S cluster that coordinates the first SAM molecule. That second 4Fe–4S cluster is proposed as the sacrificial donor (as in lipoate assembly above) of the sulfur atom that gets incorporated at C_2 of adenine (MiaB) or C_3 of Asp88 in the ribosomal S12 protein (Rim O).

The mechanisms would then be as shown in Figure 9.18B.[21] The electron input into the first 4Fe–4S cluster results in the canonical homolytic cleavage of SAM_1 to generate dA•. In the MiaB active site this initiates adenine C_2–H homolytic cleavage to create dA–H (observed product from SAM_1) and the adenine C_2 radical. This would be the radical that destroys the $[4Fe–4S]_2$ by one-electron chemistry and sulfur atom transfer, creating a C_2–SH bond. This nucleophilic thiol, in the 2-thio-6-isopentnyly adenine intermediate,

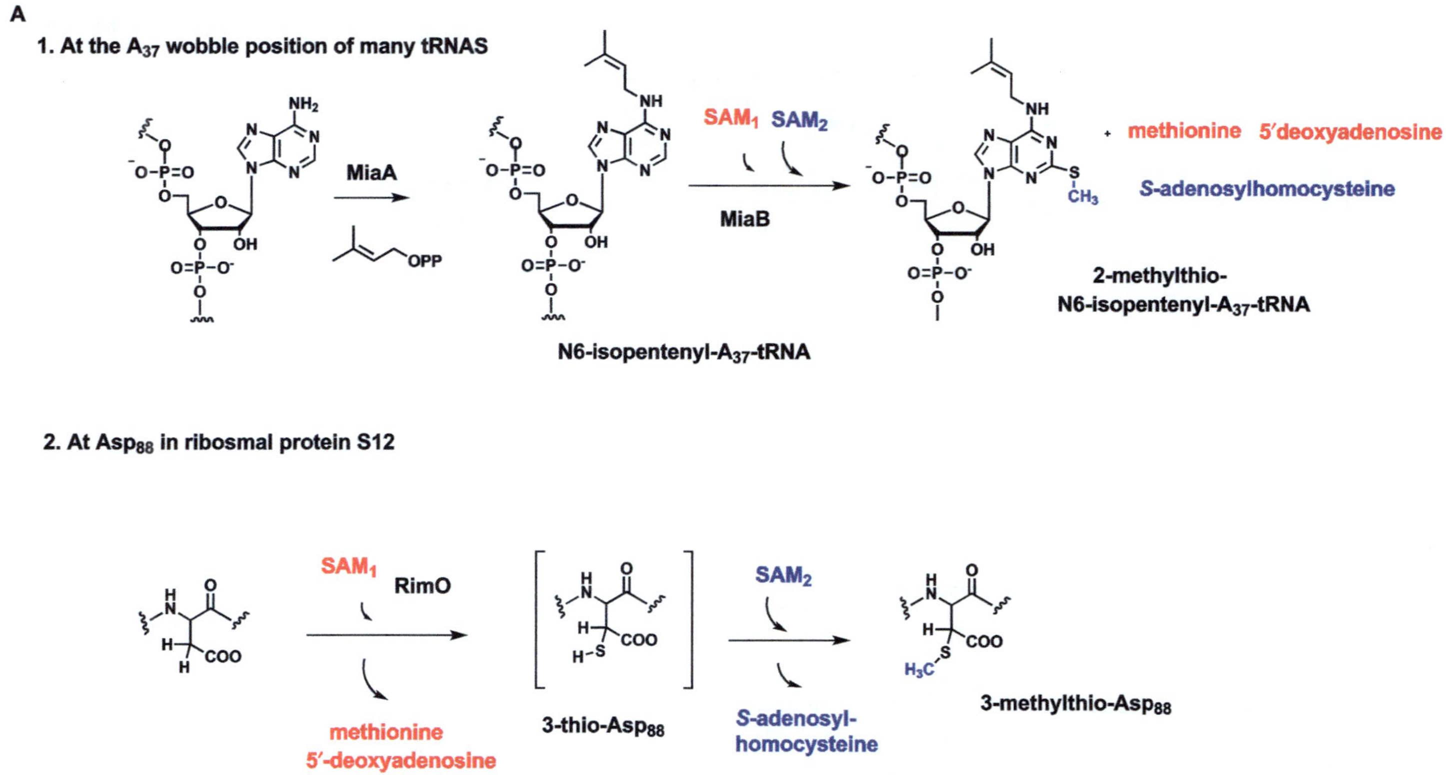

Figure 9.18 Two-step methylthiolation of a tRNA and a bacterial S12 ribosomal protein involves two SAMs and two 4Fe–4S clusters. (A) Reaction stoichiometry for *N*-6-isopentenyl-2-metylthio adenine$_{37}$-tRNA formation by MiaB; and of 3-thiomethyl-Asp88 in S12 protein by RimO, two members of a thiomethylation radical SAM enzyme subclass. (B). Thiol transfer by one-electron pathways involves sacrificial donation of a sulfur atom from the $[4Fe{-}4S]_2$ cluster in each enzyme. SAM$_1$ is cleaved to methionine and 5′-deoxyadenosine, while SAM$_2$ is the $[CH_3]^+$ donor to the cosubstrate thiol intermediate.

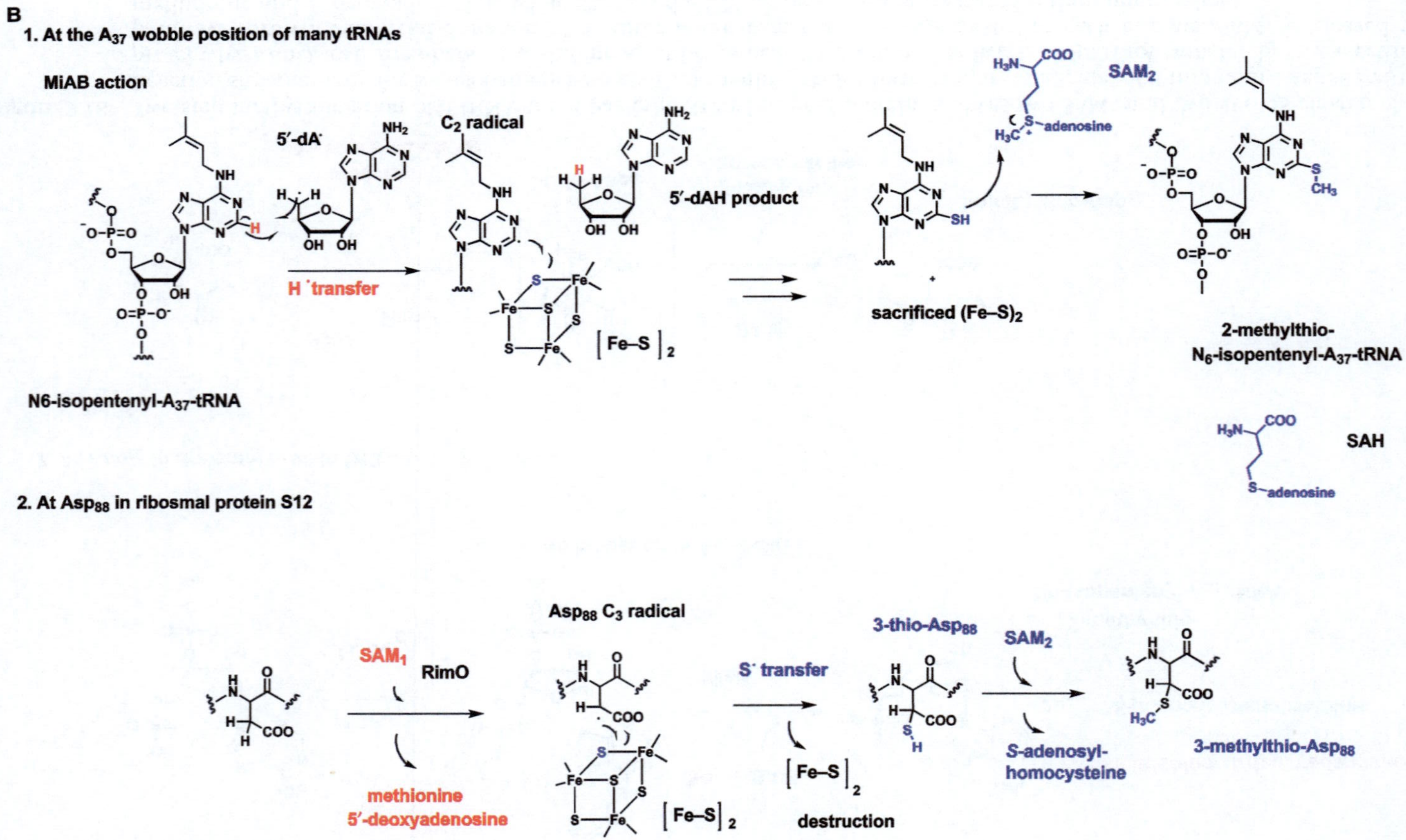

Figure 9.18 Continued.

then attacks the electrophilic CH_3 group of SAM_2 to create the observed methylthiol isopentenyl adenine-tRNA product.

In the RimO active site the transiently formed dA• must be positioned close to the C_3–CH_2 center of the Asp88 side chain of S15 protein to abstract one of the two prochiral hydrogens as an H•. Then, the resultant Asp88 C_3–CH• must again be able to extract one of the sulfur atoms of $[4Fe–4S]_2$, causing its decomposition. The 3-thio-Asp88 form of protein S12 is then the nucleophile towards SAM_2, effecting $[CH_3]^+$ transfer to produce the methylthiol-Asp88 residue.

These two enzyme examples cap off the tour through radical SAM enzymatic chemistry and show the many ways organisms have evolved to generate and control one-electron reaction pathways starting from *S*-adenosylmethionine and iron–sulfur clusters.

References

1. D. W. Bak, T. Bechtel, J. A. Falco and E. Weerapana, Cysteine reactivity across the subcellular universe, *Curr. Opin. Chem. Biol.*, 2019, **48**, 96–105.
2. L. B. Poole, The basics of thiols and cysteines in redox biology and chemistry, *Free Radical Biol. Med.*, 2015, 148–157.
3. M. Kasanmascheff, W. Lee, T. U. Nick, J. Stubbe and M. Bennati, Radical transfer in E. coli ribnucleotide reductase: a NH2Y731/R411A alpha mutant unmasks a new conformation of the pathway residue 731, *Chem. Sci.*, 2016, **7**, 2170–2178.
4. K. A. Shialer and J. Broderick, Gycyl radical activating enzymes: structure,mechanism, and substrate interactions, *Arch. Biochem. Biophys.*, 2014, **546**, 64–71.
5. S. C. Peck, K. Denger, A. Burrichter, S. M. Irwin, E. P. Balskus and D. Schleheck, A glycyl readical enzyme enables hydrogen sulfide production by the human intestinal bacterium *Biophila wadsworthia*, *Proc. Natl. Acad. Sci. U. S. A.*, 2019, **116**, 3171–3176.
6. C. T. Walsh and T. Wencewicz, *Antibiotics, Challengs, Mechanisms,Opportunities*, ASM Press, Washington DC, 2016.
7. P. Rabe, J. Kamps, C. J. Schofield and C. T. Lohans, Role of 2-oxoglutarate oxygenases and isopenicllin N synthase in beta lactam bosynthesis, *Nat. Prod. Rep.*, 2018, **35**, 735–756.
8. H. Tarhonskaya, A. Szollossi, I. K. Leung, J. T. Bush, L. Henry, R. Chowdhury, A. Iqbal, T. D. Claridge, C. J. Schofield and E. Flashman, Studies on deacetoxycephalosporin C synthase support a consensus mechanism for 2-oxoglutarate dependent oxygenases, *Biochemistry*, 2014, 2483–2493.
9. C. T. Walsh and Y. Tang, *The Chemical Biology of Human Vitamins*, Royal Society of Chemistry Press, London, 2018.
10. C. Walsh and Y. Tang, *Natural Product Biosynthesis: Chemical Logic and Enzymatic Machinery*, Royal Society of Chemistry, London, 2017.

11. J. B. Broderick, B. Duffus, K. S. Duschene and E. M. Shepard, Radical S-adenosylmethionine enzymes, *Chem. Rev.*, 2014, **114**, 4229–4317.
12. P. A. Frey, A. Hegeman and G. H. Reed, Free radical mechanisms in enzymology, *Chem. Rev.*, 2006, **106**, 3302–3316.
13. M. Ludwig and R. G. Matthews, Structure-based perspectives on B12 enzymes, *Annu. Rev. Biochem.*, 1997, **66**, 299–313.
14. J. Sancho, Flavodoxins: sequence, folding, binding, function, and beyond, *Cell. Mol. Life Sci.*, 2006, **63**, 855–864.
15. W.-B. Jin, S. Wu, Y.-F. Xu, H. Yuan and G.-L. Tang, Recent advances in HemN-like radical S-adenosylmethionine enzyme-catalyzed reactions, *Nat. Prod. Rep.*, 2020, **37**, 17–28.
16. M. R. Bauerle, E. Schwalm and S. J. Booker, Mechanistic diversity of radical SAM-dependent methylation, *J. Bio. Chem.*, 2015, **290**, 3995–4002.
17. S. J. Booker, R. Cicchillo and T. L. Grove, Self sacrifice in radical S-adenosylmethionine proteins, *Curr. Opin. Chem. Biol.*, 2007, **11**, 543–552.
18. M. I. Mclaughlin, N. Lanz, P. J. Goldman, K.-H. Lee, S. J. Booker and C. L. Drennan, Crystallographic snapshots of sufur insertion by lipoyl synthase, *Proc. Natl. Acad. Sci. U. S. A.*, 2016, **113**, 94346–94350.
19. J. E. Cronan, Assembly of lipoic acid on its cognate enzymes: an extraordinary biosynthetic pathway, *Microbiol. Mol. Biol. Rev.*, 2016, **80**, 429–450.
20. H. L. Hernandez, F. Pierrel, E. Elleingand, R. Garcia-Serres, B. H. Hunyh, M. K. Johnson, M. Fontcave and M. Atta, MiaB, a bifunctionl radical S-adenosylmethionine enzyme involved in thiolation and methylation of tRNA contains two essential[4Fe-4S] clusters, *Biochemistry*, 2007, **46**, 5140–5147.
21. M. Atta, S. Arragaain, M. Fontecave, E. Mulliez, J. F. Hunt, J. D. Luff and F. Forouhar, The methylthiolation reaction mediated by the radical SAM enzymes, *Biochim. Biophys. Acta*, 2012, **1824**, 1223–1230.
22. B. P. Anton, Saleh, J. S. Benner, E. A. Raleigh, S. Kasif and R. J. Roberts, RimO, a MiaB-like enzyme, methylthiolates the universally conserved Asp88 residue of rinosomal protein S12 in Eschericia coli, *Proc. Natl. Acad. Sci.*, 2008, **105**, 1826–1831.

CHAPTER 10

Sulfuryl Transferases and Sulfatases

In Part I we examined the chemical logic and enzymatic machinery for biological redox reactions on inorganic sulfur. As the planetary atmosphere gradually became oxygenated, a combination of abiotic and biotic oxidations of abundant low-valent sulfur (elemental sulfur and sulfide ions) to the fully oxidized sulfate dianions occurred. Thus, today dissolved sulfate is the most abundant of the biologically utilizable forms of sulfur.

In comparison to the transfer of electrophilic $-PO_3^{2-}$ groups from ATP by hundreds of phosphotransferase (kinase) enzymes, there are many fewer sulfotransferases.[1] However, the molecular logic is parallel: transfer of an electrophilic sulfuryl group from a donor molecule with a side chain sulfuric anhydride linkage. Likewise, in comparison to the phosphotransferases that put $-PO_3^{2-}$ groups on nucleophilic atoms of cosubstrates and phosphohydrolases (phosphatases) that remove them hydrolytically, so there are sulfatases[2] to hydrolytically reverse the effects of sulfotransferases (Figure 10.1).

10.1 Sulfotransferases

We have noted that the most abundant form of sulfur in aqueous environments is the sulfate dianion and that it is a key terminal electron acceptor in anaerobic sulfate-reducing bacteria and archaea. As in bacterial cells, sulfate enters eukaryotic cells by way of plasma membrane sulfate transporters.

We discussed in Chapter 2, that the first step in sulfate activation for reductive bacterial metabolism is the enzymatic conversion to the mixed phosphoric–sulfuric anhydride adenosine-5′-phosphosulfate (APS). In eukaryotes, APS is typically subjected to an additional enzymatic phosphorylation at the 3′-OH of the ribose moiety to give 3′-phosphoadenosine-5′-phosphosulfate (PAPS)[3] (Figure 10.2). The extra phosphoryl group

Chemical Biology No. 12
The Chemical Biology of Sulfur
By Christopher T. Walsh

Published by the Royal Society of Chemistry, www.rsc.org

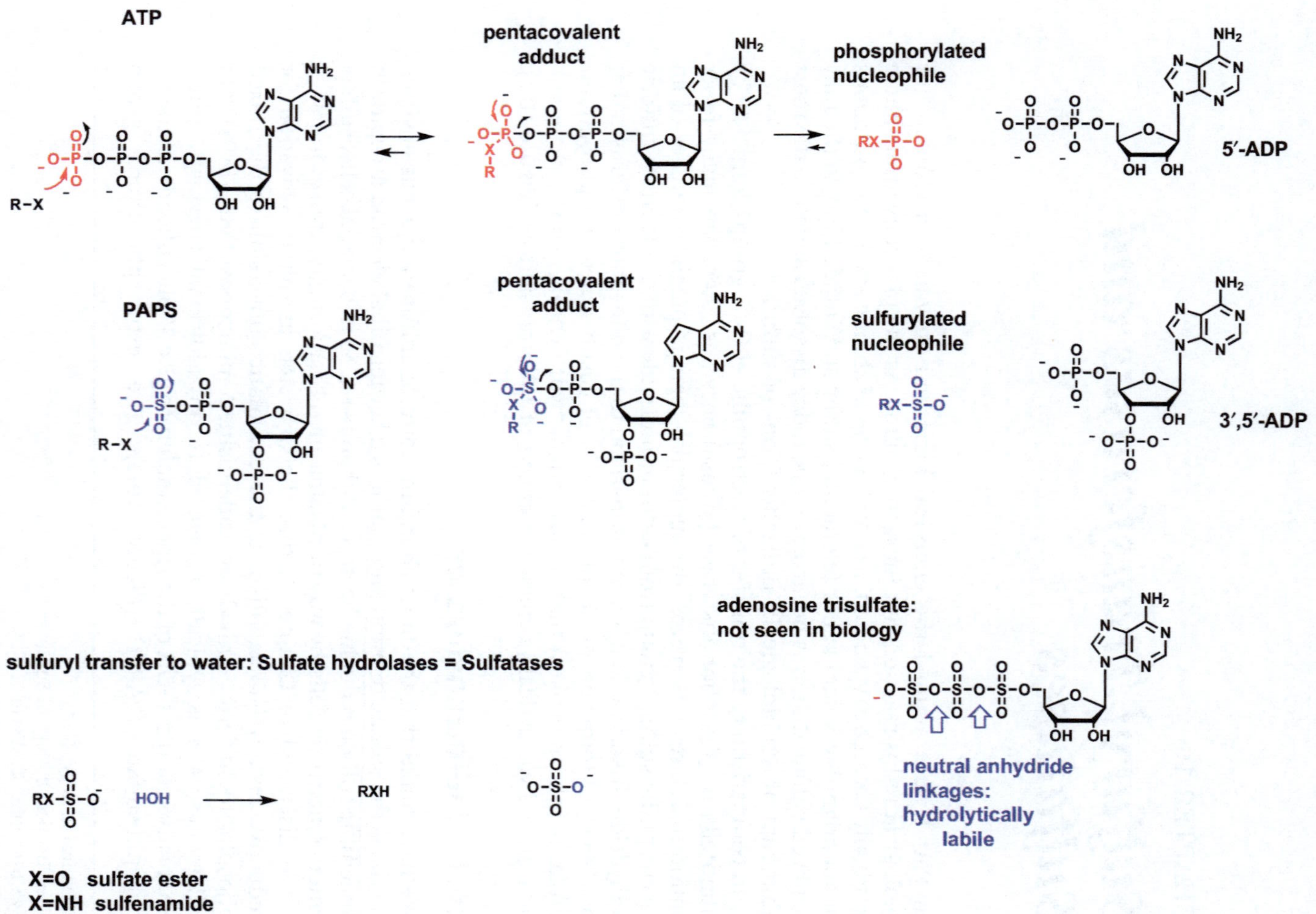

Figure 10.1 Phosphoryl transfers *vs.* sulfuryl transfers. ATP *vs.* 3′-phosphoadenosine-5′-phosphosulfate (PAPS) for thermodynamic activation and kinetic stability–lability. Cosubstrate nucleophile sulfurylation *vs.* phosphorylation.

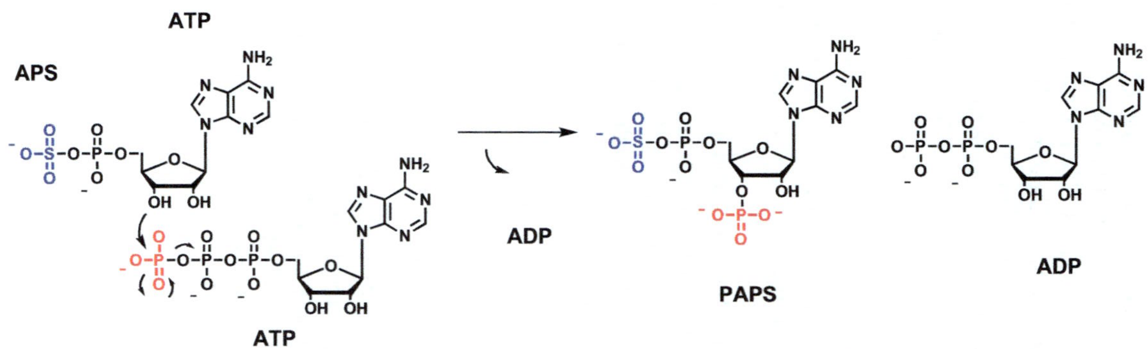

Figure 10.2 The enzyme APS kinase catalyzes attack of the 3′-OH of adenylyl sulfate on the electrophilic γ-phosphorus of the ATP side chain. PAPS is the kinetically preferred sulfuryl donor substrate.

presumably provides binding energy and specificity for binding to sulfotransferases, perhaps in analogy to the comparable late stage addition of a PO_3^{2-} group at the ribose 3′-OH to finish the biosynthesis of coenzyme A.

In contrast to the sulfur in hydrogen sulfide and its hydrosulfide anion where the electron rich sulfur ($S = -2$ oxidation state) is nucleophilic, the sulfur atom in free sulfate, and in APS and PAPS ($S = +6$), is electron-deficient and reacts only as an electrophile. PAPS is cosubstrate for a variety of sulfotransferases as an "SO_3" donor: This is termed *sulfuryl* transfer and also sometimes sulfonyl transfer or colloquially (but in fact incorrectly since it is an SO_3^- not an SO_4^{2-}) sulfo transfer.

The process is a group transfer equivalent to kinases and other phosphoryl transferases where it is an electrophilic $-PO_3$ group (phosphoryl), not a $-OPO_3$ group transfer (Figure 10.1).[4] Thus, sulfuryl/sulfo transferases transfer SO_3^- groups from PAPS to cosubstrate nucleophilic O, N (and rarely S) nucleophiles. Unlike phosphoryl transfers that are ubiquitous in cellular metabolism, sulfuryl transfers are relatively restricted in number and metabolic intersection: they number in the hundreds to thousands rather than hundreds of thousands.

It may be worth noting that the APS–PAPS mixed phosphoric–sulfuric acid side chains have two anionic charges at neutral pH as electrostatic barriers to rapid hydrolysis. The corresponding "all sulfuric" anhydride chains would be neutral at the alpha and beta positions in a putative adenosine trisulfate (Figure 10.1), would be susceptible to rapid hydrolysis, akin to that of neutral acetic anhydride in water, and so not suitable as a diffusible kinetically stable sulfuryl donor. This may be the genius of APS–PAPS: only one mixed phosphoric–sulfuric anhydride linkage.

10.1.1 Nonprotein Substrates for Sulfotransferases

In eukaryotes there are sulfotransferase family members in the cytoplasm and a distinct family subset in the Golgi compartment of the secretory pathways.[1] In humans there are some 50 genes encoding sulfotransferases (compared with approximately 550 human kinases genes), 13 predicted to be cytoplasmic and 37 membrane-associated. *Inter alia*, it indicates a rich and extensive role for sulfotransferases in the secretory compartments of cells. The 37-member class will be summarized after examining some cytoplasmic examples of sulfated metabolites.

Examples of cytoplasmic sulfotransferase substrates, collected in Figure 10.3A, include estrone-3-sulfate. Estrone-3-sulfate and the corresponding estradiol-3-sulfate are not active as hormones. Instead, in the presence of sulfatases (see next section) they can serve as reservoirs to release the active hormone forms[5] (Figure 10.3B).

Additional steroids known to be sulfurylated are cholesterol, pregnenelone and dehydroepiandrostenedione. Crystal structures are available for some of the steroid alcohol sulfuryl transferases, revealing the binding site of the PAPS sulfur donor. The logic appears to add $-SO_3^-$ groups onto the

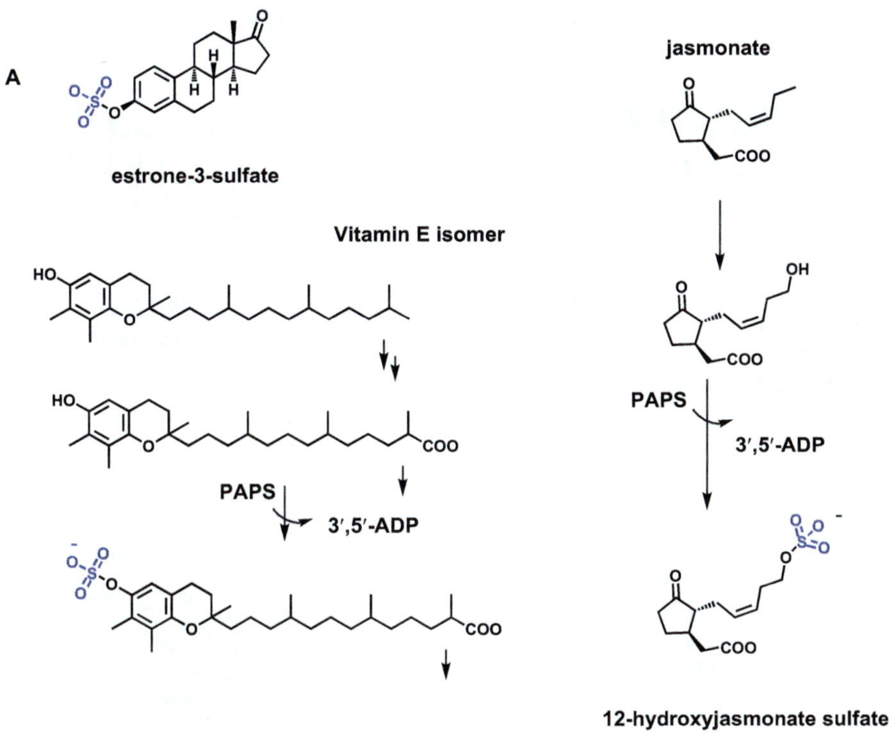

Figure 10.3 (A) Cytoplasmic substrates for *O*-sulfation (sulfuryl transfer). Estrone to estrone-3-sulfate. Two plant lipid-soluble metabolites of vitamin E and the plant hormone jasmonate, respectively, undergo *O*-sulfation to increase aqueous solubility. (B) Cyclical sulfation (formally sulfurylation) and desulfation (desulfurylation) of estradiol: the sulfate ester is an inactive, reversible reservoir for active steroid hormone. (C). Tandem double sulfation of saxitoxin to gonyatoxins by a pair of *N*- and then an *O*-sulfuryl transferases.

alcohol groups to make sulfate esters, more water soluble to allow circulation to different tissues. They are taken up into cells by anion transporter membrane proteins and then cleaved hydrolytically by sulfatases back to the unsulfated hydrophobic active forms of steroids as shown in Figure 10.3B.

Plants also have sulfuryl transferase enzymes often used in *O*-sulfurylation to increase the water solubility of hydrophobic metabolites. Two examples from plant metabolism are also shown in Figure 10.3. Lipid soluble vitamin E undergoes oxygenation of a terminal methyl group to the 13-carboxylate.[6] That can be sulfurylated to give the granddaughter metabolites the vitamin E 13-*O*-sulfate, much more water-soluble than the starting vitamin E. Analogously, the plant signaling metabolite jasmonic acid can undergo enzymatic oxygenation at C12 and then sulfurylation to impart water solubility to the deactivated jasmonate scaffold.[7] Each of these two enzyme sequences removes the active form of vitamin E or jasmonate, respectively.

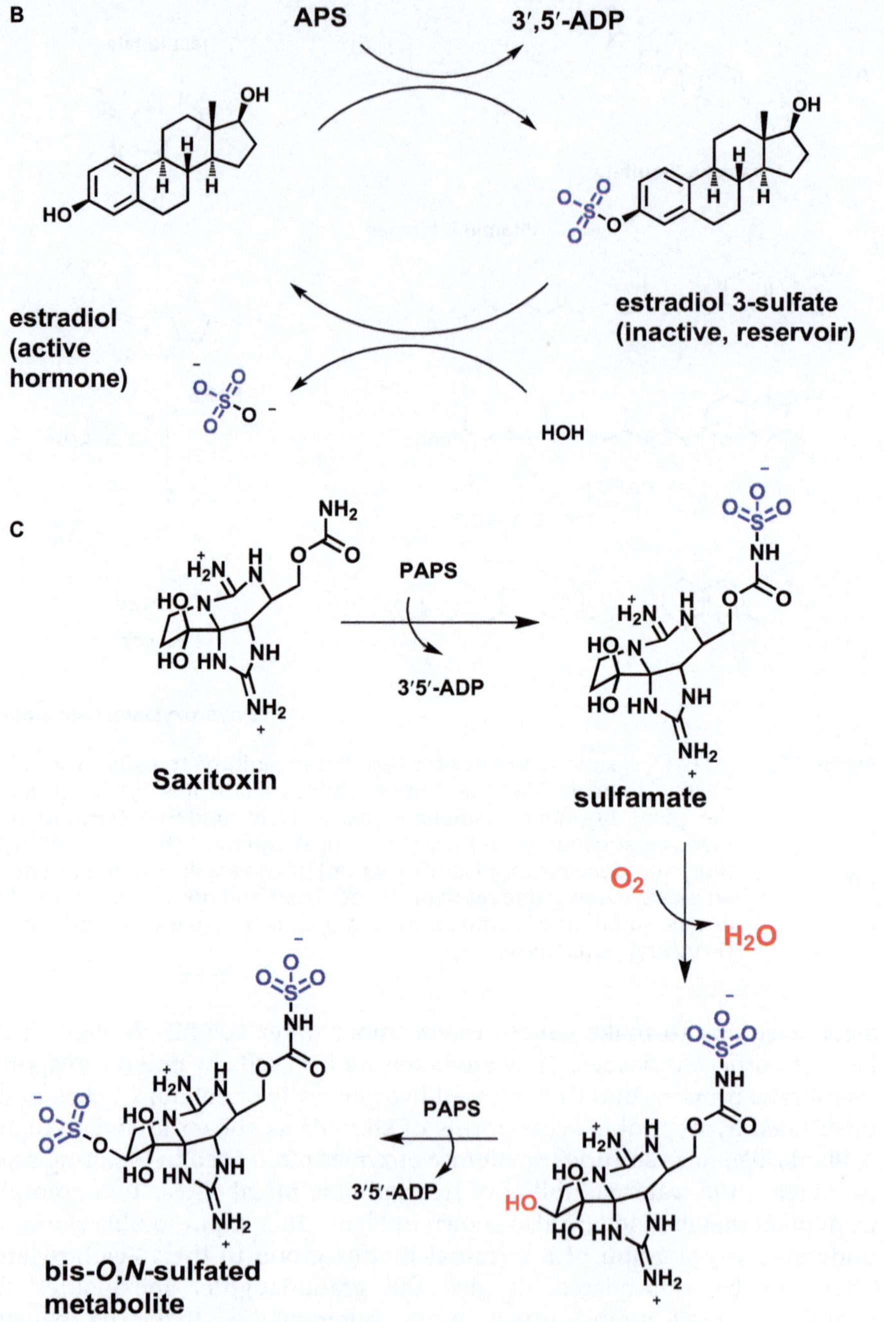

Figure 10.3 Continued.

In addition to these physiological substrates, *O*-sulfation of hydrophobic xenobiotic molecules is a typical metabolic strategy for detoxification of drug metabolites. The sulfate anionic group, once installed on a hydroxy functionality of the abiotic molecule, is a water-solubilizing group that helps partition a greater fraction of the drug metabolite into the aqueous phase for excretion by the kidney. Alternate strategies for adding hydrophilic groups in hepatic drug metabolism include glucuronidation. In Chapter 5 we summarized the acute pathophysiological response to overdoses of acetaminophen. Enzymatic $-SO_3^-$ transfer to the –N–OH acetaminophen metabolite leads to subsequent nonenzymatic loss of sulfate as the iminoquinone electrophile forms, resulting in toxification when GSH reserves are overwhelmed (Figure 5.12).

The saxitoxin class of paralytic toxins from shellfish that target mammalian sodium channels with nanomolar potency are often isolated from dinoflagellates as a mixture of unsulfated and *O*- and *N*-sulfated forms (Figure 10.3C). Two sulfuryl transferases that act sequentially, separated by oxygen insertion by a Rieske type oxygenase, have recently been characterized from dinoflagellate producers.[8] The saxitoxin family bears cationic charges from a guanidinium group and the addition of sulfuryl groups will change the net charge on this already hydrophilic scaffold.

The *N*-sulfuryl transferase acts first on the carbamate nitrogen of saxitoxin to yield the *N*-sulfamate derivative. The carbamate nitrogen is a weak nucleophile and must be activated in the enzyme complex with bound PAPS. Then, a Rieske-type iron–sulfur cluster variant (two Cys and two His residues coordinating the 4Fe–4S cubane structure) hydroxylates this *N*-sulfamate scaffold. The second sulfuryl transferase is the third enzyme in this bis-sulfation pathway. It adds an SO_3^- group to that newly introduced -OH group, creating the *N*–*O*-bis-sulfated saxitoxin scaffold. The sulfamate and sulfate ester groups presumably affect potency on voltage-gated sodium channels, and also in uptake and storage of the toxins by a comparable sulfation–desulfation cycle in the producing dinoflagellates.

10.1.2 Sulfated Disaccharide Repeats of Glycosaminoglycans as Extracellular Ligands for Receptor Protein Tyrosine Phosphatases

The different repeat sulfated disaccharide units of the glycosaminoglycan chains of heparin sulfate and chondroitin sulfate (see next section) have opposite physiological effects. These include axonal regeneration following injury and also in normal central nervous system (CNS) development. By the use of synthetic sulfated oligosaccharides of the heparin sulfate (HS) and chondroitin sulfate (CS) small repeat (Figure 10.4) fragments, Sakamoto *et al.*[9] identified the extracellular domains of two transmembrane protein tyrosine phosphatases, expressed on neuronal cell surfaces, that act as receptors (protein tyrosine phosphatase receptors; PTPRs).

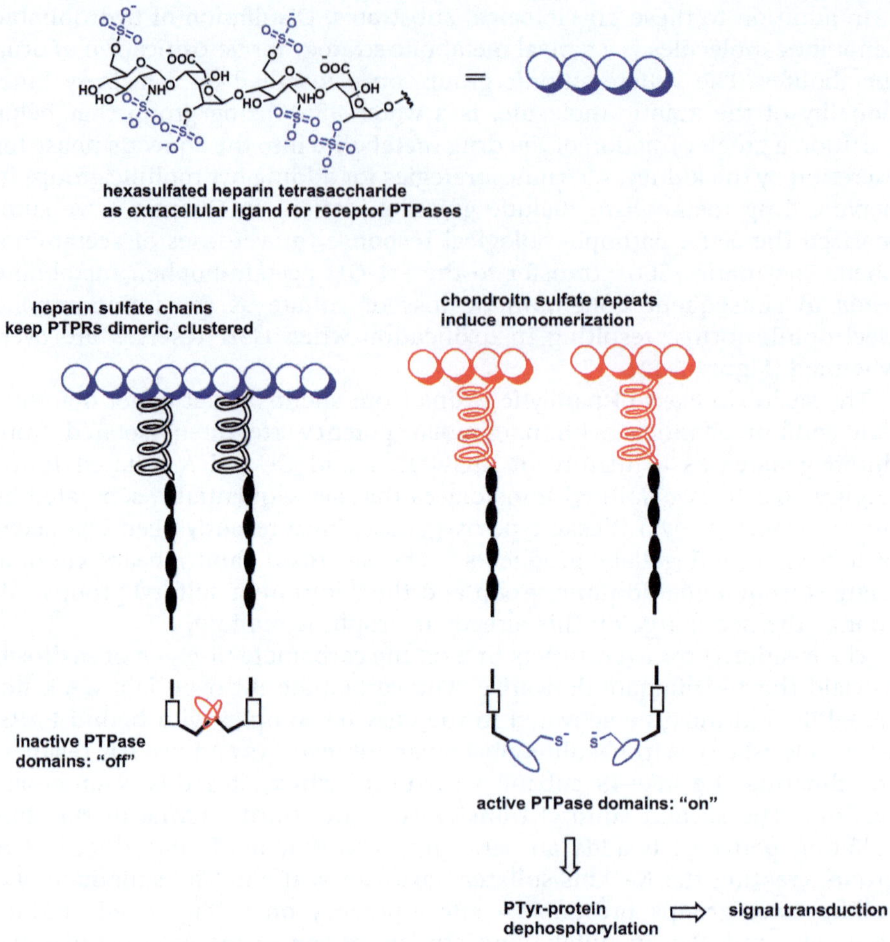

Figure 10.4 Heavily sulfated heparan sulfate tetramer as ligand to maintain transmembrane tyrosine phosphatase PTPRσ in a dimeric, inactive state.

The architecture of PTPRs involves an extracellular domain that recognizes ligands (in this case the sulfated disaccharide repeats of both HS and CS), the transmembrane domain and then an intracellular domain that has the catalytic activity for hydrolytic dephosphorylation of protein phosphotyrosine residues. As we will note in Chapter 12, such PTPases generate covalent enzyme–Cys-*S*–PO_3^{2-} intermediates in every catalytic cycle. The other relevant feature of receptor PTPase, sketched in Figure 10.4, is that such PTPRs are generally inactive as homodimers or higher oligomeric clusters, but active once dissociated by extracellular ligands into monomers.

Sakamoto *et al.*[9] concluded from their binding studies and subsequent effects in autophagic responses of axons, that heparin sulfate repeats keep the PTPases in an "off" state (favoring axon regeneration) while CS sulfated domains induce monomerization, by turning on the catalytic activity of the

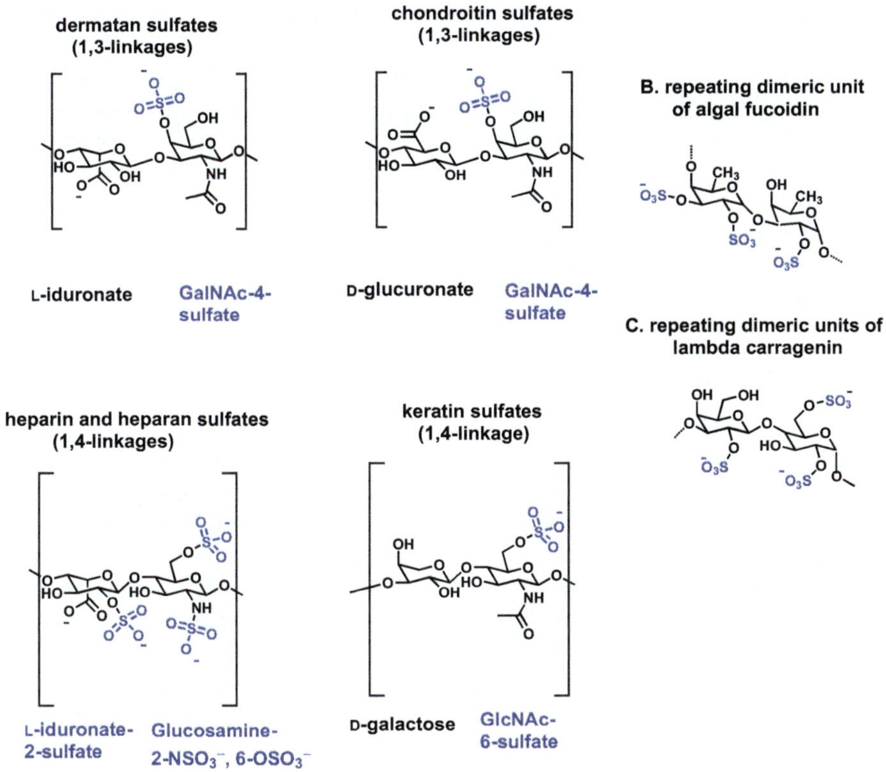

Figure 10.5 Four classes of repeat sulfated disaccharide units in extracellular mammalian proteoglycans: dermatan sulfates and chondroitin sulfates are in 1,3-hexose linkages while heparin and heparan sulfates along with keratin sulfates have the hexose residues in 1,4-linkages. *N*-acetylgalactosamine can be sulfated at the 4- or 6-hydroxyl group, L-iduronate at the 2-OH. Glucosamine can be *O*-sulfated at the 6-OH and/or *N*-sulfated at the 2-amino group, for an unusual *sulfamate* linkage.

cytoplasmic PTPase domain (see Figure 10.5 for structures). The active PTPases then remove phosphoryl groups from partner and client proteins, setting in motion signaling cascades, in this case leading to inhibition of autophagy.

The overall proposal is that the sulfation patterns in heparin sulfate *vs.* chondroitin sulfate, in the microenvirons of the extracellular matrix in the CNS, determine the fate of axons and regeneration post some injury. If this is generalizable it offers new insights into examining ligands affecting PTPR activation states.

10.2 Protein *vs.* Proteoglycan Substrates for Sulfuryl Transfer

The second class (37 in humans) of PAPS-dependent sulfuryl transferases are transmembrane enzymes, located in the Golgi complex (presumably anchored in that compartment by specific retention signals) of cells.[1] These enzymes encounter[10] protein substrates moving through the secretory pathway, whether to remain at the plasma membrane cell surface or be secreted. A significant fraction of transiting proteins are glycoproteins, either *N*-glycosylated on asparagine side chains or *O*-glycosylated on serine or threonine residues. A number of *O*-glycoproteins with complex oligosaccharide chains known as glycosaminoglycans undergo monosulfurylation or disulfurylation on repeat disaccharide units during passage to the extracellular matrix.[11] We consider the proteoglycans first and then proteins undergoing predominantly tyrosine residue *O*-sulfation.

10.2.1 Proteoglycan *O*- and *N*-Sulfation

A set of proteins with multiple serine residues, encoded by approximately 50 human genes, are the typical substrates for building related sets of *O*-linked glycan repeat units which are variably sulfated. The glycan repeats have a glucosamine or galactosamine monomer and usually a hexuronic acid (C6 carboxylate). These are categorized as glycosaminoglycan chains and encompass chondroitins, dermatans, heparans and heparin. They are significantly sulfated with disaccharide blocks that can be high in sulfation, low in sulfation, or completely des-sulfo forms. Each block is likely to have a different function in its architecture and in interactomes.[12,13] The proteoglycans are thus termed chondroitin sulfates, dermatan sulfates or heparan sulfates based on anatomical and historical assessment of function.[14] Heparins are variants of heparan sulfates with higher sulfate content, which tend to be stored in mast cell granules, and get released for local anticoagulant functions. Figure 10.5 shows typical sulfate-containing repeat units in dermatan sulfates, chondroitin sulfates, heparan and heparin sulfates and keratin sulfates. The repeat blocks in dermatan sulfates and chondroitin sulfates are 1,3-linked, the keratin sulfates and the heparan and heparin sulfates are 1,4-linked. The linkage regiospecificities impose quite distinct geometries on 1,3- *vs.* 1,4-linked glycan chains All but keratins have either a D-glucuronate or L-iduronate monomer unit as an anion at physiological pH. Likewise, any sulfate groups will add to the negative charges of these glycosaminoglycan chains.

To date the examples of enzymatic substrates for PAPS-dependent sulfuryl group transfer have been alcohols, forming *O*-sulfate ester products. The *N*-sulfo glucosamine introduces an *N–S* bond and the chemical grouping is a sulfamate not a sulfate ester. (Unlike in the case of the saxitoxin sulfamates above, the attacking nitrogen in these polymers is a nucleophilic primary amine). The particular sulfuryl transferase isoform that targets the glucosamine residue nitrogen uses the amine's nucleophilic property as the attacking species on PAPS.

The various proteoglycans can be secreted fully into extracellular media, can remain as outward-facing cell membrane proteins, or can be stored in secretory granules (*e.g.* heparins).[13] The full biological distributions and extracellular matrix and tissue functions of glycosaminoglycans are beyond the scope of this sulfo-centric focus.[14] We note only in passing some of the critical physiological roles of these anionic polymers: chondroitin sulfates in cartilage, bone and heart valves; heparan sulfates in basement membranes; dermatan sulfates in skin, blood vessels, tendons and lung structures; keratin sulfates in cornea and cartilage (*the medical biochemistrypage.org*).

Clearly, the sulfation of glycosaminoglycans is of central importance in tissue architectures and functions. As we will note in the sulfatase section, some of the almost 50 known lysosomal storage diseases arise from defects in removal of the sulfate groups during degradation of these structures during apoptosis and tissue remodeling.[15]

Typically, the repeat disaccharide units of Figure 10.5 are laid down after a core tetrasaccharide is built off a serine side chain in proteins. This core connects the long glycosaminoglycan chains to the protein cores. The number of glycan chains in such proteins can vary from one to dozens, even hundreds, (like bristles on a hair brush) depending on serine content of a given protein and accessibility to the glycosylation machinery.[14] Figure 10.6 shows the typical core tetrasaccharide linker between the serine alcohol side chain and the repeat disaccharide units of the long, linear glycosaminoglycan chains.

A proteoglycan that has been secreted into extracellular cartilage mesh might have 30 keratin-type glycosaminoglycans and 100 chondroitin sulfate chains on one protein core, with each glycosaminoglycan chain built of 100 or so disaccharide unit repeats. Such a single proteoglycan molecule could, amazingly, contain approximately 10 000 hexose units, if even 20% of them were sulfated, that would be 2000 sulfated hexoses or hexuronic acids. This gives a sense of the high metabolic flux of the several cooperating Golgi-associated transmembrane sulfuryl transferases, working to form mature glycosaminoglycan chains during passage through that organelle. The numbers also give a sense of why defects in degradative processing of this sulfocarbohydrate load can be so deleterious, (*vide infra*).

Another set of oligosaccharide frameworks where high density *O*-sulfation occurs are in cell wall components of marine algae. Part of this occurrence may arise from the uniquely high availability of dissolved sulfate ions (approximately 25–35 mM) in seawater. Figure 10.5 shows the repeat sulfated disaccharides in fucoidine and carregenin fragments. These are edible compounds from macroalgae and highlight the need for digestive sulfatases.[16,17]

10.3 Protein Sulfotransferases

The section just concluded dealt with a class of secreted glycoproteins where all the sulfuryl transfers placed $-SO_3^-$ groups on the oligosaccharide or polysaccharide chains, not on the protein residue side chains. Protein residue sulfurylations are considered here. Of the 37 predicted membrane-associated human sulfuryl transferases, two have been associated with protein sulfurylation.

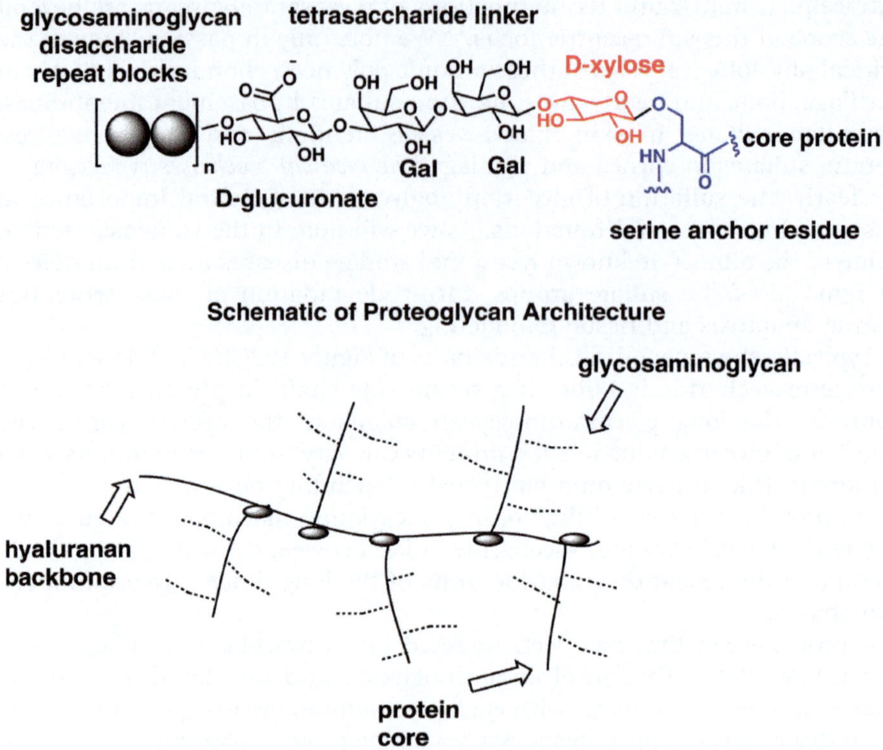

Figure 10.6 Schematics of proteoglycan architecture at two different magnifications. The upper structure shows the common tetrasaccharide core (that shows these are *O*-linked glycoproteins) and then the repeat disaccharide units, with differentially sulfated blocks are attached. The lower panel shows a low-resolution sketch of these multiple glycosaminoglycan chains, each in *O*-serine linkage through the tetrasaccharide spacer, to serine side chains in the core protein.

The best studied amino acid residue undergoing sulfurylation is the phenolate anion form of some tyrosyl residues on a subset of transiting proteins some of which undergo sulfation.[18] There is mass spectrometry-based evidence for serine and threonine residue *O*-sufurylation, in analogy to the highly prevalent *O*-phosphorylations, but they have been less well studied to date[19] and the respective sulfotransferases have not been characterized (Figure 10.7A). About 500 protein sulfurylation sites had been mapped by 2011, with the likelihood that they might represent 10% of the total predicted sites.

These constitute posttranslational modifications of tyrosyl residues and qualify for inclusion in Part III. On the other hand, since a wide variety of both protein and non-protein substrates, from low molecular weight

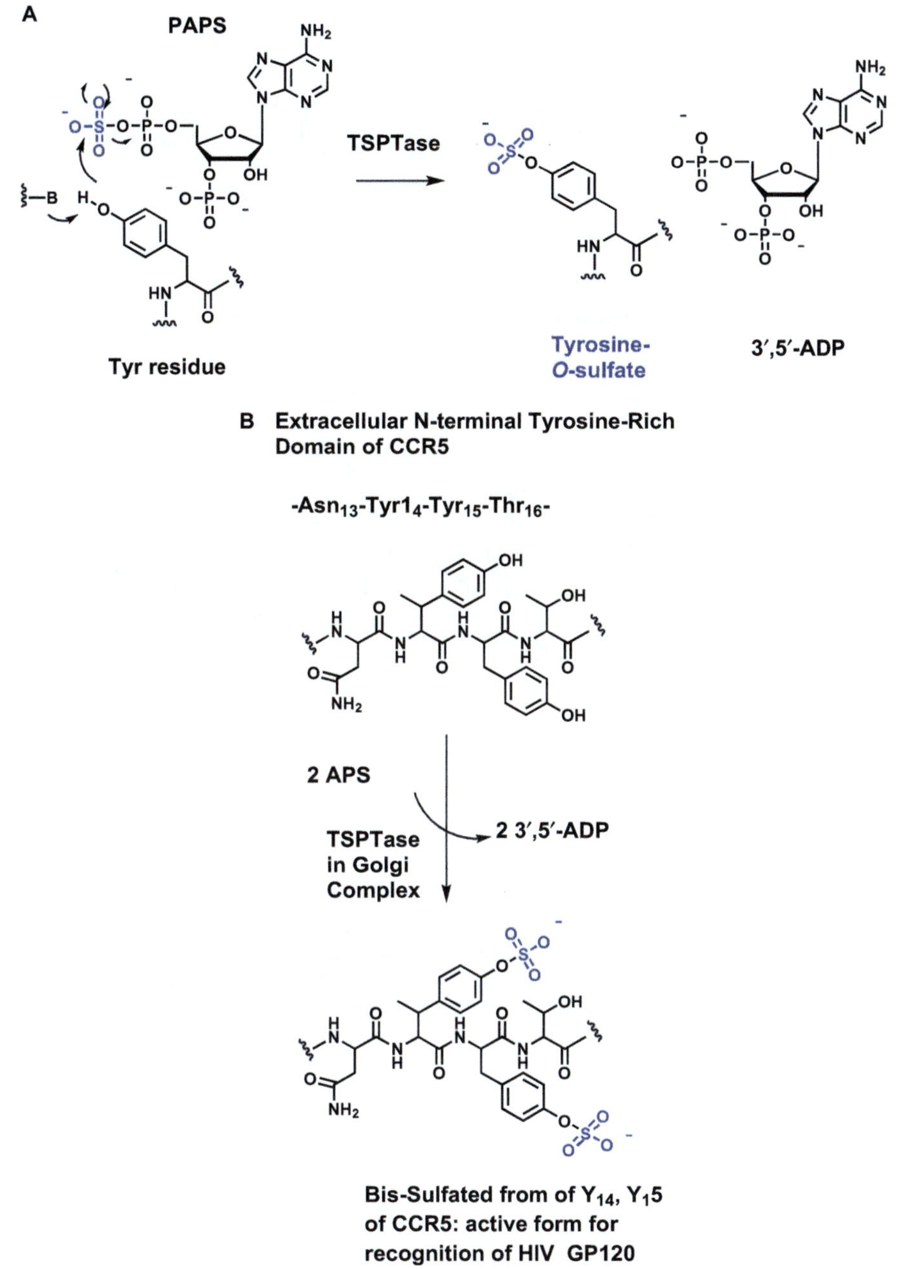

Figure 10.7 Continued on next page

steroids, lipids, monomeric carbohydrates to glycosaminoglycans and polysaccharides, can undergo enzymatic sufuryl transfers, the protein sulfurylations are examined here (Figure 10.7A).

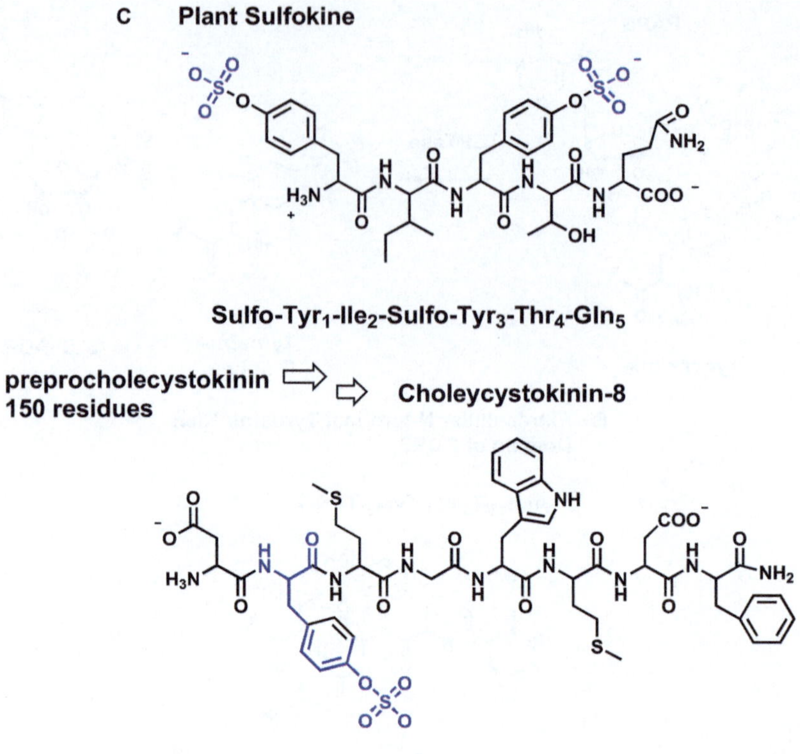

Figure 10.7 (A) Reaction stoichiometry for a protein tyrosine sulfuryl transferase using PAPS as sulfuryl donor to the phenolate side chain of a tyrosine residue, yielding a tyrosyl-*O*-sulfate residue. (B) Schematic for the protein CCR5, a seven-transmembrane-domain G-protein-coupled receptor (GPCR), and its extracellular N-terminal domain with four tyrosyl residues: Tyr3, Tyr10, Tyr14 and Tyr15. The disulfated Tyr14, Tyr15 pair are necessary for interaction with HIV G120. (C) Plant sulfokine is a bis-sulfated pentapeptide and CCK8 is a monosulfated octapeptide that are each active biological forms as ligands for receptors.
(B) Reproduced from ref. 21 with permission from National Academy of Sciences, Copyright 2002.

It is the phenolic-OH of tyrosyl residues that yield Tyr-*O*-sulfate negatively charged side chains. Among the most intriguing is the membrane receptor protein for the related chemokines macrophage inflammatory protein (MIP)-1α, MIP-1β and regulated on activation, normal T cell expressed and secreted (RANTES). The receptor is a member of the G-protein-coupled receptors (GPCRs) that are conditional guanosine triphosphate (GTP) hydrolases when an external ligand binds to the outer surface of the receptor. The specific receptor is known as chemokine receptor 5, (CCR5). It has garnered particular attention as the receptor for macrophagotropic strains of HIV by way of interaction of glycoprotein 120 when presented by CD4. It is known that CCR5 receptors with deletions in the second extracellular loops are resistant to HIV entrance.[20]

The N-terminal 18 residues of CCR5 are external to the seven transmembrane domains (Figure 10.7B). The *N*-terminal methionine is thought to be removed cotranslationally *via* hydrolysis by methionine aminopeptidase action so the mature protein starts with an N-terminal aspartate, although it is numbered from the original Met_1 residue. There are four tyrosine residues in this extracellular *N*-terminal tail, at residues 3, 10, 14 and 15. Multiple sulfated forms of the tyrosine-rich N-terminus were detected, leading to studies[21] with a peptide of residues 2–15. All four of the tyrosine residues could be *O*-sulfated in the order Tyr-14, Tyr-15, Tyr-10 and Tyr-3. It is likely that bis Tyr-OSO_3^- at residues 14 and 15 are required for HIV GP120 recognition and subsequent viral entry.

Analogously, enterovirus 71, the causative agent of some forms of hoof and mouth disease, uses P-selectin glycoprotein ligand 1 (PSGL-1)[22] as a cell-surface receptor. Again, a cluster of up to three anionic tetrahedral sulfated tyrosine residues are involved in virus–cell surface recognition as preamble to virus entry into cells. In that virus–host cell interaction an extracellular loop of PGSL-1 from residue 42 to 61 has three tyrosines at Tyr46, Tyr48 and Tyr51. It seems that the bis-sulfated forms of Tyr46 and Tyr51 may be necessary for high affinity interaction of EV71 virus.

The tyrosyl protein sulfuryl transferases (TSPTs) in mammals exists as two isoforms, as transmembrane enzymes, localized to the *trans*-elements of the Golgi complex. Estimates of their net activity are that as much as 1% of tyrosyl residues in proteins that pass through the Golgi in the secretory pathway may become sulfated.[23] Additional protein substrates that have been confirmed, include cholecystokinin, gastrin and proteins of the blood coagulation cascade, including two proteases factors V and VII and the substrate protein fibrinogen.[24] Despite some kinetic evidence interpreted in favor of a covalent sulfoenzyme intermediate, X-ray structural evidence is consistent with TPST-2 acting an in a direct transfer as suggested in Figure 10.7A.

The secreted peptide hormones that are sulfated on tyrosyl residues reflect this mode of sulfurylation during organelle transit For example, plant sulfokine is a bis-Tyr-*O*-sulfate-containing pentapeptide[25] (Figure 10.7C). It is a ligand, only when bis-sulfated, for the membrane receptor phytosulfokine receptor 1(PSKR1) and thereby drives growth in *Arabidopsis thaliana.*[25] The neighboring dityrosyl sulfates appear to be a recognition unit, with their tetrahedral doubly anionic proximity grouping.

CCK-8 (Figure 10.7C) is an active octapeptide form of the mammalian hormone cholecystokinin. The tetrahedral sulfate monoanion side chain provides geometric and electrostatic recognition elements for partner proteins in metabolism. Cholecystokinin is one of the gastrin family of hormones. It emanates from a 150-residue preprohormone nascent protein, subject to a set of protease cleavage steps that yield a set of smaller CCK peptide fragments. The smallest is CCK-8. For it to bind functionally to its membrane receptor on target cells in stomach, gall bladder, pancreas and liver, Tyr-2 must be sulfated[26] (Figure 10.7C).

10.4 Sulfatases

Given the diversity of sulfated metabolites: steroids, lipids, many types of oligosaccharide chains, along with sulfo-tyrosyl residues in proteins, it is not surprising that there are classes of sulfoester (and sulfamate) hydrolases to cleave the sulfate ester group hydrolytically to give the inorganic sulfate dianion and the corresponding product alcohol[27–29] (Figure 10.8). Sulfatase action is illustrated here for the sulfated heparan repeat unit (and specifically for hydrolysis of estradiol-3-sulfate back to estradiol in Figure 10.3). Bioinformatic data bases predict over 4000 members of the sulfate monoester hydrolases (sulfatases) across diverse proteomes.

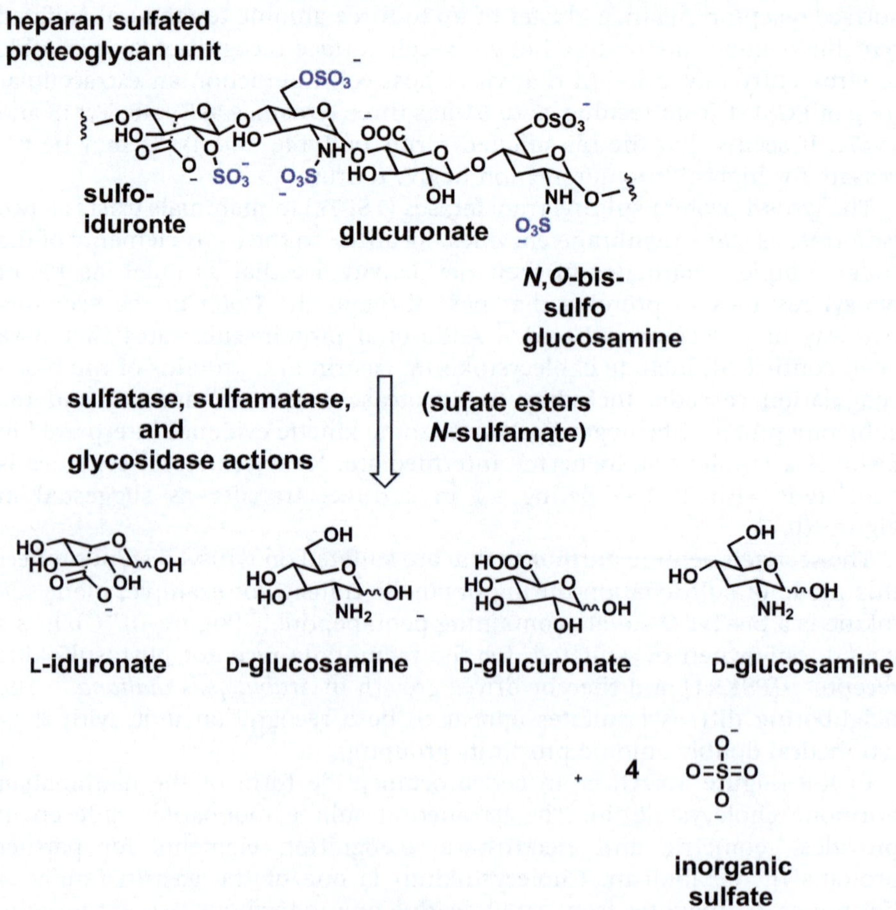

Figure 10.8 Sulfatase and glycosidase action on a sulfate-rich core of the heparin sulfate glycosaminoglycan region of that proteoglycan.

Microbes have the capacity to hydrolyze sulfate ester bonds in a variety of alkyl sulfate ester structures as well as some aryl sulfate esters. For example, the microbial hydrolysis of the surface-active sodium dodecyl sulfate is shown in Figure 10.9.

10.4.1 Metalloprotein Sulfatases

Sulfatases may have evolved three different times independently.[30,31] One set of microbial alkyl sulfatases have a binuclear zinc array, bridged by an intervening water molecule as the catalytic apparatus (Figure 10.9). They have become of interest for some biotechnology applications because some sulfate esters are positioned to undergo S–O bond cleavage while others undergo C–O bond cleavage as the sulfate group departs. For S–O cleavage there is retention at the alcoholic carbon of the released product. But for C–O bond cleavage, one sees inversion at a chiral alcohol center[30] (Figure 10.9).

A second sulfate-ester cleaving enzyme, restricted in distribution is a nonheme mononuclear iron-based monooxygenase. Rather than carrying out simple hydrolysis of an alkyl sulfate ester substrate, this enzyme performs an oxygenative cleavage to sulfate and the corresponding aldehyde (Figure 10.9). This is most likely radical chemistry and delivery of an $OH^{\bullet}$ equivalent from a high valent $Fe^{IV}{=}O$ intermediate to form the indicated tetrahedral adduct on the way to the aldehyde and sulfate dianion products (Figure 10.9).

10.4.2 Formylglycine (Fgly)-Containing Sulfatases

By far the most abundant class of sulfatases, particularly in higher organisms, are a related enzyme set that require a posttranslational modification of one active site residue to go from inactive nascent enzyme to active mature catalyst.[32,33] The posttranslational modification involves oxidative conversion of a cysteine (or sometimes a serine-CH_2OH) residue at the CH_2SH or CH_2OH side chain to the –CH=O aldehyde. That modified residue is formally 3-oxoalanine but known more familiarly as formylglycine (Fgly) (Figure 10.10).

A schematic mechanism for conversion of the proenzyme cysteine side chain into the active, formylgycine (Fgly)-containing active catalyst is shown in Figure 10.10. An oxidant (*e.g.* HOOH) is proposed to convert the cysteine thiol to the sulfenate, a known oxidative process (see Chapter 5). Now, rather than go on to be captured by a neighboring thiol to form a disulfide, the sulfenate undergoes an elimination of water. This can be written as occurring through the enethiol, which is the less stable tautomer of the thioaldehyde [we noted a proposed enethiol intermediate during the decarboxylation step in coenzyme A biosynthesis (Chapter 5)]. In turn, water addition to the thioaldehyde can lead to expulsion of H_2S and formation of the formylglycyine side chain.

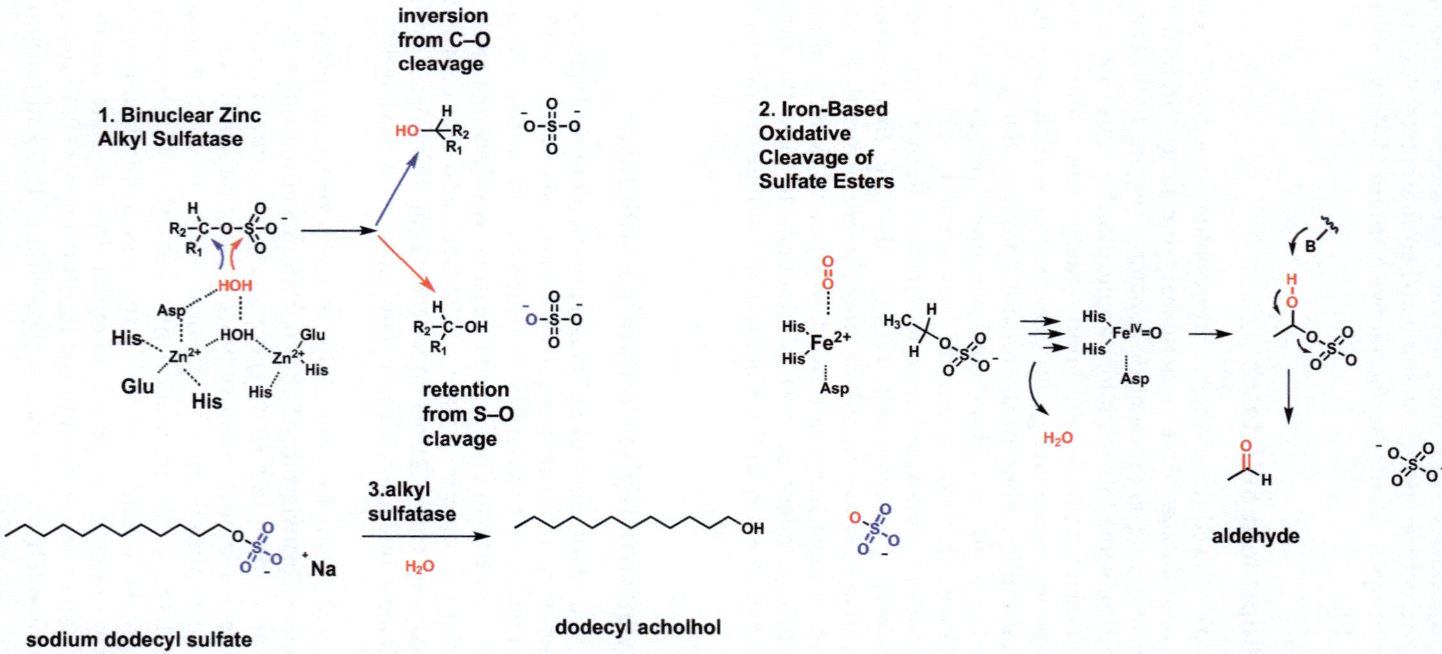

Figure 10.9 Three different classes of alkyl (and aryl) sulfatases: (1). Binuclear zinc alkyl sulfatases can give inversion (C–O bond cleavage) or retention (O–S bond cleavage) from sulfate esters to the alcohol product. (2). There are a few nonheme iron-based oxygenases that cleave sulfate esters with oxygenation as shown to yield aldehydes and inorganic sulfate. (3) Alkyl sulfatases are largely active after posttranslational modification of a Ser or Cys side chain to a formylglycine residue.

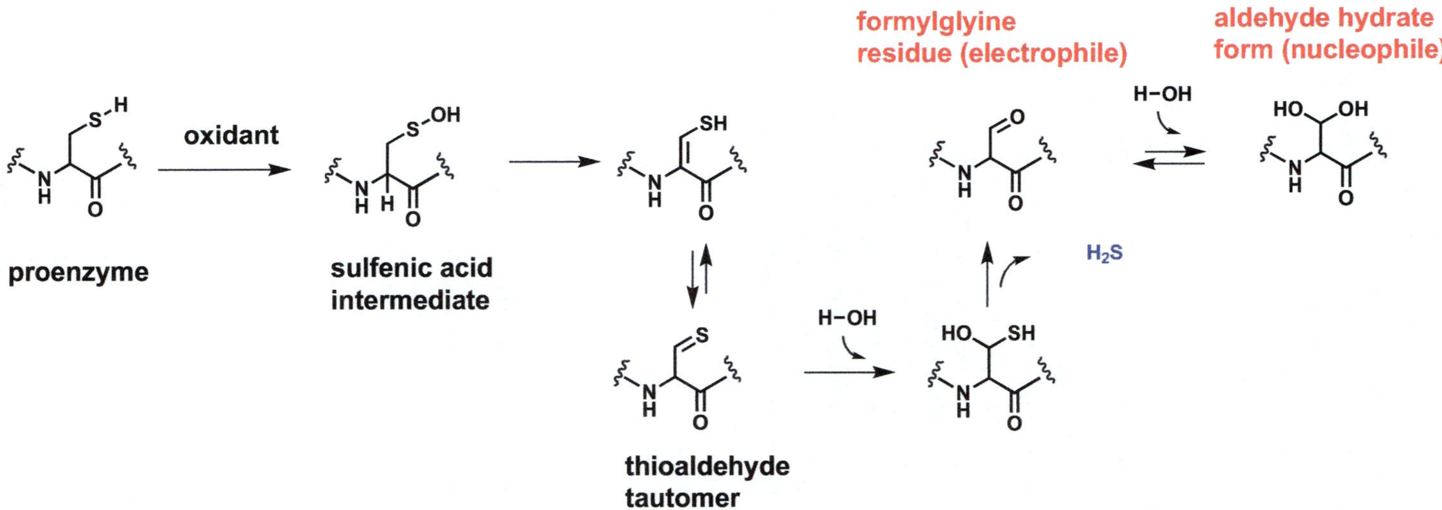

Figure 10.10 The family of sulfatases with an active site formylglycine residue arise from posttranslational radical SAM-mediated oxidative desulfuration of a cysteinyl residue to the formylglycine residue as shown. The reaction starts by oxygenation of the Cys thiolate to the sulfenate, loss of water to the enethio–thioaldehyde tautomeric mixture followed by water addition to the thioaldehyde side chain. Note the sulfur departs from the tetrahedral adduct as HS^-. Of consequence mechanistically for the catalytic activity is the tendency for aldehyde carbonyls to add water and form the gem diol–tetrahedral aldehyde hydrate form.

From the point of view of addition to the chemical inventory of protein side chains, this oxidative process has introduced an electrophilic aldehyde into the active site environment. Like most aldehydes, it can readily add water and be in equilibrium with measurable fractions of the gem diol–aldehyde hydrate form (Figure 10.9). Note that the gem diol form is potentially nucleophilic at oxygen while the aldehyde form is electrophilic at the carbonyl group, so one gets both reactivity modes in the same modified side chain.

The Fgly-containing active forms of sulfatases are also typically activated by Ca^{++} ions. A reaction scheme for sulfate ester hydrolysis is shown in Figure 10.11, involving the Fgly side chain. There has been debate as to whether catalysis is initiated by the electrophilic aldehyde or the nucleophilic gem diol form of the Fgly residue in the enzyme active site.

Here, we show the gem diol as nucleophile attacking the electrophilic central sulfur core of the sulfate group in an enzyme-bound substrate molecule. This generates a transient pentacovalent sulfur atom adduct to the Fgly residue. Decomposition of that adduct in the forward direction is the ester bond-cleaving step, expelling ROH as the leaving group as observed first product. The enzyme is left at this point in a covalent sulfate ester intermediate stage. Now intramolecular participation of the other –OH of the Fgly gem diol is pictured to cleave off inorganic sulfate as second product and return the enzyme to the electrophilic aldehyde form of the Fgly reside, ready for the next catalytic cycle.

One measure of the ubiquity of the Fgly-containing family of mammalian sulfatases is that an inactivating mutation in the Fgly-forming posttranslational oxidase, is pleiotropic for 12 sulfatases.[15] They are all inactive because of lack of this Fgly-forming catalyst. This is a known multiple sulfatase deficiency human phenotype.

10.5 Lysosomal Sulfatase Deficiencies

Many of the sulfatase variants are concentrated in lysosomes of cells, along with other hydrolytic enzymes (proteases, nucleases, glycosidases, phosphatases), consistent with the role of lysosomes and their phagolysosomal precursors as organelles of macromolecular digestion. When the capacity of lysosomal enzymes cannot keep up with the ingested load of material, a variety of secondary pathologies ensue, including organ failures, neurological deficits and early deaths. Some 40–50 lysosomal storage diseases have now been described that represent unwanted accumulation, typically of macromolecular substrates, arising from insufficient hydrolytic breakdown activity, often reflecting a mutation to enzyme deficiency.[34]

Two such syndromes are due to deficiency in a particular sulfatase. Hunter's syndrome is cased by a defect in L-iduronate sulfatase. As a consequence, lysosomes fill up with uncleaved glycosaminoglycans bearing the iduronate-sulfate moieties in chondroitin sulfate chains. Sanfilippo syndrome type A is an autosomal recessive neurodegenerative syndrome whose

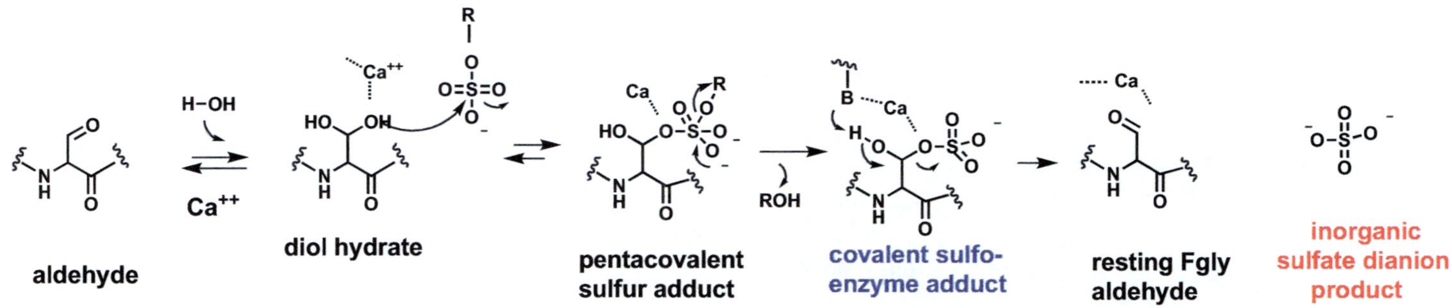

Figure 10.11 The catalytic mechanism for Fgly-containing sulfatases is proposed to go through the carbonyl hydrate form of the Fgly side chain. The oxyanion of the gem diol attacks the electrophilic sulfur of the sulfate ester group of bound substrate. Decomposition of that adduct releases the alcohol product and generates a covalent enzyme adduct between the Fgly hydrate and the sulfuryl group. Internal displacement of the sulfate as the Fgly hydrate equilibrates with the free aldehyde form to release inorganic sulfate.

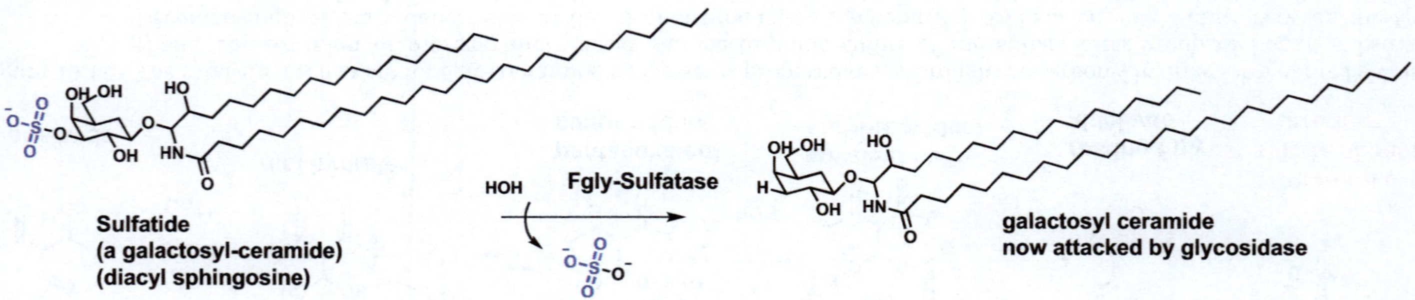

Figure 10.12 Defect in arylsulfatase A leads to accumulation of the sulfate ester-containing brain 3-*O*-sulfo-glucosyl-ceramide in the lysosomes of almost all cells with subsequent systemic pathologies.

molecular defect is lack of activity of the sulfamidase required to cleave the N–S bond in the N–SO_3^- glucosamine-2-sulfate units in both heparan- and heparin-sulfate glycosaminoglycan chains.

Analogously, a defect in the lysosomal enzyme named arylsufatase A leads to toxic accumulation of the sulfated ceramide lipid, galactosylceramide-3-*O*-sulfate, known as sulfatide (Figure 10.12) (named in 1884). Typically, sulfatide makes up to 10–20% of the myelin sheath in myelinated neurons. The disease phenotype is metachromatic leukodystrophy. Sulfatide accumulation occurs in all tissues, with functional pathologies early in the nervous system.

In consideration of sulfatase gene multiplicity to deal with the many low- and high-molecular-weight sulfated metabolites, bacteria that feed on algal cell walls may be the diversity–redundancy winners. We noted the sulfated hexose units in fucoidines and carageenins in Figure 10.3B. Up to 75 encoded sulfatases are annotated in those bacterial genomes to enable them to get to the hexose monomers to feed into energy-yielding glycolytic pathways.[17]

References

1. M. Negishi, L. Pedersen, E. Petrochenko, S. Shevtsov, A. Gorokhov, Y. Kakuta and L. G. Pedersen, Structure and Function of Sulfotransferases, *Arch. Biochem. Biophys.*, 2001, **390**, 149–157.
2. S. R. Hanson, M. Best and C. H. Wong, Activity, inhibitors, and synthetic utility, *Angew. Chem., Int. Ed.*, 2004, **43**, 5736–5763.
3. J. van den Boom, D. Helder, S. R. Martin, A. Pastore and J. W. Mueller, 3′Phosphoadenosine-5′phosphosulfate (PAPS) synthases: naturally fragile enzymes specifically stabilized by nucleotide binding, *J. Biol. Chem.*, 2012, **287**, 17645–17655.
4. C. T. Walsh, B. Tu and Y. Tang, Eight kinetically stable but thermodynamicallay activated molecules that power cell metabolism, *Chem. Rev.*, 2018, **118**, 1460–1494.
5. J. W. Mueller, L. Gilligan, J. Idkowiak, W. Arlt and P. A. Foster, The regulation of steroid action by sulfation and desulfation, *Endocr. Rev.*, 2015, **36**, 526–563.
6. Q. Jiang, H. Freiser, K. V. Wood and X. Yin, Identification and quantitation of novel vitamin E metabolites, sulfated long-chain carboxychromanols, in human A549 cells and in rats, *J. Lipid Res.*, 2007, **48**, 1221–1230.
7. S. K. Gidda, O. Miersch, A. Levitin, J. Schmidt, C. Wasternack and L. Verin, Biochemical and molecular characterization of a hydroxyjasmonate sulfotransferase from *Arabidopsis thaliana*, *J. Biol. Chem.*, 2003, **278**, 17895–17900.
8. A. L. Lukowski, N. Denomme, M. E. Hinze, S. Hall, L. L. Isom and A. R. Narayan, Biocatalytic detoxfication of paralytic shellfish toxins, *ACS Chem. Biol.*, 2019, **14**, 941–948.

9. K. Sakamoto, T. Ozaki, Y.-C. Ko, C.-F. Tsai, Y. Gong, M. Morozui, Y. Ishikawa, K. Uchimur, S. Nadanaka, H. Kitagawa, M. M. L. Zuleta, A. Bandaru, J.-i. Tamura, S.-C. Hung and K. Kadomatsu, Glycan sufation paterns defne autophagy flux at axon tip via PTPR-cortactin axis, *Nat. Chem. Biol.*, 2019, **15**, 699–709.
10. I. Wijeeskara, R. Pangestuti and S.-k. Kim, Biological activities and potential health beenfits of sulfated polysaccharides derived from marine algae, *Carbohydr. Polym.*, 2011, **84**, 14–21.
11. G. Jiao, G. Yu, J. Zhang and H. S. Ewart, Chemical structures and bioactivities of sulfated polysaccharides from marine algae, *Mar. Drugs*, 2011, **9**, 196–223.
12. J. R. Couchman and C. Pataki, An introduction to proteoglycans and their localization, *J. Histchem. Cytochem.*, 2012, **60**, 885–897.
13. R. V. Iozzo and L. Schaefer, Proteoglycan form and function: a comprehensive nomenclature of proteoglycans, *Matrix Biol.*, 2015, **42**, 11–55.
14. V. H. Pomin and B. Mulloy, Glycosamiglycans and Proteoglycans, *Pharmaceuticals*, 2018, 27.
15. M. P. Cosma, S. Peppe, I. Annunziata, R. Newbold, M. Grompe, G. Parenti and A. Ballabio, The multiple sulfatase deficiency gene encodes an essential and limiting factor for the activity of sulfatases, *Cell Chem. Biol.*, 2003, 445–456.
16. T. R. Gemmill and R. Trimble, Overview of N- and O-linked oligosaccharide structures found in various yeast species, *Biochim. Biophys. Acta*, 1999, **1426**, 227–237.
17. W. Helbert, Marine polysaccharide sulfatase, *Front. Mar. Sci.*, 2017, **9**, 839.
18. K. L. Moore, The biology and enzymology of protein tyrosine O-sulfation, *J. Biol. Chem.*, 2003, **278**, 24243–24246.
19. K. F. Medzihradsky, Z. Darula, E. Merlson, M. Fainzilber, R. J. Chalkley, H. Ball, D. Greenbaum, M. Bogyo, D. R. Tyson, R. A. Bradshaw and A. L. Burlingame, O-Sulfonation of Serine and Threonine, *Mol. Cell. Proteomics*, 2004, **3**, 429–440.
20. M. Farzan, T. Mirzabekov, P. Kolchinsky, R. Wyatt, M. Cayabyab, N. P. Gerard, C. Gerard, J. Sodroski and H. Choe, Tyrosine sulfation of the amino terminus of CCR5 facilitates HIV-1 entry, *Cell*, 1999, **96**, 667–676.
21. C. Siebert, M. Cadene, A. Sanfiz, B. T. Chait and T. P. Sakmar, Tyrosine sulfation of CCR5 N-terminal peptide by tyrosylprotein sulfotransferses 1 and 2 follows a discrete pattern and temporal sequence, *Proc. Natl. Acad. Sci. U. S. A.*, 2002, **99**, 11031–11036.
22. Y. Nishimura, T. Wakita and H. Shimizu, Tyrosine sulfation of the amino terminus of PSGL-1 is critical for enterovirus 71 infection, *PLOS Pathog.*, 2010, **6**, e1001174.
23. L. M. Danan, Z. Yu, P. J. Ludden, W. Jia, K. L. Moore and J. A. Leay, Catalytic mechanism of Golgi-resident human tyrosyl protein

sulfotransferase 2: a mass spectrometry approach, *J. Am. Soc. Mass Spectrom*, 2010, **21**, 1633–1642.
24. M. J. Stone, S. Chuang, HouX, M. Shoham and J. Z. Zhu, Tyrosine sulfation: an increasingy recognized post-translational modification of secreted proteins, *New Biotechnol.*, 2009, **25**, 299–317.
25. F. Ladwig, R. Dahlke, N. Sturwohldt, J. Hartmann, K. Harter and M. Sauter, Phytosulfokine regulates growth in Arabidopsis through a response module at the plasma membrane that includes cycic nucleotide gated channel 17 H+-ATPase and BAK1, *Plant Cell*, 2015, **27**(6), 1718–1729.
26. O. Chaudrhi, C. Small and S. Bllom, Gastrointestinal hormones, *Philos. Trans. R Soc. London B: Biol. Sci.*, 2006, **361**, 187–209.
27. D. Hosh, Three-dimensional structures of sulfatases, *Methods Enzymol.*, 2005, **400**, 273–293.
28. D. Ghosh, Human sulfatases: a structural perspective to catalysis, *Cell. Mol. Life Sci.*, 2007, **64**, 2013–2022.
29. T. Barbeyron, L. Brillet-Guegue, W. Carre, C. Carriere, C. Caron and M. Czjek, Matching the diversity of sulfated biomolecules; creation of a classification database for sulfatases reflecting their substrate specificity, *PLoS One*, 2016, **11**(10), e0164846.
30. M. Toesch, M. Schoeber and K. Faber, Microbial alkyl- and arylsulfatases: mechanism, occurrence, screening and stereoselectivities, *Appl. Micribiol. Biotechnol.*, 2014, **98**, 1485–1496.
31. E. M. Shepherd and J. Broderick, in *Comprehensive Natural Products II*, **Volume 4**, Elsevier, 2010.
32. D. Roeser, A. Preusser-Junze, B. Schmidt, K. Gasow, J. G. Wittman, T. Dierks, K. von Figura and M. G. Rudolph, A general binding mechanism for all human sulfatases by the formylglycine-generating enzyme, *Proc. Natl. Acad. Sci. U. S. A.*, 2006, **103**, 81–86.
33. M. J. Appel and C. Bertozzi, Formylgycine a post-translationally generated residue with unique cataytic caabilties and biotechnology applications, *ACS Chem. Biol.*, 2015, **10**, 72–84.
34. E. F. Neufed, Lysosomal storage diseases, *Annu. Rev. Biochem.*, 1991, **60**, 257–280.

CHAPTER 11

C–S Bond Formations in Primary and Secondary Metabolism

The major focus in all the chapters of this volume is on how physiologically accessible sulfur chemistry enables and underlies many pathways of primary metabolism. Primary metabolic pathways are defined as those that occur in every cell. (Our emphases have spanned aspects of both prokaryotic and eukaryotic sulfur metabolic chemistry). Secondary metabolic pathways are conditional, on in some cells, but not in others. While one could apply the conditional term to differentiation and developmental stages in higher eukaryotes, secondary metabolism usually encompasses low molecular weight compounds, often of remarkable scaffold complexity, that are produced under certain conditions by plants, fungi and bacteria.[1]

We take a brief look at some C–S bond-forming strategies in a few conditional biosynthetic pathways to compare and contrast molecular logic, after pointing out the logic for the four to five major C–S bond-forming strategies and protein machineries in primary metabolism.

11.1 C–S Bond-forming Strategies in Primary Metabolic Pathways

Organosulfur scaffolds that have featured prominently in metabolic transformations in the several chapters of Part II include cysteine, homocysteine as precursor to cystathionine, methionine and *S*-adenosylmethionine. The central metabolites coenzyme A and glutathione are derivatives of cysteine with no new C–S bonds formed in their biosynthesis. To understand the scope and functions

Chemical Biology No. 12
The Chemical Biology of Sulfur
By Christopher T. Walsh

Published by the Royal Society of Chemistry, www.rsc.org

of organosulfur metabolites we summarize the mechanisms that enzymes use to create the C–S bonds that define an organosulfur scaffold.

As we have noted earlier (Chapter 4), the C–S bonds in cysteine, cystathionine and homocysteine are all formed *via* the agency of the pyridoxal-phosphate (PLP) coenzyme in a specific enzyme active site (Figure 11.1). The heteroaromatic aldehyde portion of the PLP acts as an electron sink in amino acid–PLP aldimine adducts, not only in facilitating initial C2 carbanion formation but also in the β-elimination of the leaving group to create the aminoacrylyl–PLP adducts that are electrophilic at C3. The plant biosynthetic pathway uses a variant of the amino acid–PLP–aldimine strategy on the homoserine scaffold to create a corresponding βγ-enaminoacid–PLP adduct as a comparable electrophile. In all three enzymatic cases, a nucleophilic inorganic or organic thiolate ion, (HS^-, cysteine-S^-, homocysteine-S^-) is the attacking species in C–S bond formation. Conversion of cystathionine on to cysteine involves C–S bond cleavage but not C–S bond formation.

The biosynthesis of methionine from homocysteine constitutes the second major C–S bond-forming strategy (Figure 11.1B). We noted in Chapter 8, that the N5-CH_3– tetrahydrofolate (THF) cosubstrate can transfer a [CH_3^+] equivalent directly, or *via* a methyl B_{12} intermediate, to the thiolate of homocysteine. The polarity of this C–S bond formation mirrors those in the preceding paragraph: organic thiolate anion attacking carbon electrophile. Here, the carbon electrophile [CH_3^+] is quite distinct from the enamino acid–PLP adducts.

The third modality for C–S bond formation is enacted by SAM synthetase (Figure 11.1C). This is the sole case of a thioether sulfur atom attacking an sp^3 carbon center (C5′ of the ribose moiety of ATP). The resultant *S*-adenosylmethionine has the permanently trigonalized sulfonium cation from that energetically uphill C–S bond-forming step. Furthermore, the C5′ of the ribose ring in ATP is an unlikely electrophile in terms of its *a priori* chemical reactivity and is a one-off process.

In contrast to the two-electron reaction manifolds of the above three routes, thiolate anion or thioether as nucleophile in a direct two-electron step, one-electron routes involving sulfur radicals and carbon radicals are responsible for generating the four molecules in Figure 11.1D. Lipoic acid and biotin are made by higher eukaryotes although biotin is not made by humans and is required in the diet as vitamin B_9. In each case there are two C–S bonds, colored in blue in the figure, formed by radical SAM enzymes. In some analogy, two famous microbial secondary metabolites that are gateways to the penicillin and cephalosporin class of antibiotics are also shown in Figure 11.1D with the C–S bonds assembled by radical processes in ring formation and ring expansion also colored in blue. These are illustrative place holders for homolytic pathways in C–S bond metabolite constructions and a bridge to an overview of C–S bond formation routes in secondary metabolism.

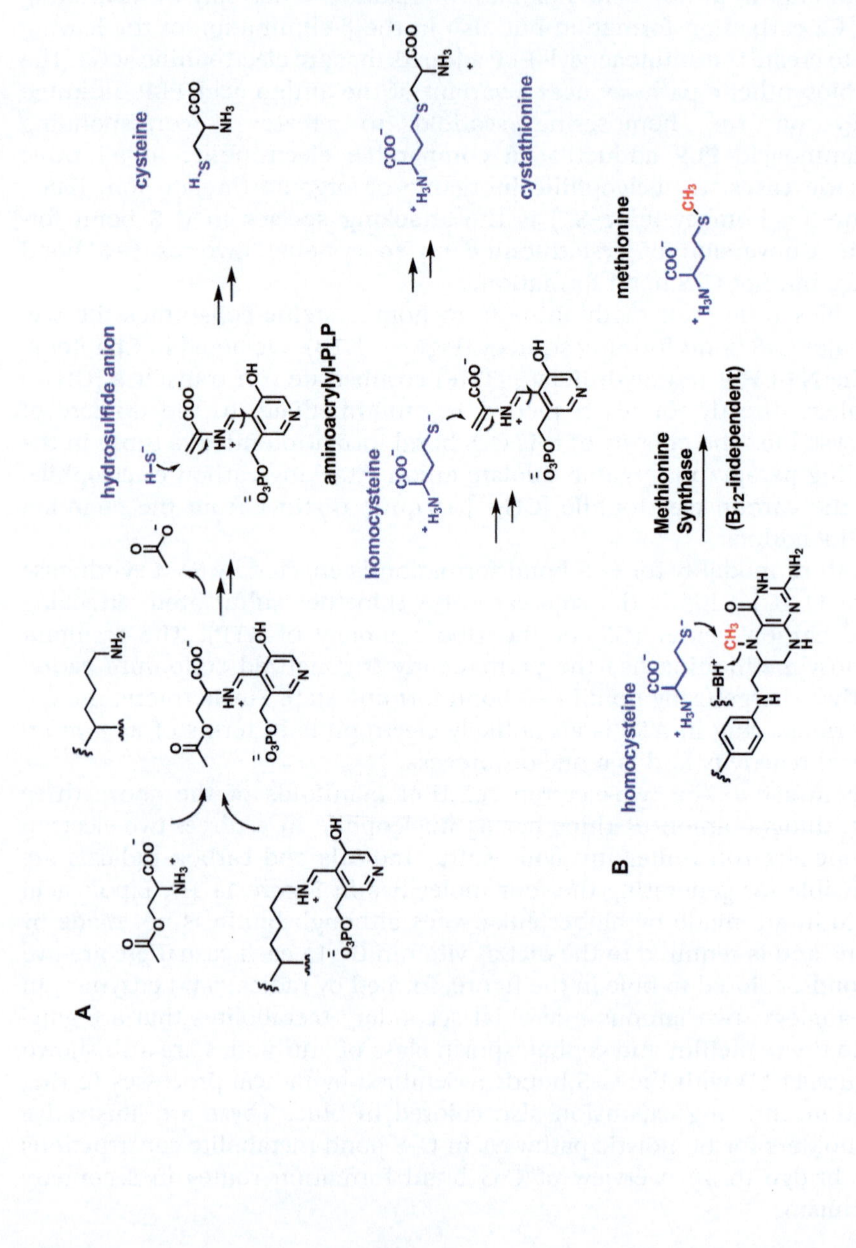
A
hydrosulfide anion
cysteine
aminoacrylyl-PLP
homocysteine
cystathionine
B
homocysteine
Methionine Synthase
(B12-independent)
methionine
N5-CH3-tetrahydro folate

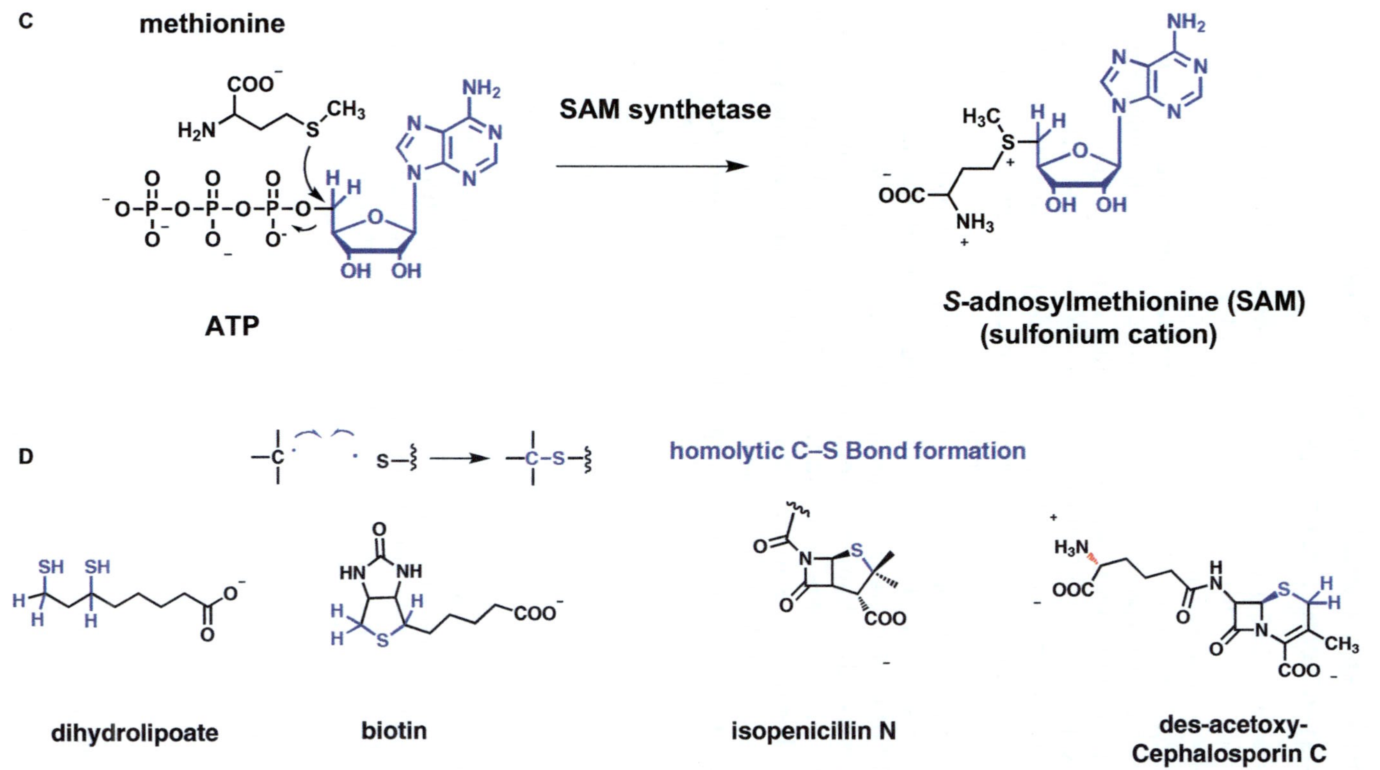

Figure 11.1 Three mechanisms of C–S bond formation in primary metabolism. (A) The formation of cysteine and cystathionine use aminoacrylyl–PLP adducts as the carbon electrophiles. (B) Methionine formation uses N5-CH_3–tetrahydrofolate, directly or indirectly in methionine synthases, as a carbon electrophile. (C) SAM synthetase uses a thioether as nucleophile and an unactivated C5′ of ATP as an unlikely carbon electrophile. (D) Metabolites spanning primary and secondary metabolism that form *via* one-electron, carbon and sulfur radical routes.

11.2 Some C–S Bond-forming Strategies in Conditional Plant and Microbial Biosynthetic Pathways

For readers who wish to delve deeply into biosynthetic pathway strategies for plant and microbial conditional sulfur-containing metabolites, a recent comprehensive review by Dunbar *et al.* in 2017 is recommended.[2] There is also a more general recent text on *Natural Product Biosynthesis: Chemical Logic and Enzymatic Machinery*.[1] In this subsection we highlight only a few of the more than 1000 known C–S bond-containing natural product scaffolds. In some cases, the C–S bond-forming enzymes are known and characterized (*e.g.* Isopeniclin N synthase see Chapter 9 and the previous section of this chapter) while in others there are proposed reaction mechanisms but the enzymes may not yet be fully characterized.

Figure 11.2, reprinted from a review by Dunbar *et al.*,[3] depicts some of the natural product scaffolds. Formation of the thiane ring in the 4,5-fused bicyclic scaffold of penicillins was noted in Chapter 9 Section 9.2 and in the previous section of this chapter, *via* one-electron reaction manifolds. Some unusual cyclic disulfides are found in tropodithietic acid and in the fungal hepatotoxin gliotoxin. The iron-chelating yersiniabactin from the plague bacterium *Yersinia pestis* has tandem five-member thiazoline and thiazolidine rings as Fe^{3+} ligands for siderophore function in iron acquisition strategies. The plant defense molecule camalexin has the fully aromatized thiazole ring version. The DNA-cleaving calecheamicin has a pendant trisulfide chain while leinamycin has an equally unusual disulfide monoxide group. The *S*-oxidations in leinamycin and amanitin presumably arise by monooxygenase actions on sulfur.

Thiacoraline, the dimerized product of a microbial non-ribosomal assembly line, has three kinds of C–S bonds, two bridging thioesters, a bridging disulfide and a pair of *S*-methylcysteine side chains,[4] reflecting three modifications of cysteine side chains by the kinds of strategies we have noted in primary metabolic transformations in previous chapters. The bisthioesters of this DNA-intercalating agent probably occur during dimerizing chain release from the assembly line's terminal thioesterase domain.[4]

11.2.1 Disulfide to Methylthioacetal

An intriguing rearrangement of a thiacoraline-related bridging disulfide of the triostin A dimer (Figure 11.3A) occurs on post-assembly line methylation.[5,6] The proposal is that SAM delivers a $[CH_3^+]$ equivalent to the disulfide. This is the first unusual step because disulfides are generally electrophilic not nucleophilic: perhaps the $[CH_3^+]$ moiety in flight from SAM is sufficiently electrophilic to transiently generate the monomethylated disulfide species. This could undergo carbanion formation at one of the adjacent CH_2 groups to form the paired − and + charge in the indicated sulfur ylid. Then, the second unusual step is proposed, a 1,2 migration of the ylid carbon to the distal sulfur as the weak $S–S^+$ bond breaks, quenching the

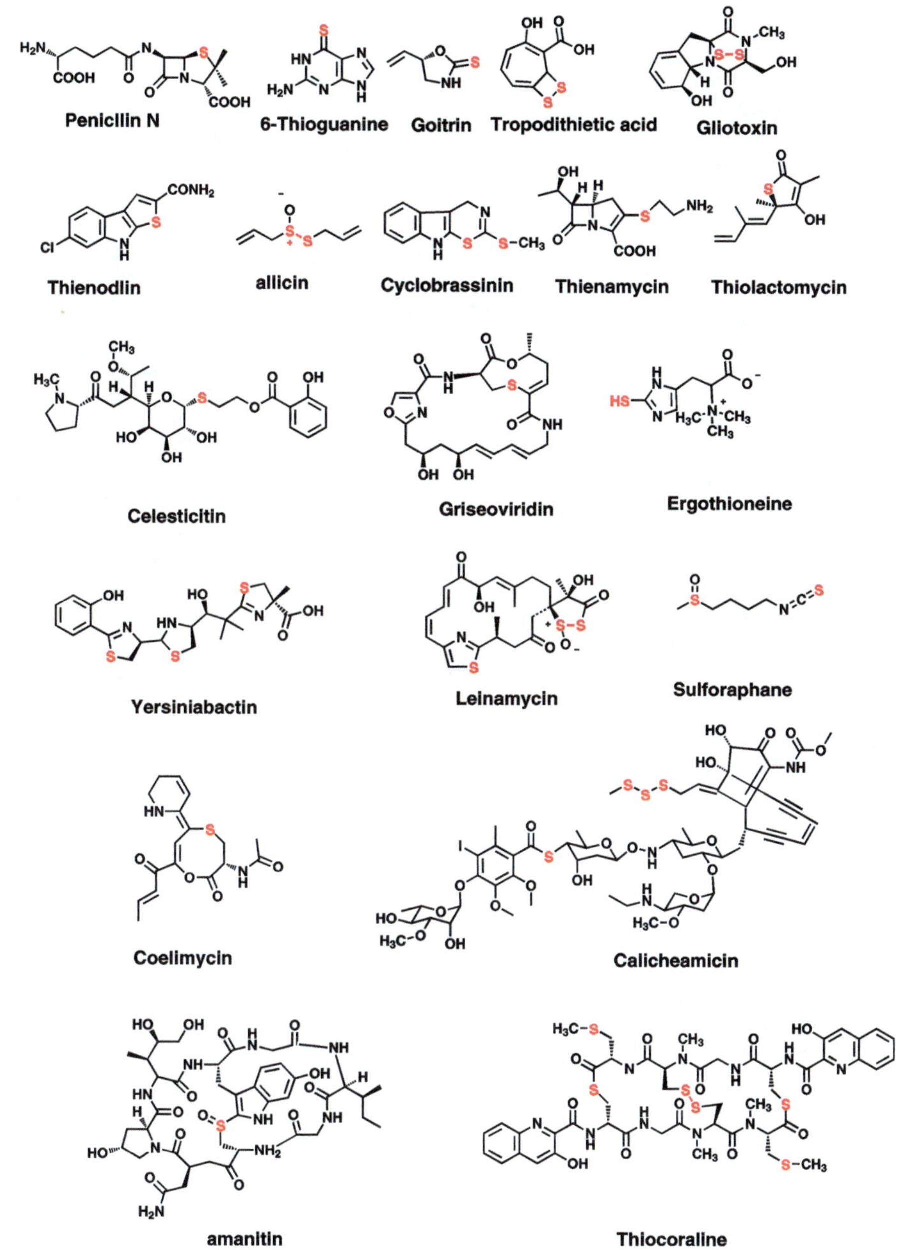

Figure 11.2 A representative set of C–S bonds in some natural product scaffolds. Reproduced from ref. 2 with permission from American Chemical Society, Copyright 2017.

charge on sulfur. This action has cleaved the initial bridging disulfide (S–S), while forming a new, tethering C–S bond. Inspection of that bridge shows that a disulfide has been converted into an *S*-methyl thioacetal linkage: a remarkable feat of sulfur chemical biology.

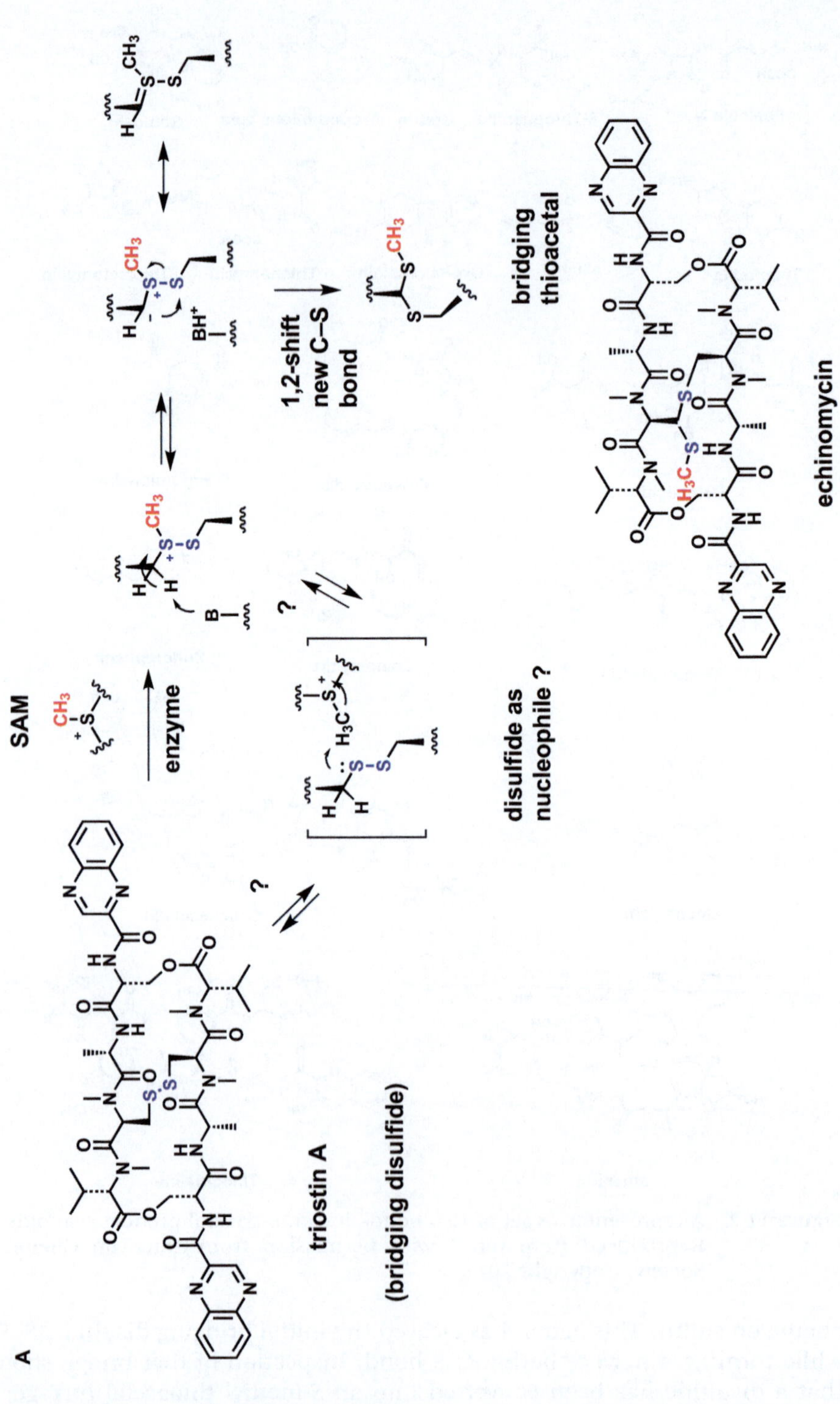
A
triostin A
(bridging disulfide)
SAM
enzyme
disulfide as
nucleophile ?
1,2-shift
new C–S
bond
bridging
thioacetal
echinomycin

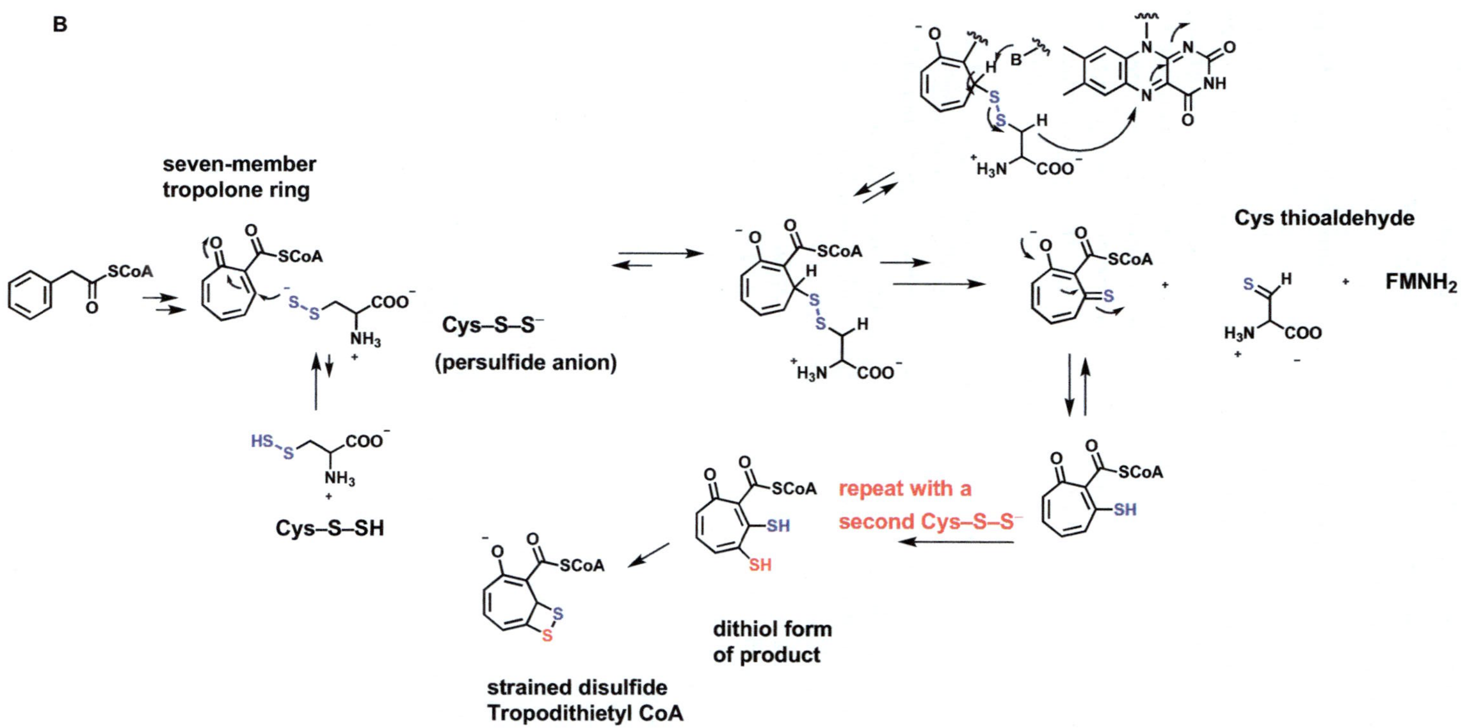

Figure 11.3 Two unusual enzymatic transformations of disulfide bonds in natural product scaffold maturations. (A) Late stage conversion of triostin A to echinomycin involves methylation of a bridging disulfide and rearrangement of the presumed nascent methylated disulfide species to bridging methyl thioacetal. (B) Formation of the four-membered fused disulfide in tropodithietic acid proposes tandem conjugate addition of cysteine persulfide anions to the tropolone system followed by oxidative fragmentation of each disulfide to a separated pair of thiocarbonyl fragments.

11.2.2 Iterative Conjugate Additions by Cysteine-persulfide Anions Followed by Oxidative Cleavage of the Resultant Disulfides

Another intriguing mechanistic proposal attends the biosynthetic proposal for the seven-membered tropolone natural products, specifically in construction of the unusual four-membered disulfide ring fused to the tropolone scaffold in tropodithietic acid.[7] The tropolones mediate symbiotic interactions between the *Roseobacter* producers and their algal consortium members. Figure 11.3B indicates that phenylacetyl-CoA is the precursor to the ring-expanded tropolone acyl CoA. The most intriguing aspect of mechanism is the as yet unproven proposal that a cysteine lyase type catalyst generates a cysteinyl–S–SH reactant. In keeping with the pK_a of cysteine persulfide, the persulfide anion would predominate (Cys–S–S^-) in neutral solution.

The persulfide anion is proposed to be the nucleophile in conjugate addition to the tripolone. A subsequent cysteine lyase type β-elimination yields the monothiol adduct. A repeat of conjugate attack by a second Cys–S–S^- persulfide anion would yield the dithiol form of the product. Ring closure to the 4,5 disulfide may or may not be enzymatically catalyzed. This inferred behavior of the cysteine persulfide anion as a nucleophile has analogies to the sulfide quinone oxidoreductase mechanism described in Chapter 3.

Once the S–S adducts from the tandem conjugate additions of the cysteine persulfide anions have formed, they are unsymmetrical disulfides and their proposed mode of breakdown is remarkable. The enzyme in question has a tightly bound flavin mononucleotide (FMN) and the proposed route for S–S bond cleavage is a further oxidation and fragmentation. The result is a pair of C=S thiocarbonyl groups, one in the tropolone scaffold, the other as the cysteine thioaldehyde (we have noted a cysteine residue thioaldehyde in Chapter 5 in discussion of the pantetheinylcysteine FAD-oxidoreductase-mediated decarboxylation).

Figure 11.4 focuses in on the two otherwise unprecedented reactions of those disulfides in the echinomycin and tropodithietic acid pathways. The first is reaction of a disulfide towards CH_3^+ as a carbon electrophile. The second is oxidative cleavage of a disulfide by a proton–hydride flavoenzyme mechanism. In the first instance one does not normally see disulfides behaving as nucleophiles. In the second instance the net two-electron oxidative fragmentation of a disulfide to two thiocarbonyl equivalents is also remarkable. The S–S bond fragmentation step is proposed to be coupled to hydride transfer to the FMN coenzyme. Both are worthy of further mechanistic scrutiny to confirm these putative mechanisms.

11.2.3 Thioglycosides and Glucosinolates

Two examples of enzymatic thioglycoside natural product formation broaden thiol-mediated group transfer chemical biology (Figure 11.5A and B). In the

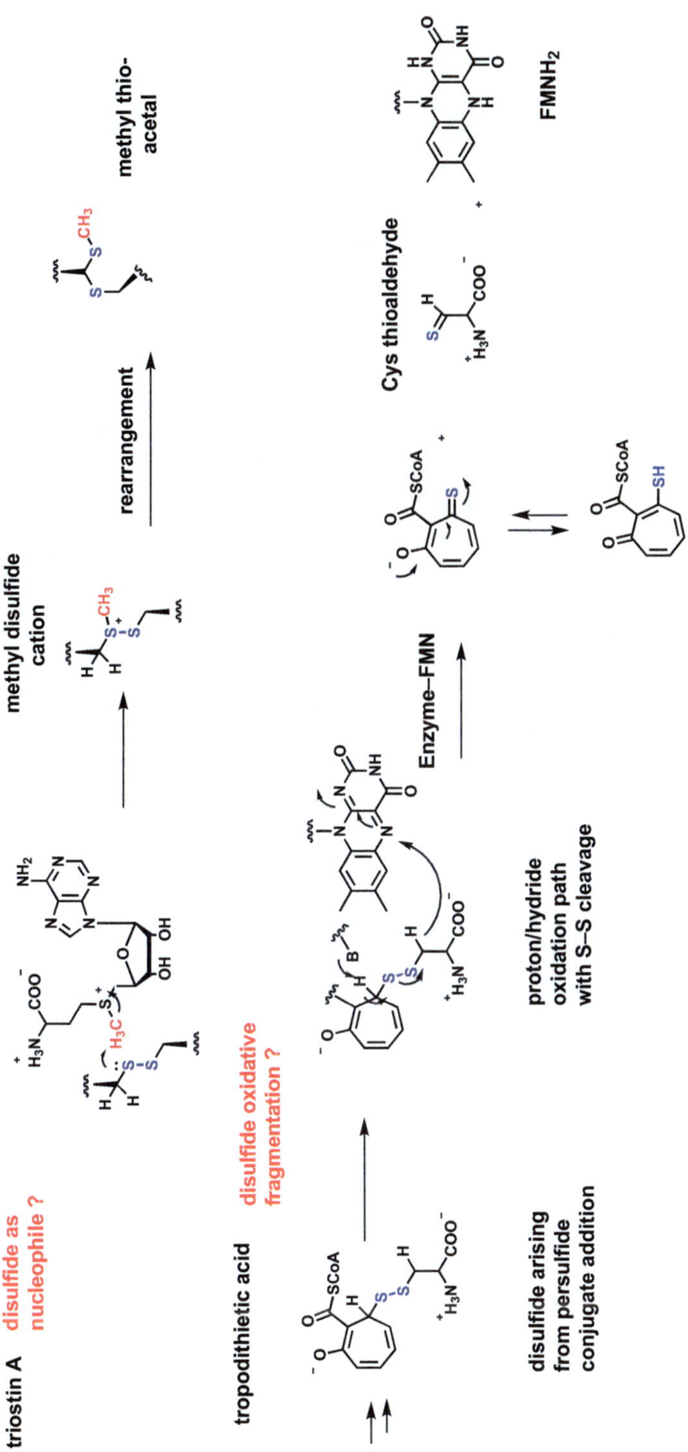

Figure 11.4 Mechanistic proposals for the rearrangement of methylated disulfides and oxidative cleavage of disulfides to two thiocarbonyl fragments.

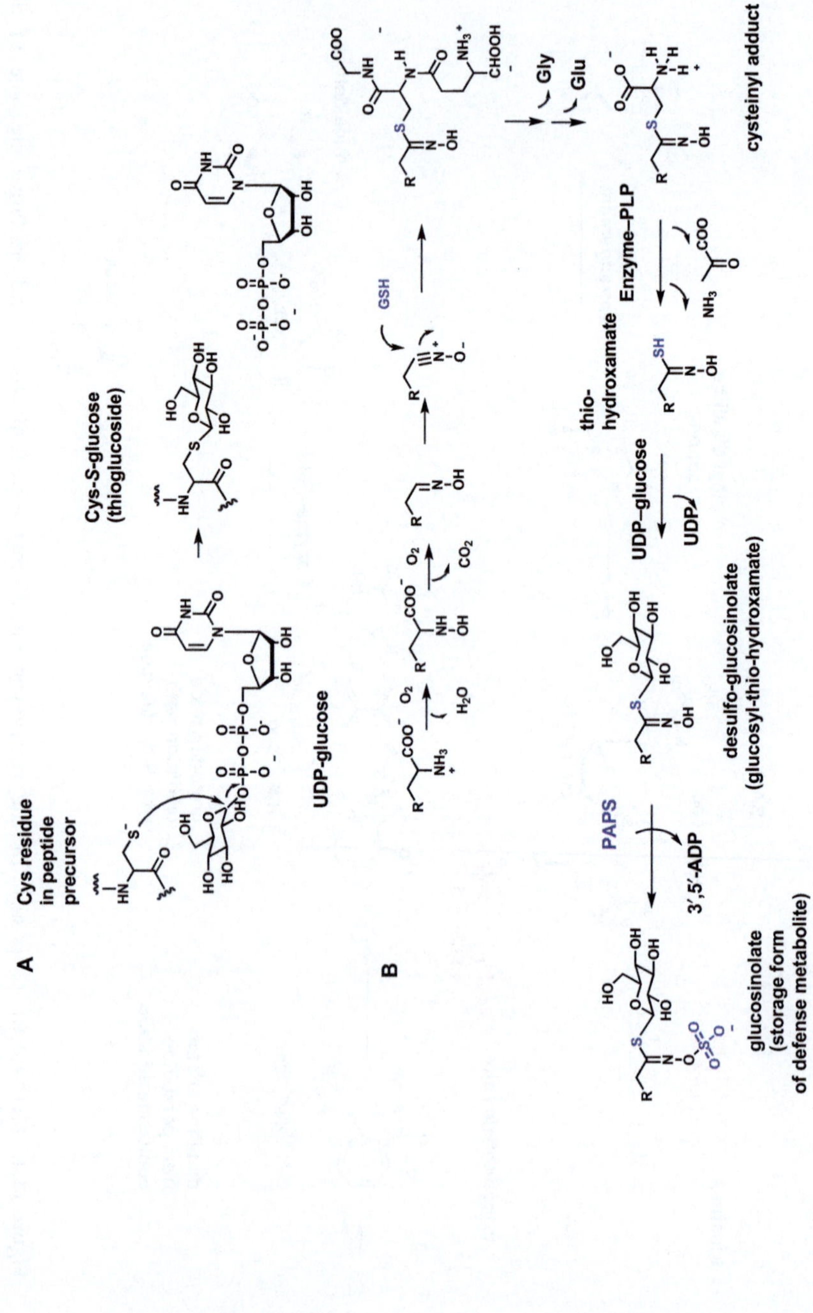
A
Cys residue in peptide precursor
UDP-glucose
Cys-S-glucose (thioglucoside)
B
GSH
Gly
Glu
cysteinyl adduct
Enzyme–PLP
thio-hydroxamate
UDP–glucose
UDP
desulfo-glucosinolate (glucosyl-thio-hydroxamate)
PAPS
3′,5′-ADP
glucosinolate (storage form of defense metabolite)

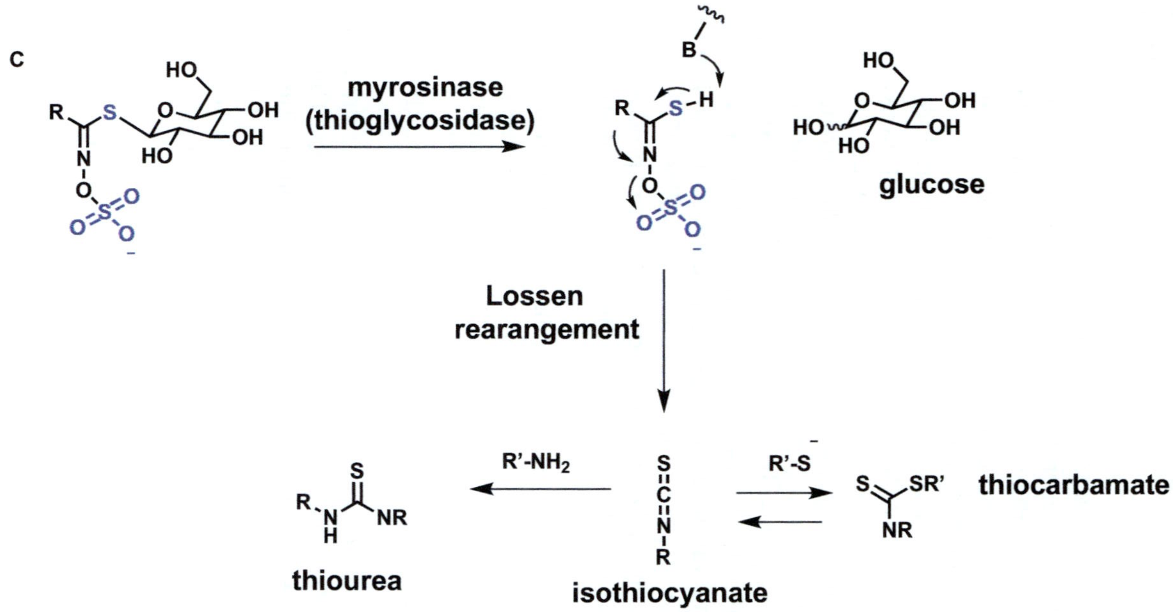

Figure 11.5 Enzymatic thioglucoside formation involves attack of a thiolate nucleophile on the electrophilic C1′ carbon of the glucosyl moiety of UDP–glucose. (A) Schematic of thioglycoside bond formation in the maturation of the peptide scaffold of sublancin. (B) Steps in the biosynthetic pathways of sulfoglucosinolates, plant storage precursor molecules to electrophilic isothiocyanates. (C) Activation of sulfoglucosinates to isothiocyanates is a two-step process. The first, enzymatic, step is catalyzed by myrosinase. The second, nonenzymatic, step involves intramolecular decomposition of the sulfothiohydroxmate enzymatic product to the isothiocyanate that can be captured by amines to form thioureas, or by thiolate anions to form dithiocarbamates.

peptide natural product sublancin, a cysteinyl side chain thiolate attacks UDP–glucose, the classic biologic donor of glucosyl groups at activated at C1′, to yield the thioacetal/*S*-glucoside linkage with inversion of stereochemistry at C1′ of the migrating glucosyl unit[8] (Figure 11.5A).

A distinct but notionally related step occurs in the biosynthesis of plant glucosinolate defensive precursors of isothiocyanates[2,9] (Figure 11.5B). Those pathways begin with cytochrome P450 *N*-hydroxylation of amino acid substrate and decarboxylation to the *N*-hydroxyimine. Further cytochrome P450 catalysis is proposed to generate transiently the nitrile oxide as electrophile for capture by the thiolate anion of glutathione. Subsequent peptidase action trims away the glycine and glutamate residues hydrolytically to generate the indicated cysteinyl-*S*-adduct.

To cleave the C3–S bond in the cysteinyl chain and produce the free thiohydroxamate, a PLP-dependent cysteine β-eliminase (of the type noted in Chapter 4), presumably creates pyruvate, ammonia and the sulfur leaving group as the thiohydroxamate. In that thiohydroxamate the –SH group is nucleophilic and can engage in thioglycoside formation analogous to the scheme in part A of this figure. The product is the desulfo-glucosinolate (Figure 11.5B) The final step in the pathway is enzymatic introduction of a second sulfur, at the oxidation state of sulfate from the standard sulfuryl donor 3′-phosphoadenosine-5′-phosphosulfate (PAPS). The sulfated form of the sulfo-glucosinolate is the end product of the pathway. Thus, sulfur at two distinct oxidation states participates in glucosinolate assembly.

When plants sense a predator attack, such a mechanical chewing on plant tissue or physical crushing of a leaf section, the enzyme myrosinase, a thioglucosidase normally segregated in a subcellular compartment separate from the sulfoglucosinolate substrate, gains physical access to substrate. The thioglycoside link is hydrolyzed, liberating free glucose and the *O*-sulfo thiohydroximate.[9]

As depicted in Figure 11.5C, intramolecular participation of the thiolate anion leads to concomitant elimination of the sulfate group. This intramolecular participation is analogous to the internal decomposition of the *O*-sulfate metabolite of *N*-hydroxyacetaminophen, described in Chapter 5. The nascent products are sulfate dianion and an isothiocyanate, electrophilic at its central carbon. Typically, the isothiocyanate is capturable by amines to give thiourea adducts that are irreversible or by thiolate anions to give dithiocarbamates that are slowly reversible. If the amine or thiol components are part of the predator, this constitutes a covalent defense mechanism.

In retrospect, the plants have made and stored a spring-loaded source of the reactive isothiocyanates in the form of the *S*-glucosylated and *O*-sulfated thiohydroxamates. The two-step unraveling of the precursor sulfoglucosinolate is set in motion by myrosinase enzymatic action. The second step, elimination of sulfate to the give the isothiocyanate is nonenzymatic. It only happens after the free thiol is produced from myrosinase action.

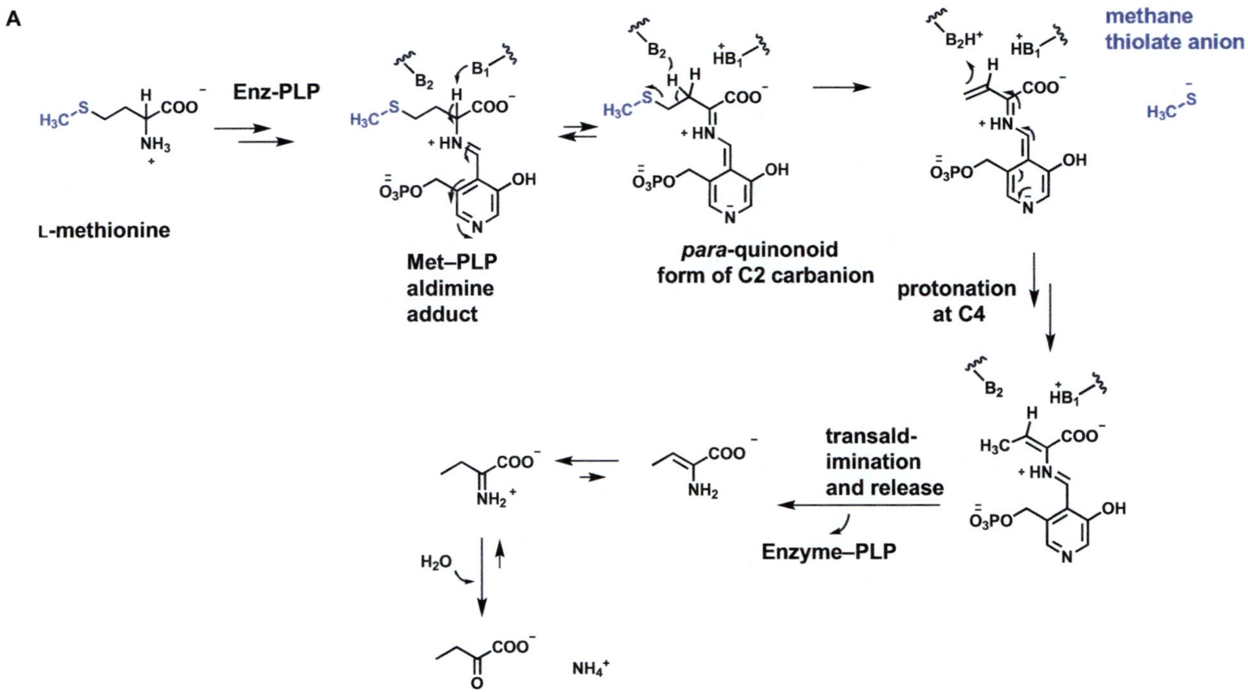

Figure 11.6 Methanethiol as a metabolic nucleophile in bacterial metabolism. (A) Bacterial methionine-γ-lyase uses its tightly bound pyridoxal-P coenzyme to fragment L-methionine to the one-carbon methane thiol and the four-carbon 2-ketobutyrate, along with ammonia. This is a γ-elimination catalyst, generating intermediate α-carbanionic–PLP and then β-carbanionic-aminoacrylyl–PLP adducts to lower the energy for C4–S bond cleavage. (B) The liberated methanethiolate anion fragment can act in *Ralstonia solnacearum* bacteria as the one-carbon thiol nucleophile in a (nonenzymatic?) conjugate addition to ralfuranone B at a late stage in ralfuranone D construction.

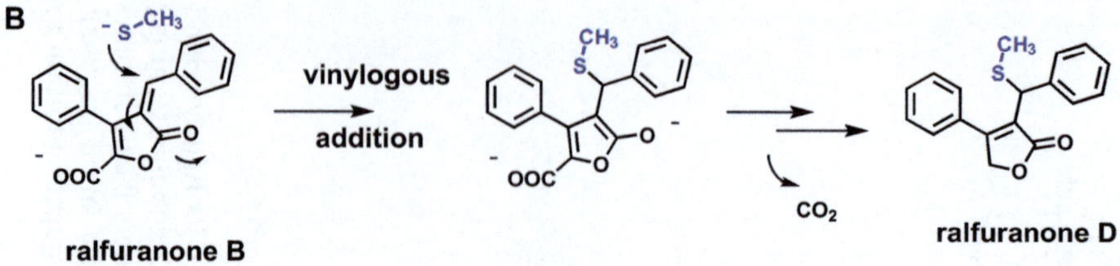

Figure 11.6 Continued.

The prior formation of the sulfate ester of the glucosinolate for storage is a necessary step for the second stage of unraveling the hidden isothiocyanate defense molecule. This strategy emphasizes the two complementary roles of the two sulfur atoms, one at the thiol and one at the sulfate oxidation state, for this two-stage chemical biology defense logic. The structure of sulforaphane in Figure 11.2 shows another natural isothiocyanate with a second, side chain sulfur at the sulfoxide oxidation state, after two-step unraveling from a sulfoglucosinolate precursor.

11.2.4 Methanethiol from Methionine-γ-lyase Activity

One final sulfur secondary metabolism example involves not cysteine or glutathione as a source of a reactive thiol nucleophile but rather methionine. Some microbes possess a PLP enzyme that acts as a γ-lyase on L-methionine[10] (Figure 11.6A). Action of the lyase yields ammonia, 2-ketobutyrate and methanethiol. It is the one-carbon methanethiol that is used as nucleophile in the bacterial (*Ralstonia*) biosynthetic pathway to ralfuranone D (Figure 11.6B). The cosubstrate, late in the pathway, is the indicated enone ralfuranone B. The double bond system is set up for conjugate addition and it is thought that the indicated vinylogous-1,4-conjugate addition by the methane thiolate anion may be nonenzymtic.[7,11] Subsequent double bond migration and decarboxylation steps yield the thioether linkage from the net 1,4-addition process in the product ralfuranone D.

Many other examples could be presented to get to the complex scaffolds of Figure 11.2. However, even this partial set of examples of the C–S bond-forming universe of conditional metabolites illustrates both the range of functional groups generated and the common logic of starting with cysteine, glutathione and methionine as building blocks to be shunted down the secondary pathways. The examples in Figures 11.3–11.6, reveal thioethers, strained disulfides, acyl thioesters, thioacetals, thioglycosides, isothiocyanates, glucosinolates, sulfate esters and persulfide anions from the common thiol (cysteine) and thioether (methionine) building blocks.

References

1. C. Walsh and Y. Tang, *Natural Product Biosynthesis: Chemical Logic and Enzymatic Machinery*, Royal Society of Chemistry, London, 2017.

2. K. L. Dunbar, D. Scharf, A. Litomska and C. Hertweck, Enzymatic C–S bond formation in natural product biosynthesis, *Chem. Rev.*, 2017, **117**, 5521–5577.
3. K. Y. Dunbar, J. Melby and D. A. Mitchell, YcaO domains use ATP to activate amide backbones during peptide cyclodehydrations, *Nat. Chem. Biol.*, 2012, **8**, 569–575.
4. F. Lombo, A. Velasco, A. Castro, F. De la Calle, A. F. Brana, J. M. Sanchez-Puelles, C. Mendez and J. A. Sala, Deciphering the biosynthetic pathway of the antitumor thiocoraline from a marine actinomycete and expression in two Streptomyces species, *ChemBioChem*, 2006, **7**, 366–376.
5. K. Watanabe, K. Hotta, A. P. Praeusth, K. Koketsu, A. Migita, C. N. Boddy, C. C. Wang, H. Oguri and H. Oikawa, Total biosynthesis of antitumor nonribosomal peptides in E. coli, *Nat. Chem. Biol.*, 2006, **2**, 423–428.
6. K. Hotta, R. Keegan, S. Ranganathan, M. Fang, J. Bibby, M. D. Winn, M. Sato, M. Lian, K. Watanabe and D. J. Rigden, Conversion of a disulfide bond into a thioacetal group during echinomycin biosynthesis, *Angew. Chem., Int. Ed.*, 2014, **53**, 8240828.
7. J. Pauly, D. Spiteller, J. Linz, J. Jcobs, C. Allen, M. Nett and D. Hoffmeister, Thioether production by the plant pathogen Ralstonia solanacearum, *ChemBioChem*, 2013, **14**, 2169–2178.
8. T. J. Oman, J. Boettcher, H. Wang, X. Okalibe and W. A. van der Donk, Sublancin is not a lantibiotic but an S-linked glycopeptide, *Nat. Chem. Biol.*, 2011, **7**, 78–80.
9. A. M. Bones and J. Rossiter, The enzymatic and chemically induced decomposition of glucosinolates, *Phytochemistry*, 2006, **67**, 1053–1067.
10. D. Sato and T. Nozaki, Methionine gamma-lyase: the unique reaction mechanism, physiologic roles and therapeutic applications against infectious diseases and cancers, *IUBMB Life*, 2009, **61**, 1019–1028.
11. B. Wackler, P. Schneier, J. M. Jacobs, J. Pauly, C. Allen, M. Nett and D. Hoffmeister, Ralfuranone biosynthesis in Ralstonia solanacearum: functional divergence in the quinone synthetase family of enzymes, *Chem. Biol.*, 2011, **18**, 354–360.

Part III: Sulfur Side Chains in Proteomes

The two proteinogenic amino acids cysteine and methionine, on incorporation into the tens of thousands of proteins in cells and tissues, have their amine and carboxylates tied up in peptide linkages, leaving the side chain thiol (cysteine) and thioether (methionine) respectively, available for functions and reactions.

The thioether side chain of methionyl residues is largely unreactive compared with the cysteine thiolate side chain. We will note in Chapter 13 that the sole posttranslational modifications of methionyl side chains are *S*-oxygenations, first to the sulfoxide and then on to the sulfone. The two-electron-oxidized sulfoxide is reversibly reduced enzymatically, while the four-electron-oxidized methionine sulfones are not reversible.

The rest of the actions of sulfur-containing amino acids in proteomes in the three chapters of Part III are thus taken up by the cysteine thiolate. The cysteinyl proteomes are largely subdivided in Chapters 12, 13, and 14. These comprise catalytic functions of cysteinyl thiolate side chains in enzymatic catalyses in Chapter 12, covalent posttranslational modifications of client proteins in Chapter 13 and exotic, often biologically restricted, cysteine morphings in Chapter 14, respectively.

In keeping with the powerful nucleophilicity of cysteinyl residue side chains, a number of enzyme families generate transient covalent adducts between an enzyme active site cysteine and a portion of a substrate molecule in a first half reaction. In the second half reaction the attached substrate fragment is typically offloaded to a nucleophilic atom in a cosubstrate.

This is the logic for approximately 100 human proteases that use cysteinyl side chains to make acyl-*S*-Cys–enzyme intermediates. In the second half reaction of proteases water is the cosubstrate. In transpeptidases it is an amine nucleophile of a cosubstrate.

The ubiquitin–deubiquitin cascades use similar acyl-*S*-cysteinyl–protein acyl transfer logic. Now the acyl group is the information-rich, folded 75-residue ubiquitin protein. These cascades are about covalent tagging of client proteins, altering function and/or fate by making one or more isopeptide bonds to lysine side chains. Hydrolysis is avoided, except when one or more of the approximately 80 deubiquitinases intervene with their cysteinyl active site nucleophiles to effect net transfer to water molecules.

A second major class of covalent cysteinyl–enzyme intermediates are found in approximately 75 protein tyrosine phosphatases where the electrophilic fragment from substrate that gets transiently attached as a covalent adduct to cysteine is the itinerant $-PO_3^{2-}$ group. The *S*-phosphoryl-cysteinyl–enzyme intermediates are then hydrolyzed in the second half reactions.

In contrast to *transient* covalent adducts to active site cysteines in certain enzyme families are the *stable* covalent adducts to cysteinyl side chains in posttranslational modifications of proteins that change their functions and/or subcellular locations. Some of the principles are exemplified in Chapter 14, starting with the two major kinds of cysteinyl side chain lipidations-*S*-palmitoylations and *S*-prenylations. The *S*-palmitoylated protein linkages are hydrolytically labile acyl thioesters, reversible by action of acyl protein thioesterases. The *S*-prenylations are hydrolytically stable alkylations and are not readily undone. Ras proteins are prototypic examples that undergo both types of cysteinyl residue lipidations.

Oxidative chemistry, purposeful or adventitious, of cysteinyl residue sulfhydryl groups may be the most commonly appreciated posttranslational modification. Sometimes these proceed through sulfenic acid (–S–OH) derivatives on the way to cystine disulfides in folded proteins. In other circumstances, such as the major disulfide-forming routes in the secretory compartments of eukaryotic cells, cystine–disulfide formations are driven by mixed disulfides with abundant protein disulfide isomerases in the endoplasmic reticulum. Chapter 13 also revisits the net oxidation of cysteinyl side chains in cysteine lyases and their partner chaperones that ferry sulfur as persulfide equivalents (–Cys–S–SH) for iron–sulfur cluster biosynthesis.

Chapter 14, the third of the thee chapters in Part III, takes up a series of cysteinyl-based transformations of proteins that are remarkable chemically but of limited biological distribution. Many of them start with the attack of a cysteinyl-thiolate residue on the immediate upstream peptide carbonyl in a folded protein substrate. The cyclic tetrahedral adduct that results can partition in the forward direction by two bond-breaking routes.

If the C–N bond is cleaved, the resonance-stabilized, unreactive peptide bond is converted to a thermodynamically activated acyl thioester linkage. In the case of the signaling protein hedgehog, this thioester is captured covalently by a molecule of cholesterol in the plasma membrane of a cell, a net autocleavage reaction. The product is an approximately 170-residue protein–*O*–cholesterol adduct that is restricted to two-dimensional diffusion in the cell membrane.

When the nucleophile capturing the acyl-*S*-protein intermediate is internal (instead of external, such as cholesterol) such as a downstream cysteinyl thiolate side chain, this comprises acyl chain migration in the process known as protein splicing. A series of steps culminate in acyl-*S* to acyl-*N* migration, reforming a peptide bond, joining upstream and downstream exteins, with expulsion of the intervening peptide sequence, known as an intein.

A second fate for the initial cyclic tetrahedral adduct, as an alternative to C–N bond cleavage, is C–O bond cleavage (see Figure 14.2). We will examine this as the route to formation of thiazolines, and then, on dehydrogenative aromatization, thiazoles. This path leads to conversion of peptide backbones into nonhydrolyzable five-membered heterocyclic thiazole rings, a stunning autocatalytic morphing of peptide bonds immediately upstream of certain cysteine (and serine) side chains in proteins.

Overall, the three chapters of Part III illustrate underlying molecular principles and uses to which the reactivity of the side chain thiol of cysteine residues (and to a much lesser extent the side chain thioether of methionyl residues) are employed to diversify proteome function.

Most obvious is the use of cysteine thiolates as catalytic nucleophiles in several enzyme families that engage in covalent catalysis during their catalytic cycles. The covalent intermediates so generated are necessarily transient, breaking down or transferring out, purposefully, in each catalytic cycle. These are distinct in both lifetimes and functions from the many cellular proteins that undergo relatively long-lived covalent posttranslational modifications. Some are reversible (*e.g.* *S*-palmitoylation), some are irreversible (*e.g.* *S*-prenylation).

In both cases, the cysteinyl thiolates, on enzymes or client proteins, act as nucleophiles. The chemistry of –cysteine–S–OH, –cysteine–S–SH and –cysteine–S–S–cysteine– (all behaving as electrophiles), and iron–sulfur clusters have distinct, albeit sometimes converging, roles and reactivities in cysteine proteomes.

CHAPTER 12

Cysteinyl Proteomes I: Metal Coordination and Catalytic Intermediates

12.1 Cysteinyl Proteomes: Introduction

On a statistically random basis the abundance of each of the 20 proteinogenic amino acids in proteomes would be 5%. For example, serine is serviced by six tRNA codons and comes in at about 8% abundance. Cysteine abundance, typically coded by UGU and UGC codons, ranges from 3.3% down to approximately 0.5% in some archaeal proteomes. In higher organisms' proteomes, cysteine residues tend to be underrepresented compared with the 5% average.[1–3] At an estimated level of 2% abundance in the human proteome, with an average size protein chain/subunit of approximately 40 kDa, that would comprise about 250 000 to 300 000 cysteines in the overall human proteome.[3] The large number both creates many possibilities for cysteine function in thousands of proteins and poses technical challenges for untargeted detection of any given subset or individual cysteine residue reaction in a proteome, a thought we return to at the end of Chapter 13.

Prior to a discussion of some of the major enzyme types that employ cysteine side chains as catalytic nucleophiles, we make a brief detour to metal cation coordination of cysteine side chains.

12.2 Coordination to Iron and Zinc

The two most abundant metal cations in cells with notable affinity for cysteine thiolate side chains of proteins are iron and zinc. The abundant monovalent cations, sodium (Na^+) and potassium (K^+) prefer oxophilic ligands as does the divalent calcium ion (Ca^{++}). Fe is a redox active

Chemical Biology No. 12
The Chemical Biology of Sulfur
By Christopher T. Walsh

Published by the Royal Society of Chemistry, www.rsc.org

transition metal, cycling between ground state Fe^{3+} and Fe^{2+} as we have noted in Chapter 6 in a discussion of the conversion of cysteine side chains to persulfides and on to 2Fe–2S and 4Fe–4S clusters.

12.2.1 Three Forms of Iron in Proteins

As we noted in Chapter 6 much of that information could be placed here in Part III, especially given the occurrence of covalent persulfide adducts to cysteine residues both in the cysteine lyases and then the series of sulfane or persulfide chaperone proteins before depositing the itinerant sulfur atoms as sulfide dianion equivalents on target Fe–S assembly proteins. We will not repeat that persulfide protein discussion here but readers may wish to refresh their insights with regard to the reach of cysteine chemistry in proteomes.[3]

Analogously, in Chapter 9 we gave several examples of cysteine residue thiol radicals in quite diverse enzyme catalytic cycles. We chose to embed that topic in the context of one-electron *vs.* two-electron reaction manifolds to emphasize the redox versatility of the thiol group, whether free or in proteins. As in Chapters 2 and 3 on inorganic sulfate biological reduction and the reverse, hydrogen sulfide biological oxidations, to illustrate the reactivity and mechanism of the available sulfur chemistry, we placed the topics in their enzymatic contexts of redox cycling. (Finally, Chapter 10 deals in significant part with protein-tyrosyl-sulfate residues: those are surely part of the sulfur proteome albeit not the cysteine proteome).

Definitionally, those topics could also be stood in their head and evaluated in the course of proteome-centered chemical biology of cysteine residues. One cannot separate the two pieces, nor the interaction of organic sulfur and inorganic sulfur functional groups in their intertwined chemical biology.

With regard to iron binding sites and arrays in proteins we have noted along the way to this point three major forms of protein-associated iron (Figure 12.1). These encompass heme groups, several of which have a cysteine thiol as the bottom axial ligand to modulate the electronics, redox potential and reactivity of the embedded central iron atom that can remain as Fe^{2+} (hemoglobin, myoglobin), interconvert between Fe^{3+} and Fe^{2+} (cytochromes), and in oxygenases traverse to $Fe^{V}{=}O$ robust oxidant equivalents (cytochrome P450s).

The second frequent platform for Fe^{2+} is the mononuclear nonheme iron array (Figure 12.1). We have met these multiple times: in isopenicillin N synthase and expandase in penicillin and cephalosporin antibiotic class assembly (Chapter 9), in the cysteine to formylglycine modification in the large family of sulfatases (Chapter 10) and in the unusual cysteine dioxygenase active site configuration (Chapter 7). This mononuclear nonheme iron class also makes high-valent ($Fe^{IV}{=}O$) oxo iron intermediates to carry out sulfur radical and then hydroxyl radical equivalent recombination chemistry.

The third iron-containing protein platform encompasses the ubiquitous 2Fe–2S and 4Fe–4S clusters (Figure 12.1) that we examined in Chapter 6. They are probably the most ancient and closest to primordial self-assembly

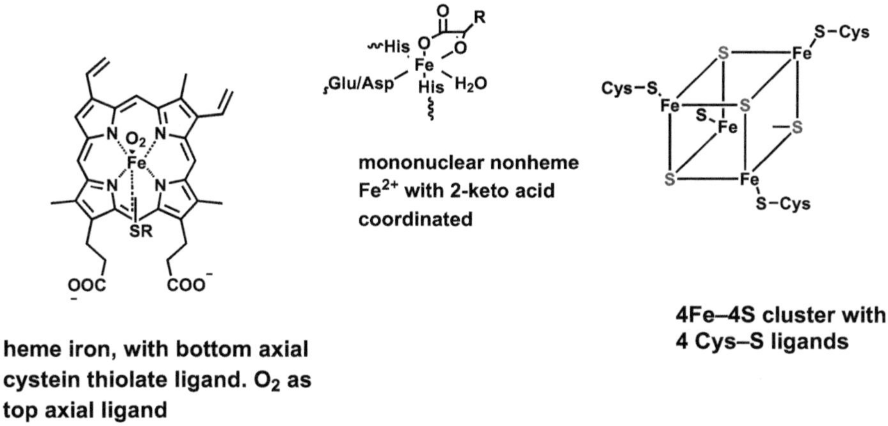

Figure 12.1 Three common forms of iron cofactors in proteins.

platforms in early anaerobic atmospheres where thiols and Fe^{2+} atoms were not intercepted by oxidizing O_2 and partially reduced oxygen species.

12.2.2 Zinc as a Pervasive Super Acid, Organizing Ligand in Proteins

Zinc, with a d10 electronic configuration, forms the positively charged $zinc^{++}$ ion. Zn^{++} is not redox active. With its 2^+ charge, it does function as a super acid. Andreini *et al.*[4] estimate that as many as 3000 human proteins may contain zinc up to stoichiometric levels. Nitrogen atoms in histidine residues and carboxylate side chains of aspartate and glutamate residues are frequent protein ligands to Zn^{2+} ions. Figure 10.9 displays a binuclear zinc active site in alkyl sulfatases, activating the water molecule that can attack alkyl sulfate esters for either C–O or O–S cleavage routes.

Cysteine thiols and their thiolate anions are also ligands for thiophilic zinc. A long known protein ligand set for a required zinc ion is in the active site of alcohol dehydrogenase (reviewed elsewhere).[5] As sketched in Figure 12.2A, the ligand set for the zinc is the imidazole ring of His66 and two thiolates from Cys46 and Cys174. The fourth ligand to zinc is the alcohol of the substrate to be oxidized by hydride transfer to cosubstrate NAD^+. The zinc anchors the alcohol substrate in proximity and orientation to the nicotinamide hydride acceptor while the alcohol hydrogen is hydrogen bonded and set up for departure as a proton in the aldehyde-forming step.

Additional examples where zinc–Cys thiolate coordinations are part of the catalytic machinery of enzymes for activation of a water molecule as a nucleophile occur in cytidine deaminase and GTP cyclohydrolase (on the way to pterin coenzyme frameworks).[5] The protein farnesyl transferases, which we discuss in the next chapter, are hybrids in that the attacking cysteine of a protein substrate becomes coordinated to the transferase active site zinc. This transient coordination is thought to lower the attacking thiol pK_a and increase its nucleophilicity.

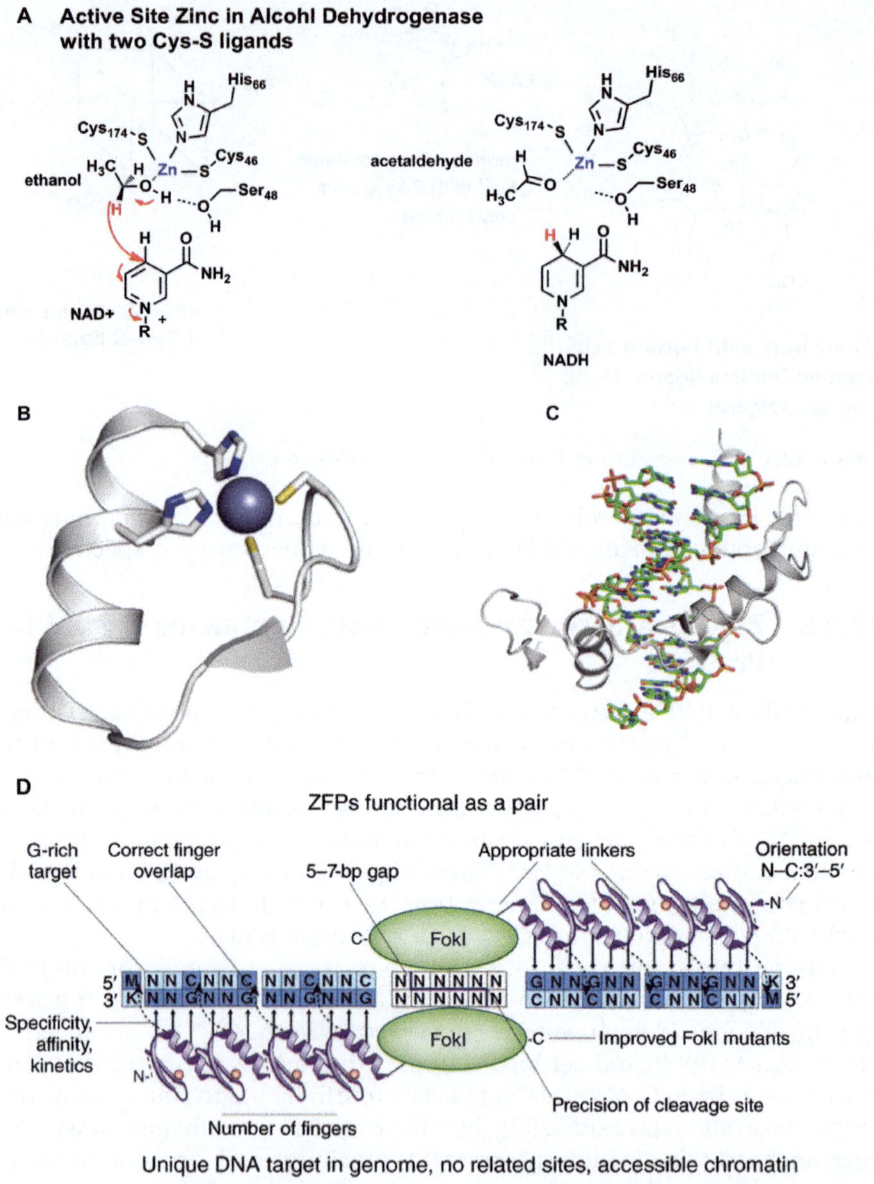

Figure 12.2 Zinc and protein thiol ligands. (A) Bis thiolate ligation to zinc ions in the active site of alcohol dehydrogenase. (B) Schematic coordination and looping in zinc finger domains (rendered from RCSB. Protein data base structure 1AAY: doi:10.2210/pdb1AAY/pdb). (C) Zinc fingers are binding elements in transcription factors that recognize DNA sequences.[5] This structure is also rendered from the same PDB1AAY file. (D) Fusion of a low specificity nuclease to zinc finger domains enables specific cuts at a desired DNA sequence.

(D) Reproduced from ref. 32 with permission from Springer Nature, Copyright 2012.

12.2.3 Zinc Finger Domains

The epitomes of Zn^{2+} as an organizing element in protein structure through coordination to cysteine side chains are zinc finger domains.[5] They are small protein domains that are loosely to fully unstructured in the absence of zinc ions. The presence of zinc recruits four protein ligands to itself in tetrahedral coordination. The canonical form of zinc fingers (Figure 12.2B) have a Cys_2 His_2 ligand set, but can go as high as three or all four cysteines as ligands.

Zinc finger protein motifs engendered significant interest when they were discovered as motifs with a short alpha helix and an antiparallel beta sheet that functioned, only in the zinc form, as DNA binding elements, often in transcription factors. The zinc fingers show up as tandem arrays to get larger coverage of specific DNA sequences. Figure 12.2C illustrates a triad of covalently connected zinc fingers with the helices engaging the major groove of a stretch of target DNA for the interaction of the transcription factor Zif268 also known as early growth factor.[6]

From a gene editing bioengineering perspective zinc fingers have been examined as a platform for DNA strand cleavage and repair. Figure 12.2D schematizes the logic where a tandem array of three zinc fingers specify a 12-nucleotide sequence in a target DNA. Fused to that triad of zinc fingers is a nonspecific nuclease, Fok1, in this sketch. The nuclease is recruited to the specific site on DNA before it makes hydrolytic cuts, in this instance, on both strands at specific base pairs.

The zinc finger architectural logic relies on the accessible thiolate anion forms of cysteine, their affinity for zinc ions, and the geometric preference of zinc for tetrahedral coordination. In turn, the thiophilic zinc ion acts as an effective organizer of unfolded regions of proteins such as transcription factors into an iterated series of loops (fingers) that can read out specific DNA base sequences to home in on particular genes. Although this is a noncatalytic role for the cysteine side chains it is a powerfully reversible route to imposing local architecture in proteins.

Zinc finger coordination requires cysteines to be in the reduced, thiol forms, not the oxidized Cys–S–S–Cys disulfide forms. In turn, this is consonant with high kevels of cellular glutathione that keeps cytoplasmic and nuclear protein thiols reduced and perhaps reflects evolution of these architectural assemblies in anaerobic organisms.

12.3 Covalent Changes in Cysteinyl Proteomes

As would be expected for the accessible reactive thiolate ion of side chains, cysteine residues participate in the widest chemical range of covalent modifications of any amino acid residues in proteomes.[3,7,8] The amine and carboxy groups are tied up in amide linkages for internal cysteine residues, leaving only the C3-thiol as an available functional group, so sulfur two-electron (nucleophilic) and one-electron (radical) chemistry dominate its proteome reactivity. Figure 12.3 depicts some of the more common reaction types of cysteinyl proteomes, although minor adducts can be consequential. Seven different two-electron manifold fates are noted.

One can divide the modifications of cysteinyl thiolate side chains into *active site chemistry* involved in enzymatic catalytic turnover machinery as distinct from modifications that alter some other property than catalysis, for example changing the intracellular address of proteins or adapting to extracellular oxidizing environments. The former we discuss in this chapter; the latter categories in the next chapter.

The categories of enzymes where active site cysteines act as nucleophiles include acyl-*S*–enzyme formation, phosphoryl-*S*–enzyme formation, conjugate addition to uracil and cytosine bases, capture of nascent thiane sulfur as persulfides and even polysulfides. The last topic has been explored in detail in Chapter 6. We will examine the other three categories here.

The one-electron reaction manifold involving cysteine sulfur radicals as key elements of the catalytic apparatuses of enzymes has been addressed in Chapter 9, *e.g.* in ribonucleotide reductase, pyruvate formate lyases and for isopenicillin *N* synthase (IPNS) and the penicillin expandase, and are not repeated here.

12.3.1 Acyl-*S*–Enzyme Intermediates: Cysteinyl Proteases

About 150 of the approximately 570 proteases in humans (collectively termed the degradome)[9] use a thiolate side chain of a cysteinyl residue as a catalytic nucleophile to attack the particular susceptible amide carbonyl of the protein or peptide substrate's bond These are collected under the terminology of "cysteinyl" proteases (Figure 12.4). All "cysteinyl" proteases engage in covalent catalysis and generate acyl-*S*–Cys–enzyme covalent intermediates in the middle of a catalytic cycle.[10–12] The tetrahedral thio-hemiaminal adduct resulting from initial thiolate attack breaks down in either the backward or forward direction. In the forward direction the amine product is released as a covalent acyl-*S*–enzyme forms between the Cys–S and the acyl portion of the protein or peptide substrate.

The second half reaction of cysteinyl proteases–peptidase–transpeptidases is hydrolysis (aminolysis in the case of transpeptidases) of the acyl-*S*–enzyme intermediate: directed attack of a water molecule in the active site on the activated thioester carbonyl. Again, a tetrahedral adduct ensues. This time, breakdown can expel the enzyme–Cys–S^- and release the acyl half of the original substrate as the carboxylate form (or a new amide product in the case of transpeptidases). The net effect is conversion of an amide bond in the protein substrate to two separate fragments, a carboxylate and an amine fragment.

This is classic *acyl thioester chemistry/chemical biology* (see Chapter 4) at work in 150 protease contexts. The alternative capture of the acyl–*S*–Cys–enzyme by an incoming amine substrate is depicted at the bottom of Figure 12.4

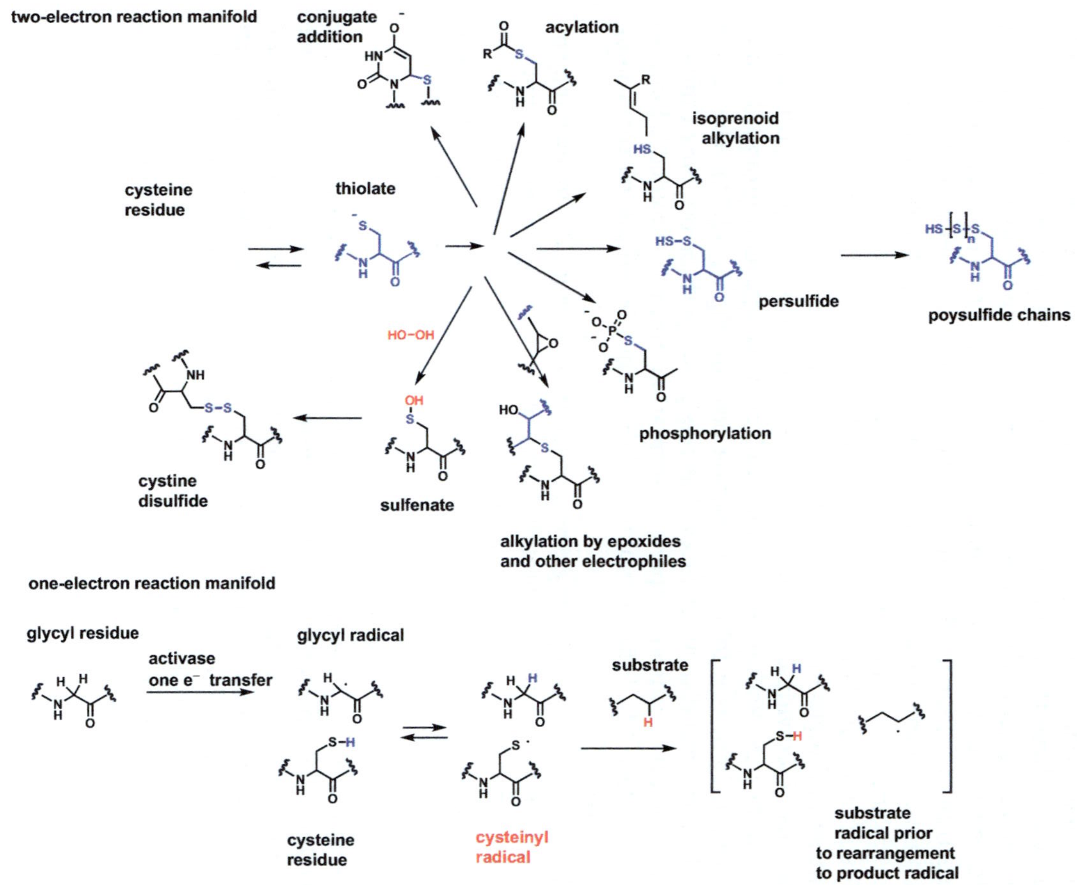

Figure 12.3 Reactions of the cysteinyl proteome: Both one-electron and two-electron reaction manifolds: some adducts are stable, others (*e.g.* some acyl-*S*–Cys–enzyme intermediates) are labile.

12.3.2 The Ubiquitinylated Proteome: Covalent Information-rich Acyl-*S*–Cys-Proteins

Another major intersection of proteomics with thiolate-enzyme catalysis is in the ubiquitinylated proteome. Two major facets underlie the logic of the ubiquitinyl acyl transfer cascade.[13,14]

The first element of logic is that the acyl-*S*–enzymes do *not* undergo hydrolysis. These thermodynamically activated acyl groups are transferred to acceptor proteins *via* iterated transfer to cysteine thiolates on partner proteins without water intervention. These are isoenergetic acyl exchanges between thiols. Then, at the late stages of the acyl transfer cascade, the acyl group is transferred to an amine, forming an isopeptide bond to a side chain lysine residue of a client protein. The conversion of acyl thioester to acyl amide is downhill thermodynamically and so biologically irreversible.

The second logical attribute underlying the ubiquitin machinery of many hundreds of individual proteins is that the acyl group undergoing transfer is *information rich*. Rather than a low molecular weight acyl group, the ubiquityl acyl chain is a folded 76-residue protein (Figure 12.5A). The folded protein has several distinct faces that recognize partner proteins differentially. The acyl transfer imparts information from the folded ubiquitin protein surface. The information is read differently by various partners along the cascades.

Mono-ubiquitylation (a single acyl chain in an isopeptide linkage) may be the signal for movement of proteins from one cell compartment to another. A single ubiquitin acyl tag is not sufficient information, or the wrong information, to be read as a degradation signal. By the time one gets to four ubiquitin chains lined up as a $[Ub]_4$ tetramer that is a signal for transport to the proteasome and degradation of the client proteins to which the tetra-ubiquitin chain is attached.

Enzyme 1 of the typical three-enzyme sequence for regiospecific ubiquitylation of client proteins makes a thioester intermediate between its active site cysteine thiolate and the C-terminal Gly_{75}-carboxylate of ubiquitin. This acyl-*S*–enzyme intermediate does *not typically undergo hydrolysis*–that would be an unwanted derailment. Enzyme 1 thus makes the 76-residue Ub–S–enzyme 1 covalent intermediate and then transfers that protein acyl group to an active site cysteine side chain in up to 40 partner proteins known as the enzyme 2 family.

The enzyme–Cys–S–Ub adduct is formed by typical ATP utilization logic. Ub–COO^- is converted to Ub–AMP as ATP is cleaved with release of PPi. The activated C-terminal Gly76 carbonyl is then captured by the active site cysteine thiolate, converting the mixed Ub–AMP anhydride into the Ub–S–enzyme 1 thioester.

Next, the information-rich 76-residue ubiquityl (acyl) group is transferred to a comparable nucleophilic thiolate side chain of a cysteine residue in any of several enzyme 2 family members, a transacylation process.[15] This is

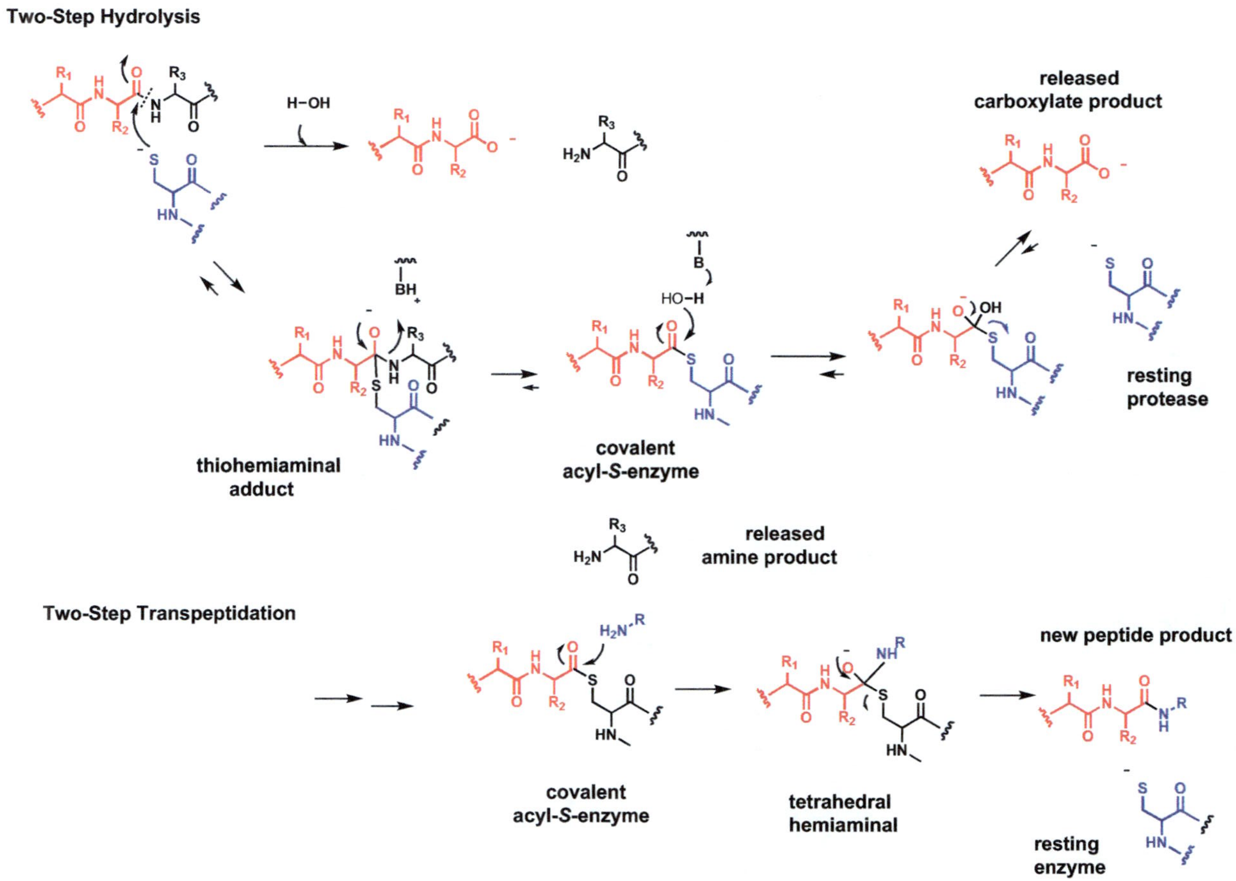

Figure 12.4 General mechanism for ''cysteinyl'' proteases: Attack of the active site cysteinyl thiolate generates a thiohemiacetal adduct in the way to a covalent acyl-*S*-enzyme intermediate. In the second half reaction the acyl-*S*-enzyme undergoes catalyzed hydrolysis. In transpeptidase catalysis an amine nucleophile from the cosubstrate outcompetes water.

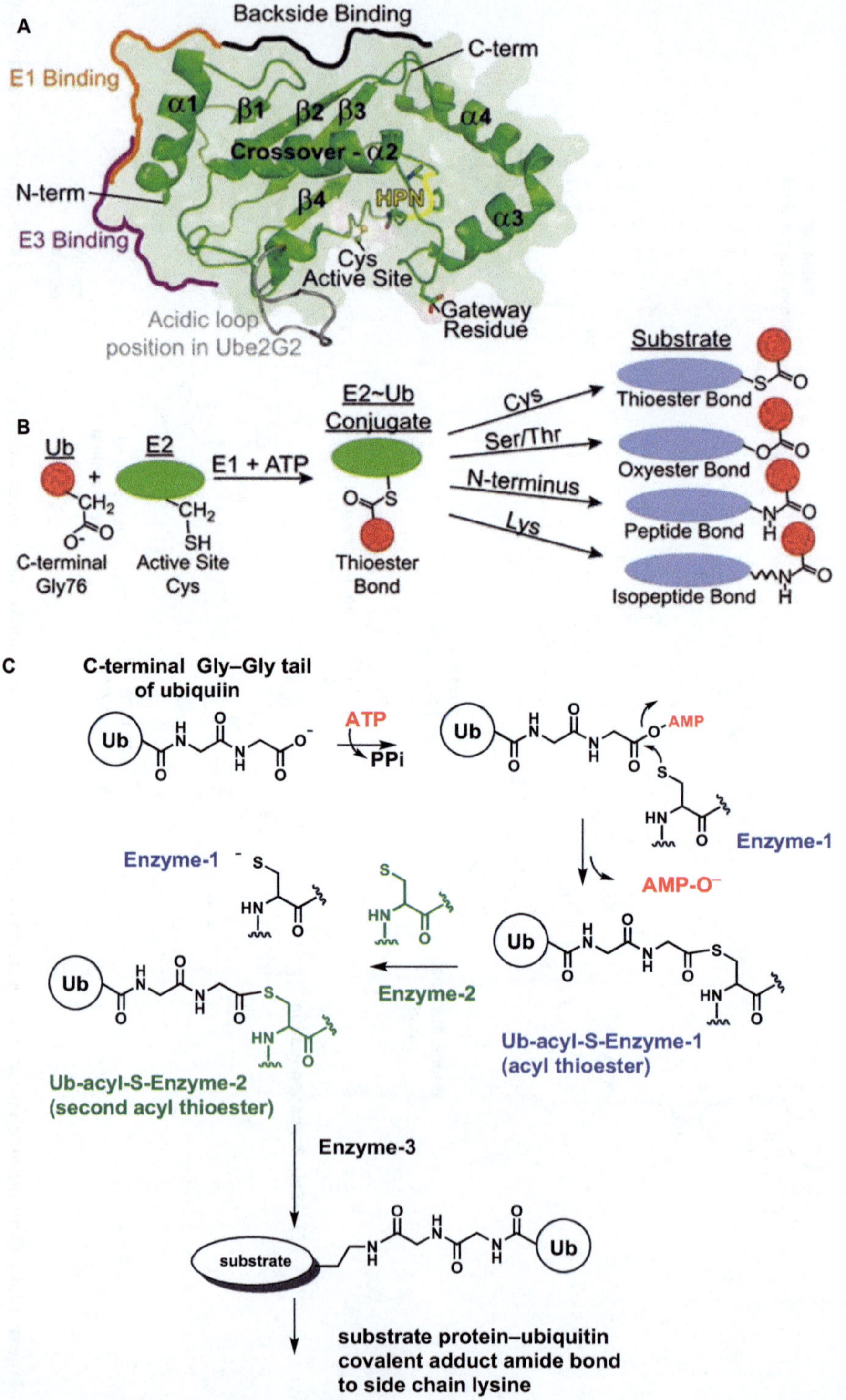

Figure 12.5 Continued on next page.

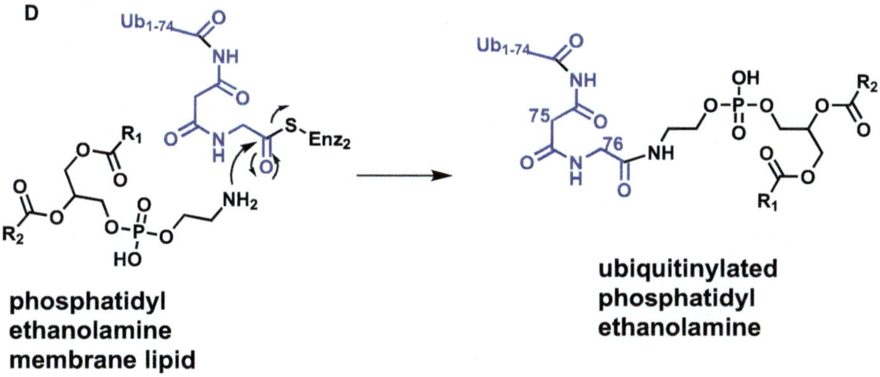

Figure 12.5 (A) The acyl group in this case is an information rich 76-residue-folded protein that conveys information to partner proteins and downstream cellular machinery. (B) Enzymes 2 transfer abilities. (C) The ubiquitylation cascade involves a series of acyl–S–enzymes on the way to an irreversible transfer of the acyl (ubiquityl) group to the lysine side chain of a client protein. (D) Transfer of a Ubiquitin moiety to the amine of a phosphatidylethanolamine lipid.
(A and B) Reproduced from ref. 15 with permission from Springer Nature, Copyright 2016.

preamble in the cascade to mono-and then polyubiquitylation of target proteins, *via* action of specifying-enzyme 3 family members.[13]

Enzyme 2 family members are often overlooked in the cascade, relegated to the role of middle men.[15] Yet, as shown in Figure 12.5B, the ubiquitinylated *S*-cysteinyl–enzyme 2 covalent adducts can deliver their information-rich 76-residue acyl group to four kinds of amino acid side chains on partner proteins. Direct transfer to a client protein, without the intervention of an E3 specifier protein, is net aminolysis: transfer to a lysine side chain to form an amide linkage as a stable isopeptide bond. There are also examples where the aminolysis occurs with transfer of the Ub-acyl group to an N-terminal amine of a client protein. A notably unusual transfer to an amine by a ubiquitinyl–S–E2 is the protein ATG3, involved in apoptotic cascades, where the cosubstrate is a phosphatidylethanolamine molecule in the maturing phagosome membrane. The product is a covalently ubiquitylated phosphatidylethanolamine (PE) headgroup[15] (Figure 12.5D), a modified lipid rather than a modified protein. The Ub–PE lipid accumulates in apoptotic vesicle membranes as a signaling agent.

The role of middle man for an E2 is net ubiquitinyl transfer to a cysteine side chain on an E3 ligase family member. This is classic thioester interchange between cysteinyl side chains. Finally, there are examples where it is the side chain hydroxyl of a serine or threonine residue in a client protein that is the kinetically competent nucleophile to undergo *O*-ubiquitylation as a (relatively) unstable oxoester is formed rather than the classical isopeptide bond to a lysine side chain (Figure 12.5C).

The faces of E2 proteins that interact with E1 donor and E3 acceptor proteins have been mapped on several E2 proteins and the binding regions are distinct but apposite (Figure 12.5A). The covalent ubiquitylated–S–E2 forms must persist in cells long enough to find their partner acceptor proteins (E3 ligases or direct client proteins) so these acyl–S–enzyme forms must be shielded from adventitious attack by water molecules. One of the roles of E3 ligases, on binding to Ubi–S–E2 acyl enzyme forms, is to induce conformational changes that increase the reactivity of the acyl–S–E2s for ubiquityl transfers onto those partner proteins.

There are hundreds of E3 protein members of the cellular ubiquitin cascades.[16,17] There are diverse in evolutionary origin and in mechanism. Some are catalytic, making yet a third ubiquityl–S–enzyme in the cascade prior to aminolytic transfer to a client protein. Others are effectively protein chaperones, bringing client proteins to the vicinity of a given ubiquityl–S–enzyme 2 for the isopeptide bond-forming aminolysis outcome. Some E3s are multicomponent with noncatalytic subunits imparting specificity, sensing, and multistate integration of cellular information to determine whether an endpoint protein is to be ubiquitylated in a given time and place within a cell.

The combinatoric possibilities are almost astronomical. They reveal logic that allows ubiquitylation of proteins for quality control and orderly monitoring of protein lifetimes as *acyl–S–enzyme cascade processes* (Figure 12.5) for controlling lifetimes of thousands of intracellular proteins.

The logical parallel to acyl–S–enzyme intermediates in the cysteinyl protease family evolution is clear. The distinctions include multiple transfers of the acyl moiety (ubiquitins) between thiol groups on enzyme 1 and the suite of enzyme 2 active site thiols and lack of hydrolytic breakdown.

Because the acyl group in ubiquitin cascades is large, folded in three dimensions and information rich, it has seven lysine side chains, any one of which can be targeted for sequential subsequent ubiquitylations. The mono- to poly-ubiquitin chains can have distinct connectivities and architectures (Figure 12.6). The linkages to their diverse client proteins are through stable amide linkages to side chain lysines. The distinct regiochemistries, for example connections to Lys48, or Lys63 or Lys11, not only give different three-dimensional architectures of the poly Ub chains, they are recognized for different functions (Figure 12.6B). A few of the possible patterns that show up on cytoplasmic domains of membrane receptors are shown in Figure 12.6C.

12.3.3 Deubiquitinases

In addition to the enzyme 2 and 3 combinations that add ubiquitin groups singly and in multiples, there are suites of isopeptidases that reverse or counterbalance the biological effects of ubiquitinylation. These are the deubiquitinases (DUBs).[18] There are almost one hundred members encoded in human genomes, reflecting the diverse tasks, the multiplicity of regiochemical linkages, and the range of Ub–client protein architectures to be recognized for initial binding prior to isopeptide bond cleavage. Because the

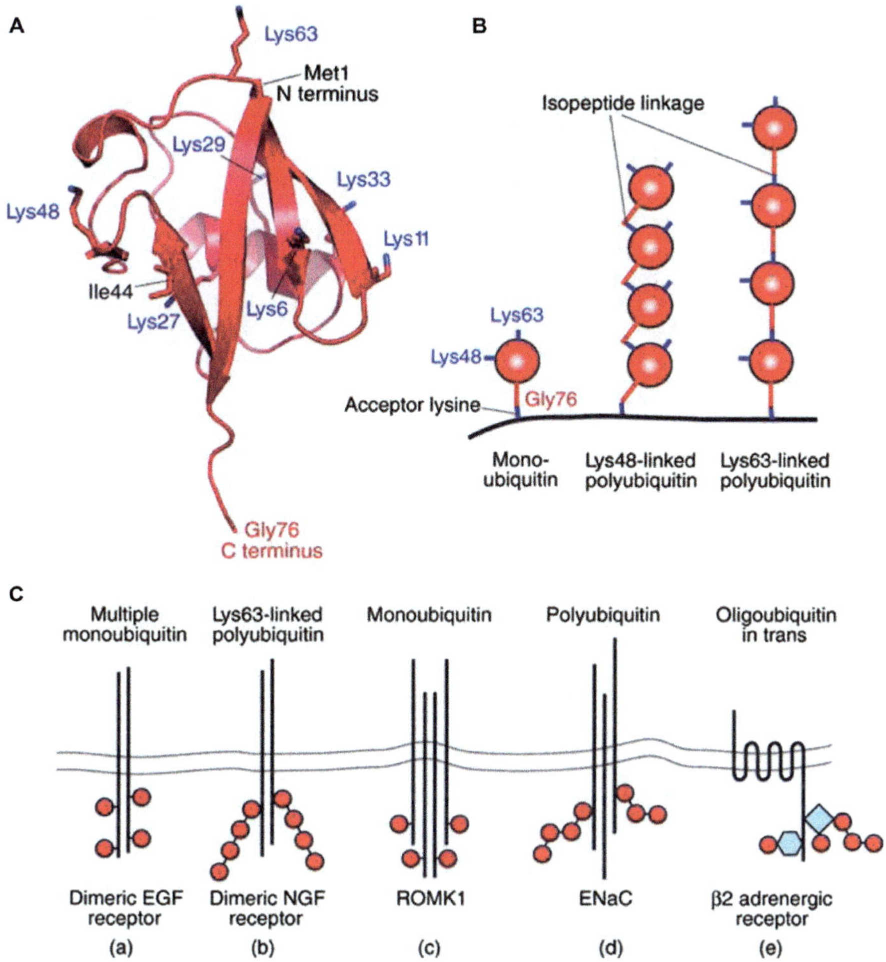

Figure 12.6 (A) Seven-lysine side chins in each ubiquitin monomer offer multiple regiochemical possibilities for specific oligomerization by E2 and/or E3 transferases. (B) Schematic of monoUb, and two polyUb chains, at K48 and K63 respectively. (C) Examples of differing degrees and arrays of Ub chain modifications on the cytoplasmic domains of five transmembrane proteins.
Reproduced from ref. 30 with permission from The Company of Biologists, Copyright 2007.

linkage of ubiquitin chains to each other and to client proteins are isopeptide bonds they are resistant to cleavage by the normal complement of host peptidases and proteases, Thus, this special class of deubiquitinylation isopeptidases are required for cleavage.

There are two mechanistic families for isopeptide bond hydrolysis (Figure 12.7). One family, minor in numbers, is comprised of metallopeptidases. The other major family is composed of active site cysteinyl peptidases, cousins

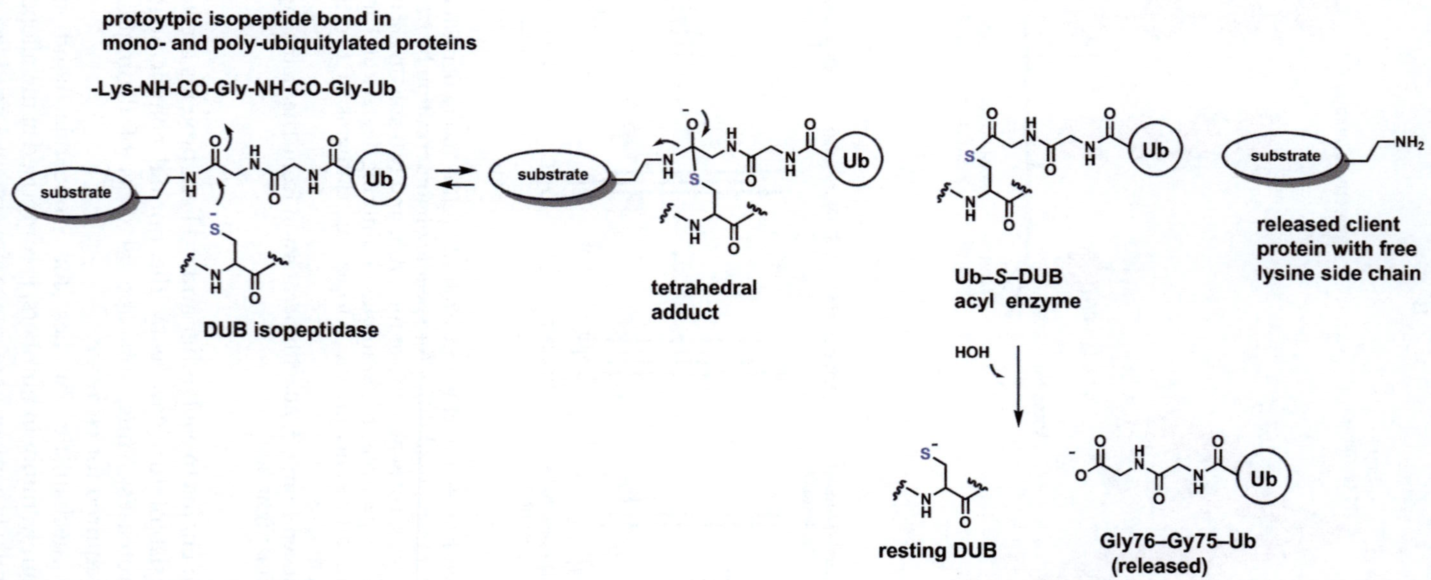

Figure 12.7 The class of deubiquitinases with active site cysteines hydrolyze the isopeptide bonds between client protein lysine side chains and the C-terminal Gly75–Gly76 of a ubiquitin protein tag. Ubiquityl-*S*-cysteinyl enzyme intermediates are involved in the disconnection mechanism.

of the cysteinyl proteases noted in Section 11.2.3. The cysteinyl isopetidases form the larger class and use acyl–*S*–Cys enzyme intermediates to transfer the acyl (ubiquityl) group to a cosubstrate water molecule.

DUBs show preferences that can be absolute for regiochemistry, for the isopeptide linkages between Ub units in polyubiquitin chains and also in the recognition of lysine side chains in client proteins, including ones that are monoubiquitylated in signaling cascades, as depicted in Figure 12.6. DUBs can also act as endo- or exo-isopeptidases in polyUb chains. The balance of ubiquitylation and deubiquitylation is of current interest not only in normal physiological homeostasis but in many pathophysiologies, from neurodegeneration to cancer.

To give calibration of the extent of the ubiquitin proteome network, there is a single E1 encoded in human genomes to use ATP to make ubiquityl-AMP and then load it as the ubiquityl–S–Cys enzyme 1 covalent intermediate. Then, there are 38 human E2 acyl transferases that serve for ubiquityl transfer by thiol interchange. In turn those E2 acyl–S–enzymes interface with some 600–1000 E3 ligases. Those can be single ligases, or catalytic subunits embedded in multiprotein complexes, or chaperones that bring client proteins to the E2 ubiquinyl transferases. Add in polyubiquitylation, with different chain lengths and regiochemistries and multi-monoubiquitylations as depicted in Figure 12.6, and almost any outcome can be engineered. Then, add the dynamics of reversal by a suite of 100 DUBs for moment to moment control of ubiquitinylation stoichiometries and subcellular locations.

The combinatoric possibilities reach astronomical dimensions and enable the information-rich UB acyl transfer cascade and its hundreds of client proteins to provide complex networks for cell decision-making. The commitment to covalent ubiquitylation and its many cellular roles probably dwarfs other facets of the cysteinyl proteome universe. This brief summary is but one of many jumping off points to the ubiquitin realm for interested readers.

12.3.4 Glutaminase Catalytic Logic

Acyl-*S*–cysteinyl thioester logic is also employed in low molecular weight metabolite interconversions. In particular glutamine, one of the 20 proteinogenic amino acids functions as a latent surrogate for nucleophilic ammonia. Ammonia is promiscuously reactive when it accrues to high levels in cells. To avoid toxicity organisms have evolved the use of glutamine as an "on demand" NH_3 precursor. Glutamine serves to liberate ammonia in the microenvirons of an active site where ammonia reacts with cosubstrates before it diffuses out to the local environment. The device for liberating ammonia from the glutamine-γ-carboxamide group is to use a cysteine side chain in the active site as a catalytic nucleophile. The central logic: hydrolysis of glutamine to glutamate proceeds *via* an acyl–*S*–Cys enzyme intermediate.

There are several glutaminase domains in metabolic enzymes that cleave the γ-carboxamide of glutamine to free ammonia and glutamate[19] (Figure 12.8). Catalysis proceeds through acyl–*S*–cysteinyl–enzyme thioester intermediates.

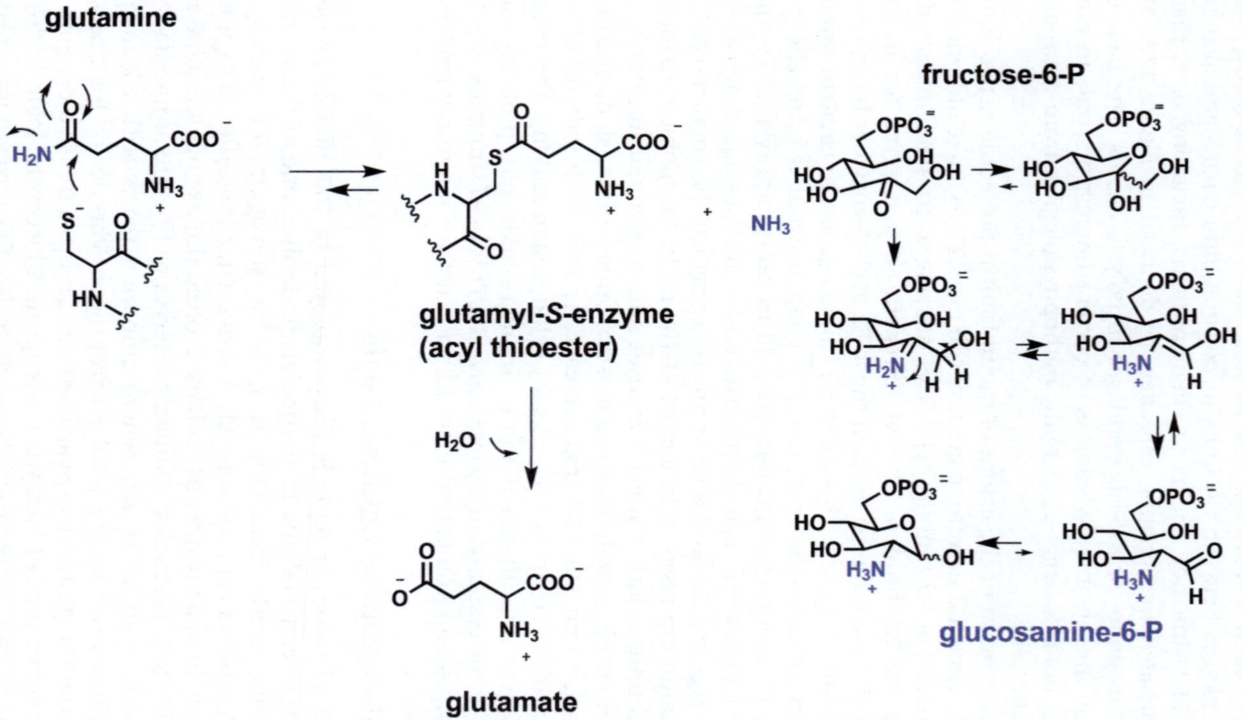

Figure 12.8 Covalent acyl–S–enzyme intermediates in glutaminase action to deliver nascent ammonia without release into general solution. Catalysis proceeds by glutamyl–*S*–cysteinyl enzyme intermediates. In the specific example the nascent ammonia is condensed with fructose-6-P followed by an aldose–ketose isomerization to glucosamine-6-P.

This strategy allows cells to use glutamine as a ready supply of ammonia, but to carry it around in an unreactive latent from in the amino acid glutamine. These catalysts can then serve up ammonia to the active site of some amination enzyme that uses the nascent ammonia before it can accumulate to toxic levels. Figure 12.8 shows enzymatic conversion of fructose-6-phosphate to glucosamine-6-phosphate, the first building block for protein *N*-glycosylations, *via* a glutamine-dependent thioester hydrolytic route for *in situ* release of nascent ammonia.

Acyl-*S*-cysteinyl enzyme catalytic machinery is widespread in prokaryotic and eukaryotic metabolic logic. Control of the lifetime and fate of the thermodynamically activated acyl–*S*–enzymes generates a broad spectrum of function. from cysteinyl proteases, to transpeptidases, to the ubiquitin cascade.

We turn next to one more example of acyl–*S*–enzyme intermediates not for degradation, but for biosynthesis. The thiol nucleophile is not the cysteine thiolate side chain itself but the terminal thiolate of covalently tethered phosphopantetheine.

12.3.5 Fatty Acid, Polyketides and Non-ribosomal Peptide Assembly Lines

In Chapter 5 we explored the activating effects of the sulfur atom in acyl thioesters on both the C1 (electrophilic) and the C2 (nucleophilic) positions of the acyl moiety of acyl thioesters. This was largely in the context of acyl-*S*–CoAs as soluble metabolites. We also noted in Figure 5.5A that fatty acid synthases contained 80–100-residue carrier proteins, either as independently folding domains or as free-standing subunits. A specific serine-side chain in each of these acyl carrier proteins can attack coenzyme A at the electrophilic phosphorus atom of the 4′-phosphopantetheine moiety to convert apo forms of carrier proteins into the phosphodiester-tethered panthetheinyl arms.[20] These are posttranslational modifications that install the phosphopantetheinyl moiety as a covalent prosthetic group and most importantly provide its terminal cysteamine thiolate as a catalytic nucleophile in the fatty acid synthase machinery.

The covalently attached phosphopantetheinylated acyl carrier protein domains/subunits act as a cysteine thiolate surrogates. In fatty acid chain elongations, acetyl CoA carboxylase generates the three-carbon malonyl CoA. A loading transacylase performs thiol exchange to give malonyl-*S*–ACP, akin to the enzyme 1–enzyme 2 acyl exchange noted above at the start of ubiquitin cascades (Figure 12.9). The orthogonal activation at C1 and C2 is revealed on subsequent decarboxylation of the malonyl-*S*–ACP to the transient C2-acetyl carbanion, needed to engage in C–C bond formation. The resulting 3-ketobutyryl acyl undergoes three enzymatic steps that convert the 3-ketone to the 3-CH_2 group: the saturated butyryl chain is generated before the next chain elongation cycle.

Thus, each cycle of C2 unit elongation and then four-electron reduction of the 3-ketogroup uses the C1 electrophilicity and the C2 nucleophilicity of the

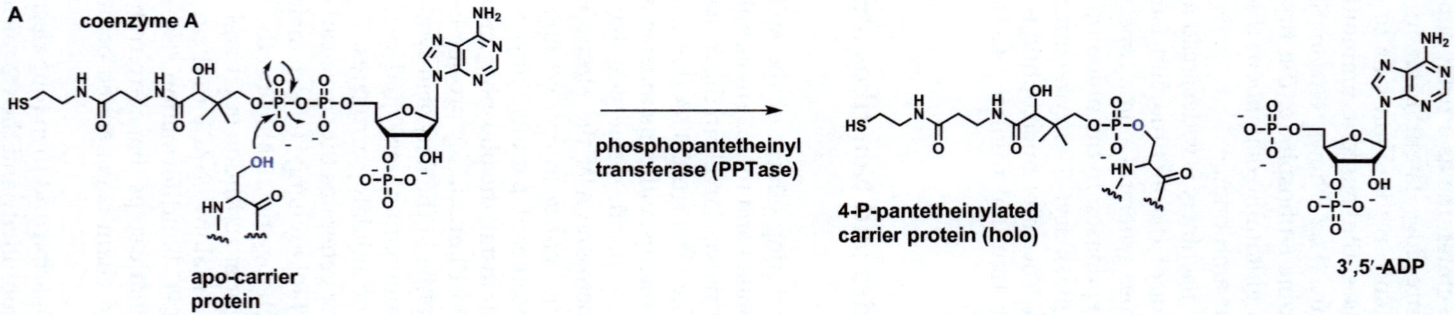

Figure 12.9 (A) A specific serine-OH side chain in acyl carrier proteins acts as nucleophile towards CoASH to transfer the phosphopantetheinyl moiety to form the holo form of the acyl carrier protein. The terminal thiol of the pantetheine prosthetic group is one of the two thiol sites used for acyl chain growth in fatty acid chain elongations. (B) Acyl exchange between thiols moves the malonyl group from malonyl CoA to malonyl–*S*–ACP. (C) C–C bond formation in chain elongation occurs *via* decarboxylative thioclaisen condensation. The resultant 3-ketoacyl–*S*–ACP is subjected to three consecutive enzyme reactions to convert the 3-keto to the 3-methylene in the butyryl–*S*–ACP example.
(A) Reproduced from ref. 31 with permission from Elsevier, Copyright 2017.

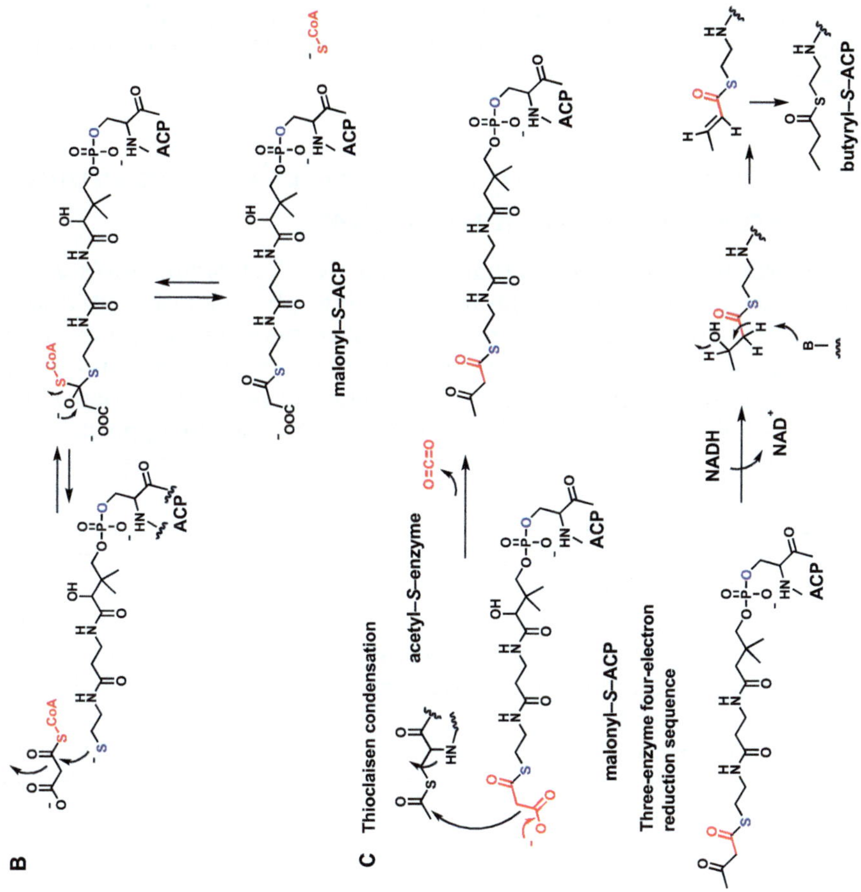

Figure 12.9 Continued.

acetyl group in the C–C bond-forming steps. In the subsequent three-enzyme four-electron reduction process, the dehydration step relies on the accessibility of the C2 carbanionic transition state, enabled in the thioester. Given the crucial role of fatty acids as energy sources in cell metabolism, this acyl thioester logic for C–C bond formation shows an additional facet of the special role of the sulfur atom in thioester chemical biology.

The same logic of using phosphopantetheinylated forms of carrier proteins is carried over to both microbial polyketide and non-ribosomal peptide assembly lines. Acyl–*S*-pantetheine and aminoacyl–*S*-pantetheinyl thioesters constitute the building block units for both polyketide C–C bond forming steps and also N–C amide bond-forming steps in non-ribosomal peptides.[21,22]

12.4 *S*-Phosphoryl–Enzyme Intermediates in Protein Tyrosine Phosphatase Action

A second large family of enzymes that use an active site cysteine thiolate as a catalytic nucleophile are the protein tyrosine phosphatases (PTPases) (Figure 12.10). The PTPases dynamically remove phosphoryl groups from tyrosine residues in the phosphotyrosyl proteome that have been previously added by protein tyrosine kinases (PTKs). There are some 90 human PTKs identified by bioinformatics, and 107 PTPases that act in opposition to control the fraction of tyrosine phenol oxygen atoms that are phosphorylated at any time and place in mammalian cells.[23]

PTPases are just one branch of enzymes that work on protein substrates that have been post-translationally modified by the action of protein kinases.[24] Analogously, the family of protein serine phosphatases balance the serine and threonine residue phosphorylations carried out by the hundreds of protein serine–threonine kinases. But those PSPases do *not* utilize covalent catalysis and do not involve cysteinyl residue side chains as nucleophiles.

Although the PTPases map into different subfamilies, the specific enzymes that have been examined to date do proceed *via* *S*-phosphoryl–Cys-covalent enzyme intermediates (Figure 12.10).[25,26] The intermediacy of these covalent S–PO_3-enzyme intermediates in hydrolase catalytic cycles is in contrast to the other large class of protein phosphatases – the serine-threonine phosphatases. In those enzyme families, phosphatase reactions appear to involve *direct attack of water* on the Ser– or Thr–OPO_3 group on target proteins. The mechanistic difference might reflect independent evolution of PTPases *vs.* PSPases and the opportunity to intervene in PTPases catalysis selectively by targeting the reactivity of the active site cysteine.

There are analogies to the "cysteinyl" proteomes in the "serine" enzyme proteome. There are acyl-*O*-seryl enzyme and phosphoryl–*O*-serine covalent protein adducts, indicating parallel development of Ser and Cys side chains as catalytic nucleophiles in the evolution different families of proteomes.

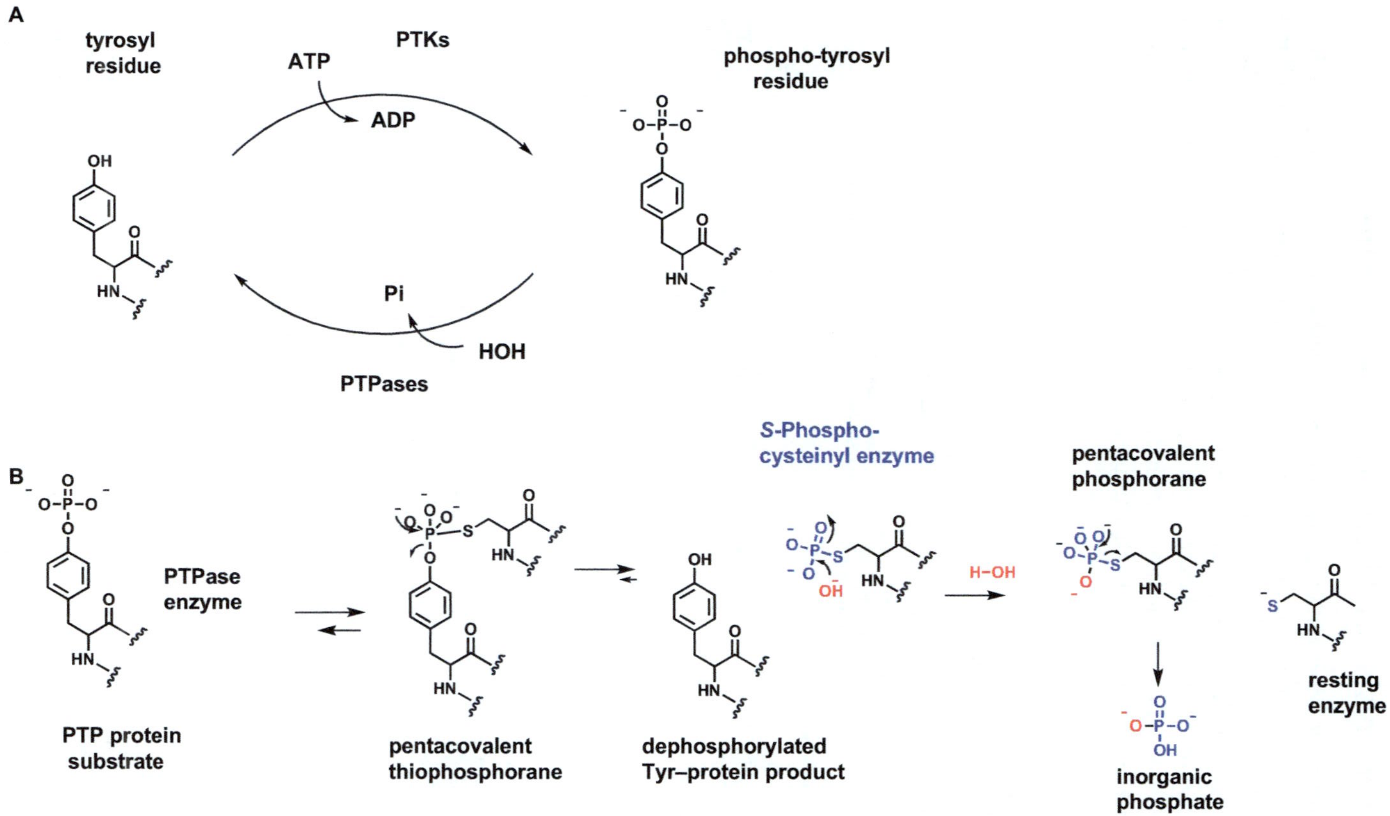

Figure 12.10 (A) Protein phosphotyrosine phosphatases balance the action of protein tyrosine kinases. (B) The catalytic mechanism of PTPases involves covalent *S*-phosphoryl enzyme intermediates, arising from nucleophilic attack of the enzyme's active site Cys–thiol on the phosphotyrosyl residues of client proteins.

12.5 Covalent Addition of Enzyme Thiolate Initiates Catalysis on Thymidylate and Cytosine

A small number of enzymes carry out formal substitutions of a C5–H by a C5–CH_3 group on the pyrimidines uracil and cytosine. The 5-methyluracil formation occurs on the free 2′-deoxymononucleotide dUMP.[27] On the other hand, the 5-methylcytosine product is formed in DNA double helices and is a key covalent modification for whether DNA is actively transcribed[28] (Figure 12.11A). Thymidylate synthase is the key enzyme that converts the uracil base found in RNA to the 5-methyluridine (=thymidine) found in DNA.

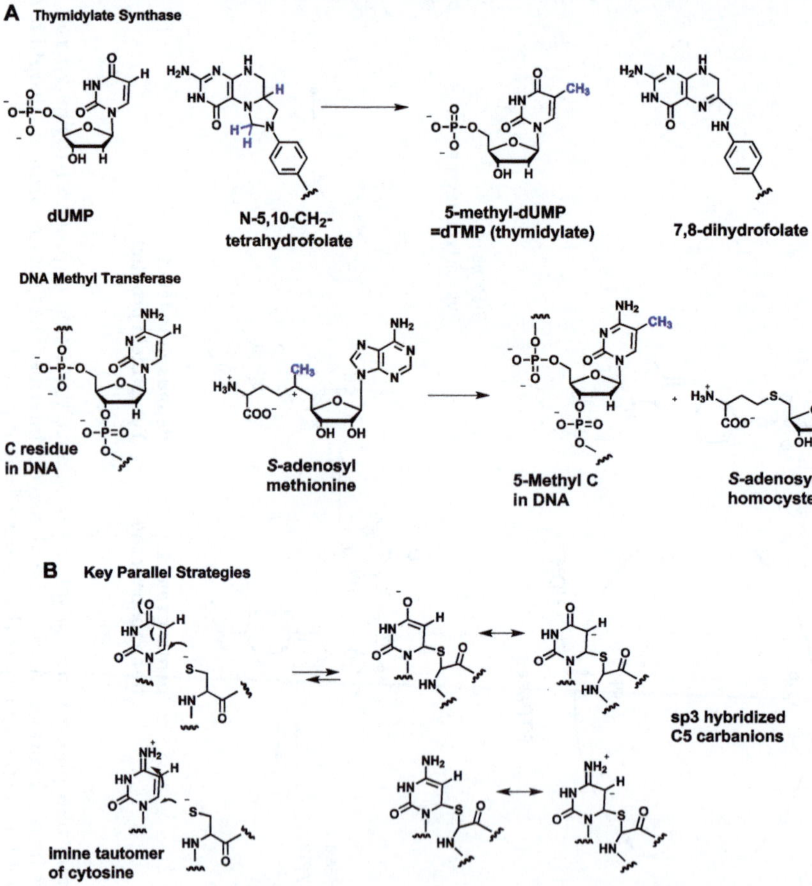

Figure 12.11 (A) Thymidylate synthase and the cytosine-selective DNA methyltransferases use parallel logic in adding an active site cysteine thiolate into the 5,6-double bond of pyrimidines in conjugate addition reactions to initiate catalysis. This action converts unreactive sp^2 hybridized C5 atoms to carbanionic sp^3 hybridized carbon nucleophiles. (B) One-carbon units are transferred from methylene-tetrahydrofolate for thymidylate and (C) from SAM for 5-methylcytosine. (D). The one-carbon units transferred during thymidylate synthase undergo two-electron reduction.

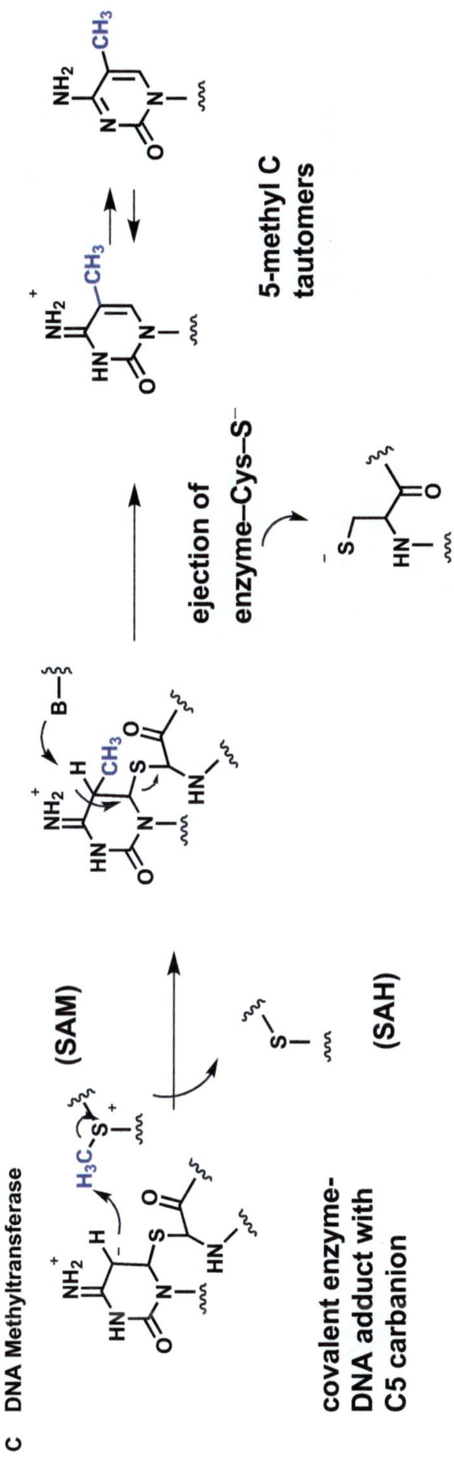

Figure 12.11 Continued on next page.

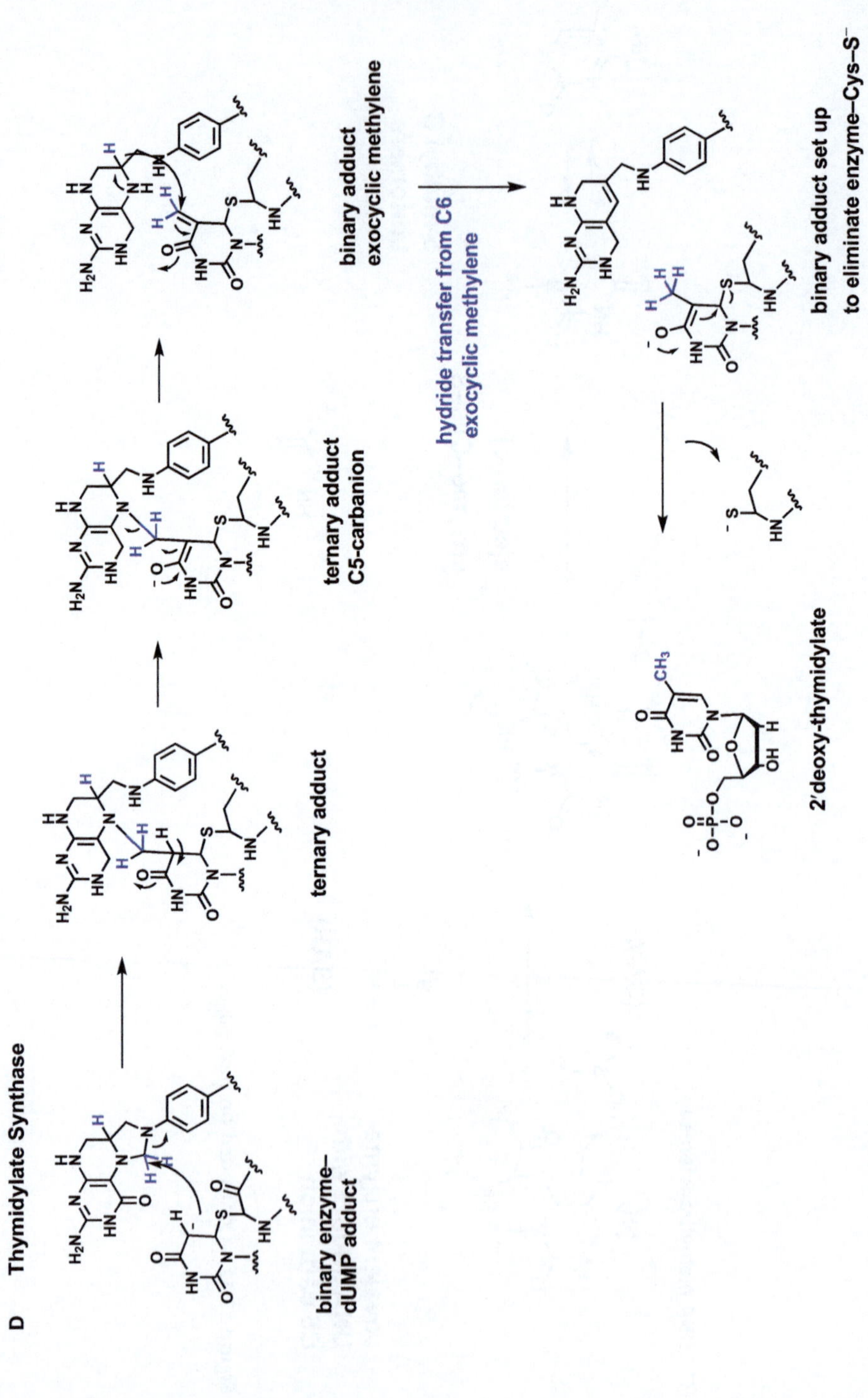

Figure 12.11 Continued.

Although each enzyme adds a one-carbon unit to C5 of their pyrimidine substrates, the C1 donors are at different oxidation states and carried by different chemical chaperones. The one-carbon donor in thymidylate synthase is N-5,10-methylene tetrahydrofolate (THF). In the 5-methyl-C-forming enzyme acting on DNA, a DNA methyltransferase, the methyl donor is *S*-adenosylmethionine.

The distinct oxidation states of the C1 fragments mandate distinct late-stage mechanisms. The methylene–THF cosubstrate is a carrier of latent formaldehyde. As it unravels from the cyclic N-5,10-CH_2– adduct to release nascent formaldehyde, an sp^2-hybridized one-carbon electrophile.

The SAM in the DNA methyltransferase catalytic cycle transfers a $[CH_3^+]$ equivalent. Both transfers require a C5-pyrimidine carbanion equivalent. Because C5 of uracil and cytosine are trigonal, sp^2-hybridized carbons there is no easy route to a low-energy carbanion in either starting case.

That is the role of the active site cysteine thiolate anion: to create a sp3 carbanionic species at C5 of the uracil ring in dUMP or at C5 of the cytosine residue in the DNA to be methylated. Both catalytic cycles begin with catalyzed addition of the active site thiolate across the 5,6-double bond, with regiospecific addition at carbon 6. This chemistry is akin to addition of glutathione thiolate to conjugated carbon electrophiles such as acrolein or the iminoquinone metabolites from acetaminophen, examined in Chapter 5.

The resultant adduct has covalently connected the enzyme to the dUMP framework, or to the DNA molecule *via* an adduct to the specific cytosine base by a thioether bond. Note that the electrons in the 5,6 double bond can be stored in the C4 ketone (uracil) or C4 imine (cytosine), as well as at C5. Delocalization means lowered energy and more stability for that C5 carbanion.

Completion of the catalytic cycle for the DNA methyltransferase is straightforward. The cytosine enzyme adduct's C5 carbanion acts as carbon nucleophile (Figure 12.11C) towards the methyl group of SAM. An S_N2 type transfer of $[CH_3^+]$ forms the C–C bond of 5-methylcytosine. It now remains to eject the initially added cysteine thiolate by reversal of the first step. This regenerates the resting enzyme form and releases the 5-methyl C DNA. Repetition of this catalytic cycle can occur millions of times per genome per cell cycle.

Completion of the thymidylate synthase catalytic cycle is more complex. The corresponding enzyme–dUMP covalent adduct with carbanionic character at C5 can attack the N-5,10–CH_2 bridge to open that ring, with directed cleavage of the C–N10 bond. This is now a ternary covalent adduct between the enzyme–dUMP framework and the folate coenzyme bridged by the CH_2 group. One mechanism for resolution of this ternary adduct is shown in Figure 12.11B. It relies on the C5–H in the pyrimidine scaffold of this ternary adduct being sufficiently acidic to be removed as a proton. This is the second consecutive carbanion to form at that carbon. Note that it is delocalized into the adjacent C4-ketone as the enolate. Formation of the carbanion leads to its use to form a transient exomethylene (=CH2) group at C5, releasing the tetrahydrofolate as that CH_2–N5 bond breaks.

The observed folate product of the thymidylate synthase reaction is 7,8-dihydrofolate. The 5,6 position is a C=N double bond. That product can be explained by transfer of a hydride from C6 of nascent THF to the exomethylene group as depicted in Figure 12.11C.

As that mechanistic dust settles, the folate is in the 7,8-dihydrofolate oxidation state, the one carbon attached to C5 of the uracil ring is now a CH_3 group. The ternary enzyme–pyrimidine–folate adduct has been cleaved to the enzyme–5-methyl-uracil adduct. The reaction completes the way the DNA methyltransferase also works: ejection of the initially added enzyme cysteinyl thiolate anion.

The thymidylate synthase second half reaction is more complex than that of the DNA methyltransferase because of the oxidation state of the itinerant one-carbon unit that becomes the C5 methyl group on the pyrimidine scaffold. In DNA methyltransferase action the donor SAM has the methyl group at the same oxidation state as in the 5-methyl C product. The actual $[CH_3^+]$ transfer is a simple S_N2 type displacement by the pyrimidine–enzyme adduct C5 carbanion.

For the thymidylate synthase reaction, the one-carbon donor, N-5,10-methylene–THF, carries around the C1 unit at the oxidation state of formaldehyde. Yet, it ends up at the formal oxidation state of methanol (CH_3 *vs.* starting CH_2) in the 5-methyluracil product. Thus, as the C1 unit undergoes a net two-electron reduction during a catalytic cycle, the folate ring system undergoes the compensatory two-electron oxidation from tetrahydro to dihydro oxidation state. (Cells subsequently spend an NADPH to reduce the dihydrofolate back to tetrahydrofolate, billions of times per cell cycle as the DNA genome is replicated.) The actual redox step is the postulated hydride transfer from C6 of the tetrahydrofolate to the exomethylene group at C5 of the pyrimidine ring.

Moving up one level in abstraction, the underlying logic in these two key DNA metabolism manipulations at C5 of pyrimidine rings is to create low-energy C5-carbanion equivalents to attack the incoming, electrophilic C1 fragments. The two enzymes use a conjugate addition strategy with a nucleophilic active site cysteine thiolate to add across the 5,6-double bond of the two pyrimidines. This action converts inert sp^2 hybridized C5 atoms to sp^3 hybridized enolate or carbanion equivalents.

12.6 Use of an Enzyme Cysteine Active Site Nucleophile to Create a Persulfide S–S Bond

A fourth set of enzymes using nucleophilic catalysis *via* the agency of an active site cysteinyl thiolate anion are the cysteine lyases, which set in motion the provision of inorganic sulfur for Fe–S cluster formation. These were discussed in detail in Chapter 6 from the perspective of the cysteine substrate undergoing conversion to alanine and the electrophilic form of inorganic sulfur. Here, we turn the perspective on its head and look at the catalytic cycle from the perspective of the enzyme cysteinyl nucleophilic thiolate anion.

The reaction manifold is unusual in several respects. Both free cysteine as substrate and enzyme active site cysteine residue are involved directly. One of the products is an S–S bond, a persulfide linkage between the active site Cys nucleophile and the sulfur atom extracted from the cysteine substrate. We noted in Chapter 6 that the formation of product alanine indicates a β-replacement pyridoxal phosphate (PLP) catalytic logic. The actual S–S bond forming step (Figure 12.12) occurs as the enzyme Cys–S^- attacks the substrate–PLP stabilized carbanion adduct. The itinerant sulfur moves without the electron pair that was in the C3–S bond. Those electrons can be temporarily stored in the aminoacrylyl–PLP framework by using the PLP scaffold transiently as a four-electron sink.

The S–S bond formation yields a persulfido enzyme adduct. It shows the versatility of one thiolate (the enzyme cysteine) as a nucleophile while the other thiol behaves formally as electrophile (substrate–PLP adduct), driven that way by hanging the PLP electron sink onto the substrate amino acid framework. The nascent enzyme–Cys–S–SH product cannot turn over again until the terminal SH of the persulfide link has been passed out to a partner protein, *e.g.* the IscU partner protein (Chapter 6). These sulfur passages to chaperones also convey the sulfur as the six-electron sulfane. When that inorganic sulfur becomes deposited on protein platforms that can build 2Fe–2S clusters and reductively dimerize them to 4Fe–4S clusters, they become instead S^{2-} inorganic sulfide ions.

12.7 Thiohemiacetal Adducts on the Way to Acyl Thioesters

Enzymatic aldehyde oxidation can also offer insights into how organisms can capture energy from the process in the guise of acylthioesters. While most aldehyde dehydrogenases use the free aldehyde as substrate, typically transferring the C1-hydrogen as a hydride equivalent to NAD^+, and so producing the resonance-stabilized free carboxylate anion, a few enzymes take a different course.

Glyceraldehyde-3-P dehydrogenase is a prime example of the strategy of generating a thiohemiacetal and then oxidizing it directly to an acyl thioester (Figure 12.13). The strategy is to add an active site cysteine thiolate into the aldehyde carbonyl of bound substrate, creating a covalent enzyme–substrate tetrahedral hemiacetal adduct (in analogy to the first step in glyoxalase 1 catalysis described in Chapter 6 for glutathione thiolate as a nucleophile). Ejection of a hydride ion to cosubstrate NAD^+ creates two thermodynamically activated species. NADH is the cellular activated electron currency, thermodynamically poised to deliver a hydride ion back to a bevy of different carbonyl groups as the dihydropyridine ring aromatizes.

The other activated species is the 3-phosphoglyceryl-*S*-Cys–enzyme adduct. That thermodynamic acyl activation is readily harvested by transfer of the acyl group to inorganic phosphate. The released product is then the

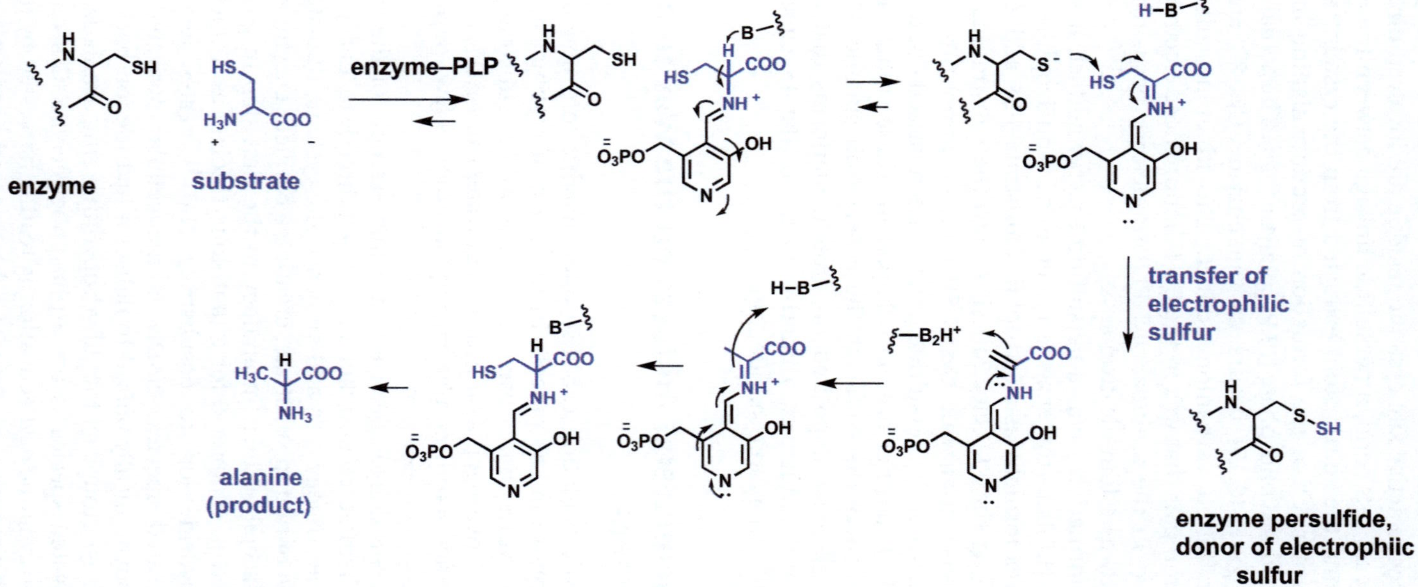

Figure 12.12 Cysteine lyases providing sulfur atoms for Fe–S cluster assembly are β-replacement pyridoxal-P-containing enzymes. They replace the departing –SH with an incoming proton to turn cysteine into alanine. The itinerant sulfur moves as an electrophilic SH equivalent, not a nucleophile.

mixed acyl phosphoric anhydride, 1,3-diphosphoglycerate. The underlying logic: convert the aldehyde substrate to the sp^3 tetrahedral thiohemiacetal substrate–enzyme adduct. Oxidize it by hydride transfer to give the acyl thioester. Release the activated acyl group as the acyl-phosphate.

A second example of this aldehyde to acyl thioester catalytic machinery occurs in the oxidation of malonyl semialdehyde by a bacillus bacterial enzyme. Cosubstrates are $NADP^+$ and CoASH. In this case the malonyl thiohemiacetal enzyme adduct gives the malonyl–*S*–enzyme.[29] That undergoes facile decarboxylation to the thioester enolate. Ketonization to acetyl–*S*–cysteinyl enzyme precedes product release as the diffusible cellular acylation currency, acetyl–*S*–CoA (Figure 12.13).

12.8 Summary of the Types of Substrate Electrophiles that Form Covalent Enzyme–Cysteinyl Adducts

In the next chapter we continue examination of cysteine proteome reactivity modes but switch focus from *catalytically central enzyme cysteine residues* to cysteines that are reactive in various proteins in both enzymatic and nonenzymatic modes. Those transformations do *not* typically involve covalent adducts between all or part of the substrate and a catalytically active cysteine nucleophile in the enzyme active site.

Before leaving the topic of covalent cysteinyl enzyme adducts that are the central topic of this chapter, we summarize in Figure 12.14 six types of substrates that undergo such covalent adduct formations in mid catalytic cycles.

Many amide-containing substrates, both high molecular weight proteins and low molecular weight amides, such as glutamine, undergo acyl transfers to form covalent acyl thioester intermediates to active site cysteines. In the second half reactions, breakdown of the acyl–*S*–enzyme species, the activated acyl groups, derived from substrates, undergo thermodynamically favored acylation of cosubstrate nucleophiles. Hydrolysis is one main outcome but transacylations are also known. These enzymes utilize the same balance of kinetic stability and thermodynamic activation in acyl thioesters that make coenzyme A such a mainstay of diffusible metabolite acyl transfers.

The most straightforward reaction logic is examined in the next chapter, in the protein *S*-palmitoylations that happen in hundreds of proteins post-translationally. The chemistry is straightforward use of an acylthioester substrate to generate an acylthioester covalent enzyme adduct to generate an acylthioester in the targeted protein cosubstrate.

The second type of substrate electrophilic carbon captured in Figure 12.14 are the conjugated enone systems in the uracil and cytosine pyrimidine bases. The relevant enzymes catalyze 1,4 additions of an active site thiolate to create covalent thioether adducts. This mechanism turns unreactive sp^2 carbon centers (C5 of the pyrimidines) into carbanionic sp^3-hybridized centers to engage in attack on one-carbon electrophiles on the way to

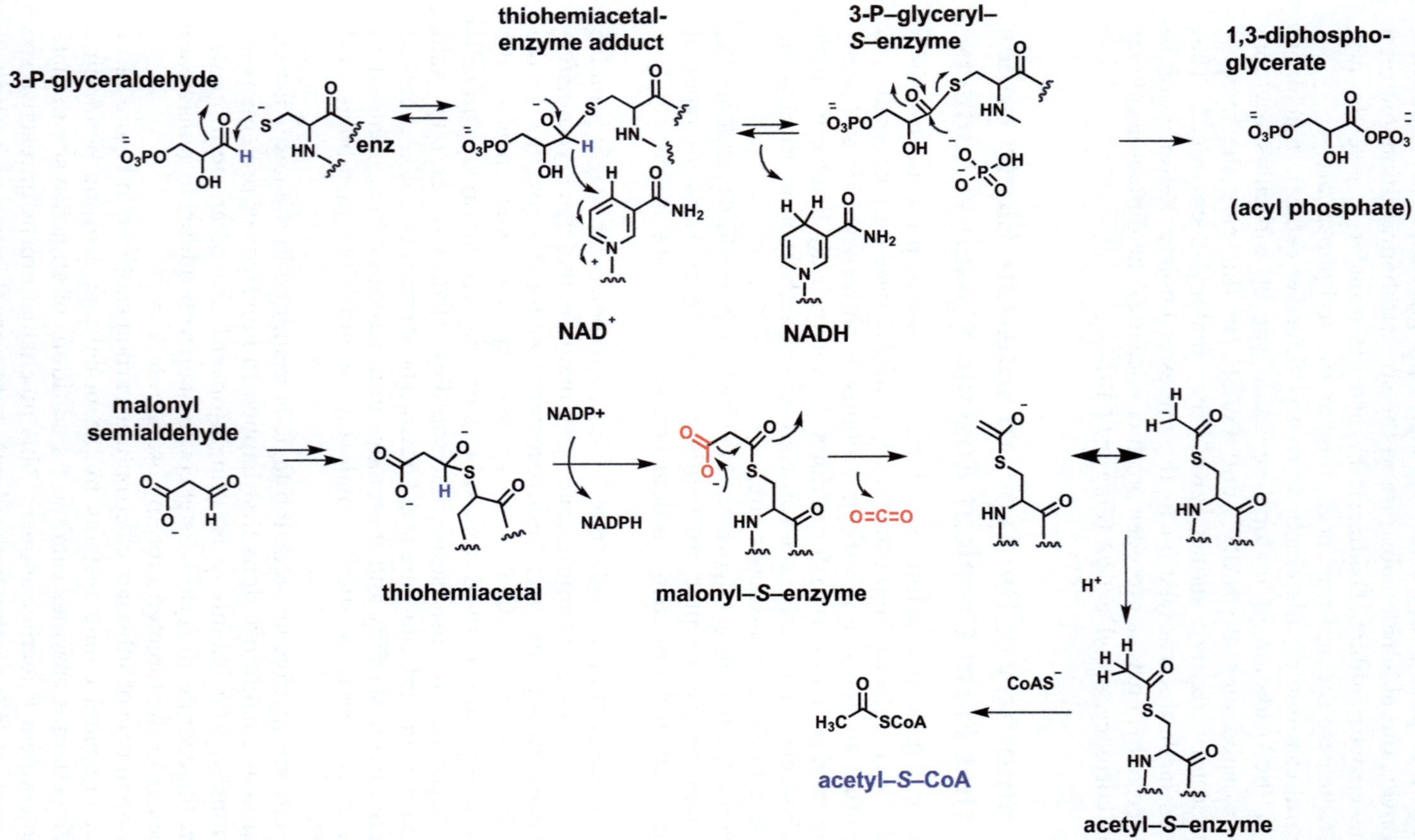

Figure 12.13 Addition of an enzyme cysteinyl thiolate to an aldehyde substrate creating the thiohemiacetal adduct is the prelude to oxidation to the thermodynamically activated acyl–*S*–Cys–enzyme intermediate. That can be released by phosphorolysis or thiolysis to the acyl phosphate or acyl CoA released products.

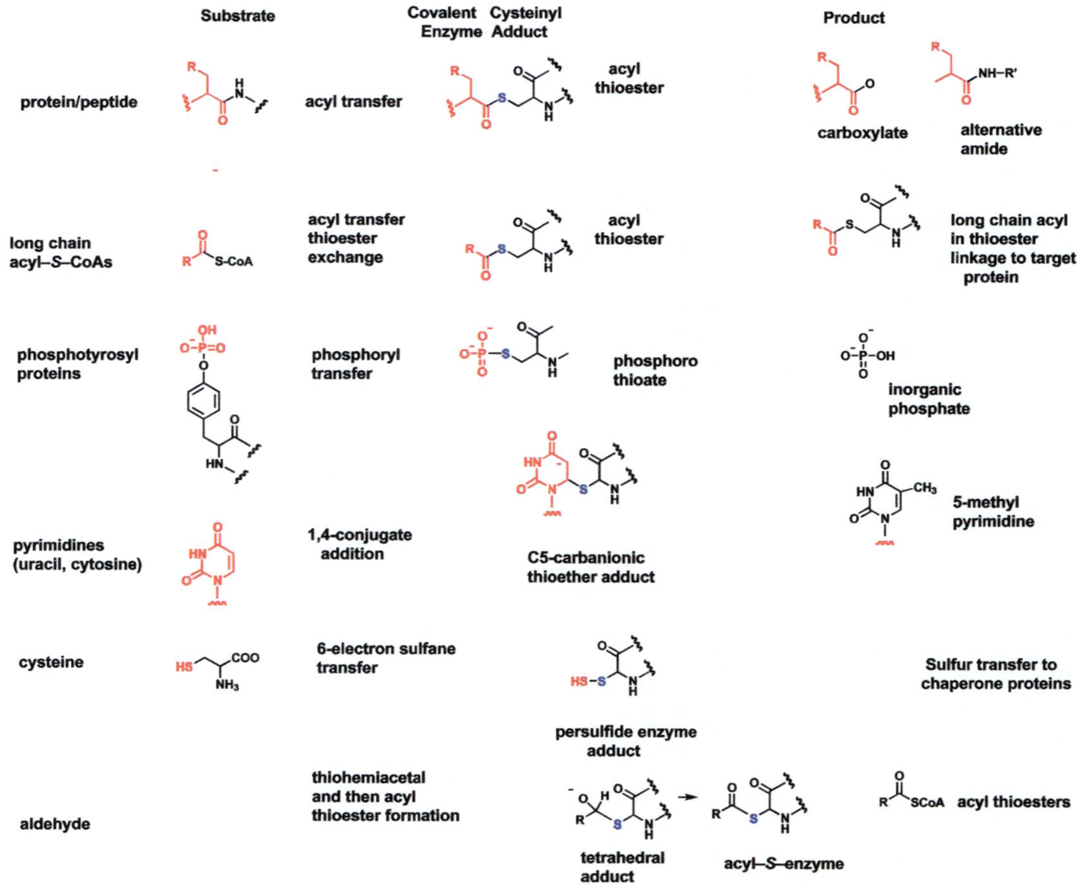

Figure 12.14 Covalent enzyme intermediates from enzyme cysteinyl thiolate nucleophilic catalysis.

5-methyl pyrimidine frameworks. This logic is key to two quite different kinds of DNA metabolic transformations. The last step for each enzyme is ejection of the initially added enzyme thiolate anion, proving the value of reversibility of conjugate addition mechanisms, and resetting the stable sp^2 hybridization at C5 and C6 of the methyl pyrimidine products.

The third type of substrate electrophile captured is the aldehyde carbonyl group, leading to an initial thiohemiacetal substrate–enzyme adduct. The choice to oxidize the thiohemiacetal rather than the aldehyde allows capture of some of the oxidation energy as the acyl-*S*-enzyme rather than the resonance-stabilized free carboxylate product.

The remaining two entries in figure 12.14 represent attacks of enzyme cysteine thiolates on inorganic cosubstrate moieties. The hundred or so protein tyrosine phosphatases that act on many signaling pathway, use the active site thiolate to attack the electrophilic phosphorus atom on the post-translationally modified protein substrates. The covalent enzyme adduct has an unusual S–P bond, in a phosphothionate linkage.

These PTPases catalyze the second half reaction by activating a water molecule to attack the still electrophilic phosphorus atom and generate released inorganic phosphate. It is not obvious why this class of protein phosphatases have evolved to utilize the intrinsic nucleophilicity of the active site thiolate as a reactant, while the equally broadly distributed protein serine phosphatases use water as a direct cosubstrate.

The sixth category of substrates attacked by enzyme cysteine nucleophiles is the small but important class of PLP-dependent cysteine lyases that are the gateways for converting the thiol of free cysteine molecules to persulfide adducts. The inorganic sulfur extracted is the ultimate source of the inorganic sulfide dianions found in hundreds of Fe–S clusters in proteins. This class is intriguing for making an enzyme S–S bond that is not a disulfide and for the ability to transfer that terminal persulfide –SH to cysteine side chains on partner proteins.

References

1. R. C. Fahey, J. Hunt and G. C. Windham, On the cysteine and cystine content of proteins. Differences between intracellular and extracellular proteins, *J. Mol. Evol.*, 1977, **10**, 155–160.
2. A. Miseta and P. Csutora, Relationship between the occurrence of cysteine in proteins and complexity of organisms, *Mol. Biol. Evol.*, 2000, **17**, 1232–1239.
3. Y.-M. Go, J. Chandler and D. P. Jones, The Cysteine Proteome, *Free Radical Biol. Med.*, 2015, **84**, 227–245.
4. C. Andreini, L. Banci, I. Bertini and A. Rosato, Counting the zinc-proteins encoded in the human genome, *J. Proteome Res.*, 2006, **5**, 196–201.
5. N. J. Pace and E. Weeringa, Zinc-binding cysteines: diverse functions and structural motifs, *Biomolecules*, 2014, **4**, 419–434.

6. A. Veyrac, A. Besnard, J. Caboche, S. Davis and S. LaRoche, The transcription factor Zif/268EGR1 brain plasticity and memory, *Prog. Mol. Biol. Transl. Sci.*, 2014, **122**, 89–129.
7. Y. M. Go, D. Duong, J. Peng and D. P. Jones, Protein cysteine functional networks according to steady state level of oxidation, *J. Proteomics Bioinf.*, 2011, **4**, 196–209.
8. D. W. Bak, T. Bechtel, J. A. Falco and E. Weerapana, Cysteine reactivity across the subcellular universe, *Curr. Opin. Chem. Biol.*, 2019, **48**, 96–105.
9. V. Quesada, G. Ordonez, L. M. Sanchez, X. S. Puente and C. Lopez-Otin, The degradome database: mammalian proteases and diseases of proteolysis, *Nucleic Acids Res.*, 2009, **37**, D239–D233.
10. J. C. Powers, J. Asgian, O. D. Ekici and K. E. James, Irreversible inhibitors of serine, cyteine, and threonine proteases, *Chem. Rev.*, 2002, **102**, 4639–4750.
11. H.-h. Otto and T. Schirmeister, Cysteine proteases and their inhibitors, *Chem. Rev.*, 1997, **97**, 133–171.
12. S. Vermas, R. Dixit and K. C. Pandey, Cysteine proteases: Modes of activation and future prospects as pharmacologic targets, *Front. Pharmacol.*, 2016, 7, 1–12.
13. W. Kim, E. Bennett, E. L. Huttlin, A. Guo, J. Li, A. Possemato, M. E. Sowa, R. Rad, J. Rush, M. J. Comb, J. W. Harper and S. P. Gygi, Systematic and quantitative assessment of the ubiquitin-modified proteome, *Mol. Cell*, 2011, **44**, 325–340.
14. Y. Shi, P. Xu and J. Qin, Ubiquitinated Proteome: ready for gobal? *Mol. Cell. Proteomics*, 2011, **10**, R110.006882.
15. M. D. Stewart, T. Ritterhoff, R. E. Klevitt and P. S. Brzovic, E2 enzymes: more than just middlemen, *Cell Res.*, 2016, **26**, 423–440.
16. C. E. Berndsen and C. Wolberger, New Inisghts into ubiquitin E3 ligase mechanism, *Nat. Struct. Mol. Biol.*, 2014, **21**, 301–307.
17. Z. Zheng and N. Shabek, Ubiquitin ligases: structure, function, and regulation, *Annu. Rev. Biochem.*, 2017, **86**, 129–157.
18. D. Komander, M. Clague and S. Urbe, Beaking the chains: structure and function of deubiquitinases, *Nat. Rev. Mol. Cell Biol.*, 2009, **10**, 550–563.
19. F. Massiere and M.-A. Badet-Denisot, The mechanism of glutamine-dependent amidotransferases, *Cell. Mol. Life Sci.*, 1998, **54**, 205–222.
20. R. H. Lambalot, A. Gehring, R. S. Fugel, P. Zuber, M. Lacelle, M. A. Marahiel, R. Reid, C. Khosla and C. T. Walsh, A new enzyme superfamily: the phosphopantetheinyl transferases, *Cell Chem. Biol.*, 1996, **3**, 923–936.
21. D. E. Cane, C. Walsh and C. Khosla, Harnessing the biosynthetic code combinations, permutations, and mutations, *Science*, 1998, **282**(5386), 63–68.
22. C. T. Walsh and Y. Tang, *Natural Product Biosynthesis: Chemical Logic and Enzymatic Machinery*, Royal Society of Chemistry, Cambridge, 2017.

23. A. Alonso, J. Sasin, N. Bottini, I. Friedberg, I. Friedberg, A. Osterman, A. Godzik, T. Hunter, J. Dixon and T. Mustein, Protein tyrosine phosphatases in the human genome, *Cell*, 2004, **117**, 699–711.
24. Z. Y. Zhang, Protein-tyrosine phophatases: biological function, structural characteristics and mechanism of catalysis, *Crit. Rev. Biochem. Mol. Biol.*, 1998, **33**, 1–52.
25. A. D. Pannifer, A. Flint, N. K. Tonks and D. Barford, Visualization of the cysteinyl-phosphate intermediate of a protein-tyrosine phosphatase by X-ray crystallography, *J. Biol. Chem.*, 1998, **273**, 10454–10462.
26. K.-l. Guan and J. Dixon, Evidence for protein tyrosine phsphatase catalysis proceeding by a cysteine-phosphate intermediate, *J. Biol. Chem.*, 1991, 17026–17030.
27. C. Carreras and D. Santi, The catalytic mechanism and structure of thymidylate synthase, *Annu. Rev. Biochem.*, 1995, **64**, 721–762.
28. G. Vilkaitis, E. Merkiene, S. Serva, E. Weinhold and S. Klimasauskas, The mechanism of DNA cytosine-5-methylation, *J. Biol. Chem.*, 2001, **276**, 20924–20934.
29. C. Stines-Chaumeil, F. Talfournier and G. Branlant, Mechanistic characterization of methylmalonate semialdehyde dehydrogenase from Bacillus subtlis, *Biochem. J.*, 2006, **395**, 107–115.
30. L. M. Traub and G. L. Lukacs, Decoding ubiquitin sorting signals for cltahrin-depemdemt endocytosis by CLASPs, *J. Cell Sci.*, 2007, **120**, 543–553.
31. A. M. Marcella, S. J. Culbertson, M. A. Shogren-Knaak and A. W. Barb, Structure, high affinity, and negative cooperativity of the E.coli holoACP: holo ACP synthase, *J. Mol. Biol.*, 2017, **429**, 3763–3775.
32. M. Isalan, Zinc-finger nucleases: how to play two good hands, *Nat. Methods*, 2012, **9**, 32–34.

CHAPTER 13

The Posttranslational Cysteinyl Proteome

In keeping with the cysteine thiolate side chain being kinetically accessible and powerfully nucleophilic, a variety of covalent modifications accrue on sets of client proteins acted on by families of posttranslational modification enzymes. Cysteine side chains are not the only ones that engage in posttranslational modifications Lysine side chain amines famously get methylated and acetylated for transcriptional control,[1] and we have referred to tyrosyl residue *O*-sulfation in Chapter 10.

However, cysteine thiolate side chains engage in perhaps the widest set of enzyme-mediated attachment of electrophilic covalent groups. Figure 13.1 depicts seven of the major covalent modifications, to be analyzed in this chapter. Chapter 14 covers some more restricted but chemically fascinating additional reactions of cysteines in organic metabolite scaffolds.

The placement of the seven modifications is arbitrary with regard to the hypothetical protein sequence and not meant to indicate that all seven are on a given protein at a given time and place in a cell. We will note, though, that C-terminal *S*-palmitoylation and *S*-geranylgeranylation do occur as double modifications on the C-terminus of one of the human Ras protein isozymes. These two lipidations are acylations and alkylations of the attacking cysteine side chains, respectively. The other five, formylglycine formation, disulfide bond formation, *S*-oxygenation to the sulfenic acid, persulfide bond formation and *S*-nitrosylation are all oxidative modifications. In Section 13.6 we also note regiospecific double oxygenation of Cys_2 residues to sulfinates in hypoxia sensing.

The two lipidations will be examined first, each of which gives a higher affinity to the modified client protein for intracellular membranes.

Chemical Biology No. 12
The Chemical Biology of Sulfur
By Christopher T. Walsh

Published by the Royal Society of Chemistry, www.rsc.org

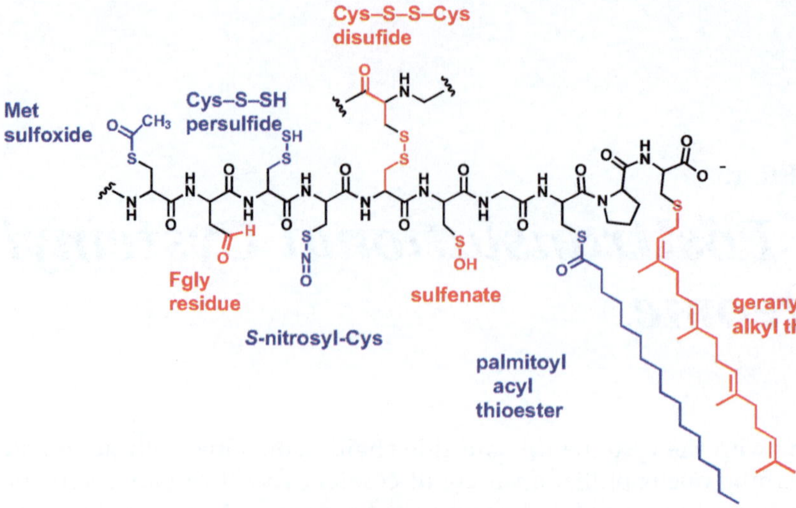

Figure 13.1 Seven posttranslational modifications of cysteinyl proteomes: formylglycine, persulfide, disulfide, sulfenate, *S*-nitrosyl, *S*-palmitoyl and *S*-geranylgeranyl covalent adducts. The sulfoxide modification of methionine is also shown.

13.1 Post-translation Lipidations of Cysteinyl Side Chains

13.1.1 Long Chain Acyl-thioesters *via* Acyl–*S*–Cys–Enzyme Intermediates

The family of posttranslational *S*-acyltransferases that add palmitoyl (or stearoyl acyl chains) (Figure 13.2) are in a sense transitional between the enzymes of the preceding chapter and the covalent modification of target proteins in this chapter. Of the preceding examples of enzyme classes that generate covalent acyl-*S*-cysteinyl–enzyme intermediates (Figure 12.14), three deal with protein or amino acid substrates: proteases, the ubiquitinyl-*S*-transferases and glutaminases. The fourth, the fatty acid synthase complexes, iteratively grow an elongating series of acyl–*S*–enzymes up to 16–18 carbons before they are released by CoASH thiolysis. This last type is closest to these palmitoyl-*S*-transferases that use palmitoyl–*S*–CoA as substrate and transfer it to one or more cysteinyl residues on target proteins.[2]

The underlying logic is to add the long chain fatty acids as hydrophobic groups that may send otherwise soluble proteins to membrane microenvironments to meet up with potential partner proteins docked at specific intracellular or plasma membranes.

The catalysts are a family of posttranslational modification acyl transferases (up to 23 in mammalian cells) that are named for their preferential use of the thermodynamically activated palmitoyl–*S*–CoA as acyl donor (Figure 13.2A). However, in plants the 18-carbon saturated stearoyl–CoA and

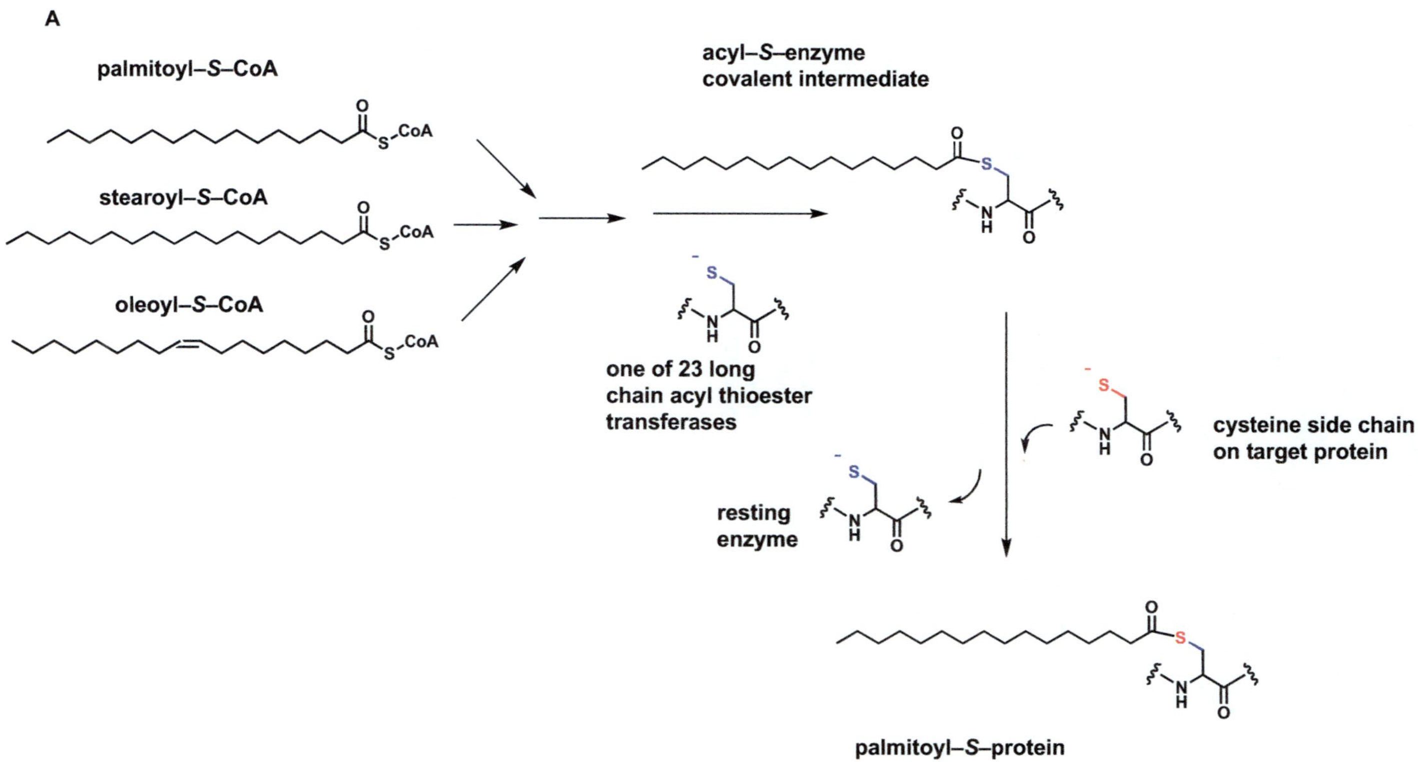

Figure 13.2 Posttranslational lipidation of proteins. (A) Long chain *S*-acylation, exemplified by *S*-palmitoylation. (B) Three classes of protein lipidation include, N-terminal glycine *N*-myristoylation, cysteinyl-*S*-palmitoylation and cysteinyl-*S*-prenylation. (C) Of the three the *S*-palmitoyl group is uniquely labile to hydrolysis by action of acyl–*S*–protein thioesterase.

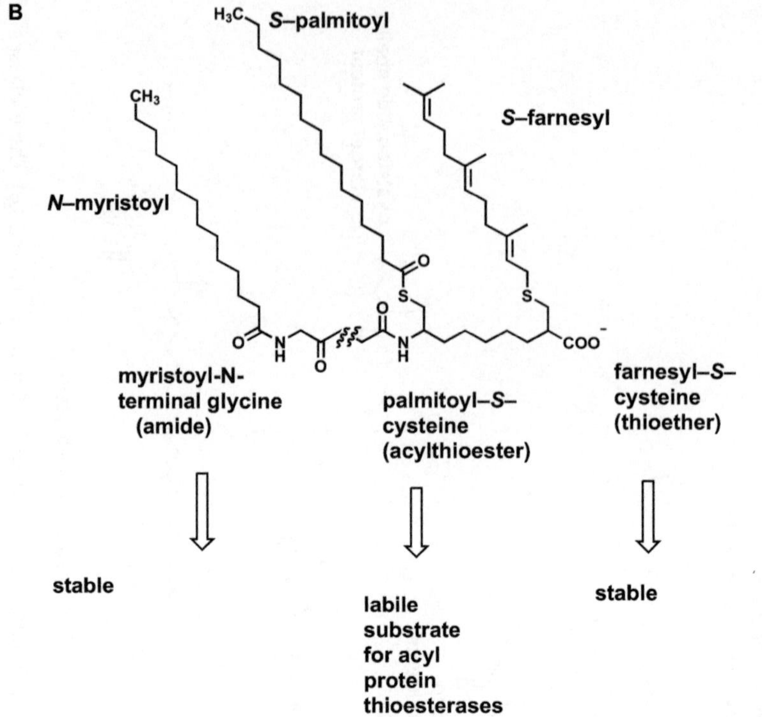

Figure 13.2 Continued on next page.

its monounsaturated oleoyl–CoA are also significant acyl donors to protein cysteinyl side chains.[3]

Mechanistic analysis has identified Cys146 in one of the well characterized mammalian transferases as a catalytic nucleophile, forming a covalent long-chain acyl–*S*–cysteinyl enzyme intermediate in mid-catalytic cycle (Figure 13.2A). This is acyl transfer between thiols (CoAS– *vs.* enzyme Cys–S–), also termed a thioester exchange. It should proceed isoenergetically, once the acyl transferase has bound the long chain acyl–*S*–CoA substrate.

Likewise, the second half reaction is a similar thioester exchange. The acyl chain moves from the enzyme active site sulfur on the cysteinyl residue to a cysteinyl thiolate on a target protein cosubstrate. The overall *S*-palmitoylation of client proteins proceeds *via* three acyl thioesters: substrate, covalent acyl enzyme, product.

The scope of the reaction is indicated by the observation that the palmitoyl–*S*–proteome has reached some 500 members in mammalian cells and up to 650 predicted sites in *Arabidopsis thaliana*. Remarkably, some 342 synaptic proteins have been identified as substrates for *S*-palmitoylation, at some 490 distinct cysteine side chains.[2] Clearly this posttranslational modification is central to synaptic function and perhaps trafficking of proteins at synapses.

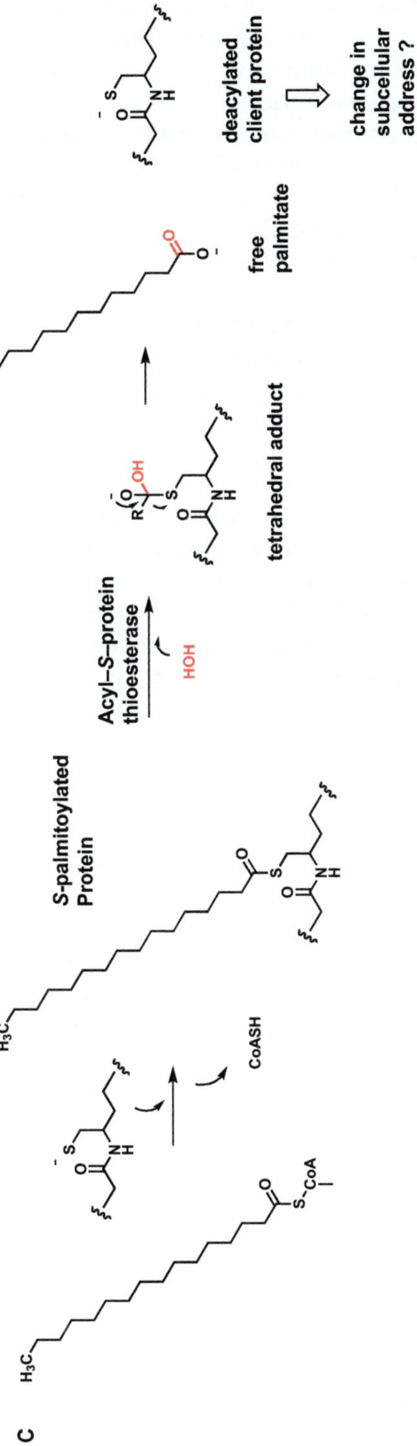

Figure 13.2 Continued.

The distribution of palmitoyl chains can range from one to four in mammalian proteins. In the 18-protein subunit complex for cellulose synthase, up to 70–100 sites of palmitoylation have been predicted: a bristling of post-translationally added acylated thioesters!

The biological readout of protein *S*-palmitoylation is typically a change in subcellular address. The hydrophobic 16-carbon acyl group decreases water solubility of the modified protein and increases its partition to hydrophobic microenvironments, specifically association with subcellular membranes and organelles. The change in subcellular address can bring proteins to membrane surfaces where partner proteins may now be able to function in complexes that transmit intracellular signals.

This posttranslational lipidation strategy is only one of three enzymatic machineries that add lipid chains to protein cores to affect subcellular address, partitioning into membranes and connecting to partner proteins, many functioning in signaling cascades.[4,5]

Figure 13.2B summarizes that two of the three modifications involve long chain fatty acylation strategies. *N*-myristoylation (14-carbon fatty acid) occurs cotranslationally on an N-terminal glycine amine, liberated as the new N-terminus by methionine aminopeptidase action. The *N*-myristoyl glycine linkage is an amide linkage and is functionally irreversible. In contrast, the *S*-palmitoylation of cysteine thiolate side chains, maintaining the thermodynamically activated acyl thioester link, is labile to hydrolysis, as we will note momentarily. The third lipidation mode also involves a cysteinyl thiolate as side chain nucleophile but the electrophilic partner is either the C_{15} (farnesyl) or C_{20} (geranylgeranyl) prenyl diphosphates. We will comment on *S*-prenylation in the next subsection and only note here that the lipid adduct is a thioether and is also stable to hydrolytic removal.

There are suites of enzymes that put the acyl group on (palmitoyl transferases) different categories of target proteins. Currently 23 mammalian palmitoyl-*S*-transferases[6] have been identified by sequence homology. They are transmembrane proteins, consistent with the lipid solubility of palmitoyl CoA and have a long zinc finger-like cysteine rich core (Chapter 12) that probably encompasses the active site. As with many other posttranslational modifications that carry one or more axis of cellular signal transduction (*e.g.* protein kinases–phosphoprotein phosphatases, and protein acetylase–acetylated protein deacetylases), there are opposing enzymes to undo the post-translational covalent mark in the target protein.

Because the acyl-*S*-cysteine[6] is thermodynamically activated and kinetically somewhat labile towards hydrolysis, the removing enzymes are hydrolytic-thioesterases, releasing the free palmitate.[6] The small family of acyl–*S*–protein thioesterases (less than five established catalysts) thus balance the protein acyl transferases.[6] The subcellular location, the accessibility of the palmitoyl–thioester–protein linkage and the specificity of the protein thioesterases will all contribute to the lifetime of acyl chains covalently attached to specific members of cell proteomes. The acyl protein thioesterases have been detected in cytoplasm and lysosomes, indicating that palmitoylated proteins that have become tightly bound to membranes may not be physically accessible except when they dissociate.

Proteins whose palmitoylation cycle have been well studied include the Ras example below, as well as endothelial cell nitric oxide synthase and the β_2-adrenergic receptor. Additional membrane receptors undergoing *S*-palmitoylation are the transmembrane receptors for serotonin, dopamine, vasopression, adenosine, opioids and acetylcholine and CB1 for cannabinoids.[7]

Among the most celebrated examples of cysteinyl residue palmitoylation are the reversible cycles of acylation of Ras protein isoforms, shuttling to and from plasma membranes as a function of their *S*-palmitoylation state[4,5] (Figure 13.3). For example, human processed R-ras protein has two cysteines in the final three C-terminal residues, one of which is *S*-acylated with the 16-carbon palmitoyl acyl group, the other alkylated with the C_{20}–geranylgeranyl group (see next section). Both act as membrane anchors. Palmitoylation of the Ras C-terminal cysteine occurs during passage through the Golgi complex and sends the modified RAS protein products to the plasma membrane.

Likewise, the approximately 30 proteins of the regulators of G protein signaling (RGS) group are reversibly palmitoylated to traverse intracellular locales and activate the G-protein-coupled signaling complexes at the inner surface of cytoplasmic membranes of animal cells.[8]

Ras proteins are also typically lipidated not only by palmitoylation (acylation; Figure 13.2A) but also by isoprenyl group alkylation, discussed next. A single lipid tether may set up enough affinity for an acylated protein to partition into but also back out of membranes. Addition of a second lipid anchor increases the probable residence time of the bis-lipidated protein in membrane compartments.

13.1.2 Protein Prenylations: Posttranslational Lipidation *via* Thioether Adducts

The second lipid anchor is often a C_{15} (farnesyl) or C_{20} (geranylgeranyl) isoprenoid group that gets attached in a thioether linkage to a cysteinyl-sulfur side chain in a client protein (Figures 13.1 and 13.3). The protein prenylations differ from *N*-myristoylations (Figure 13.2B) and are similar in one sense to *S*-palmitoylations in the use of cysteineyl thiolates in targeted client proteins as nucleophiles. However, the prenylations are irreversible, unlike the palmitoylation events, because the prenylations generate thioether linkages rather than thioesters. In terms of abundance, there are estimates that prenylated proteins can comprise up to 2% of mammalian proteomes (The LipidWeb.org: proteolipids and protein lipidation).[9]

The prenyl donors utilized are the C_{15}-farnesyl-PP and the C_{20}-geranylgeranyl-PP sesquiterpene (C_{15}) and diterpene (C_{20}) metabolites (Figure 13.3A), respectively. They are sequential key intermediates in both linear isoprenoid and cyclic terpenoid frameworks, including all the steroid metabolites.

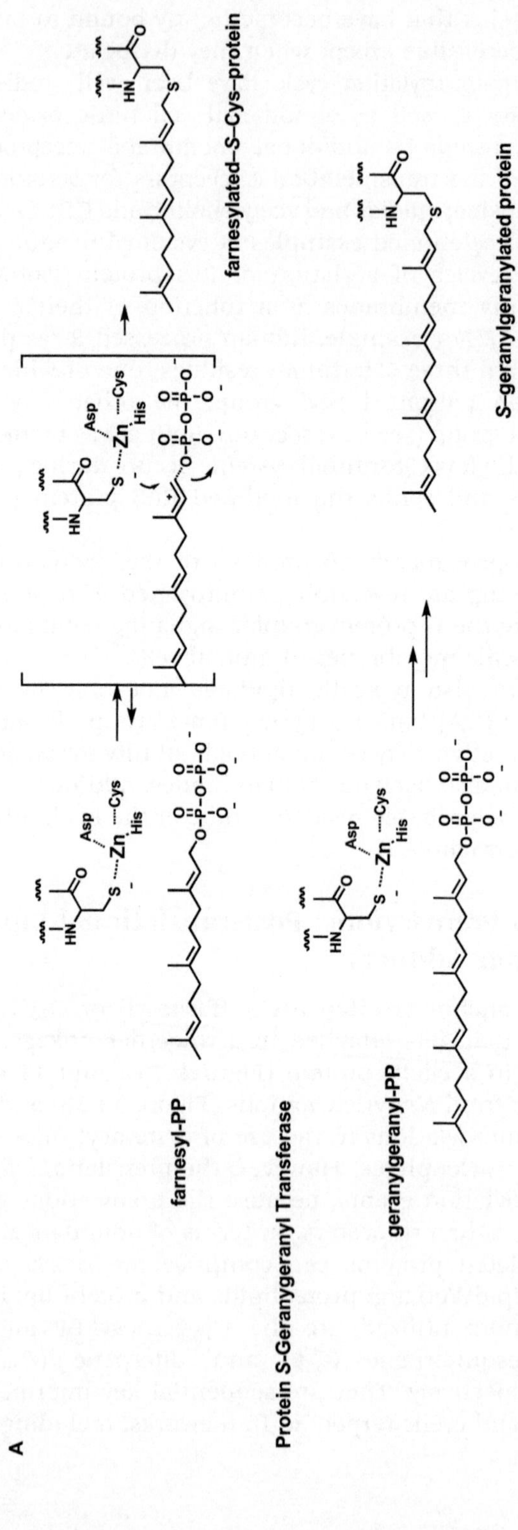
A
Protein S-Geranygeranyl Transferase
farnesyl-PP
farnesylated–S–Cys–protein
geranylgeranyl-PP
S–geranylgeranylated protein
Asp
Zn
Cys
His

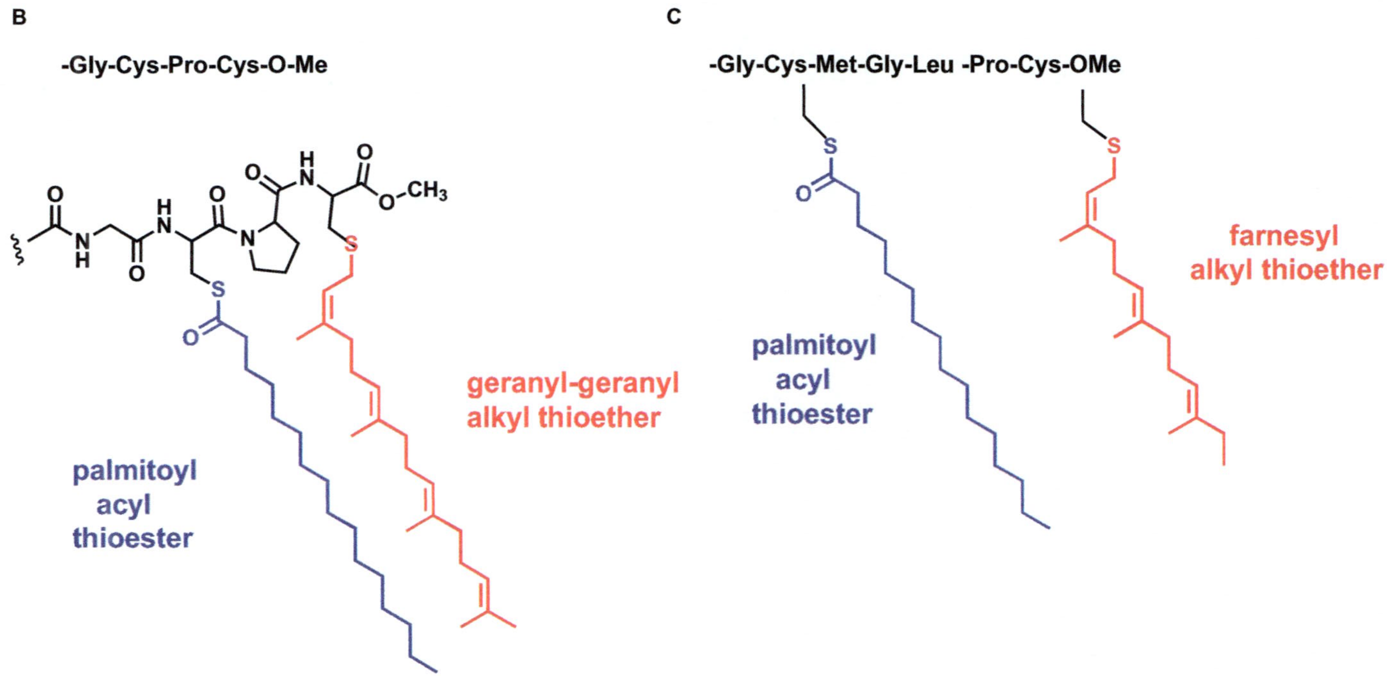

Figure 13.3 Prenylation of ras proteins at cysteines near the C-termini. (A) The cysteine thiolate side chain can be prenylated with a C_{15} farnesyl side chain by a farnesyl–*S*–protein prenyltransferase or can be *S*-geranylgeranylated by a C_{20} selective prenyltransferase. (B) The C-terminus of human R-ras has two lipid anchors, a palmitoyl–*S*–acyl thioester and a geranylgeranyl alkyl thioether. (C) The C-terminus of human N-ras also has two lipid anchors, a C_{18} acyl thioester and a C_{15} prenyl thioether. Both R- and K-ras have the C-terminal carboxylate covered up as a neutral oxo-methylester, increasing the lipophilicity of this region of each protein.

All the different length prenyl-diphosphates serve as allyl cation equivalents in enzymatic alkylation reactions. Early dissociation of the C1–OPP bond to nascent PPi and the delocalized, stabilized C1 allyl cations provide the carbon electrophiles for capture by cosubstrate nucleophiles.[10] These reactions are prefigured by the action of glutathione to capture the C_{20} geranylgeranyl chain, producing geranylgeranyl glutathione, noted in Chapter 5 as a germinal center signaling metabolite.[11]

There are three forms of prenyl-*S*-protein transferases. One uses farnesyl-PP as substrate and the other two use the C_{20} geranylgeranyl-PP isoprenoid as cosubstrate (Figure 13.3A). This small posttranslational enzyme family has a zinc divalent cation in the active site (see Section 12.2) that coordinates the target protein thiolate in the ternary complex of enzyme–prenyl diphosphate–target protein. The coordination presumably sets the geometry for attack and may even activate the attacking thiolate side chain.

Figure 13.3B and C show now classic cases of prenylation of C-terminal cysteines of R- and N-ras protein isoforms.[12] Note that both *S*-palmitoylation and *C*-geranylgeranylation or farnesylation occur on nearby cysteine sulfurs to create a densely hydrophobic patch. The double lipidations, of two types, are key for trafficking of Ras proteins to cell membranes. The enzymatic hydrolysis of the palmitoyl thioester, leaving the prenyl thioether chain intact is thought to modulate the partition of Ras isoforms on and off membranes.

Prenylations occur in cell cytoplasmic compartments.[13] Then, as the modified Ras proteins move through the endoplasmic reticulum, a committed protease removes the last three residues of a CAAX terminus, liberating the prenylated cysteine as the new carboxy terminus. This form of the protein is then the substrate for the SAM-dependent carboxymethyl ester synthase, noted briefly in Chapter 8. This final state of R-Ras, after undergoing four types of posttranslational modifications (palmitoylation, geranylgeranylation, C-terminal peptidase trimming and C-terminal carboxylate methylation) is shown in Figure 13.3B. In some cases, there is no obvious functional distinction conferred by a C_{15}- *vs.* C_{20}-length prenyl chain, while in others it may affect both affinity for a membrane niche and/or interaction with partner proteins.

We have noted above the chemical contrast between C_{16} palmitoyl group acylations and C_{15} or C_{20} prenylation of cysteine thiolates is the stability of the posttranslational modifications. Acyl thioesters are thermodynamically activated, kinetically labile to hydrolysis, and reversible by action of the family of acyl–*S*–protein thioesterases. In contrast, the prenyl–thioether linkages are chemically stable, functionally irreversible in cells and thus not dynamically removed. Study of the fate of farnesylated cysteinyl proteins (*e.g.* Ras), indicates lysosomal protease-mediated hydrolysis all the way to free farnesyl-cysteine. Then, a dedicated flavoprotein carries out a net oxidative cleavage, yielding back free cysteine and the farnesyl aldehyde.[14,15] One can write the mechanism shown in Figure 13.4, which is reminiscent of the enethiol route in the oxidative decarboxylation of phosphopantetheinyl cysteine in the coenzyme A biosynthesis pathway, discussed in Chapter 5.

A

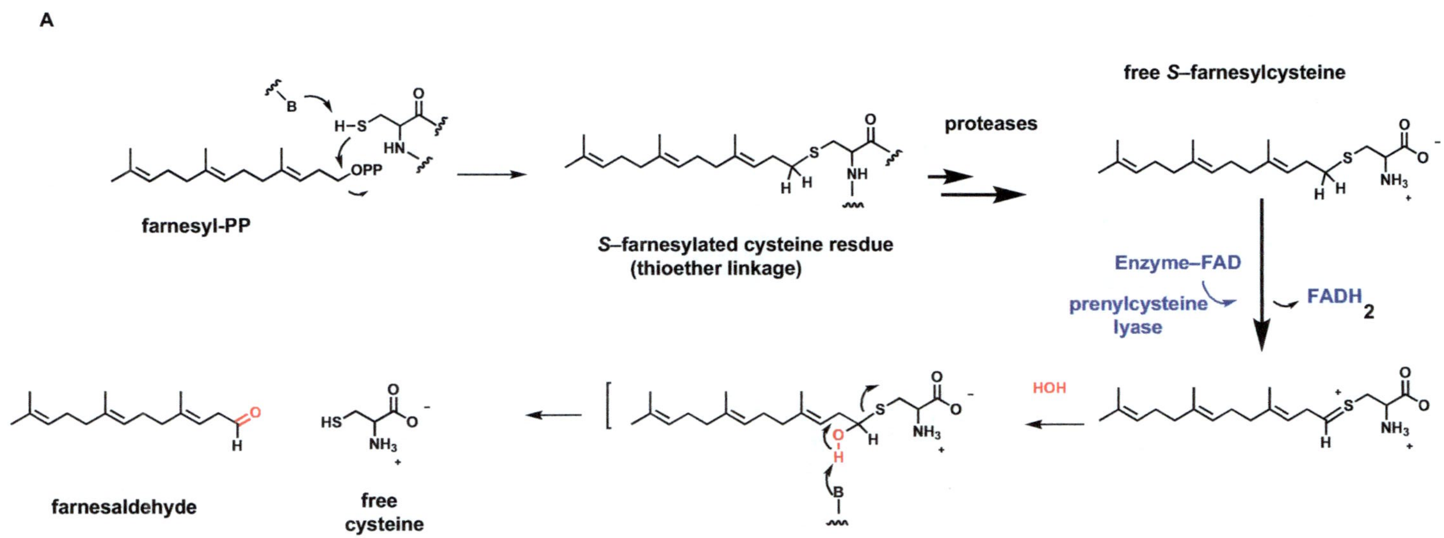

Figure 13.4 (A) Metabolism of prenylated-Cys–protein residues. Proteolytic cleavage to free prenyl–*S*–cysteines and then oxidative release of free cysteine and the prenyl aldehyde by a flavoprotein lyase. (B) Four types of oxidized sulfur in cysteine derivatives in proteomes: cystine disulfide, cysteine persulfide, cysteine sulfenic acid and *S*-nitroso-cysteine.

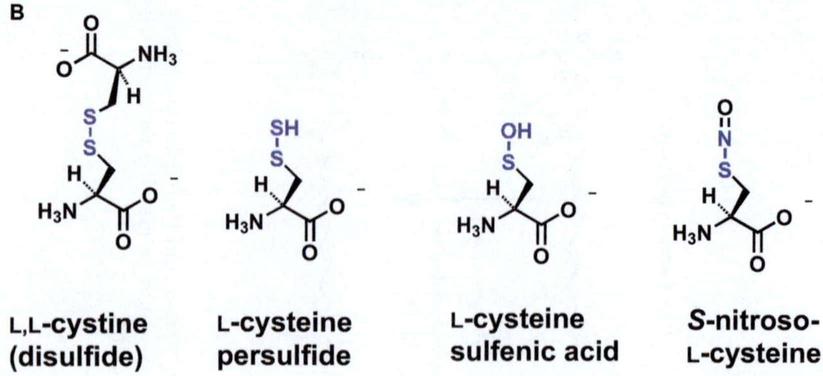

Figure 13.4 Continued.

The virtues and liabilities of reversible acylation *vs.* irreversible prenylation of cysteinyl residues in proteins to be lipidated and sent to different subcellular addresses may reflect distinct intracellular functions for the two type of cysteine residue modifications.

13.2 Cystine Disulfides

The oxidative dimerization of two cysteinyl residues within or between proteins to form cystine disulfide linkages is probably the most widely recognized posttranslational covalent modification of proteins in oxidizing environments[16–18] (Figures 13.1 and 13.4). The disulfide bond is typically formed by one cysteine side chain acting as a nucleophile (thiolate) and the other as an electrophile. Disulfide bond formation is an oxidative transformation as are the other three transformations (persulfide, sulfenic acid and *S*-nitrosylation) that will also be summarized in later sections of this chapter.

The cystine disulfide bond in proteins, within subunits and between subunits, is typically a reflection of a local oxidizing environment. Within cells, most cysteines are reduced, with one or two exceptions noted below. A sample of cysteine residue abundance in 32 intracellular mammalian proteins indicated a 1.2% abundance of cysteine, substantially below the 5% level one might have expected for a random usage of the 20 proteinogenic amino acids.[19] In comparison, in 34 extracellular mammalian proteins the cysteine residue abundance, much of it in cysteine disulfide linkages, was 4.1%, raising the prospect that during evolution of aerobic (*vs.* anaerobic) organisms, that cysteine may have been selected for proteins that would spend their lifetimes in oxidizing microenvironments where the disulfide bridges provide architectural elements that may aid folding, function and stability. In turn, that would suggest the posttranslational abundance of cysteine cross links in plasma membrane and extracellular proteins was a post global oxygenation development. If so, the history of proteomic architecture may have been quite different in early anaerobic organisms.

13.2.1 Keeping Cytoplasmic Cysteinyl Side Chains Reduced

With the notable exception of a few cytoplasmic enzymes, such as glutathione reductase and thioredoxin reductases (see Chapter 9) which maintain active site disulfides for mechanistic purposes, most cytoplasmic proteins have cysteine side chains in the reduced, thiol forms. Certainly, they can undergo transient oxidation, *e.g.* by hydrogen peroxide, to sulfenic acids[20,21] and then on to the disulfides. However, in the presence of millimolar reduced glutathione, they are brought back to the thiol oxidation states by thiol–disulfide interchange processes. As exemplified in Figure 13.5, dithiol–disulfide exchange chemistry is both isoenergetic and, even though bimolecular, presumably reasonably fast. With glutathione presumably at over 100-fold in molar excess over any transient disulfide-containing cytoplasmic protein that would be returned to the reduced state.

This logic and machinery succeeds as long as the cells can convert the oxidized glutathione disulfide back to two molecules of reduced glutathione. We noted in Chapter 5 that cells spend a molecule of NADPH for every molecule of GSSG *via* action of glutathione reductase (present in cytoplasm and in mitochondria). In Figure 5.17 we pointed out two features of glutathione reductase (and also cytoplasmic thioredoxin reductase) that are essential to this logic. One is the presence of bound FAD flavin coenzyme in the oxidized state of the enzyme at rest. The other is the occurrence of a rare Cys–S–S–Cys disulfide pair in the active site, within bonding distance of the FAD, on the opposite face from the NADPH binding site.[22]

Electrons literally pass from NADPH into the top side of FAD and then out from the bottom side of the $FADH_2$ to the active site disulfide[23] (Figure 13.5B). From the resultant dithiol form of the reduced enzyme they pass out to reduce glutathione. The sketch in Figure 13.5B is from a 1985 paper on glutathione reductase that schematizes that layout of the electron flow in the active site.[24] That electron flow occurs in three stages by three distinct mechanisms. NADH transfer of two electrons to oxidized FAD is *via* hydride ion transfer to N5 of FAD. Electron transfer from $FADH_2$ to the active site Cys–Cys disulfide proceeds by covalent thio–4a–FAD adduct (Figure 13.6B). The transfer of electrons from active site dithiol to bound GS–SG occurs by the dithiol-disulfide interchange process described in Figure 13.5A.

We describe the chemistry of the flavin-4a–thiol adduct as the covalent conduit of electron transfer between the riboflavin-based coenzyme and the active site dithiol or disulfide. That logic is also central in the ERO1 and QSOX enzymes involved in disulfide formation during protein transit through the secretory system noted in the next section.

One additional mitochondrial protein in which a disulfide bond is in the resting state of an enzyme is found in enzyme 2 of the three-enzyme complexes found in a small family of α-keto acid dehydrogenases. These catalyze oxidative decarboxylation of substrates such as pyruvate to acetyl CoA, CO_2 and NADH and represent a central energy harvesting strategy in metabolism. Rather than a Cys–S–S–Cys protein-based disulfide the lipoamide acyltransferase enzyme 2

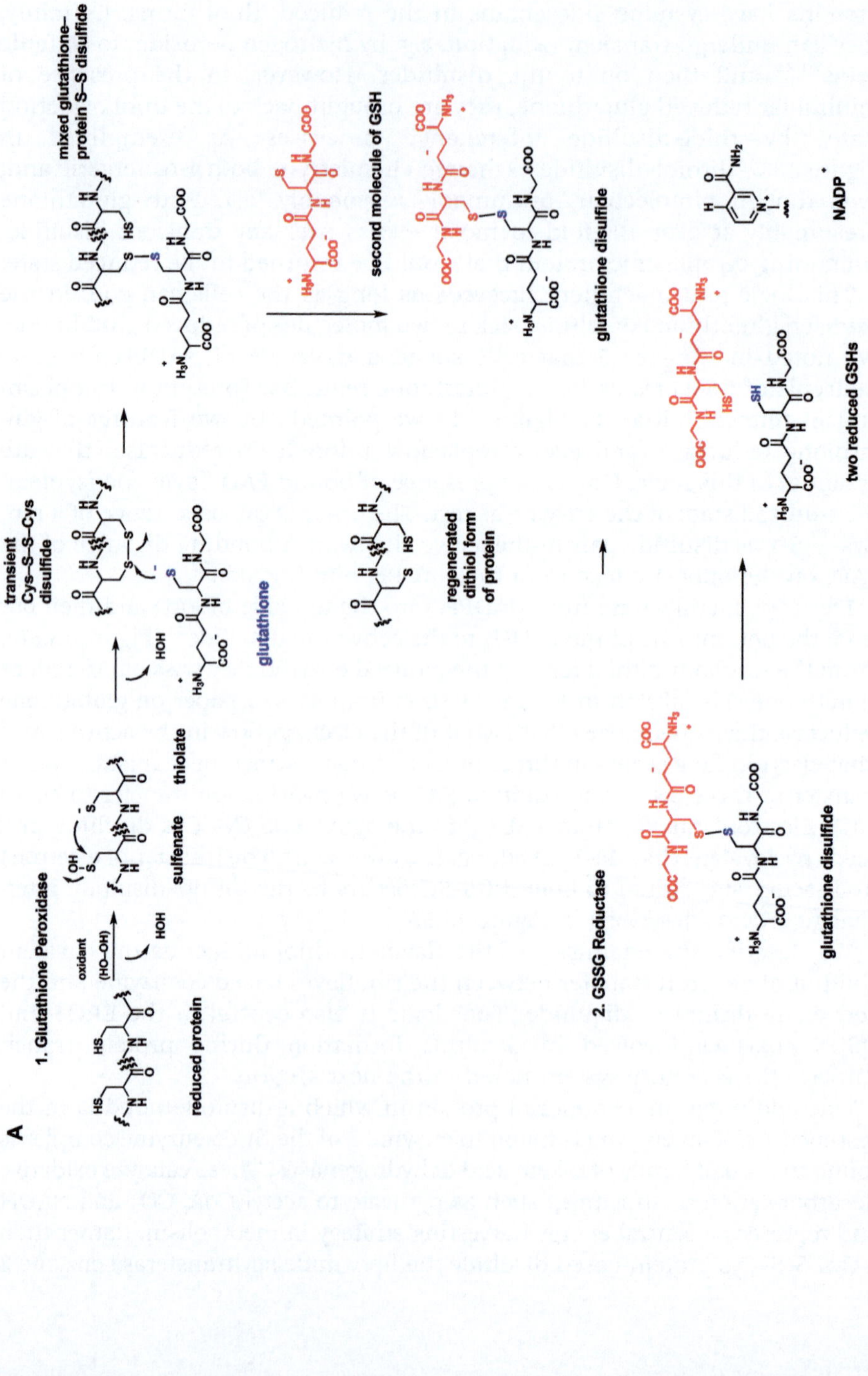

Figure 13.5 Continued on next page.

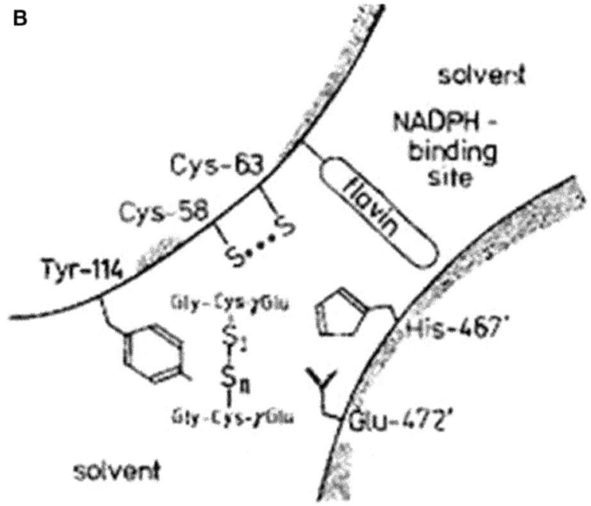

Figure 13.5 (A) Schematic of oxidation of a cysteinyl thiolate in a protein by H_2O_2 leads to the two-electron-oxidized sulfenic acid. That may exist indefinitely or be captured by a neighboring cysteinyl thiol to give the protein cystine disulfide. In reducing microenvironments such as cell cytoplasms with high GSH levels, GSH will reduce the protein cystine disulfide back to the dithiol form, *via* the mixed protein–Cys–S–SG disulfide on the way to forming the oxidized GS–SG dimer. GS–SG is then returned to two molecules of GSH by action of glutathione reductase. (B) A schematic of the active site groups in glutathione reductase showing NADPH binds on one side of the FAD with the reducible active site disulfide on the other side. In turn GS–SG binds external to that active site disulfide for linear electron flow from NADPH to GS–SG;
(B) Reproduced from ref. 59 with permission from John Wiley and Sons, Copyright 2005 Federation of European Biochemical Societies.

family members contain an oxidized lipoamide (Chapter 9) tethered covalently to a lysyl side chain.[25] We noted in Figure 9.17, that the lipoamide disulfide is the electrophilic acceptor for a migrating C2 carbanion in a hydroxyethyl–thiamin–PP adduct in enzyme 1 of the complex. When the resultant C2-acetyl group is transferred out to CoASH by thiol exchange the dihydrolipoamide form of enzyme 2 must be reoxidized. That is the task of enzyme 3 in the complex, a dihydrolipoamide dehydrogenase. Enzyme 3 operates on a similar principle to glutathione reductase, but in the reverse direction. Again, 4a-thiol flavin covalent adducts (Figure 13.6B) are at the heart of that molecular logic and redox machinery for electron transfer.

13.2.2 Peroxiredoxins

An unusually reactive family of proteins for transformation of dithiols to disulfides are six *peroxiredoxin* isoforms.[26] As the name implies these proteins react with hydrogen peroxide to reduce it by two electrons to two molecules of water. The bimolecular rate constants for peroxiredoxin

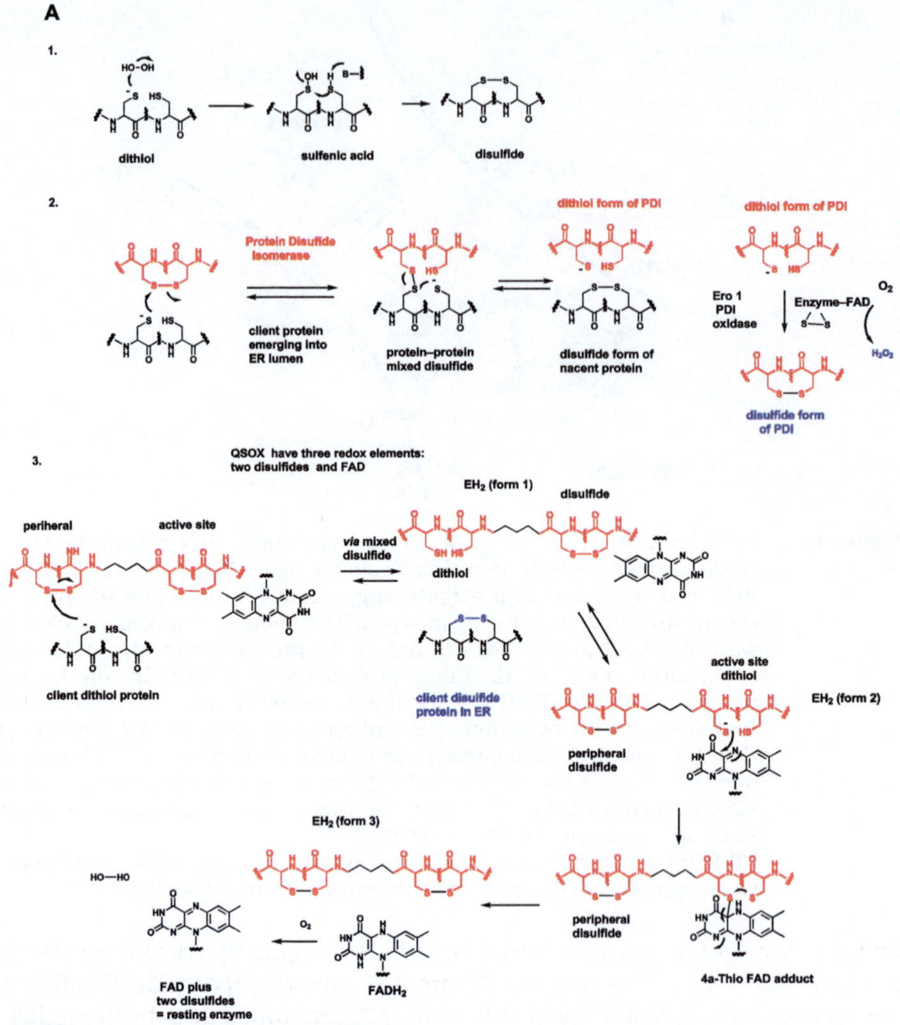

Figure 13.6 Continued on next page.

thiolate attacking HOOH can approach the diffusional upper limit[21,27] of 10^8–10^9 $M^{-1}\,s^{-1}$. In contrast many other protein thiols react at slow to negligible rates with H_2O_2 and also with O_2 and so are kinetically stable in their presence (*e.g.* CoASH and GSH as small molecule thiols).

The resulting peroxiredoxin sulfenic acid is captured by an adjacent thiol to release the second water molecule as the peroxiredoxin disulfide forms. Return of peroxiredoxin to the active dithiol form requires input of a two-electron reductant. That comes *via* reduced thioredoxin and ultimately from NADPH as it donates electrons *via* thioredoxin reductase.

More generally, as noted in the next section, the process of protein dithiol–disulfide exchange within and between proteins can be terminated or directed

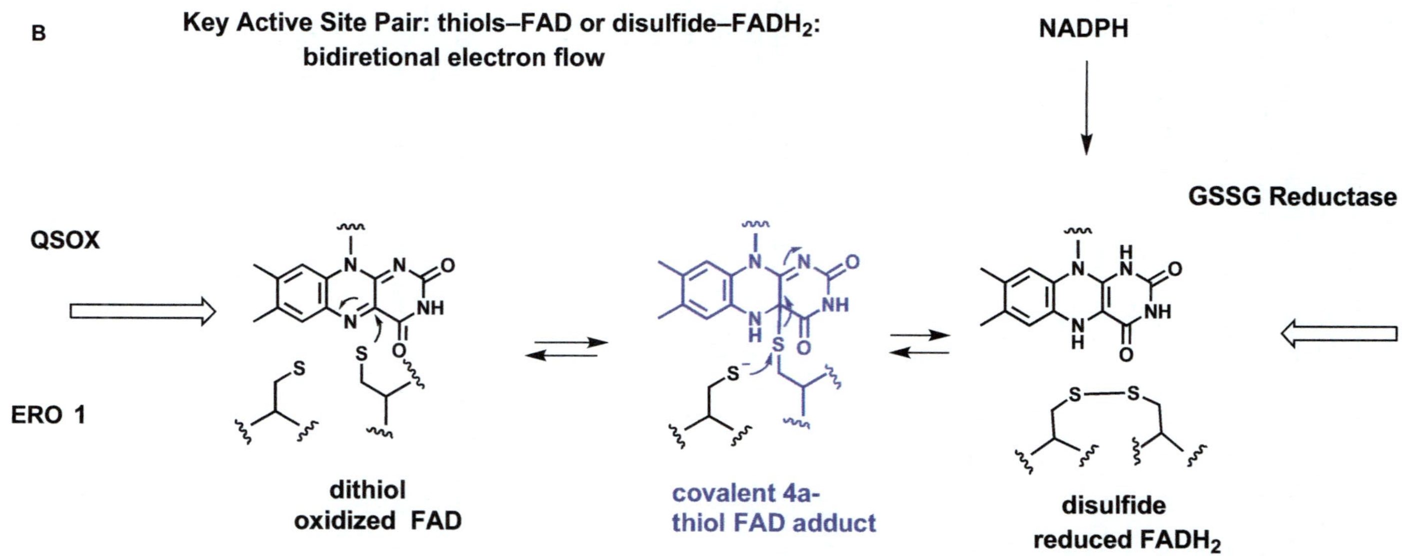

Figure 13.6 Routes for conversion of protein dithiols to disulfides. (A) Three routes: 1. *via* sulfenic acid intermediate; 2. by protein disulfide isomerase (PDI)-mediated dithiol–disulfide exchange in the endoplasmic reticulum; 3. by direct oxidation of client protein dithiols by QSOX flavoprotein oxidase. (B). Key redox pairing of FAD and active site dithiol or disulfide. GSSG reductase works from right to left. ERO1 and QSOX work from left to right.

by input or exit of two electrons. The input from NADPH *via* thioredoxin reductase and reduced thioredoxin leads to electron pressure to favor dithiol accumulation. In the oxidizing microenvironments of the secretory compartments, the cosubstrate is O_2. Its reduction to H_2O_2 by two enzymes discussed below removes electrons *via* H_2O_2 formation and so puts pressure on accumulation of disulfide forms. The first strategy applies in the cytoplasm. The second strategy applies in the endoplasmic reticulum and Golgi complex.

The preroxiredoxins are the workhorses for H_2O_2 reduction in reducing cellular compartments. From Figure 13.7, one can see that the net reaction stoichiometry is for a molecule of NADPH to be spent reducing H_2O_2 to two water molecules. Of course, NADPH is not reacting directly with peroxide. The catalysis requires first the peroxiredoxin and then thioredoxin and finally thioredoxin reductase: three proteins in the cascade.

Amidst all the thiol–disulfide interchanges, the net pull on the equilibrium is the NADPH reduction. In the secretory pathway the net pull is O_2 reduction to H_2O_2, putting pressure on accumulating protein disulfide isomerases in the oxidized disulfide form.

13.2.3 Sulfiredoxin

Not surprisingly, occasionally the –Cys–S–OH sulfenate intermediate gets over-oxidized to the sulfinic acid (Figure 13.7B) (see Chapter 7 for sulfenate and sulfinate forms of free cysteine). Historically, the sulfinic oxidation state was thought to be irreversibly inactivated enzyme. Then, in 2003 a small protein named *sulfiredoxin*, for its ability to reverse the *sulfinate* to thiol by reduction, was discovered.[28] This is a net four-electron reduction and consumes one molecule of ATP.

The sulfiredoxin has a thiolate that attacks the γ-$PO3^-$ of ATP to form a rare *S*-phosphoryl cysteinyl enzyme residue (Figure 13.7B and C). This is reminiscent of the *S*-phosphoryl enzyme intermediates in protein tyrosine phosphatase action noted in Chapter 10 but this one forms by attacking the terminal phosphoryl group of ATP. The phosphoryl–*S*–sulfiredoxin acts as chaperone and donor to the sulfinate residue oxyanion to create a mixed sulfinic–phosphoric anhydride.[29] That has set up the S–O bond in the S–OPO_3 group for low energy cleavage as an intersubunit Cys–SH from the peroxidredoxin dimer attacks. This creates a disulfide monoxide species.[30] (It is also known as a thiosulfinate.) In turn, that can be attacked by a thiol to regenerate a tautomer of the normal sulfenate form. Isomerization to the sulfenate gets back to a normal intermediate in the peroxiredoxin cycle. To get all the way back to the thiol form consumes another two thiol molecules.

When the dust settles, one ATP has been cleaved and four thiols have been expended to return this key inactive sulfinate form of peroxiredoxins to the starting thiol form. As noted in Figure 13.7 *three unusual enzyme cysteine-based intermediates* have been formed and dissipated: the thiophosphoryl–*S*–sulfiredoxin, the mixed sulfinic–phosphoric anhydride and the disulfide monoxide, both in the peroxiredoxin undergoing reductive resuscitation.

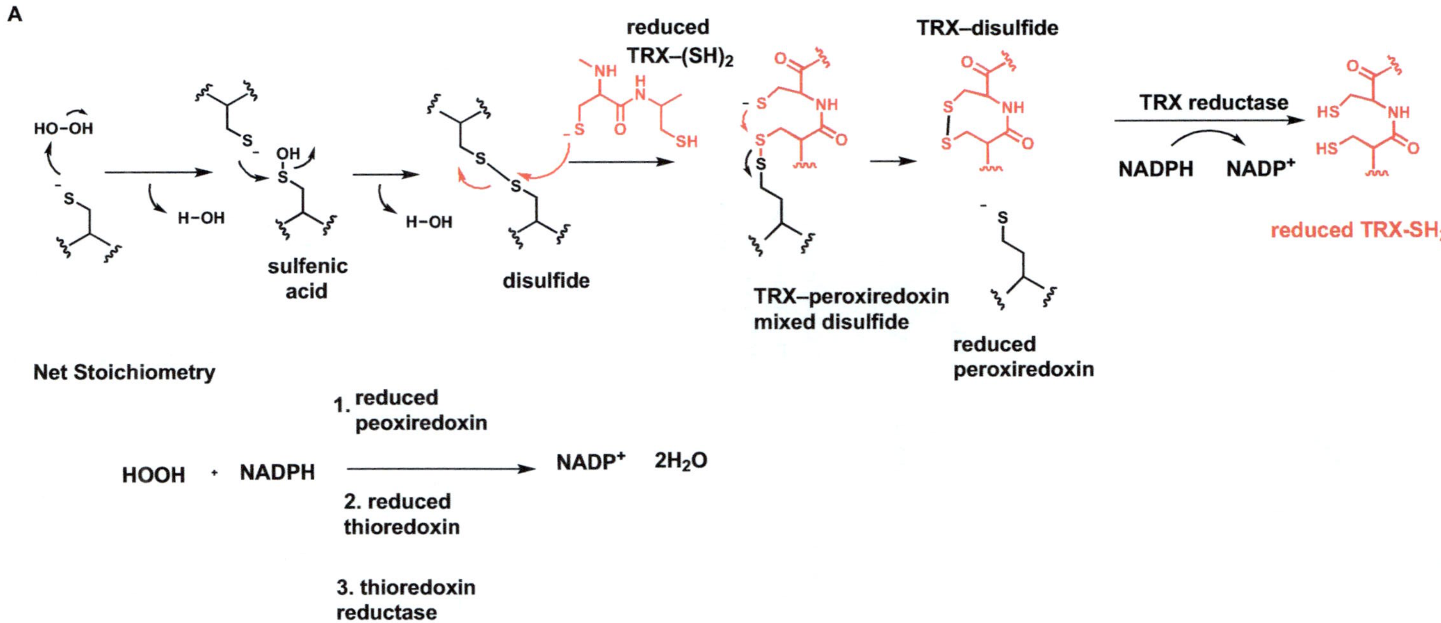

Figure 13.7 Continued on next page.

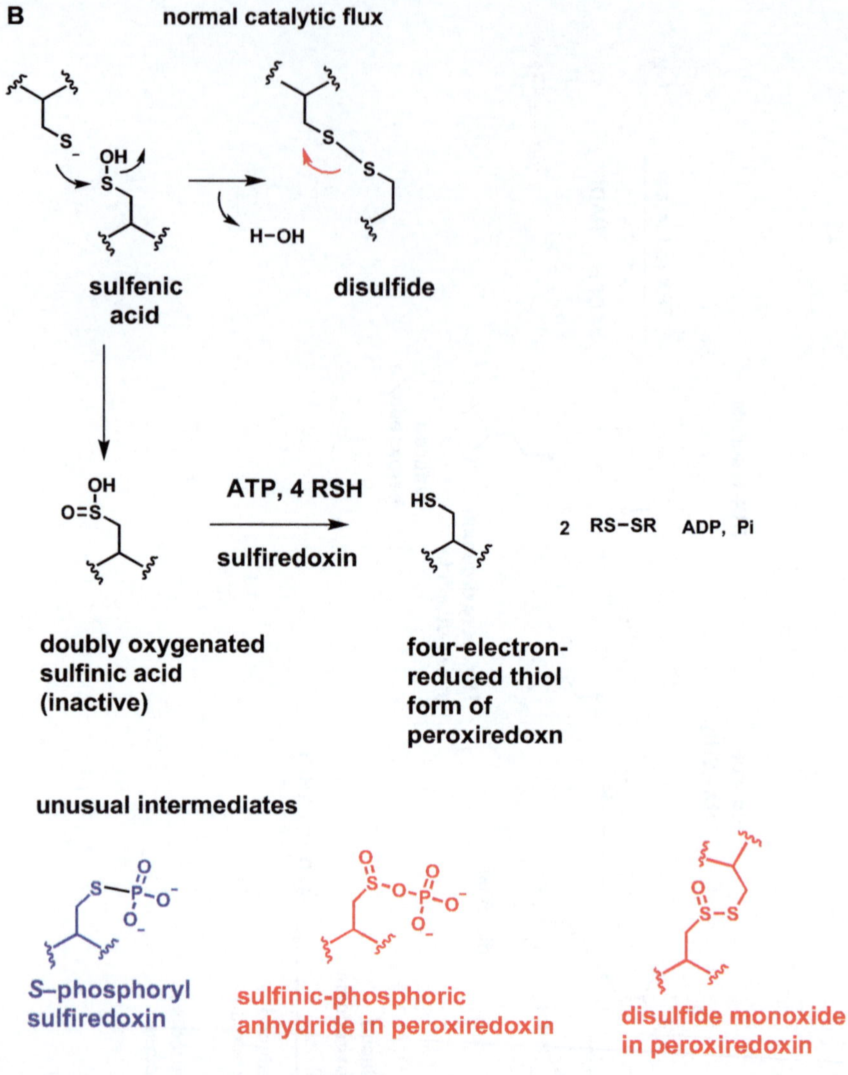

Figure 13.7 Continued on next page.

It was originally thought that sulfiredoxin showed exclusive specificity for reductive repair of cysteineyl sulfinic acid forms of inactive peroxiredoxins and that these were unique occurrences of sulfinic acid residues. More recent investigations and detections of protein sulfinic acids also show their reductive repair to Cys–thiol side chains (making them much more prevalent in proteomes than had previously been realized) to the thiol forms by sulfiredoxin action with ATP and RSH cosubstrates.[21]

Sulfiredoxin has also been identified as a catalyst for denitrosylation of a *S*-nitrosylcysteine residue in peroxidredoxin-2.[31] Phosphoryl transfer from the S–PO_3–sulfiredoxin to the nitrosyl nitrogen would set up elimination of

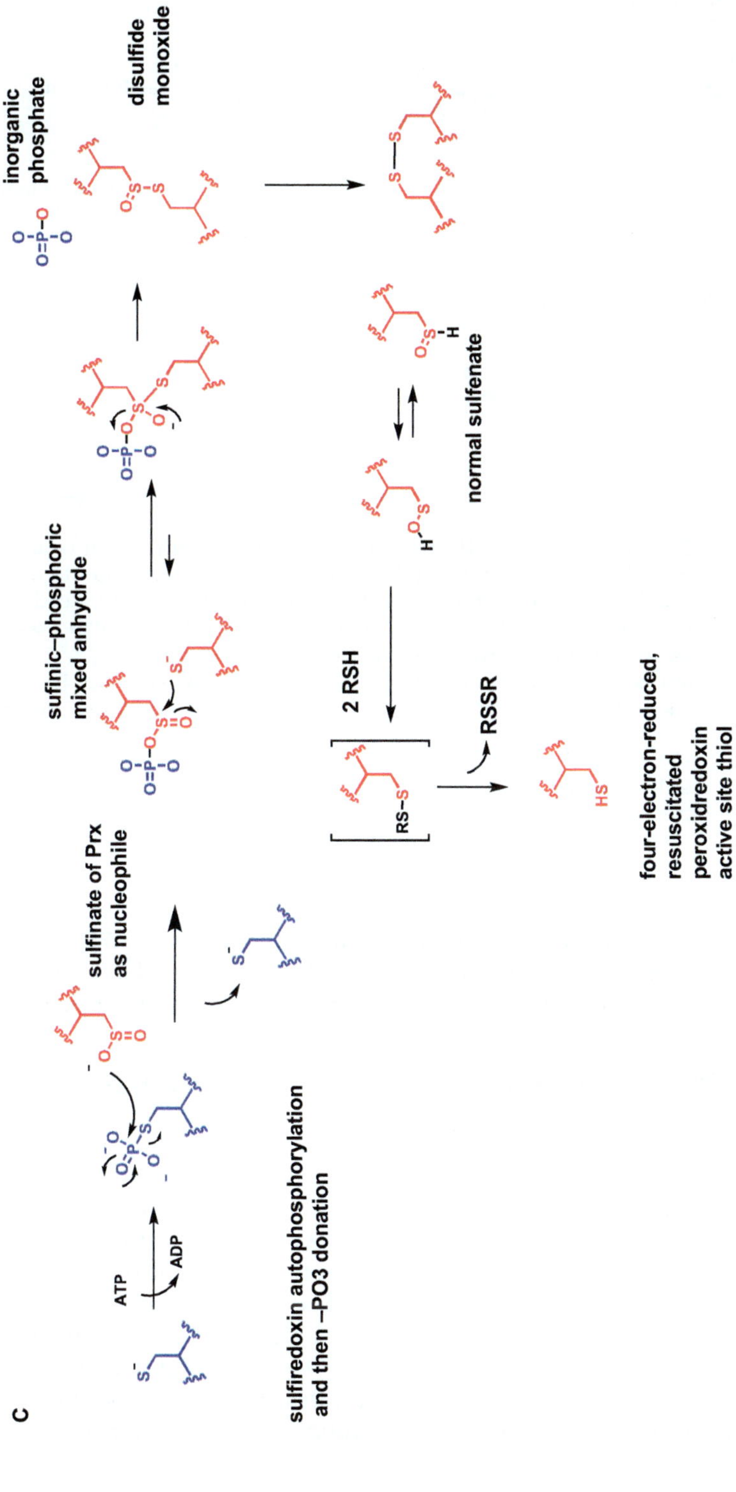

Figure 13.7 Continued on next page.

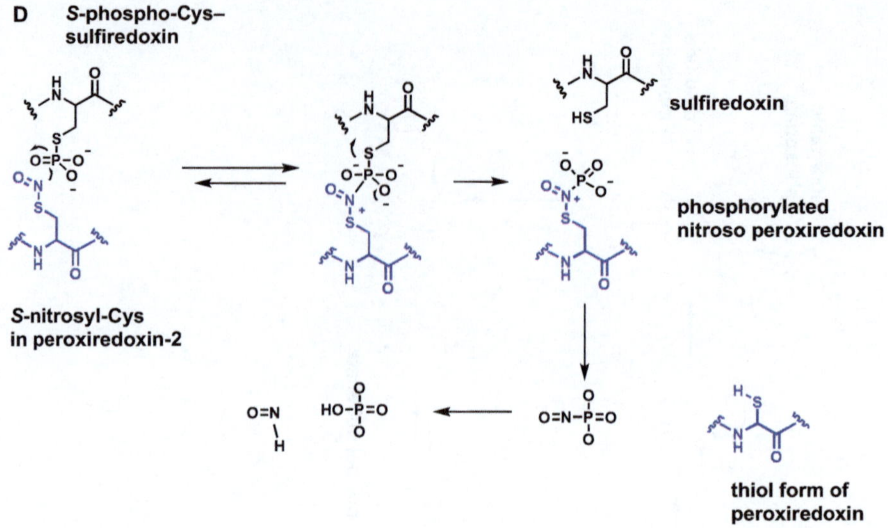

Figure 13.7 Peroxiredoxin catalytic cycle and resuscitation after over-oxidation. (A) Peroxiredoxin normal catalytic cycle proceeds *via* sulfenic acid intermediate. The oxidized form of peroxiredoxins is re-reduced by thioredoxin and thioredoxin reductase. (B) Peroxiredoxins can occasionally be over-oxidized from the sulfenic acid to the doubly oxygenated sulfinic acid. Re-reduction consumes ATP and four thiol equivalents *via* action of sulfiredoxin. (C) The catalytic mechanism for sulfiredoxin or peroxiredoxin resuscitation involves three unusual covalent intermediates: *S*–phosphocysteinyl–*S*–sulfiredoxin, the sulfinyl–phosphoric anhydride in peroxiredoxin and the disulfide monoxide form of peroxiredoxin (D) *S*- phosphorylation of nitrosylated peroxiredoxin to effect denitrosylation of the cysteinyl residue.

HNO and return to the peroxiredoxin thiol state (Figure 13.7D). In parallel thioredoxin-1 has been regarded as a protein denitrosylase, in that case presumably *via* a nucleophilic Cys thiolate as the NO transfer agent.

13.3 Protein Machinery for Disulfide Formation in Oxidizing Compartments

In contrast to the great bulk of cytoplasmic, nuclear and mitochondrial proteins that have cysteine thiols in the reduced thiol forms, proteins that enter the secretory pathways of cells begin to undergo oxidative conversion to disulfides.[32] Glutathione as a small tripeptide has some diffusional access to the endoplasmic reticular lumen but gets diluted out both by oxidation to the disulfide and by mass action as proteins move further into distal regions of the endoplasmic reticulum. They track to the *cis*- and then *trans*-Golgi apparatus, to secretory vesicles and granules, to plasma membrane localization and/or to secretion into extracellular media and matrix.

As depicted in Figure 13.6, there are three routes by which thiols of transiting proteins get oxidized to disulfides in secretory pathways (Figure 13.6A). One is *via* the sulfenate route that we have noted previously. This occurs to drain metabolically-generated HOOH from the endoplasmic reticulum (ER).[17,33]

Perhaps more important is the second route involving multiple forms of protein disulfide isomerases (PDIs). There are some 15 members of the PDIs in the secretory pathway, many in the ER.[33] It is thought that some engage proteins cotranslationally as they are extruded across the ER membrane by attached ribosomes. In aggregate, PDIs reach millimolar concentrations in the ER compartment, enough to act as effective disulfide-oxidizing and disulfide-isomerizing catalysts. The structure of the multifunction protein disulfide oxidase, dithiol oxidase 57 kilodalton ER protein ERp57[34,35] is shown in Figure 13.8A.

There are two thioredoxin-like domains (designated as regions a and a′) with a pair of cysteines that can shuttle between dithiol and disulfide oxidation states.[35] Regions b and b′ can act as chaperone domains for binding hydrophobic regions of nascent proteins extruded into the ER. As shown in Figure 13.8B, PDIs in the oxidized, disulfide state are initial electrophilic partners for reaction with client partner thiolate nucleophiles. Mixed disulfide formation is then on the way to the reduced PDI and the initial disulfide regioisomer of the oxidized protein (Figure 13.8B).

PDIs can also act as disulfide isomerases. Initial disulfide bond regioisomers may lock nascently oxidizing proteins extruded into the ER in non-native conformations. PDIs can then react, in their reduced, dithiol states, again to generate mixed disulfides as intermediates. Presumably, if this process is iterated enough times, the native, thermodynamically favored disulfide oxidation state of oxidized proteins will emerge. This process can be lengthy, especially in proteins with multiple disulfide bonds (see Figure 13.9). Iterative folding cycles may be partly determined by folding propensities of secondary and then tertiary structural elements in proteins as they fold to a condensed state. The availability of PDIs in both dithiol and disulfide oxidation states and chaperone proteins, at appropriate concentrations, are required to drive bimolecular events.

A necessary part of the protein disulfide isomerase folding cycles, with nascent and maturing proteins in the ER and Golgi complexes, is the ability to convert dithiol forms of PDIs back to the oxidized, disulfide entities. To avoid so much build up of PDI dithiol forms that the corresponding oxidation of client proteins dithiols to disulfides does not grind to a halt, there must be one or more ER-resident PDI oxidases. That is the role of the ER oxidase ERO in yeast and ERO1α- and β-isoforms in mammalian cell.[36]

The molecular logic of ERO1 action utilizes the pair of redox elements FAD and an active site disulfide in analogy to the glutathione reductase active site, in particular the use of a 4a-thio–FAD adduct to transfer electrons between organic coenzyme and active site disulfide (Figure 13.6B). In the ERO1 cases, the dihydro $FADH_2$ is reoxidized by O_2, generating H_2O_2 as product. In

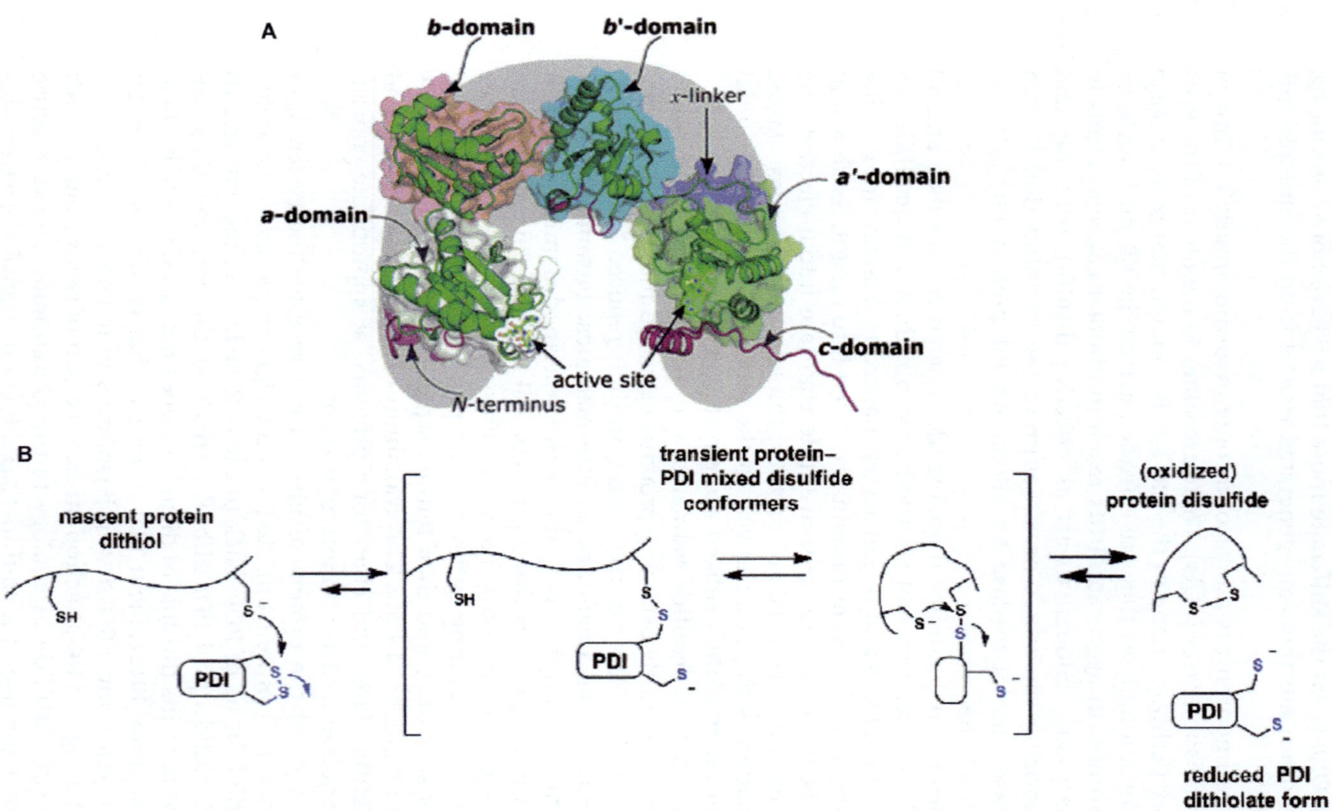

Figure 13.8 Protein disulfide isomerases (PDIs). (A) The 57 kDa endoplasmic reticular protein, ERp57, is one of the several PDIs in the ER initial compartment of the cellular secretory pathway. It has two thioredoxin-like domains (a and a′, each with a dithiol–disulfide pair) connected by linkers to two presumptive chaperone/protein binding domains (b and b′), (B) Schematic of interaction of the predominant disulfide form of a PDI with the dithiol form of a newly imported ER protein, showing disulfide formation through a mixed PDI–S–S–client protein disulfide intermediate.
(A) Reproduced from ref. 60, https://doi.org/10.1073/pnas.1618985114, with permission from National Academy of Sciences, under the terms of a CC BY 4.0 license, https://creativecommons.org/licenses/by/4.0/.

turn, this hydrogen peroxide load accelerates the nonenzymatic route 1, *via* sulfenic acid intermediates.

Studies on ERO1 and several PDI proteins demonstrate that ERO1 reduces PDI disulfides. but not the disulfides of emerging nascent client-protein disulfides. This is a protein–protein recognition optimization to funnel the electrons from reduced forms of PDIs out to O_2.

The third line of Figure 13.6A reveals that there is a small family of protein dithiol oxidases, QSOXs, resident in the ER and other secretory pathway sites, that *do* oxidize client protein dithiols directly.[37,38] The proteins are related to and named for a quiescin protein overproduced in lung tissue under calm conditions. One of the family members was fully characterized as a flavoprotein dithiol oxidase.[37] Again, its redox inventory is a bound FAD and an active site disulfide, furthering the centrality of this redox pair as a general device for dithiol–disulfide redox molecular logic (Figure 13.6B).

Whereas Ero1 oxidizes PDI dithiol forms but not emergent proteins with new S–S bonds, QSOXs work in the same direction but with altered protein dithiol specificity. They oxidize dithiol forms of proteins transiting the secretory pathway, but do not oxidize PDI dithiol forms. Catalytic turnover studies indicated that the quiescin dithiol oxidase could introduce 1000 disulfide bonds per minute into peptides corresponding to ER proteins.

One of the intriguing features of QSOX as well as some other flavoenzymes dealing with thiol redox metabolism is the presence not only of an active site disulfide but an additional peripheral disulfide.[22] The peripheral disulfide is often the docking site for interaction with client protein dithiols. As depicted in Figure 13.6A line 3, this gives QSOX three different forms of enzyme that has been reduced by two electrons. The electrons can sit in the peripheral dithio pair (EH_2, form 1), they can sit in the reduced $FADH_2$ (EH_2 form 2) or in the active site dithiol pair (E to EH_2, form 3).

These three EH2 forms are interconvertible, some *via* thiol–flavin adducts. The addition of the second, peripheral dithiol–disulfide pair extends the reach of this molecular logic. For example, it is the architecture used by bacterial mercuric ion reductases that use NADH to reduce FAD and then the two sets of thiols–disulfides to chelate and reduce Hg^{2+} to Hg^0 that volatilizes from the local microenvironment around the bacteria.[39]

These two O_2-reducing oxidases, ERO 1 and QSOX, run two parallel pathways of maintaining a balance of PDI buffering capacity as nascent proteins are oxidized to disulfides as they fold or refold in the endoplasmic reticular lumen. The directional arrows in Figure 13.6B show that Ero1 pushes the equilibrium from left to right when PDI-dithiols are the substrates. QSOX pushes the equilibrium in the same direction as nascent dithiol proteins (but not reduced forms of PDIs) are oxidized. GSSG reductase pushes the equilibrium from right to left and is not an oxidase. Instead of dumping electrons into O_2, it receives two electrons from NADPH. These observations illustrate the elegant chemical versatility of the FAD/dithiol–disulfide pairing as redox elements that can be run in either direction.

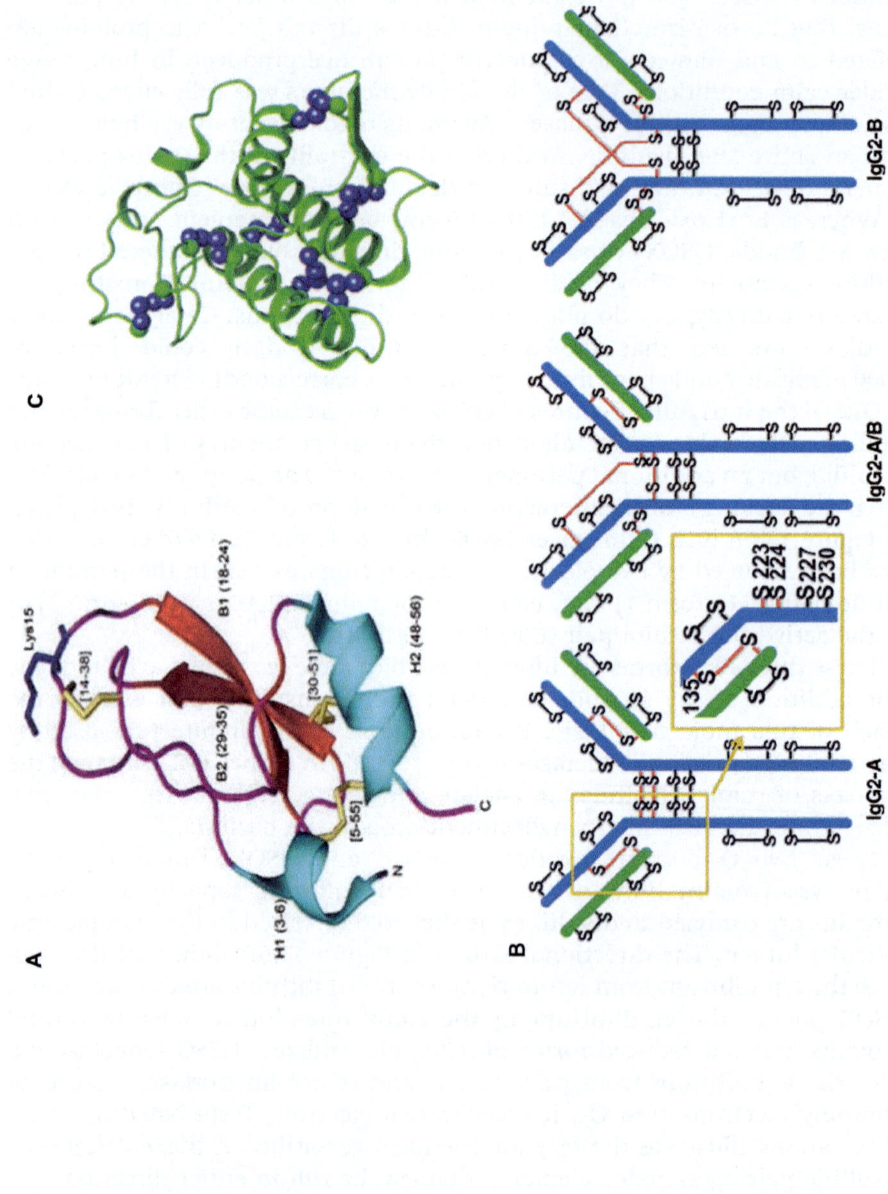
A
Lys15
[14-38]
B1 (18-24)
B2 (29-35)
[30-51]
H2 (48-56)
[5-55]
H1 (3-6)
N
C
B
IgG2-A
IgG2-A/B
IgG2-B
135
S 223
S 224
S227
S230
C

13.4 Examples of Protein Cys–S–S–Cys Disulfide Bonds

An article as long as all these chapters could justifiably be written on the occurrence and functions of disulfide bonds: enabling posttranslational modifications in the oxidative folding of proteins in transit through the endoplasmic reticulum of cells, misfolding *vs.* rearrangement of disulfide links and the many recurring disulfide-containing repeat motifs of proteins such as antibodies, LDL receptor and EGF receptor domains. The interested reader might start with a recent review by Fass and Thorpe.[18] Some 40% of mammalian proteins are predicted to be secreted or sent to the plasma membrane of cells and most of these have conformations stabilized by disulfide cross links, so disulfides may well be the most abundant modification type in the secreted and membranous cysteine proteome.[40]

Small protein toxins, such as conotoxins secreted from cone snails, the disintegrin proteins from snake venoms and the antimicrobial defensin peptides typically have one or more cystine disulfides that stabilize compact conformations required for biological activities.

Small proteins that function as protease inhibitors such as bovine pancreatic trypsin inhibitor also have architectures imposed by disulfide bonds. Figure 13.9A shows that the 57 BPTI residue protein has three Cys–Cys disulfides connecting residues 5 and 55, 14 and 38, and 30 and 51. Note that although Cys51 and Cys55 are close in primary structure they are not in disulfide connection in the native folded BPTI structure.

The vast repertoire of human antibodies contain multiple disulfide bridges within the light chain and the heavy chain and between the chains, creating S–S linkages covalently connecting the two light (L) and two heavy (H) chains in the heterotetramer L_2H_2 structures (Figure 13.9B) that circulate in oxidizing extracellular milieus.[41] The disulfide patterns are nuanced and distinct between the two distinct forms of the IgG2 antibody class shown. Figure 13.9B shows four intra-subunit disulfide and four inter-subunit disulfides in the heavy chains. Correspondingly, there are two intra-light chain disulfides and one inter-chain disulfide linking each light chain to a heavy chain, near the

Figure 13.9 Cystine disulfide bonds in proteins. (A) Three intrachain disulfide bonds stabilize the folded architecture of the BPTI (bovine pancreatic trypsin inhibitor), a small secreted protein that complexes with the protease trypsin and inactivates it by forming a tight long-lived complex. (B) Antibodies, secreted by B cells have heavy and light chains in an L_2H_2 heterotetramer that are long lived in extracellular solutions.[62] The extensive network of intrachain and interchain disulfide bonds stabilize the antibodies in folded, native conformations. (C) Structure of 14 kDa Ca^{++}-phospholipase A2. The eight disulfide bonds are colored in blue and anchor the helical core of the secreted enzyme.

variable, specificity-conferring ends. Clearly the disulfide code is the chief feature determining the structure of the extracellular mature antibodies.

The Ca^{++}-dependent secreted phospholipase A2 family, small (approximately14 kDa) proteins of about 130 residues, can have up to eight cystine disulfide bonds connecting the helical core of these phospholipid hydrolases (Figure 13.9C) to impose architecture.[42] At the other end of size and complexity, the extracellular glycoprotein mucins that provide a protective barrier between the gastrointestinal (GI) tract microbes and the outer GI layer of epithelial cells can have up to 200 cysteines that are oxidized in disulfide linkages to provide the structural barrier features.[18]

In archaea, CXXC tetrapeptide sequences are common,[43] probably reflecting adjacent dithiolate metal-binding sites, perhaps indicating an evolutionary selection for those motifs that occurred under anaerobic conditions. These may have been precursors of disulfides once pO_2 levels arose, replacing a superacid metal such as Zn^{2+} by a covalent S–S bond.

An example of the interconversion of protein cysteine thiols to disulfides as a response to environmental conditions is observed in the transcription factor PecS of the plant pathogenic bacterium *Pectobacterium atrosepticum.* PecS in the bacterial cytoplasm is ordinarily a non-crosslinked dimer that acts as transcriptional repressor of certain regulons of genes.[44] As the bacteria invade a plant apoplast, the plant responds with a burst of reactive oxygen species as defense.

The reactive oxygen species oxidize the side chain of Cys45 in one of the dimer subunits to a sulfenate that is captured by the Cys45′ in the other dimer subunit to form a disulfide bond. This new S–S bond deforms the dimer conformation sufficiently to lower affinity for its target DNA. Now, it can be displaced by the bacterial RNA polymerase to transcribe previously repressed genes. This leads to enhanced expression of virulence genes for the bacteria to react to a suddenly hostile microenvironment.

In this case the Cys45 thiolate has acted as a redox sensor. The subsequent formation of the inter subunit dimer causes a conformational change that alters affinity for its target.[44] This is a paradigm for regulatory function for the dithiol–disulfide interconversion in other cases and may have been a driver in evolution if zinc-based dithiol organization became replaced by disulfides. In Section 13.6.2 we will note similar cysteine residue *S*-oxygenation, in that case dioxygenation to the sulfinate, as a molecular oxygen sensor.

13.5 Nitrosylation of the Cysteinyl Proteome

In Chapter 5 we discussed the origin of NO from five-electron oxidation of the terminal nitrogen of arginine by a family of nitric oxide synthases. We also noted that glutathione at *circa* two to three orders of magnitude higher concentrations than even the most abundant cellular proteins, was likely to be an efficient trap for excess NO, forming *S*-nitroso glutathione (GSNO). GSNO, once formed, can be reductively cleaved to hydroxylamine or can act on its own as a soluble nitrosylating reagent. In that role it can participate both in transfer of NO to cysteinyl–thiol side chains in proteins and in the

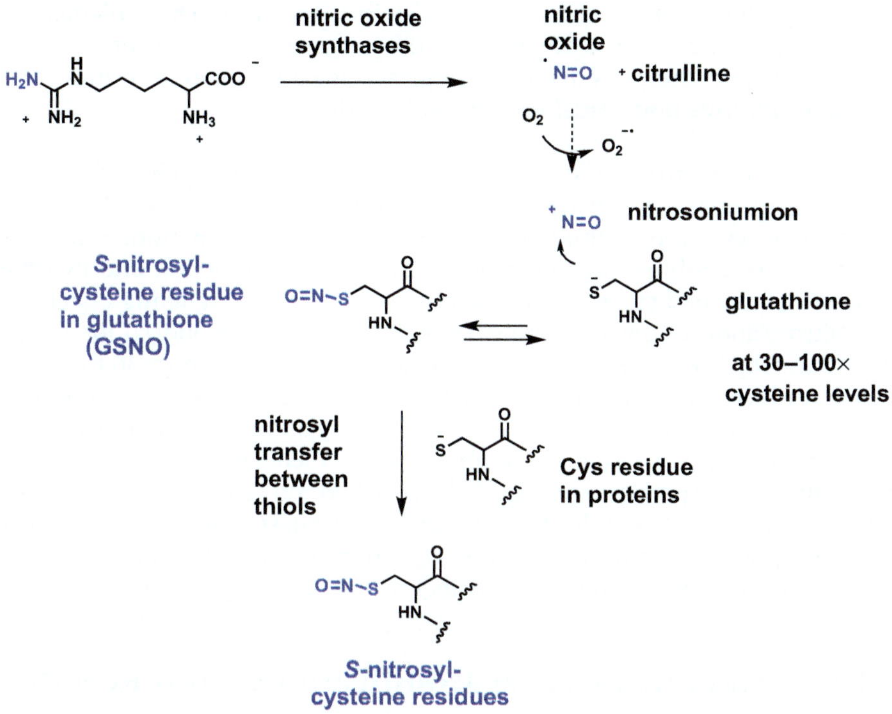

Figure 13.10 The nitric oxide radical, produced by several isoforms of tissue nitric oxide synthases is used as a short-lived signaling molecule, *e.g.* in smooth muscle relaxation. As a radical species, $NO^{\bullet}$ can be captured by glutathione, presumably with most flux through prior oxidation of NO to the nitrosonium ion. GSNO (glutathione-*S*-nitrosyl) can be a vehicle for removal of NO by enzymatic reduction to GSH hydroxylamine or it can serve as a NO donor to cysteine side chains (*via* sulfur radicals?). In the back direction free GSH can also remove NO groups from nitrosylated cysteines.

back direction Thus, GSH can remove nitrosyl groups from cysteinyl residues (Figure 13.10). Analogously, thioredoxin can undergo *S*-nitrosylation and also act as a nitrosyl transfer agent.

A variety of effects of *S*-nitrosyl-cysteine formation are observed, in target proteins affecting either structure and/or function, raising the question of whether protein cysteinyl side chain nitrosylations should be considered purposeful or stochastic regulation.[45] Some 177 proteins undergoing cysteine-*S*-nitrosylation were tabulated in 2013,[46] with a follow up analysis that detected more than 1000 *S*-nitrosylated cysteine residues in almost 650 proteins from mouse tissues. In their 2013 review Gould and coworkers[46] identified 30 proteins from mouse heart that are *S*-nitrosylated. They also noted:

> 'We propose that, in cellular proteomes, distinct and separate clusters of proteins that share cysteine residues within similar microenvironments

show preferential reactivity toward specific modifiers. The biological selectivity is then a consequence of the enzymatic process or chemical modifier and the presence of strategically located cysteine residues that guide selective post-translational modifications.'

In one sense nitrosylation has been termed the prototypic redox-based signaling mechanism[45] with a set of protein machinery devoted to redox changes on cysteine residues that affect protein conformations and functions. The reversibility, by glutathione, by sulfiredoxin action and by other routes is presumed to be important for reversal of any signaling effects.

S-Nitrosylation of one cysteine side chain in the mitochondrial long chain acyl-CoA dehydrogenase improved catalytic efficiency by almost 30-fold, indicating an increase in metabolic capacity of this step. In contrast, the human brain serine racemase is inhibited by *S*-nitrosylation. This PLP-dependent enzyme racemizes L-serine to D-serine which then serves as a co-agonist for *N*-methyl-D-aspartate receptors[47,48] enriched on glial cell membranes. *S*-Nitrosylation of Cys13 in human serine racemase shuts down catalytic activity[48] (Figure 13.10). The *S*-nitrosylation proteome continues to be an evolving and dynamic arena for physiological and pathophysiological functions.

13.6 Cysteinyl Residue Modifications from Reactive Oxygen Species

13.6.1 Protein Sulfenates and Sulfenamides

Figure 13.4B notes four types of oxidative modifications to cysteinyl residue side chains. In addition to formation of disulfides and *S*-nitroso adducts, there are sulfenates and persulfides. The S–OH sulfenate functional groups have been noted several times in the earlier chapters, most often as intermediates on the way to disulfides. If neither reduced glutathione nor internal cysteine residues are in position to capture the sulfenate, the sulfenates may persist.[49] That is the case for protein phosphatases such as PTPase 1B (Cys215), MAP kinase, the lipid phosphatidyl inositol triphosphate (PI3) kinase, and also the PTPase Pten (Cys431), examples of cysteine side chain oxygenations that abrogate activity.[16]

Salmeen *et al.*[50] noted that sulfenate formation in PTPase 1B then led to intermolecular reaction with the immediate downstream peptide N–H bond to form a cyclic sulfenamide (Figure 13.11) that is reversible. One interpretation is that the sulfenamide is a reversible protecting group to keep the sulfenate from over-oxidation to the inactive sulfinate residue.

Perhaps the best studied case of a specific functional effect of conversion of a cysteinyl residue to a sulfenic acid group are the effects of H_2O_2 on the intracellular portion of the transmembrane epidermal growth factor receptor (EGFR).[51] On ligation of EGF and relatives to the external domain of EGFR, there is an increase in tyrosine residue autophosphorylation in the

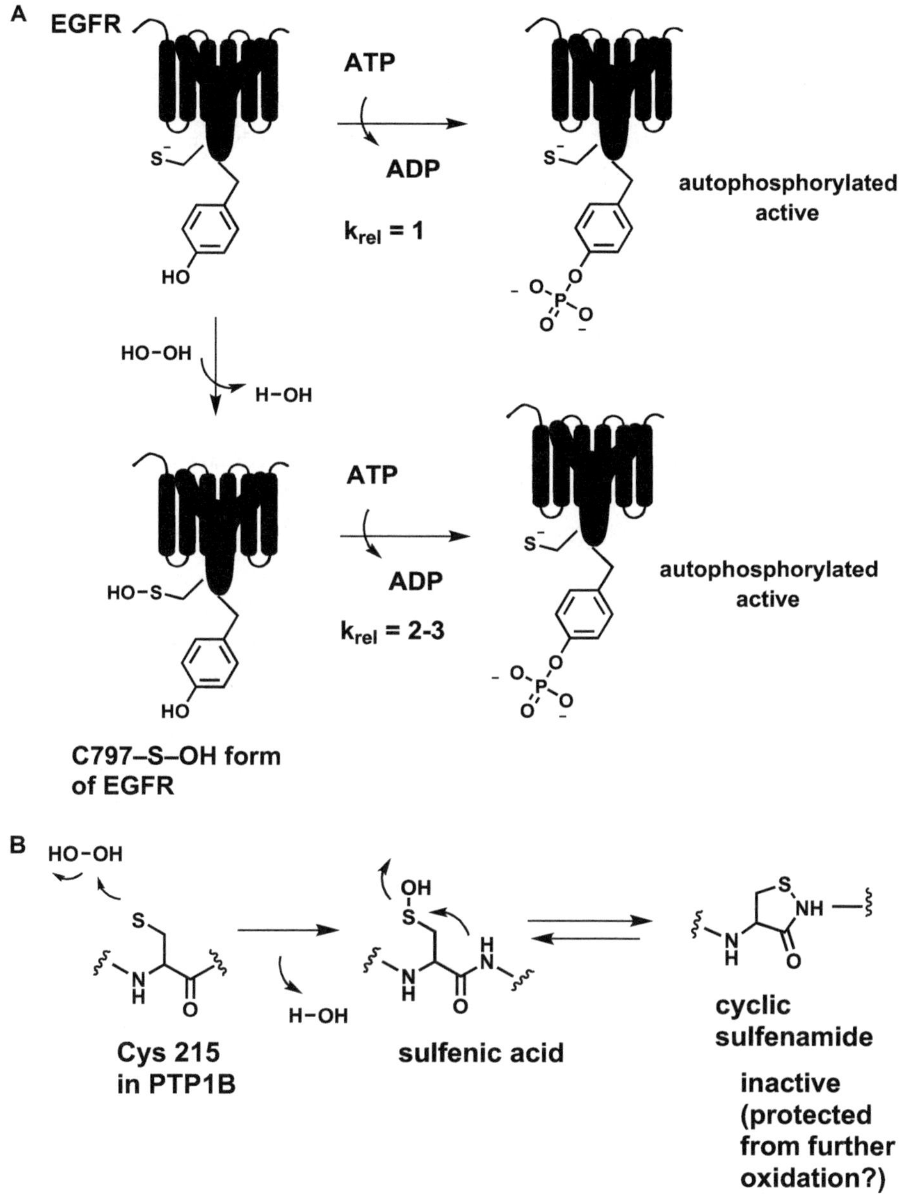

Figure 13.11 Sulfenate formation in proteins. (A) EGFR undergoes H_2O_2-mediated oxygenation of Cys797 specifically at the edge of the ATP binding site of the EGFR-intracellular tyrosine kinase domain. The resultant sulfenic acid is twofold to threefold more active for the approximately 12 minutes that the sulfenate survives reduction back to starting EGFR. (B) Sulfenylation of the catalytic Cys 215 in the protein tyrosine phosphatase PTP1B leads to reversible formation of the cyclic sulfenamide.

internal tyrosine kinase domain and consequently in the phosphoryl transfer to tyrosine residues on partner proteins.

One of the indications of rise in intracellular [H_2O_2] as a signaling agent is the increase in oxidation rate of some cysteinyl residues to sulfenates. Kate Carroll's research group has determined that EGFR Cys797 at the junction of the ATP binding site is rapidly oxidized to the sulfenic acid.[51] It turns out the pK_a of that Cys–SH for dissociation to the thiolate is 5.5, a thousand-fold lower than the pK_a for free cysteine residues. Furthermore, they observed that the Cys797–SOH group resulted in a twofold to threefold increase in EGFR autokinase and *trans* kinase activity. On average the half life of the sulfenate form of EGFR is about 12 minutes after an H_2O_2 burst.[51] Presumably the cysteine sulfenic acid residue then is captured by way of a disulfide intermediate.

13.6.2 Dioxygenation of Cys_2 Residues in Certain Proteins Is a Signal for Sequential Arginylation, Polyubiqitinylation and Rapid Degradation During [O_2] Sensing

A restricted but intriguing case of purposeful cysteine residue irreversible dioxygenation occurs in hypoxia sensing pathways. We have discussed the nonheme iron cysteine dioxygnase (CDO) family previously in both Chapter 7 and Chapter 12. In Chapter 7 we noted the unusual catalytic capacity to carry out *S*-dioxygenation of the thiolate side chain of free cysteine molecules as a detoxification strategy against high levels of free thiolate metabolite and also as the first step in taurine biosynthesis. In Chapter 12 we observed that cysteine dioxygenases generate an active site thioether cross link between a tyrosyl aromatic ring and a Cys side chain thiol, probably by radical pathways without affecting catalytic efficiency in subsequent cycles.

In the context here, we note the ability of certain CDOs in plants and related aminoethanethiol (cysteamine) dioxygenases (ADOs) in humans to serve as rapid [O_2] sensors in hypoxic responses.[52,53] The canonical well-studied hypoxic response logic and machinery in humans involves hydroxylation of a pair of prolyl (and asparaginyl) residues in the transcription factor hypoxia inducible factor 1 alpha (HIF1α).[54,55] This is a protein-based titration of [pO_2] levels in cells *via* Km values for O_2 for the responsible protein monooxygenases. In the absence of sufficient [O_2], hydroxylation decreases, the deoxygenated HIF1α transcription factor subunit is rapidly ubiquitinylated by a ubiquitin E3 ligase (Chapter 12) and sent to the proteasomes for proteolytic degradation.

The absence of HIF1α now allows derepression and expression of many genes that comprise the hypoxic response. Thus, this is an [O_2] sensor strategy that is protein-based, involves transcription factor degradation and will require new protein synthesis to run the hypoxic response, with the attendant delay for the time to make new mRNA and proteins.

The CDO–ADO-mediated oxygen sensing response uses parallel logic in that [pO_2] levels are titrated as Km values for O_2 for cysteine residue dioxygenases.[53] In these cases, the substrate to be deoxygenated is not free cysteine but rather N-terminal cysteine residues and the *S*-dioxygenation is irreversible, targeting the protein for proteolytic degradation.

The actual process has four sequential steps (Figure 13.12). (1) Mammalian proteins start with N-terminal methionyl residues. These are often removed cotranslationally as they emerge from the ribosome by methionyl aminopeptidases. (2) Now, what had been Cys_2 residues have become N-terminal and are recognized by plant CDOs or human ADOs by the mechanism for iron-based *S*-dioxygenation outlined in Chapter 7. (3). The *S*-dioxygenated N-terminal Cys-residue-containing proteins are then recognized as substrates for N-terminal arginylation by an arginyl transferase. (4). At that point the doubly modified *N*-arginylated-Cys_2–sulfinate–protein is recognized as a client by a specific E3 ligase that polyubiquitinates the modified protein. This leads to its transport to the proteasomes and rapid degradation of that client protein.

Among the substrate Cys_2-proteins in humans are regulator of G protein signaling 4 and 5 (RGS4/5) and the cytokine interleukin 32 (IL32). The RGS proteins interact with a series of G-proteins and decrease the half-lives of GTP-bound states by accelerating hydrolysis. In turn this sets in motion a variety of hypoxia responses. Because this Cys-N-terminal degron pathway does not require a transcriptional response for new protein synthesis, it is more rapid than the canonical HIF1α machinery and the two distinct time regimes may complement each other.

It is worth passing note that this Cys N-terminal degron pathway involves and requires four sequential posttranslational modifications of potential client proteins for them to act as O_2 sensors: $methionyl_1$ removal, Cys_2 thiolate dioxygenation; Cys sulfinate N-terminal arginylation; and polyubiquitination by an E3 ligase. The actual O_2 sensing step is the CDO–ADO dioxygenation catalyst.

13.7 Cysteine Persulfide Residues

The fourth covalent type of cysteine side chain oxidative conversion in proteomes is formation of Cys–S–SH persulfides. The distribution of persulfides is likely to be more restricted than the other three oxidative modifications of cysteine residues. On the other hand persulfide side chains are found in glyceraldehyde 3-P dehydrogenase; other enzymes that undergo sulfenylation are also subject to activating effects have been reported.[56,57] Additional Cys–SSH side chains have been detected in protein disulfide isomerases, nuclear factor kappa B (NFkB), heat shock protein (Hsp) 70 and several other proteins.[58]

Proteins that carry persulfide residues are thought to function predominantly in pathways to form Fe–S clusters in target proteins. These include the partner proteins to the cysteine lyases that generate the first S–SH link in the lyase active sites, as well as any downstream chaperone proteins that may intervene on the way to the several dozen proteins that acquire

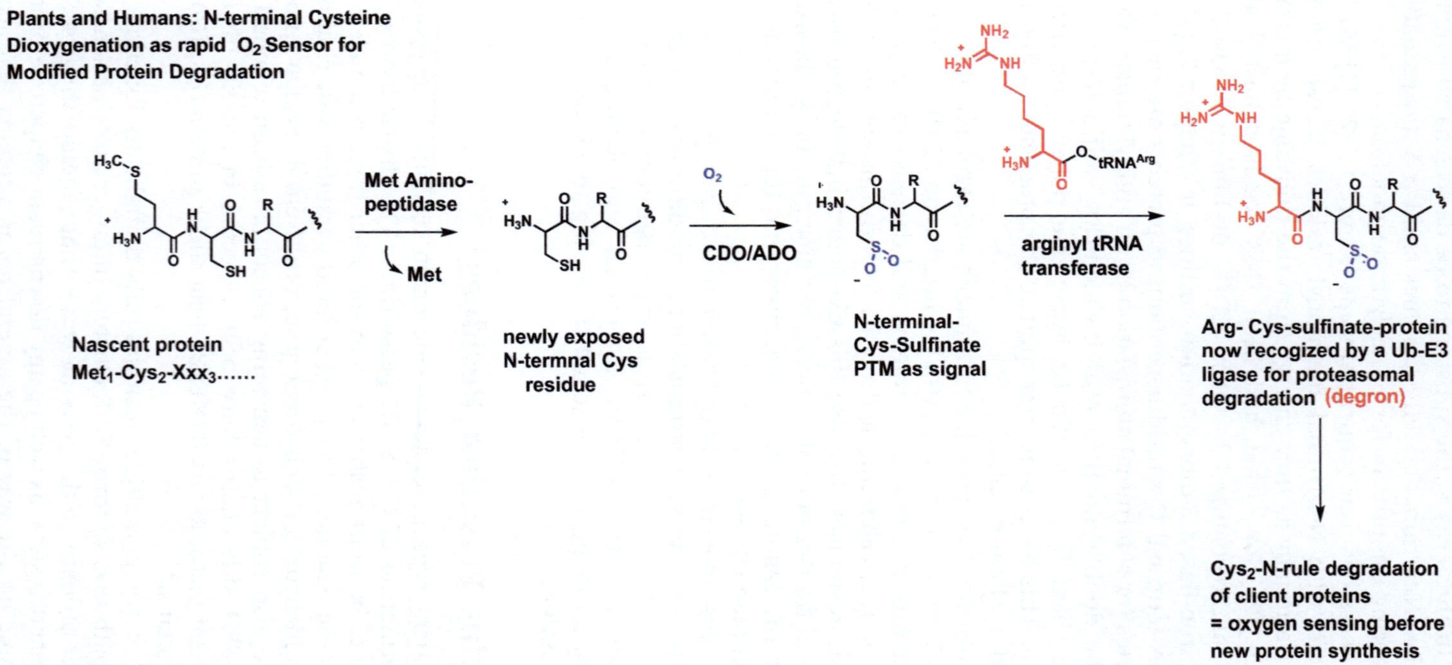

Figure 13.12 Four steps of posttranslational oxygenative processing of Cys_2 in protein *N*-degrons: (1) proteolytic removal of Met_1; (2) *S*-dioxygenation of the new N-terminal Cys side chain thiol by O_2; (3) N-terminal arginylation by a Cys-sulfinate-recognizing arginyl transferase; (4) polyubiquitinylation by an E3 ligase sends the modified client proteins for proteasomal degradation.

2Fe–2S and 4Fe–4S clusters. The mechanism and reactivity of persulfide formation and transfer have been extensively analyzed in Chapter 6.

13.8 Methionine Residue *S*-Oxidations and Reversibility

As noted in Figure 13.1, along with autoxidation of the cysteinyl thiol side chain of protein residues to sulfenic acids, the thioether sulfur atom in methionyl residues is also subject to autoxidative modifications. Two stages of oxidation can occur. The first generates methionine sulfoxide groups and the second is conversion of the divalent thioether of a methionyl side chain from tetravalent sulfoxide to hexavalent sulfone oxidation state at sulfur (Figure 13.13A).

For decades it was known that such methionyl residue oxidations in proteins could occur nonenzymatically and were associated with aging pathology and loss of function of some of the proteins thus autoxidized. More recently (see reviews ref. 55 and 56), an enzyme dedicated to methionyl residue oxidation in actin filaments was discovered and shown to play a role in actin polymer dynamics *via* oxygenation of two Met residues to the R-sulfoxides. We take up the nonenzymatic aspects first.

13.8.1 Nonenzymatic Methionyl Residue *S*-Oxidation

The two diastereomeric methionyl sulfoxide forms can occur by two-electron mechanisms, *e.g.* by oxidative interaction with hydrogen peroxide or perhaps by one-electron oxidative routes. A nonenzymatic route showing reduction of H_2O_2 (to one H_2O product molecule) as a methionyl side chain thioether sulfur acts as nucleophile and is oxidized by two electrons to the sulfoxide is shown in Figure 13.13A. A subsequent further two-electron oxidation of sulfoxide to sulfone can occur and that turns out to be irreversible and not repairable reductively in cells.

Nonenzymatic formations of methionyl sulfoxides generate both the *R*- and *S*-isomers since there is no chirality control. There are surveillance enzymes for reducing each of the *R*-sulfoxide and *S*-sulfoxide back to the starting methionyl residue oxidation states. Defects in the reductases correlate with aging, indicating that accumulating methionyl sulfoxide (and even more so the accumulation of sulfones) marks lead to some of the proteins being dysfunctional. The *R*-specific methionyl sulfoxide reductase (Msr) is the more extensively studied of the two (*R*-specific *vs.* *S*-specific) catalysts. It uses a pair of active site cysteines in the reductase catalytic cycle. In fact, the underlying logic of these Msr enzymes show the confluence of proteomic cysteine thiol oxidation and methionine thioether oxidation.

A catalytic cycle begins with one of the active site cysteine residues attacking the electrophilic sulfur of the bound methionine sulfoxide residue in a protein substrate. That attack yields an S–S bond in what is a disulfide-*S*-

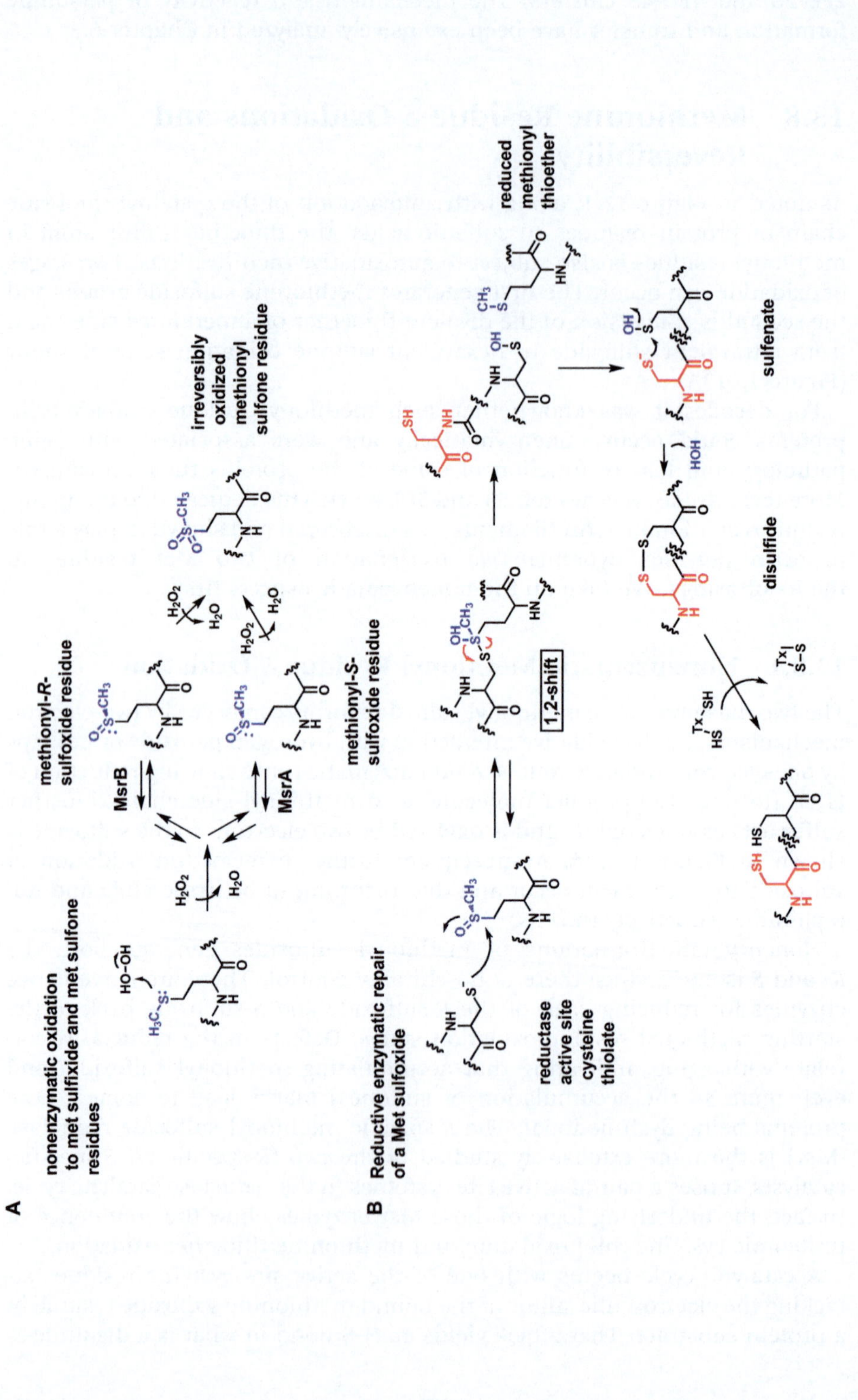
A
nonenzymatic oxidation to met sulfixide and met sulfone residues
methionyl-R-sulfoxide residue
MsrB
MsrA
methionyl-S-sulfoxide residue
irreversibly oxidized methionyl sulfone residue
B
Reductive enzymatic repair of a Met sulfoxide
reductase active site cysteine thiolate
1,2-shift
reduced methionyl thioether
sulfenate
disulfide

oxide intermediate as shown in Figure 13.13B. Now, the most unusual mechanistic facet as proposed, a 1,3-OH shift reaction. The S–S bond breaks as the OH group moves, with its pair of electrons from the methionyl sulfur to the cysteinyl-enzyme's sulfur atom. The result is regeneration of the methionyl residue thioether in the protein cosubstrate that, now that it has been repaired, can depart the reductase active site. Meanwhile the reductase has a sulfenate form of its active site cysteine. This undergoes attack by a second cysteine thiolate residue, in the anticipated expulsion of water and formation of the reductase active site Cys–Cys disulfide.

The methionine sulfoxide has been reduced by two electrons while the enzyme has been oxidized by two electrons. Before the reductase can initiate a next catalytic cycle, its active site disulfide must be reduced by an external reductant. That is the protein thioredoxin. So, the overall stoichiometry is oxidation of the dithiol form of the redox protein thioredoxin to the disulfide as protein methionine sulfoxide residues are reduced. NADPH-dependent thioredoxin reductase flavoenzymes in turn regenerate the dithiol form of thioredoxins.

In contrast to reductive rescue of the methionine sulfoxide side chain, the further oxidation to methionyl sulfone is biologically irreversible. To the extent that methionyl sulfoxide intermediates are successfully intercepted by these sulfoxide reductases, the probability of progress on to the irreversible sulfone oxidation state is lowered.

13.8.2 Methionine Residue Enzymatic Sulfidation in Actin

Actin is abundant, dynamic and central to the biology of muscle and non-muscle cells, where its assembly from globular (G) actin monomers to polymerized actin filaments (F-actin) occurs. The F-actin and G-actin states create sets of interactomes that have distinct mechanical and signaling properties in many cell types. The control of actin polymerization and depolymerization dynamics are imposed at several levels, from partner protein binding and enzymatic activity to a host of posttranslational, reversible modifications, including lysine acylations, SUMOylations, ubiquitinylations of lysine side chains, arginylation of glutamyl residues and, most intriguingly, enzyme-mediated chiral sulfoxidation of methionines at positions 44 and 47.

Figure 13.13 (A) Autoxidation of methionyl thioether side chains in proteins can occur nonenzymatically to generate either the *R*- or *S*-sulfoxide stereochemistry. Further oxidation yields the sulfone irreversibly. Enzymes exist (MsrA and MsrB) to effect reduction of either the *S*- or the *R*-methionyl sulfoxides and return the proteins to their original functionalities. (B) The mechanism of Msr involves attack by a Cys residue thiolate in the reductase active site to form a series of intermediates, including a *mixed Cys–Met disulfide monoxide* that is proposed to undergo *a 1,2 shift with fragmentation* to methionine and the cysteine sulfenate that is ultimately reduced by disulfide formation and reduction at the expense of NADPH and thioredoxin.

We eschew discussion of the other (non-sulfur) actin posttranslational modifications (PTMs) here to focus on the first example of methionyl sulfoxidation for controlled functional purposes on a specific protein target.

A dedicated flavoprotein monooxygenase with both catalytic and noncatalytic domains, known as molecule interacting with CasL (MICAL), specifically oxygenates two methionine residues to the corresponding sulfoxide residues. These are a purposeful complement to the nonenzymatic oxidations of proteins noted above (Figure 13.13A) that have been thought to be adventitious. In contrast, the *S*-oxygenation (Figure 13.14) of these two residues in actin is a purposeful, major contributing input to accelerated depolymerization, a net severing of the F-actin filaments. This is particularly relevant during cytokinesis in cells.[56] Specifically, the *S*-oxygenation of Met44 and/or Met47 is thought to interfere with actin monomer interactions that lead to homopolymerization of the G-actin subunits. Thus, polymerization is slowed and net disassembly or filament severing occurs.

Consistent with the reversibility of most signaling posttranslational modifications of proteins, the sulfoxidation of these two residues is specifically counteracted by the MsrB enzyme that recues the *R*-isomers of methionyl sulfoxide residues but not the *S*-isomers. This susceptibility defines the stereochemistry of MICAL as an enzyme creating only the *R*-sulfoxide diastereomer at the thioether sulfur of methionyl residues 44 and 47 in the F-actin substrate. F-actin–G-actin dynamics are important in axonal guidance in neurons and also in hippocampal mossy fiber connections, to name just two attributes of MICAL–MsrB-mediated regulation in the nervous system.[55] Figure 13.14C shows that the balance of MICAL and MsrB activity controls the level of actin methionyl sulfoxidation and in turn the depolymerization dynamics of F-actin back to G-actin monomers. This panel also notes that, for MsrB to act in multiple turnovers, a molecule of NADPH is oxidized by thioredoxin reductase to convert the MsrB nascent product disulfide back to the active dithiol form. In the three-enzyme cycle MICAL–MsrB–Trx reductase, two NADPH equivalents are consumed for this PTM cycle to run so it is energetically costly to cells.

The strategy for controlled methionyl residue enzymatic *S*-oxygenation by MICAL is to use readily available cellular NADPH to reduce the tightly bound FAD coenzyme in its active site. The dihydro $FADH_2$ state is oxygen-reactive and probably forms the flavin-4a-peroxy adduct shown in Figure 13.13A. The terminal oxygen of that flavin peroxide is electrophilic and capturable by the methionine-44 and -47 thioether sulfur atoms acting as nucleophiles. The initial result is cleavage of the weak peroxy O–O bond, transfer of one oxygen atom to the sulfur and leaving one oxygen atom from O_2 as the flavin–hydroxy adduct. That decomposes to H_2O and reoxidized FAD. Meanwhile, the sulfoxide residue can be written either as the S^+–O^- contributor or the more familiar S=O rendering of the *R*-sulfoxide diastereomer. The charged resonance contributor emphasizes the partial negative charge on oxygen as one of the disrupting elements of these two methionine PTMs. It also draws some formal

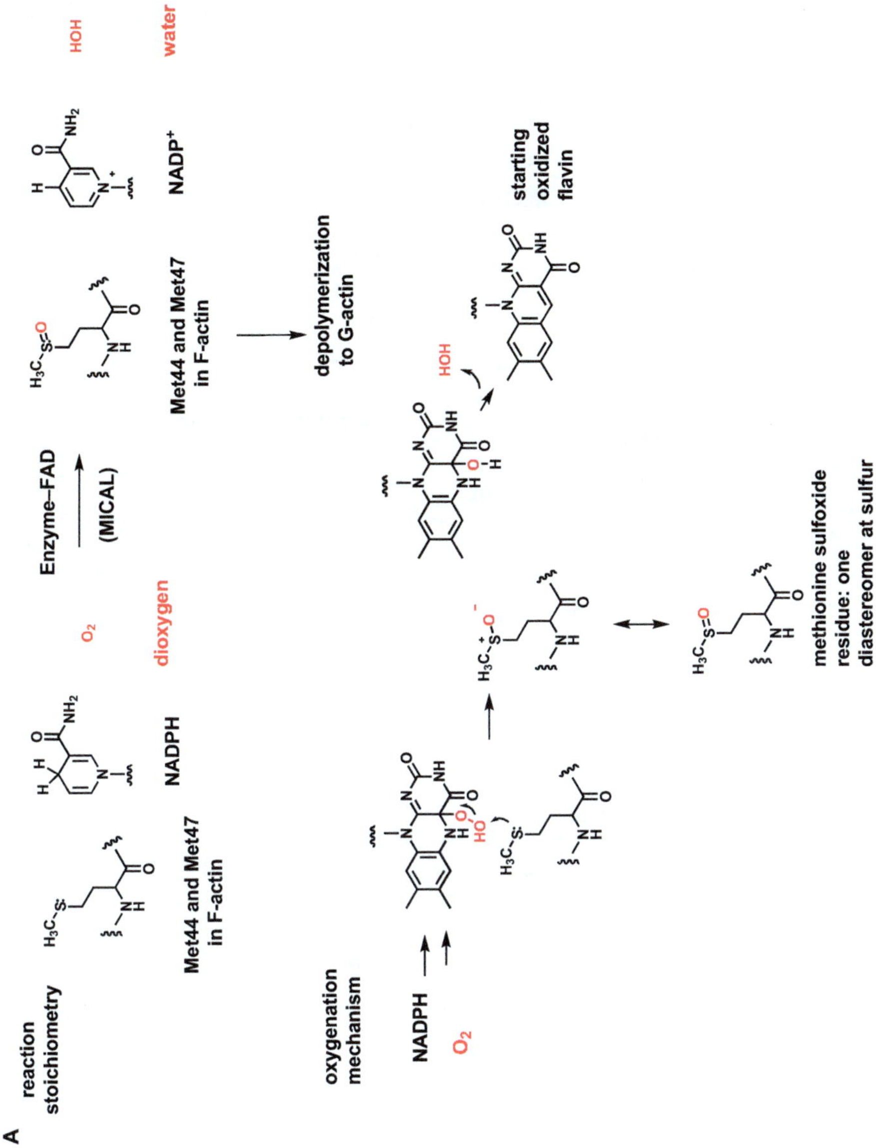

Figure 13.14 Continued on next page.

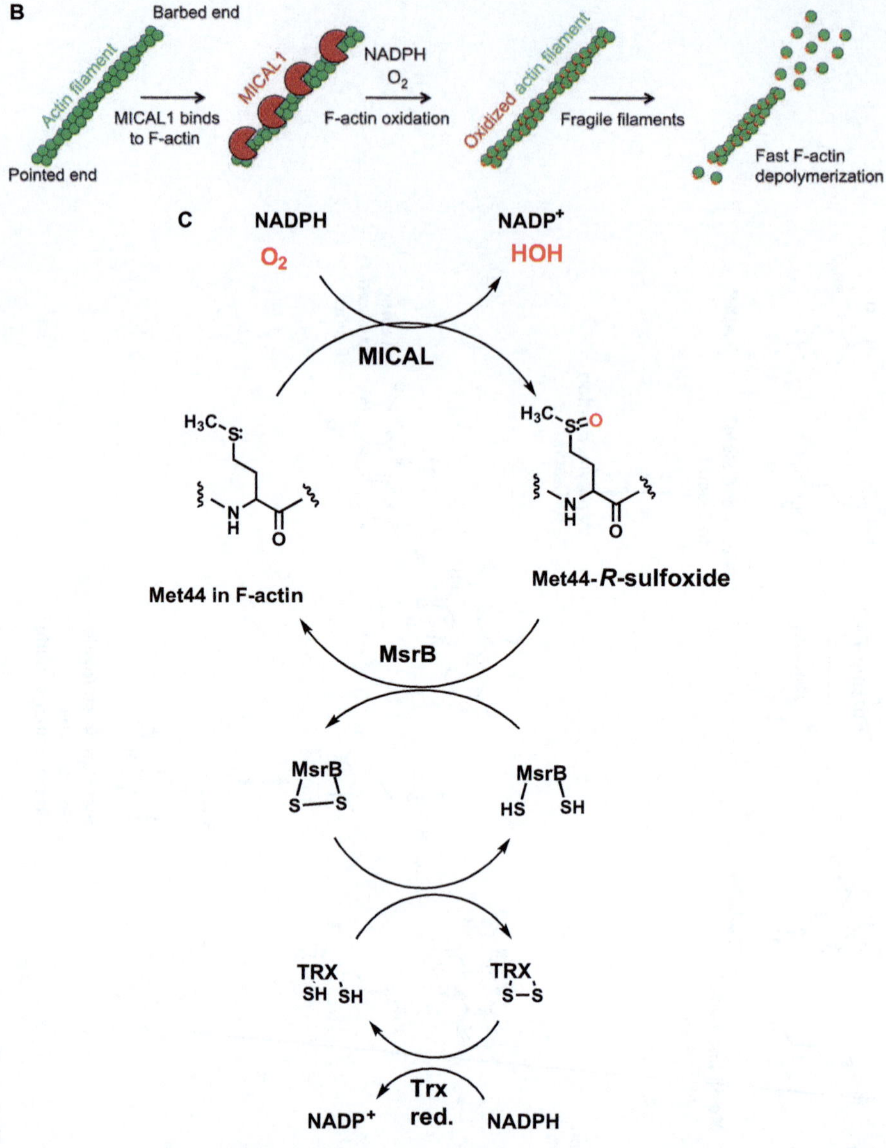

Figure 13.14 (A) Enzymatic oxygenation of Met44 and Met47 thioether side chains in actin to the *R*-sulfoxides by the FAD-containing monooxygenase MICAL. The covalent FAD-4a–OOH is proposed as the proximal donor of electrophilic oxygen to the thioether sulfur atom. (B) Schematic of acceleration of F-actin depolymerization by MICAL-mediated *S*-oxygenation of F-actin Met44and Met47 to generate a destabilized protein polymer. (C) A regulatory posttranslational sulfoxidation cycle: Actin chiral sulfoxidation by MICAL is opposed or reversed by MsrB. In turn MsrB is reduced by thioredoxin and thioredoxin reductase. Two NADPH molecules are spent: one to drive each direction. (B) Reproduced from ref. 64 with permission from The Company of Biologists, Copyright 2017.

comparison to the trigonal sulfonium ion in *S*-adenosylmethionine (where the S=C forms are not significantly populated).

To date, there are two examples of that thioether sulfur in methionine acting as a biologically relevant nucleophile. The first occurs on free methionine in the SAM synthetase reaction, that creates *S*-adenosylmethionine, with sulfur acting as a nucleophile on an sp^3 carbon atom of ATP. The MICAL enzyme provides the second catalyzed example. The enzyme acts not on free methionine, but on two methionyl residue side chains in actin, dictated by the protein–protein specificity of MICAL for its actin substrate. This is sulfur acting as a nucleophile towards bound flavin–OOH. It may well be that additional examples of enzymatic and/or reversible methionyl residue sulfoxidations for regulatory purposes remain to be discovered as a more generalized post-translational modification strategy.

References

1. A. J. Bannister and T. Kouzarides, Regulation of chromatin by histone modifications, *Cell Res.*, 2011, **21**, 381–395.
2. M. Blanc, F. David, L. Abrami, D. Migiozzi and F. Armand, Swiss Palm: Protein palmitoylation data base, *F1000Res*, 2015, **4**, 261.
3. M. R. Linder and R. Deschennes, Palmitoylation:policing protein stability and traffic, *Nat. Rev. Mol. Cell Biol.*, 2007, **8**, 74–84.
4. C. Aicart-Ramos, R. Valero and I. Rodriguez-Crespo, Protein palmitoylation and subcellular trafficking, *Biochim. Biophys. Acta*, 2011, **1808**, 2981–2994.
5. S. Eisenberg, A. Laude, A. J. Beckett, C. J. Mageean, V. Aran, M. Hernandes-Valladares, Y. I. Henis and I. A. Prior, The role of palmitoylation in regulating Ras localization and trafficking, *Biochem. Soc. Trans.*, 2013, **41**, 79–83.
6. E. Conibear and N. Davis, Palmitoylation and depalmitoylation at a glance, *J. Cell Sci.*, 2010, **23**, 4007–4010.
7. V. S. Naumenko and E. Ponimaskin, Palmitoylation as a functional regulator of neurotransmitter receptors, *Neural Plast.*, 2018, **2018**, 5701348.
8. J. L. Osterhout, A. Waheed, A. Hiol, R. J. Ward, P. C. Davey, L. Nini, J. Wang, G. Milligan, T. L. Z. Jones and K. M. Druey, Palmitoylation Regulates Regulator of G-protein Signaling RGS16 Function, *J. Biol. Chem.*, 2003, **278**, 19309–19316.
9. M. J. Blanden, K. Suzako, E. R. Hildebrandt, D. S. Hardgrove, M. Patel, W. P. Saunders, M. D. DiStefano, W. K. Schmidt and J. L. Houghland, Efficient farnesylation of an extended C(x3)X sequence motif expands the scope of the prenylated proteome, *J. Biol. Chem.*, 2017, **293**(8), 2770–2785.
10. M. Wang and P. Casey, Protein prenylation:unique fats make their mark on biology, *Nat. Rev. Cell Biol.*, 2016, **17**, 110–122.
11. E. Liu, F. Wolfreys, J. R. Muppidi, Y. Xu and J. G. Cyster, S-Geranylgeranyl-L-glutathione is a ligand for human B cell confinement receptor, *Nature*, 2019, **568**, 245–248.

12. D. K. Simanshu, D. Nissely and F. McCormick, Ras proteins and their regulators in human disease, *Cell*, 2017, **170**, 17–33.
13. S. Maurer-Stroh, M. Koranda, W. Benetka, G. Schneider, F. L. Sirota and F. Eisenhaber, Towards complete sets of farnesylated and geranylgeranylated proteins, *PLoS Comput. Biol.*, 2007, **3**, e66.
14. W. R. Tschantz, J. A. Digits, H.-J. Pyun, R. M. Coates and P. J. Casey, Lysosomal prenylcysteine lyase is an FAD-dependent thioether oxidase, *J. Biol. Chem.*, 2001, **276**, 2321–2324.
15. W. R. Tschantz, L. Zhang and P. J. Casey, Cloning, expression,and celllar localization of prenylcysteine lyase, *J. Biol. Chem.*, 1999, **274**, 35802–35808.
16. C. Klomsiri, P. Karplus and L. B. Poole, Cysteine-based redox switches in enzymes, *Antioxid. Redox Signaling*, 2011, **14**, 1065–1077.
17. N. J. Bulleid, Disulfide bond formation in the mammalian endoplasmic reticulum, *Cold Spring Harbor Perspect. Biol.*, 2012, **4**, a013219.
18. D. Fass and C. Thorpe, Chemstry and enzymology of disulfide cross-linking in proteins, *Chem. Rev.*, 2018, 1169–1198.
19. R. C. Fahey, J. Hunt and G. C. Windham, On the cysteine and cystine content of proteins. Differences between intracellular and extracellular proteins, *J. Mol. Evol.*, 1977, **10**, 155–160.
20. M. L. Conte and K. Carroll, The redox biochemistry of protein sulfenylation and sulfinylation, *J. Biol. Chem.*, 2013, **288**, 26480–26488.
21. J. Yang, K. Carroll and D. C. Liebler, The expanding landscape of the thiol redox proteome, *Mol. Cell. Proteomics*, 2016, **15**, 1–11.
22. P. T. Chirvers, K. Prehoda and R. T. Raines, The CXXC motif: a rheostat in the active site, *Biochemistry*, 1997, **36**, 4061–4066.
23. D. S. Berkholz, H. Faber, S. N. Savvides and P. A. Karplus, Cataytic cycle of human glutathione reductase near 1 Agstrom resolution, *J. Mol. Biol.*, 2008, **382**, 371–384.
24. E. Pai and G. Schulz, The catalytic mechanism of glutathione reductase as derived from x-ray diffraction analyses of reaction intermediates, *J. Biol. Chem.*, 1983, **258**, 1752–1757.
25. A. S. Harkansson and A. Smith, Enzymatic characterization of dihydrolipoamide dehydrogenase from *Streptococcus pneumoniae* harboring its own substrate, *J. Biol. Chem.*, 2007, **282**, 29521–29530.
26. H. A. Woo, W. Jeong, T. S. Chang, K. J. Park, S. J. Prk, J. S. Yang and S. G. Rhee, Reduction of cysteine sulfinic acid by sulfiredoxin is specific to 2-cys-peroxiredoxins, *J. Biol. Chem.*, 2005, **280**, 3125–3128.
27. K. G. Reddie and K. Carroll, Expanding the functional diversity of proteins through cysteine oxidation, *Curr. Opin. Chem. Biol.*, 2008, **12**, 746–754.
28. S. G. Rhee and H. Woo, Multiple functions of preoxiredoxins: peroxidases, sensors, and regulators of the intracellular messenger H_2O_2 and protein chaperones, *Antioxid. Redox Signaling*, 2010, **15**, 781–794.
29. T. J. Jonsson, M. Murray, L. C. Johnson and W. T. Lowther, Reduction of cysteine sulfinic acid in peroxiredoxin by sulfiredoxin proceeds through

a sulfinic phosphoric ester intermediate, *J. Biol. Chem.*, 2008, **283**, 23846–23851.
30. X. Roussel, A. Kriznik, C. Richard, S. Rahuel-Clermont and G. Branlant, Catalytic mechanism of sulfiredoxin from *Saccharomyces cerevisiae* passes through an oxidized disulfide sulfiredoxin intermediate that is reduced by thioredoxin, *J. Biol. Chem.*, 2009, **284**, 33048–33054.
31. C. R. Sunrico, A. Sultan, T. Nakamura, N. Dolatabadi, J. Parker, B. Shan, X. Han, J. R. Yates III, E. Masliah, R. Ambasudhan, N. Nakanishi and S. A. Lipton, Role of sulfiredoxin as a peroxiredoxin-2 denitrosylase in human iPSC-derived dopaminergic neurons, *Proc. Natl. Acad. Sci. U. S. A.*, 2016, **113**, E7564–E7571.
32. P. Kosuri, J. Aegre-Cebollada, J. Feng, J. M. Sanchez-Ruiz, A. Holmgren and J. M. Fernandez, Protein folding drives disulfide formation, *Cell*, 2012, **151**, 794–806.
33. L. Ellgaard and L. Ruddock, The human protein disulfide isomerase family: interactions and fuctional properties, *EMBO Rep.*, 2005, **6**, 28–32.
34. E.-M. Frickel, P. Frie, M. Bouveier, W. P. Stafford, A. Helenius, R. Glockshuber and L. Ellgaard, Rp57 is a multifunctional thiol-disulfide oxidoreductase, *J. Biol. Chem.*, 2004, **279**, 18277–18287.
35. G. Kozlov, P. Maattanen, J. G. Schrag, S. Pollock, M. Cygler, B. Nagar, D. Y. Thomas and K. Gehring, Crystal structure of the bb' domans of the protein disulfide isomerase ERp57, *Structure*, 2006, **14**, 1331–1339.
36. E. Zito, ERO1: a protein disulfide oxidase and H_2O_2 producer, *Free Radicals Biol. Med.*, 2015, **83**, 299–304.
37. C. Thorpe, K. Hoober, S. Raje, N. M. Glynn, J. Burnside, G. K. Turi and D. L. Coppock, Sulfhydryl oxidases: emerging catalysts of protein disulfide bond formation in eukaryotes, *Arch. Biochem. Biophys.*, 2002, **405**, 1–12.
38. E. J. Heckler, P. Rancy, V. K. Kodali and C. Thorpe, Generating disulfides with the quiescin sulfhydryl oxidases, *Biochim. Biophys. Acta*, 2008, **1783**, 567–577.
39. B. Fox and C. Walsh, Mercuric reductase: Purification and characterization of a transposon-encoded flavoprotein contaning an oxidation-reduction active disulfide, *J. Biol. Chem.*, 1982, **257**, 2498–2503.
40. J. Chiu and P. Hogg, Allosteric disulfide: sophisticated molecular structures enabling flexible protein regulation, *J. Biol. Chem.*, 2019, **294**, 2949–2960.
41. H. Liu and K. May, Disulfide bond structures of IgGs, *MAbs*, 2012, **4**, 17–23.
42. E. A. Dennis, Diversity of group types, regulation and function of phospholipase A2, *J. Biol. Chem.*, 1994, 13057–13060.
43. D. E. Fomenko and V. Gladyshev, Identity and function of CxxC-derived motifs, *Biochemistry*, 2003, **42**, 11214–11225.
44. D. K. Deochand, A. Pand, J. K. Meariman and A. Grove, Redox sensing by PecS from the plant pathogen *Pectobacterium atrosepticum* and its effect on gene expresssion and the conformation of PecS-bound promoter DNA, *Biochemsitry*, 2019, **58**, 2564–2575.
45. J. S. Stamler, S. Lamas and F. C. Fang, Nitrosylation. The prototypic redox-based signaling mechanism, *Cell*, 2001, **106**, 675–683.

46. N. Gould, P.-T. Doulias, M. Tenopoulou, K. Raju and H. Ischiropoulos, Regulation of protein function and signaling by reversible cysteine-S-nitrosylation, *J. Biol. Chem.*, 2013, **288**, 26437–26479.
47. Z. Yng-Luan, Y. Zhao and H. Mori, Role of D-serine in the mammalian brain, *Brain Nerve*, 2007, 725–730.
48. F. Marchesani, S. Bruno, G. Paredi, S. Raboni, B. Campanini and A. Mozzarelli, Human serine racemase is nitrosylated at multiple sites, *Biochim. Biophys. Acta*, 2018, **1866**, 813–821.
49. V. Gupta and K. Carroll, Sulfenic acid chemistry detection and cellular lifetime, *Biochim. Biophys. Acta*, 2014, **1840**, 847–875.
50. A. Salmeen, J. Anderson, M. P. Myers, T. C. Meng, J. A. Hinks, N. K. Tonks and D. Barford, Redox regulation of protein tyrosine phosphatase 1B involves a sulphenyl-amide intermediate, *Nature*, 2003, **423**, 769–773.
51. T. H. Truong, P. Ung, P. B. Palde, C. E. Paulsen, A. Schlessinger and K. S. Carroll, Molecular basis for redox activation of epidermal growth factor receptor kinase, *Cell Chem. Biol.*, 2016, **23**, 837–848.
52. D. A. Weits, B. Giuntoli, M. Kosmacz, S. Parlanti, H.-M. Hubberten, H. Riegler, R. Hoefgen, P. Perata, J. T. van Dongen and F. Licausi, Plant cysteine oxidases control the oxygen-dependent branch of the N-end-rule pathway, *Nat. Commun.*, 2014, **5**, 3425.
53. N. Masson, T. Keeley, B. Giuntoli, M. D. White, M. L. Puerta, P. Perata, R. J. Hopkinson, E. Flashman, F. Licausi and P. J. Ratcliffe, N-terminal cysteine dioxygenases transduce responses to hypoxia in animals and plants, *Science*, 2019, **365**, 65–69.
54. A. C. R. Epstein, J. M. Gleadle, L. A. McNeill, K. S. Hewitson, J. O'Rourke, D. R. M. Mole, M. Mukherji, E. Metzen, M. I. Wilson, A. Dhanda, Y.-M. Tian, N. Masson, D. L. Hamilton, P. Jaakkola, R. Barstead, T. Hodgkin, P. H. Maxwell, C. W. Pugh, C. J. Schofield and P. J. Ratcliffe, *Cell*, 2001, **107**, 43–54.
55. M. Ivan, K. Kondo, H. Yang, W. Kim, J. Valiando, M. Ohh, A. Salic, J. M. Asara, W. S. Lane and W. G. Kaelin Jr, HIFalpha targeted for VHL-mediated destruction by proline hydroxylation: Implications for O2 sensing, *Science*, 2001, **292**, 464–468.
56. C.-M. Park, L. Weerasinghe, J. J. Day, J. M. Fukuto and M. Xian, Persulfides: current knowledge and challenges in chemistry and chemicalbioloy, *Mol. BioSyst.*, 2015, **11**, 1775–1785.
57. R. K. Yadav, M. Martinove, V. Vitvitsky, J. Seravalli, R. Wedmann, M. R. Filiopvic and R. Banerjee, Biosynthesis and reactivity of cysteine persulfides in signaling, *J. Am. Chem. Soc.*, 2016, **138**, 289–299.
58. E. G. Mueller, Trafficking in persulfides: delivering sulfur in biosynthetic pathways, *Nat. Chem. Biol.*, 2006, **2**, 185–194.
59. L. R. Krauth-Seigel, R. H. Schirmer and S. Ghisla, FAD analogs as prosthetic groups of glutathione reductase: proerties of the modified enzyme species and comparison with the active site structure, *Eur. J. Biochem.*, 1985, **148**, 335–344.

60. R. P. P. Neves, P. A. Fernandes and M. J. Ramos, Mechanistic insights on the reduction of glutathione disulfide by protein disulfide isomerase, *Proc. Natl. Acad. Sci. U. S. A.*, 2017, **114**, E4724–E4733.
61. H. Zeiger, Simulation studies reveal the role disulfide bonds play in protein folding, Phys.org. report 17 August 2015, 2015.
62. L. Jones, H. Zhang, W. Cui, S. Kumar, J. Sperry, J. Carroll and M. Gross, Complementary MS methods assist conformational characterization of antibodies with altered S-S bonding networks, *J. Am. Soc. Mass Spectrom.*, 2013, **24**, 835–845.
63. M. Qin, W. Wang and D. Thirumalai, *Proc. Natl. Acad. Sci. U. S. A.*, 2015, **112**, 11241–11246.
64. S. Frémont, G. Romet-Lemonne, A. Houdusse and A. Echard, *J. Cell Sci.*, 2017, **130**, 1509–1517.

CHAPTER 14

Cysteinyl Proteomes: Exotic Modifications

This third of three chapters on the reactivity of cysteine side chains in proteomes covers a diverse set of reactions. The chemistry can be quite illuminating but generally occurs in more restricted settings than the catalytic and posttranslational modifications described in Chapters 12 and 13.

14.1 Thioethers

Microbial proteomes, notably bacterial and fungal, contain examples of unusual chemistry in the posttranslational modifications of proteins and peptides derived by limited proteolysis of nascent proteins.[1] We do not attempt an encyclopedic coverage here, except to note that cysteinyl radicals are involved in different classes of enzyme catalytic cycles, some noted in Part II, and those radicals can occasionally crosslink to other electron-rich amino acid side chains.[2]

The two examples that occur in hundreds to thousands of peptides derived from microbial protein precursors are thioether formation and heterocyclic thiazole formation (Figures 14.1 and 14.3).

Thioether formation between cysteine side chain thiolates and dehydroalanine residues arising from Ser or dehydrobutyrines from Thr residues create nonhydrolyzable thioether cross links, termed lanthionines or methyllanthionines, respectively.[1,3] The molecules bearing these bifunctional amino acid residue crosslinks are termed lanthipeptides and are typically matured as proteolytic fragments of precursor proteins that have undergone the enzyme-catalyzed cross linkage.[1]

The chemical logic is always to use the cysteine thiolate as an internal sulfur nucleophile to capture the electrophilic group arising from net dehydration of serine or threonine residues. There are subclasses of lanthipeptide synthases

Chemical Biology No. 12
The Chemical Biology of Sulfur
By Christopher T. Walsh

Published by the Royal Society of Chemistry, www.rsc.org

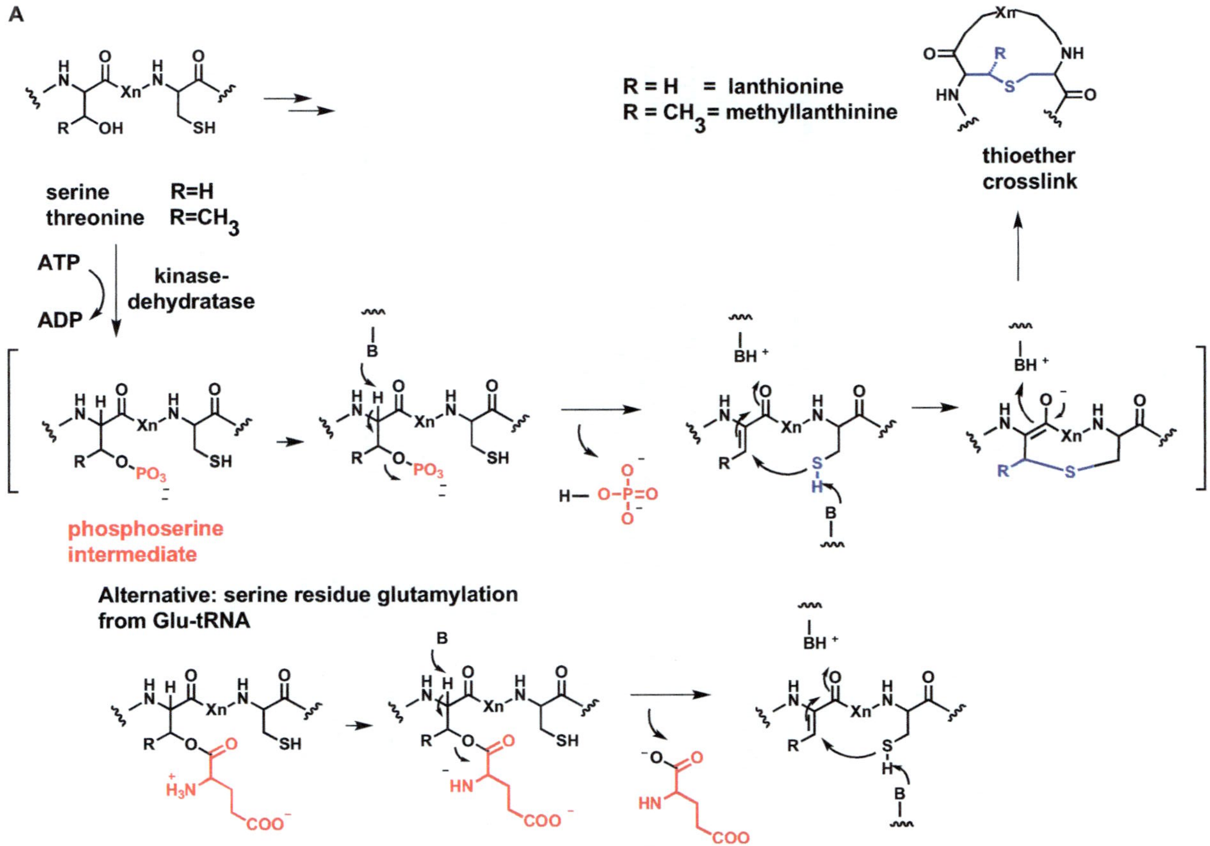

Figure 14.1 Posttranslational lanthionine thioether formation. (A) Formation of lanthionine and methyllanthionine residues by activation of the side chain –OH by phosphorylation (or glutamylation) and dehydration to dehydroalanine and dehydrobutyrine residues. Subsequent capture by cysteine side chain thiolates generates the thioether linkages. (B) Five post-translationally introduced thioether linkages in the lanthipeptide antibiotic nisin. The Ala–S–Ala residues originate as Cys and Ser residues while Ala–S–Abu residues originate as Cys–Thr residues. Note that directionality of some thioether linkages run in either N to C or C to N-directions *vis a vis* the peptide chain.

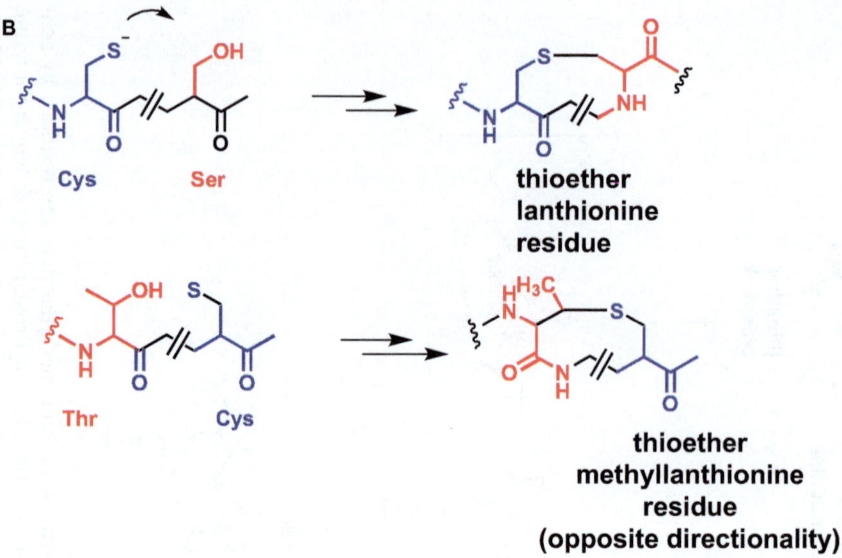

Figure 14.1 Continued.

that have distinct evolutionary histories, but all need to convert the $-C_3$–OH group of particular Ser and Thr residues to leaving groups, downstream of an information-conferring leader peptide. Sometimes the ser–OH or Thr–OH side chain is phosphorylated (Figure 14.1B), sometimes glutamylated[4] and then subjected to action of dedicated dehydratases to create the dehydroalanine and dehydrobutyrine residues as electrophiles.

The tendency to undergo conjugate addition is sufficient for capture by the strong thiolate nucleophiles, but in general not by water or other, adventitious, oxygen or nitrogen nucleophiles. In certain circumstances then, the dehydroalanine or dehydrobutyrine residues persist because there are no thiolate side chains that can reach them in the ever more restricted space as multiple thioether crosslinks form.

When the phosphorylation route is used, this amounts to a *protein serine kinase* action. Usually, phospohoserine residues are stable until reversed by phosphatase action. In the lanthipeptide case, however, the dehydratase domain of the kinase–dehydratase actually carries out αβ-elimination of the elements of inorganic phosphate across the C_2–C_3 carbons of the phosphoserine residues, to dehydroalanine (or dehydrobutyrine if a phosphothreonine residue were the intermediate). As noted above, those dehydro side chains may persist or may be intermediates on the way to thioether dervatives.[4]

Then, the thioether synthases catalyze attack of particular cysteine side chain thiolates on particular electrophilic dehydroalanine (or dehydrobutyrine) residues to establish definite regiochemistry to the thioether cross bridges. Each thioether introduces rigidity into the peptide scaffold, fixes conformational constraints and can generate high-affinity mature peptide-derived scaffolds for interaction with cellular targets. The food preservative lanthipeptide nisin with five thioether bridges has served as an antibiotic for over 50 years.[4] After thioether bond formation, the adduct is reprotonated at the C2 carbon of the Ser- or Thr-derived residue and typically yields a D-Ala or D-aminobutyryl (Abu) three-carbon or four-carbon chain.

The 34-residue mature nisin structure, schematized in cartoon fashion in Figure 14.1, reflects proteolytic trimming from a preprotein, with five subsequent dehydrations and five thioether bond formations. Of the five residues dehydrated four are threonines (yielding the Abu) and one is a serine residue (yielding the Ala in Figure 14.1B).

The regiochemistry of the thioether formations is dictated by the timing of the net dehydrations and the orientations of cysteine thiolate side chains to the nascent four dehydrobutyrines and one dehydroalanine residue. Note that the four methyl lanthionine thioethers in nisin run retrograde, in a C to N-direction, compared with the N to C direction of the main peptide chain. Those thioethers have resulted from a downstream cysteine thiolate attacking an upstream dehydrobutyrine residue. The final structure is highly knotted and compact and targets the lipid II component of bacterial cell walls with nanomolar affinity.

14.2 Attack by a Cysteine Side Chain on the Immediate Upstream Peptide Carbonyl Group

A small set of newly folded single-chain proteins are capable of self-modification reactions.[5] Among the more celebrated examples are the conversion of the precursor form of green fluorescent proteins to the mature protein with the eponymous green fluorophore that assembles as a three-residue stretch of Ser65–Tyr66–Gly67 rearranges to the conjugated fluorogenic heterocycle.[6]

Here, we note proteins with specific internal cysteine residues that can lead to three alternative fates: to protein chain fragmentation, to thiazole ring formation without peptide backbone cleavage and to replacement of a specific amide carbonyl with a thio carbonyl group. (Figure 14.2). They occur in a limited number of biological contexts but show the remarkable power of a cysteine side chain to attack the upstream peptide carbonyl group.

14.2.1 Autoproteolysis and Capture of a Thioester Intermediate by the Alcohol Group of Cholesterol

One of the most remarkable fates of a secreted protein, known as hedgehog, is its autoproteolysis at a specific internal residue and capture of the acyl portion of the protein fragment by a molecule of cholesterol in the membrane[7,8] (Figure 14.3).

The process starts when a secreted, folded protein molecule at the plasma membrane surface achieves a conformation where an internal cysteine side chain can attack the immediate upstream carbonyl of the preceding peptide

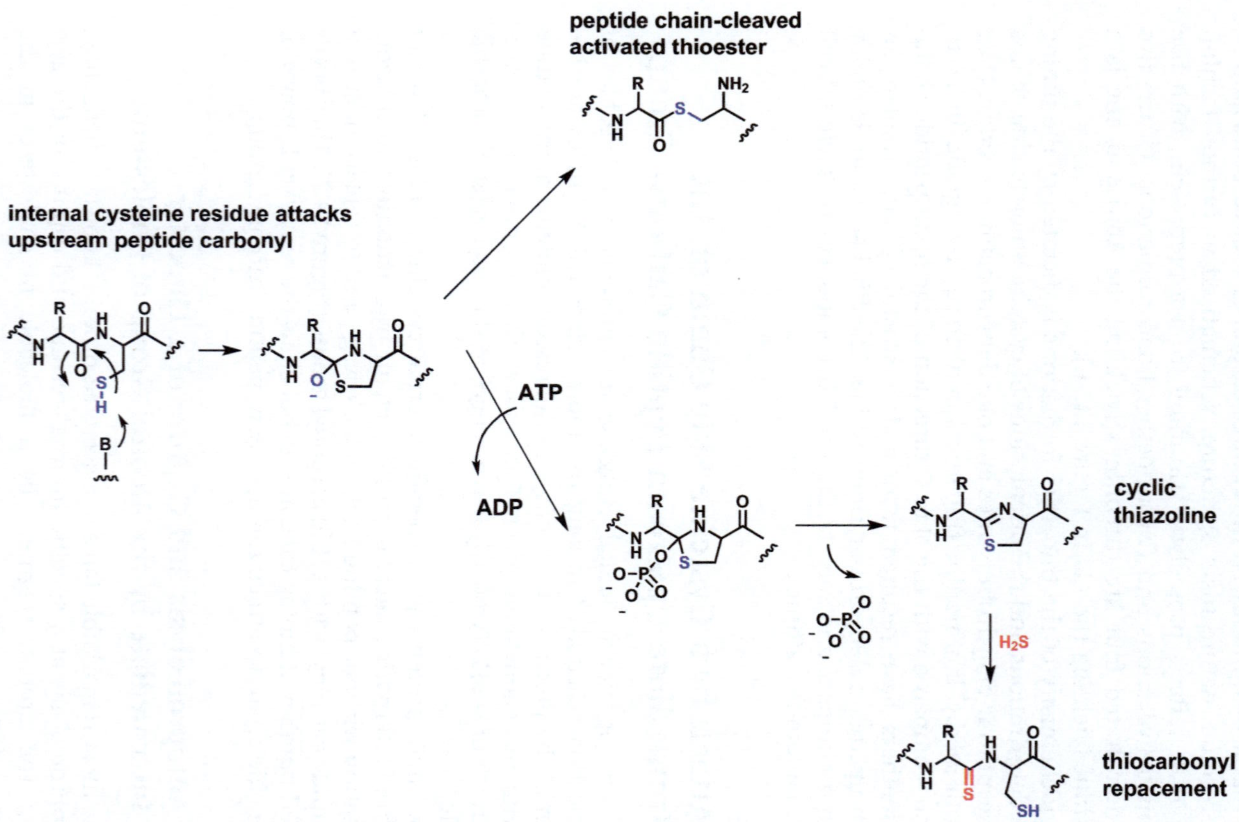

Figure 14.2 Three outcomes from attack of a cysteine thiolate side chain on the immediate upstream peptide carbonyl group. The initial tetrahedral adduct can undergo C–N bond cleavage and yield the acyl-*S*-cysteinyl enzyme intermediate with release of the downstream amine fragment. C–O bond cleavage of the tetrahedral adduct occurs after phosphorylation from ATP. This can lead to the cyclic thiazoline ring or subsequently to the thiocarbonyl group from attack by external H_2S.

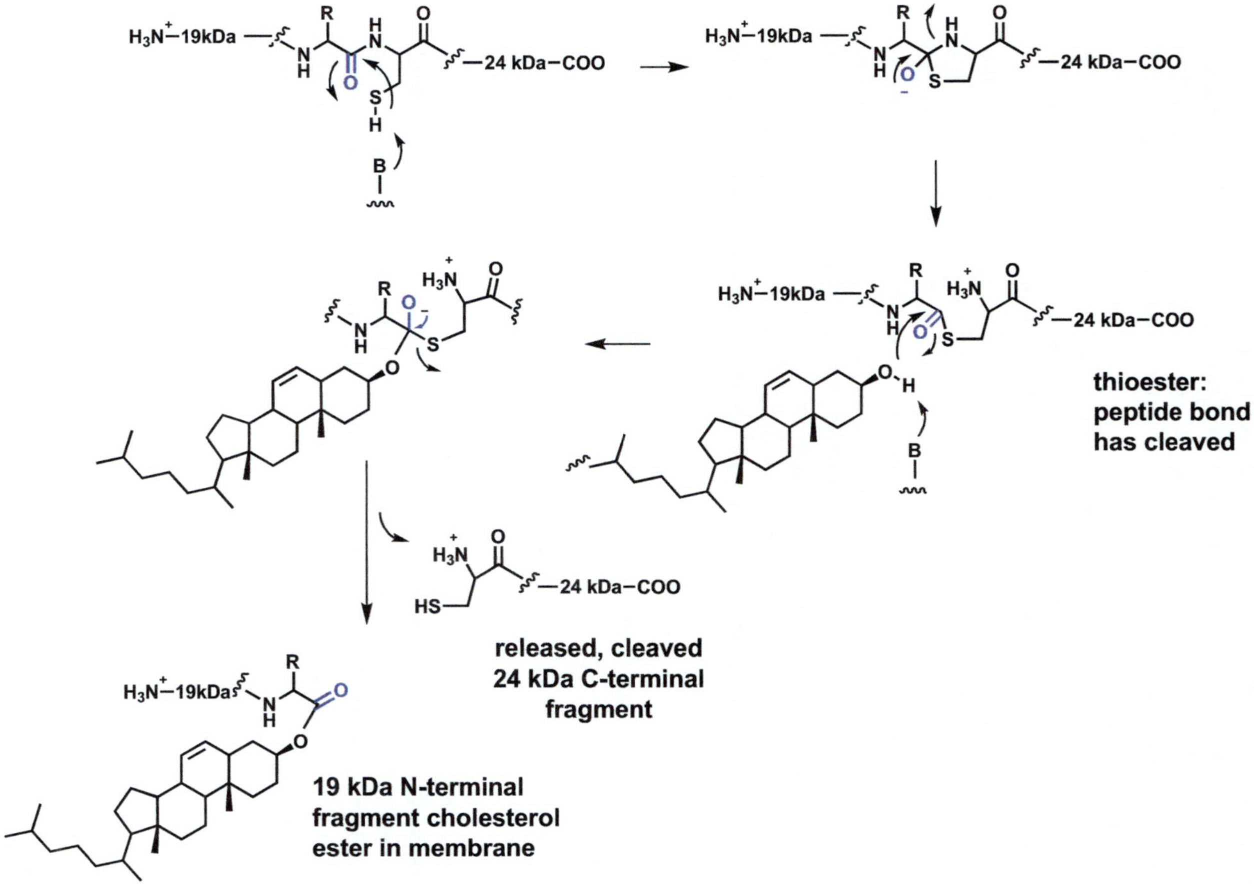

Figure 14.3 Protein automodification at an internal cysteine side chain. Maturation of the membrane protein hedgehog involves peptide bond cleavage to an acyl thioester and specific capture by the 3-β-OH of cholesterol to anchor the N-terminal protein fragment in the membrane as a cholesterol ester.

bond. This forms a transient tetrahedral carbonyl addition adduct. It can reverse to the starting peptide bond. If it goes forward the result is a species where the peptide bond has cleaved and been replaced by a thioester linkage.

Remarkably, the C–N peptide bond that is the primary covalent backbone in proteins has been broken. In its place is a thermodynamically activated acyl thioester. That upstream acyl group is now much more activated for transfer to a cosubstrate nucleophile than when the resonance-stabilized low energy amide bond was intact.

This autoproteolytic process occurs in the maturation of the hedgehog signaling protein.[7] The 45 kDa single chain protein cleaves itself at an internal cysteine residue but in a net *esterolysis* rather than hydrolysis. The attacking cosubstrate nucleophile is the 3-β–OH of a cholesterol molecule embedded in the plasma membrane, not a water molecule. The resultant covalently tethered N-terminal 19 kDa hedgehog fragment is a protein–oxoester bond to C3 of cholesterol The covalent attachment to the cholesterol moiety confines the lipidated protein ester to the plane of the plasma membrane for two dimensional diffusion to its partner proteins in morphogenesis, cell patterning and growth.[8,9]

The oxoester linkage between the 19 kDa protein fragment and the cholesterol is sufficiently stable for the lifetime of the modified hedgehog to function in its signaling roles.

14.2.2 Thiazole and Oxazole Formation

A different set of over 100 microbial proteins are subjected to posttranslational modifications at both cysteinyl and seryl residues that convert side chains and also peptide backbones into five-membered heterocycles, thiazoles and oxazoles, embedded within the peptide structures[5,10–13] (Figure 14.4).

The first step in such enzymatic morphing of peptide backbones is exactly the same as the above autoproteolysis, but the tetrahedral adduct is captured and diverted to heterocycle formation (Figure 14.3). The process again uses the nucleophilicity of both the Cys–SH and Ser–OH side chains to attack the immediate upstream peptide carbonyl to yield the same initial tetrahedral adduct.

The fate is now directed away from C–N bond cleavage instead to C–O bond cleavage by phosphorylation of that tetrahedral oxyanion by a phosphoryl group from ATP (Figure 14.4), lowering the energy barrier for C–O bond cleavage.[14] This arises by action of the tetrahedral oxyanion as a nucleophile on the electrophilic γ-phosphorus of ATP. Elimination of that phosphate by neighboring group participation of the nitrogen electron pair creates the cyclic imines, thiazoline and oxazoline respectively. The figure shows tandem conversion of a serine residue to an oxazoline and a cysteine residue to a thiazoline ring: two five-membered heterocycles that have altered both side chains and peptide backbone.

Finally, a companion flavoenzyme dehydrogenase acts to further desaturate those initial dihydro-heterocycles to the fully heteroaromatic thiazole and oxazole rings[15] (Figure 14.4). These cyclizations have morphed the peptide

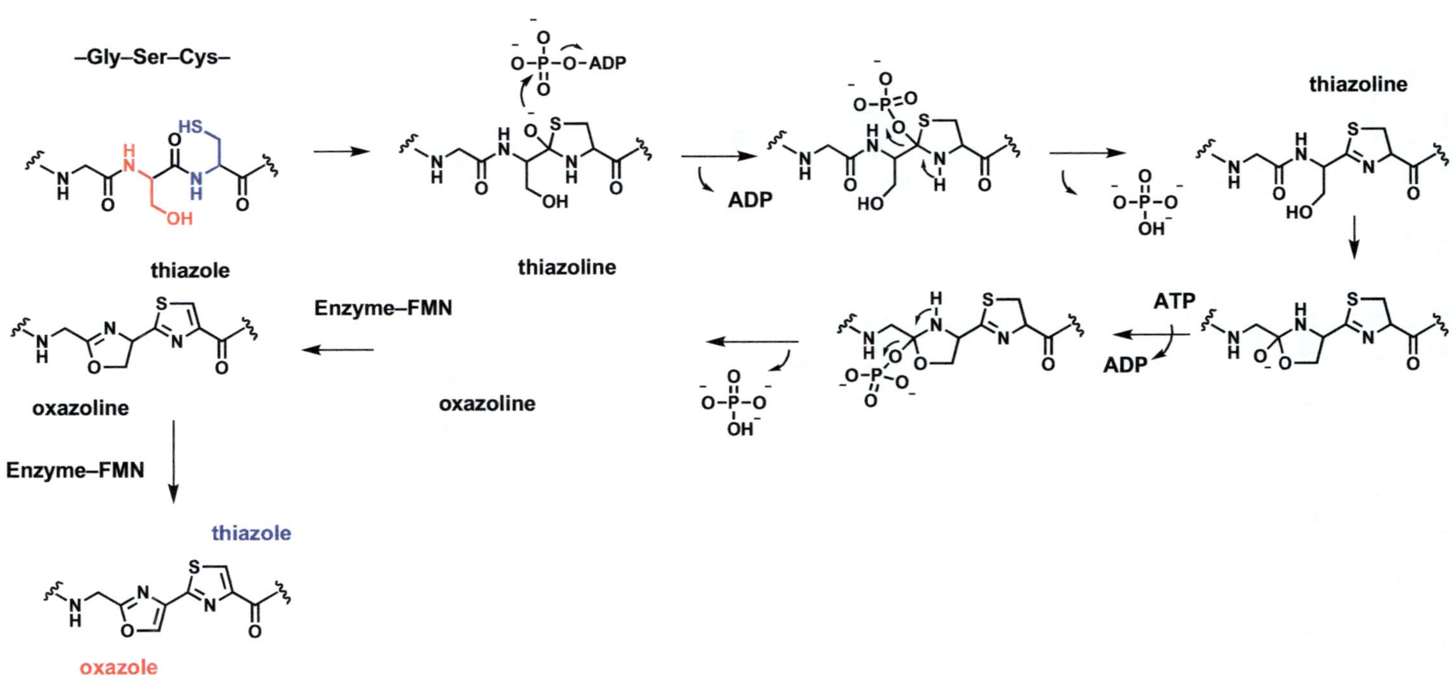

Figure 14.4 Conversion of an internal Gly–Ser–Cys tripeptide sequence to a tandem oxazole–thiazole modified peptide backbone. C–O bond cleavage of the tetrahedral adducts is driven by *O*-phosphorylation. Thiazoline and oxazoline rings are desaturated to the heteroaromatic oxazole and thiazole heterocycles by a flavin mononucleotide enzyme.

backbones into rigid, non-hydrolyzable ring structures. Multiple tandem cyclizations convert the floppy, acyclic, proteolytically unstable peptide precursors into compact, stable, densely functionalized, peptide-based natural products.

14.2.3 Thiocarbonyl Group Formation

A small number of microbial peptides, hydrolysis products of larger protein precursors, have one or more thiocarbonyl groups replacing the carbonyl group of peptide backbones (Figures 14.3, 14.5A and B). These are now known to arise by an intermolecular variant of the thiazole-forming post-translational modifications just detailed.

In this instance the early step leading to attack on the specific residue's upstream carbonyl is mediated by an HS^- (inorganic sulfide anion, shown in red in Figure 14.5) substrate molecule, rather than the thiolate side chain of a cysteine residue. The resultant tetrahedral adduct is acyclic rather than cyclic as in Figure 14.5. Attack of that adduct's oxyanion on the γ-phosphoryl group of

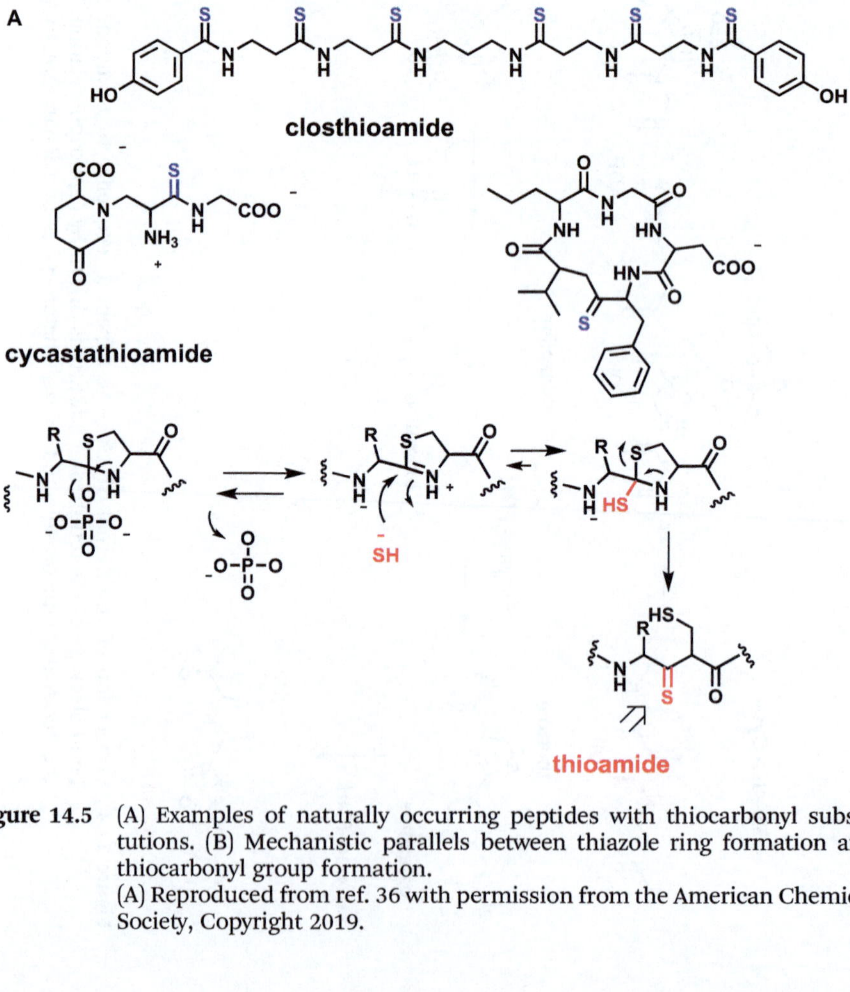

Figure 14.5 (A) Examples of naturally occurring peptides with thiocarbonyl substitutions. (B) Mechanistic parallels between thiazole ring formation and thiocarbonyl group formation.
(A) Reproduced from ref. 36 with permission from the American Chemical Society, Copyright 2019.

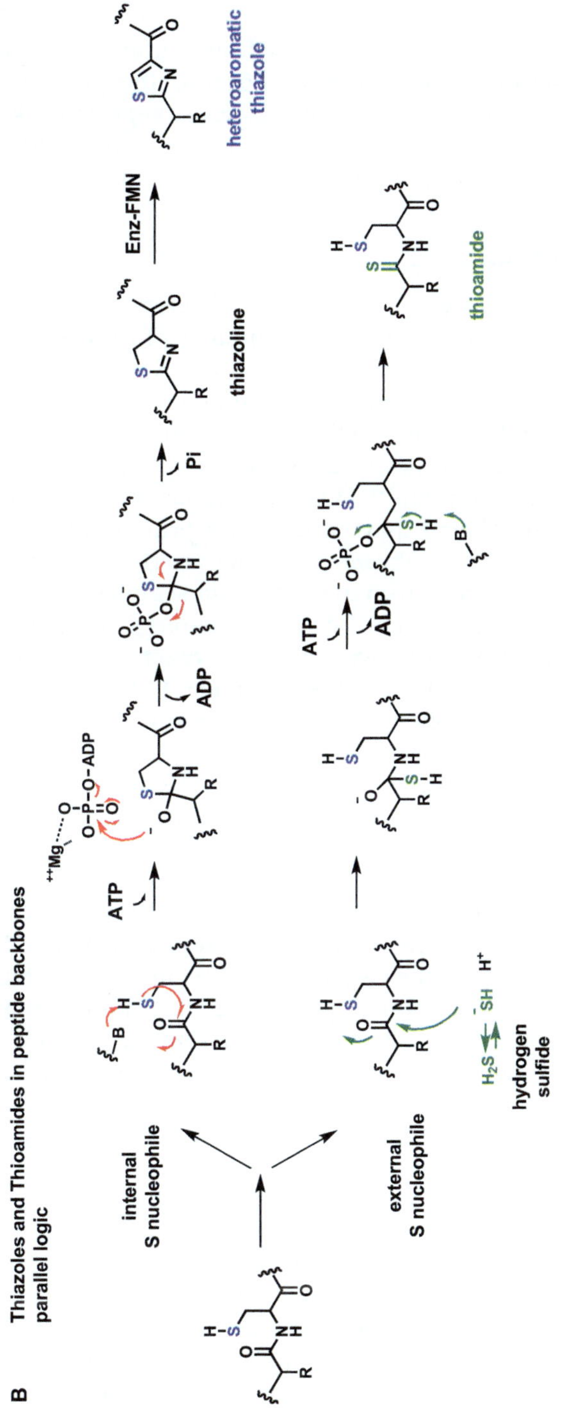

Figure 14.5 Continued.

ATP gives the phosphorylated adduct, now biased for C–OPO^{2-} bond cleavage in the next step.

As depicted in Figure 14.5, a lone pair from sulfur can act as an internal nucleophile, generating a C=S double bond as the product thioamide is formed. These transformations, autoproteolysis, peptide backbone thiazole formation and thioamide bond formation, show that both thiolate and inorganic sulfide sulfur nucleophiles can morph peptide bonds in proteins.

14.2.4 Protein Splicing

Simple eukarya, such as yeast, archaea and bacteria carry out protein splicing reactions,[16,17] catalytic capacity presumably of value in protein domain shuffling in evolution, but lost in multicellular eukaryotes. This has analogies to RNA splicing[18,19] in that an intervening region of sequence (RNA or protein) is looped out and the two flanking regions joined together, either in restored phosphodiester linkage (RNA splicing) or peptide linage (protein splicing). Unlike the abundant RNA splicing capacity of current multicellular organisms, protein splicing may have been a primordial autocatalytic capacity that has been lost in evolution. In analogy to the RNA splicing language of exons and introns—the intron to be split out seamlessly—the few families of proteins that are spliced have exteins separated by an intein that gets spliced out[16,17] (Figure 14.6). Two schematic diagrams of the intein expulsion and extein ligations are depicted in Figure 14.6.

In the discussion that follows there are three key residue side chains. Two are cysteine residues and their nucleophilic thiolate side chains are internal reactants. The first cysteine is at the C-terminus of the upstream extein-1 and will initiate the peptide bond to acyl thioester step that begins the autocatalytic cycle. The second cysteine is at the downstream boundary of intein–upstream boundary of exein-2 (Figures 14.6 and 14.7). It will attack the initial acyl thioester, thereby making a second acyl thioester intermediate, dragging the extein-1 past the intein sequence. The third key residue is an asparagine at the C-terminus of the intein to be removed. It acts as an internal nucleophile on its upstream peptide bond, converting it to a cyclic pyroglutamate residue. That cleaves the intein–extein-2 peptide bond and liberates the intein as illustrated in Figures 14.6 and 14.7. The use of the asparagine carboxamide nitrogen, a very weak nucleophile, as a peptide chain-cleaving agent in real time may be the most surprising feature of the whole protein splicing process, chemically.

The protein splicing chemistry is a variant of the autocleavage chemistry initiated by serine or cysteine side chain attack on the upstream peptide carbonyl that has been elaborated in the previous two sections. Figure 14.7A shows the initial upstream attack by a cysteine residue at the extein-1–intein boundary on the upstream peptide carbonyl. The tetrahedral adduct is now directed to autocleavage (much like the case of hedgehog autocleavage) of the peptide backbone to a thioester by C–N bond cleavage. Recall that to get to either the heterocyclic ring (thiazoline) or thiocarbonyl outcome,

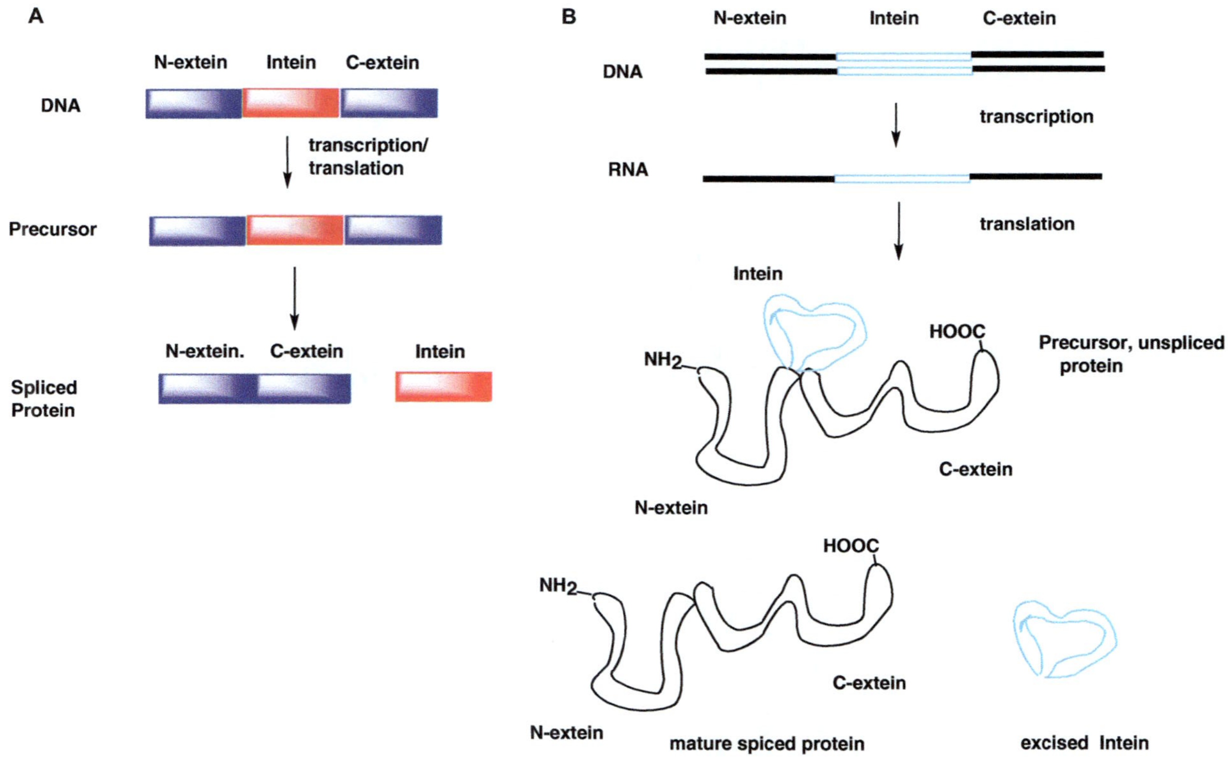

Figure 14.6 (A) Schematic in two dimensions of protein splicing to remove an intein sequence from a preprotein. The right-hand cartoon (B) indicates that the precursor protein is in a native, folded state for protein splicing to occur. The reaction stoichiometry for protein splicing joins an upstream extein to a downstream extein (both in blue) with hydrolytic excision of the intervening protein intein sequence (red). Note that at the extein–intein upstream boundary, a sentinel cysteine gets left within the intein. At the downstream intein–extein boundary an intein C-terminal asparagine is cyclized to a pyroglutamate ring structure on release.
(A left) Reproduced from ref. 35 with permission from Elsevier, Copyright 2013. B) Adapted from ref. 37 with permission from Springer Nature, Copyright 2010, The Author(s).

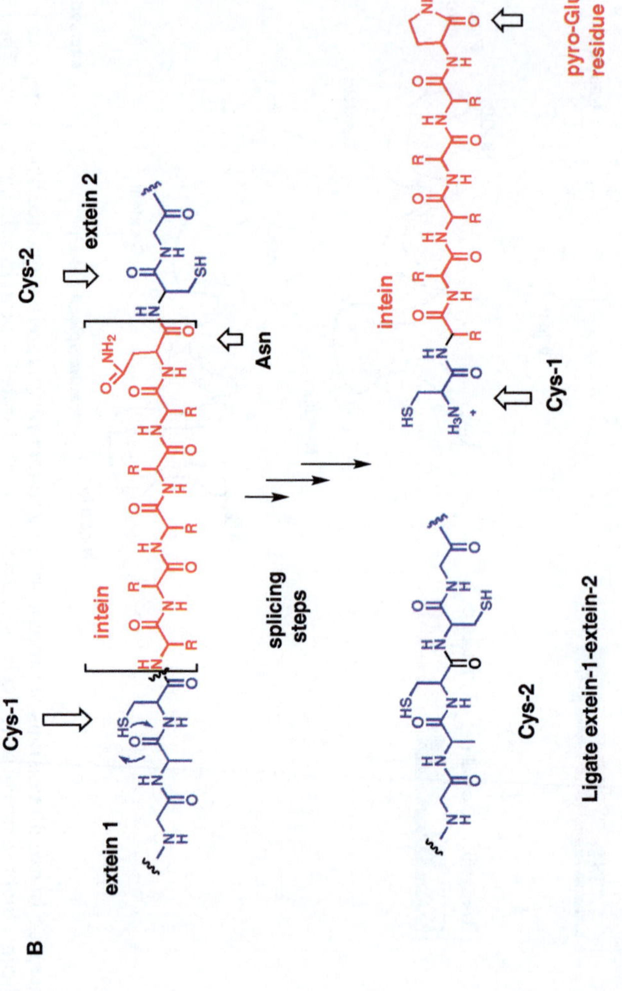

Figure 14.6 Continued.

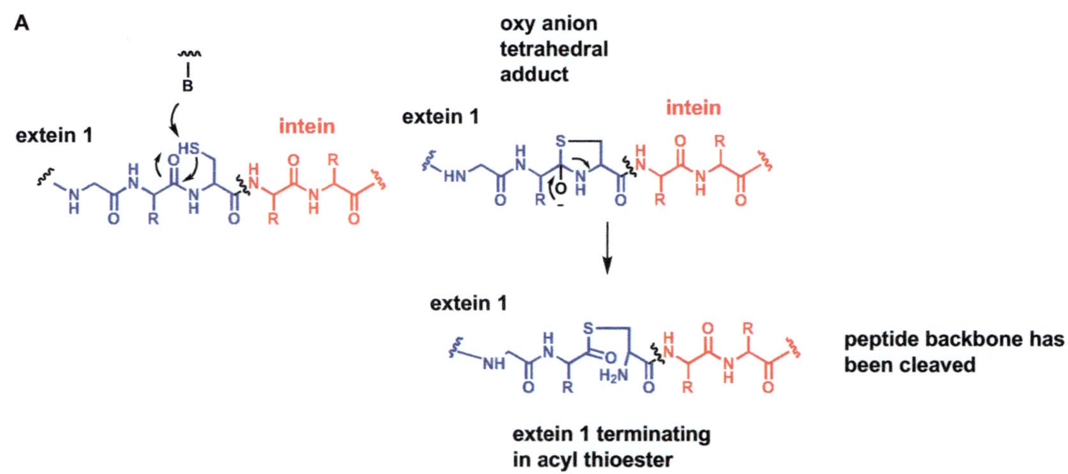

Figure 14.7 Details of the protein splicing reaction mechanism. (A) The cysteine at the upstream extein–intein boundary attacks its adjacent peptide carbonyl and cleaves the peptide backbone with thioester bond formation. (B) The cysteine thiolate side chain at the downstream intein–extein boundary acts as intramolecular nucleophile for acyl transfer, creating a lariat structure. This is resolved by intramolecular attack of the indicated Asn carboxamide N on the boundary peptide carbonyl. This releases the modified intein and gives a ligated extein–extein still as acyl thioester. *S*- to *N*-acyl shift restores the peptide backbone and completes the protein splicing ligation.

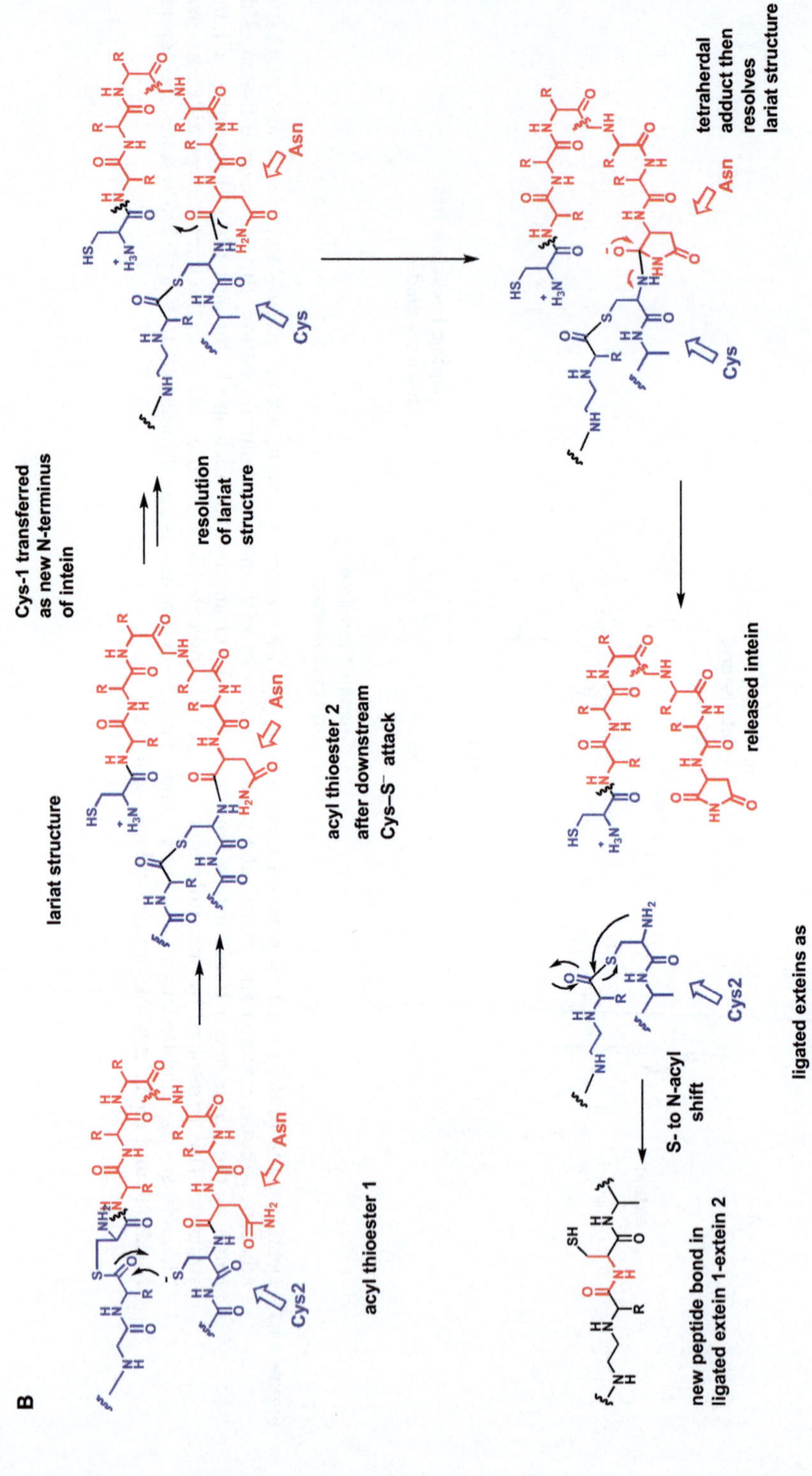

Figure 14.7 Continued.

O-phosphorylation is needed to lower the energy to favor the alternative C–O bond cleavage route heterocyclic ring.

At this stage in the protein splicing process the peptide backbone at the extein-1–intein junction has been cleaved and the extein-1 is an acylthioester activated thermodynamically for acyl transfer. The architectures of self-splicing proteins brings a downstream cysteine residue, at the intein–extein-2 border in proximity to the upstream acyl thioester (Figure 14.7B). Acyl transfer proceeds *via* the usual tetrahedral adduct intermediate with decomposition–resolution in the forward direction.

In the protein splicing cases, it is the altered fate of the peptide chain-cleaved upstream acyl thioester (from cysteine residue) or oxoester (from serine residue), not the initial tetrahedral adduct itself. These acyl-*S*– and acyl-*O*–protein intermediates have long enough lifetimes to be captured by a downstream Cys or Ser (not shown) side chain (Figure 14.7B). These constitute internal acyl transfers and set up branching or lariat structures as a result.

Note that Cys-1 at the C-terminus of extein-1 gets left behind as the new N-terminus of the intein fragment to be split out. The downstream Cys-2 will be the marker residue for the locus of joining of extein-1 to extein-2.

Resolution of the thioester–oxoester linkages in the branched, lariat structures involves participation of the side chain of the asparagine residue that is at the C-terminus of the intein to be excised. The asparagine amide NH_2 attacks its own immediate downstream peptide carbonyl group. The amide linkage becomes a tetrahedral adduct that resolves on decomposition to a cyclic residue (pyroglutamate) that is at the C-terminus of the now free intein that can diffuse from the active site. The extein is released as a separated protein fragment with Cys-1 from extein-1 at the N-terminus and the pyroglu residue at the C-terminus.

Extein1 and extein 2 are at this point joined by a thioester (Cys involvement) or oxoester bond (Ser involvement). The intein–extein-2 boundary cleavage has generated a primary amino group that can serve for acyl *S* to *N* transfer. That *S*- to *N*-acyl shift (*O*- to *N*-shift if an Ser residue is involved) is depicted as the last step of (cysteine-based) protein splicing, converting the thermodynamically activated thioester bond to the stabilized amide–peptide bond in the final spliced protein. In nature these four coordinated steps occur in the minute time frame but selection for rapid kinetics has increased the turnover number.[20,21]

The steps, in summary: (1) amide to thioester, (2) thioester 1 to thioester 2, (3) resolution of the lariat structure by internal pyroglu formation; (4) thioester to amide. The first and third steps break the initial extein-1–intein peptide bond and then remake the extein-1–extein-2 peptide bond, respectively. The second step is acyl transfer between two thiols and creates the lariat structure. Step 3 cleaves the intein–extein-2 peptide bond, releasing the modified intein and the ligated extein-1–extein-2 joined as a thioester. The process is presumably driven by the final favored *S*- to *N*-acyl shift, reforming a peptide link between the two exteins.

Protein splicing, however limited its current biological distribution, displays the power and versatility of the side chains of cysteine (and serine) as nucleophiles and the kinetic and thermodynamic interplay between thioester, (oxoester), and amide linkages in terms of rates and energetics, and the value of *S*- to *N*-acyl shift chemistry in protein contexts. It seems a bit clunky but must have had value in the course of protein evolution to still exist in extant microbes.

14.3 Cysteine–Tyrosine Thioether Automodification

There are a small set of enzymes, that undergo automodification reactions of one or more residues within or near the active site, that affect activity. In one sense the large family of protein serine kinases could fit this definition. Most of the phosphorylations are activating, inducing an electrostatically-driven conformational rearrangement. However, the great bulk of the Ser–Thr and Tyr protein kinases that get autophosphorylated are modified *in trans*, by another protein kinase molecule (often within a dimer) rather than *in cis*.

A true *in cis* automodification occurs in the active site of the cysteine dioxygenase. That enzyme was discussed in Chapter 7 as the first enzyme sending cysteine down the taurine biosynthetic pathway, and then in Chapter 13 in the context of a molecular oxygen sensor. The enzyme acts by inserting both oxygen atoms of a dioxygen molecule cosubstrate into the thiol side chain of cysteine to create the nonnucleophilic, dioxygenated sulfinate group. Figure 14.8 shows that in a presumptive early catalytic turnover cycle, the dioxygenase creates a covalent link between a cysteine and tyrosine residue just outside the first coordination sphere of the active-site iron.

The figure depicts a native enzyme molecule part of the way through an early catalytic cycle, as the high valent $Fe^{IV}{=}O$ intermediate. This is a radical generator and most likely can mistakenly abstract an $H^{\bullet}$ atom from the adjacent Cys–SH residue. This yields a sulfur radical on that Cys side chain, demonstrating a one-electron reaction manifold. The sulfur atom is proposed to do one-electron chemistry with the adjacent electron rich tyrosine side chain. This forms the covalent Cys–S–Tyr adduct, after another Tyr-derived $H^{\bullet}$ atom transfers to the Fe^{III} to create resting Fe^{II}. As it happens, the autocatalytic Cys–Tyr crosslink is not notably deleterious to activity and may even stabilize productive active site geometry for subsequent dioxygen transfer catalytic cycles.[22]

14.4 Capturing the Nucleophilic Thiol Side Chains by Foreign Electrophiles in Cysteinyl Proteomes

As befits the most nucleophilic of the side chains in the 20 amino acid residues in proteomes, the thiol side chain of cysteine is capturable by a wide variety of electrophilic groups both in natural molecules and in synthetic organic scaffolds. These mimic the thiolate nucleophilic chemistry of reduced glutathione described in Chapter 5. The hepatic necrosis from overdoses of acetaminophen that have overcome the GSH intracellular hepatic chemical protectant capacity arises from conjugate addition of the cysteine resides in hepatic proteins to the iminoquinone metabolite as in Figure 5.12.

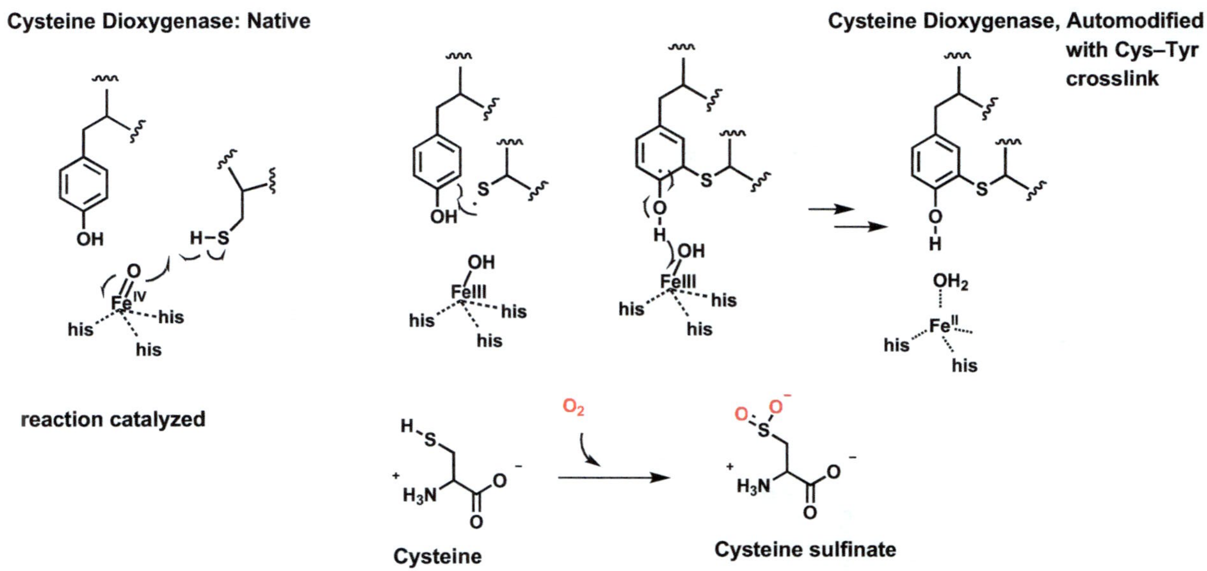

Figure 14.8 Cysteine dioxygenase-mediated autocatalytic adduct formation between an active site cysteine and a tyrosine side chain probably proceeds *via* a thiyl radical (S•) route.

The literature on both cysteine and serine protease inactivating groups has been comprehensively reviewed by Powers *et al.*,[23] summarizing how embedding aldehydes, halomethylketones, epoxides, vinyl sulfones, unsaturated ketones and acrylamides (Figure 14.9), in peptide and/or polyketide backbone

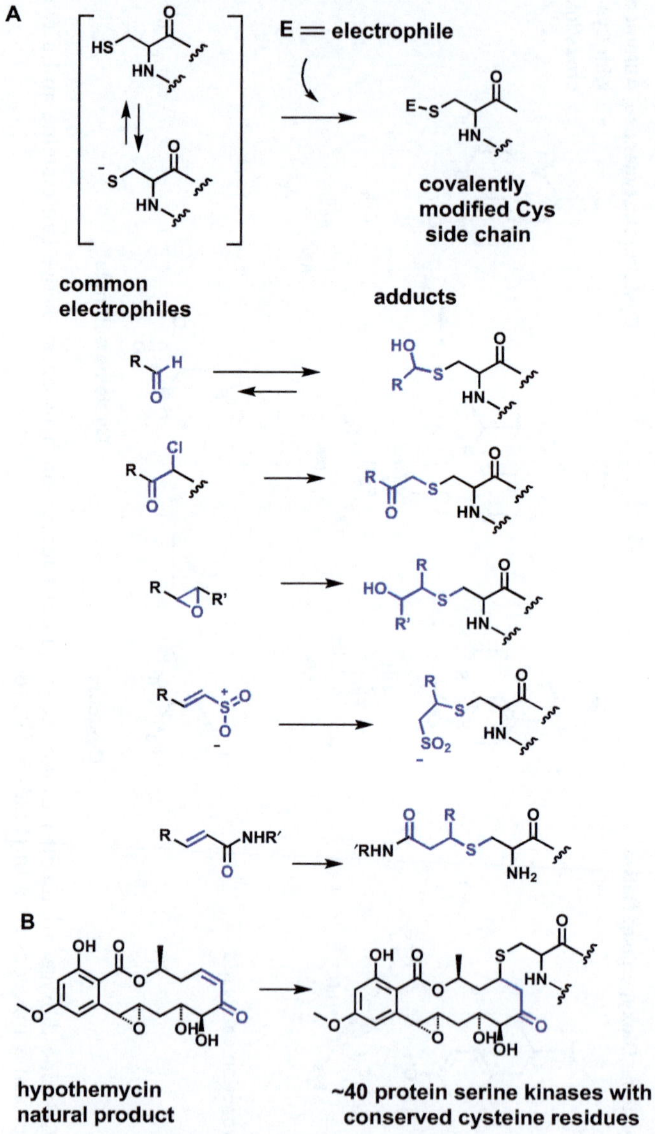

Figure 14.9 Common electrophiles for covalent modification of cysteinyl thiol side chains in proteomes. (A) Aldehydes, halomethyl ketones, epoxides, vinyl sulfones and acrylamides. (B) Conjugate addition of thiolate side chain of a cysteine residue near the active site of protein kinases to the enone moiety of the natural polyketide lactone hypothemycin.

contexts could be used to generate chemoselective inactivators for one or more target cysteinyl proteases.[24,25]

Cravatt and coworkers, among others, have recently undertaken a systems biology approach to defining the great bulk of cysteine residues within proteomes that are capturable by electrophiles (such as substituted acrylamides) in fragment-based screens and have reported that approximately 700 cysteine side chains are capturable.[26] These approximately 700 residues offer the prospect for selective modification, once the electrophilic groups are titrated for intrinsic reactivity and embedded within scaffolds to confer initial recognition and binding specificity.

The universe of cysteine side chains capturable by electrophiles is expected to encompass not only catalytic cysteines in enzyme active sites, but additional residues that may be in distant, allosteric sites.[27] Some may also be neighbors to active sites where covalent modification could block access of substrates to the active site region, in analogy to the effect of Ser520 acetylation in cyclooxygenases that is the molecular basis of the therapeutic effects of aspirin.

14.4.1 Protein Serine Kinases with Conserved Non-catalytic Cysteine Residues Near the Active Sites

Such contexts occur in some 40–50 members of the 550-member human kinome superfamily of enzymes, including protein tyrosine kinases.[24] That set has a conserved cysteine residue apposite to the aspartyl residue whose side chain carboxylate coordinates the magnesium ion that is always part of the Mg–ATP substrate in kinase actions.

Early studies with the natural product polyketide hypothemycin[24] revealed irreversible, time-dependent loss of catalytic activity of a set of protein kinases that contained the conserved cysteine residue, while those lacking that Cys residue were unaffected. Hypothemycin has two potential electrophiles, the epoxide and the αβ-unsaturated ketone group. Subsequent structural analyses indicated the conjugated enone as the culprit electrophile (Figure 14.9).

The thiol capture strategy has then been incorporated into potent synthetic ATP mimetic ligands for protein kinases of therapeutic interest. Two such molecules,[28,29] FRF0, a K-Ras inactivator, and WZ4002, an irreversible inactivator of mutant forms of epidermal growth factor receptor (EGFR) which arise during treatment of cancer patients, are shown in Figure 14.10. Both molecules bear conjugated acrylamide moieties as the electrophilic warheads that come into play after potent and selective binding to the target protein kinases.

Cravatt's group used their system-wide proteome screening approach to examine how specific such protein kinase covalent inhibitors actually are within tissues.[30–32] In particular, they analyzed the approved third-generation cancer drug AZD9291 (osimertinib), one of the acrylamide-bearing molecules that targets the predominant clinical Thr to Met EGFR mutant, that evolves under chemotherapy pressure. A highly related probe molecule with a terminal

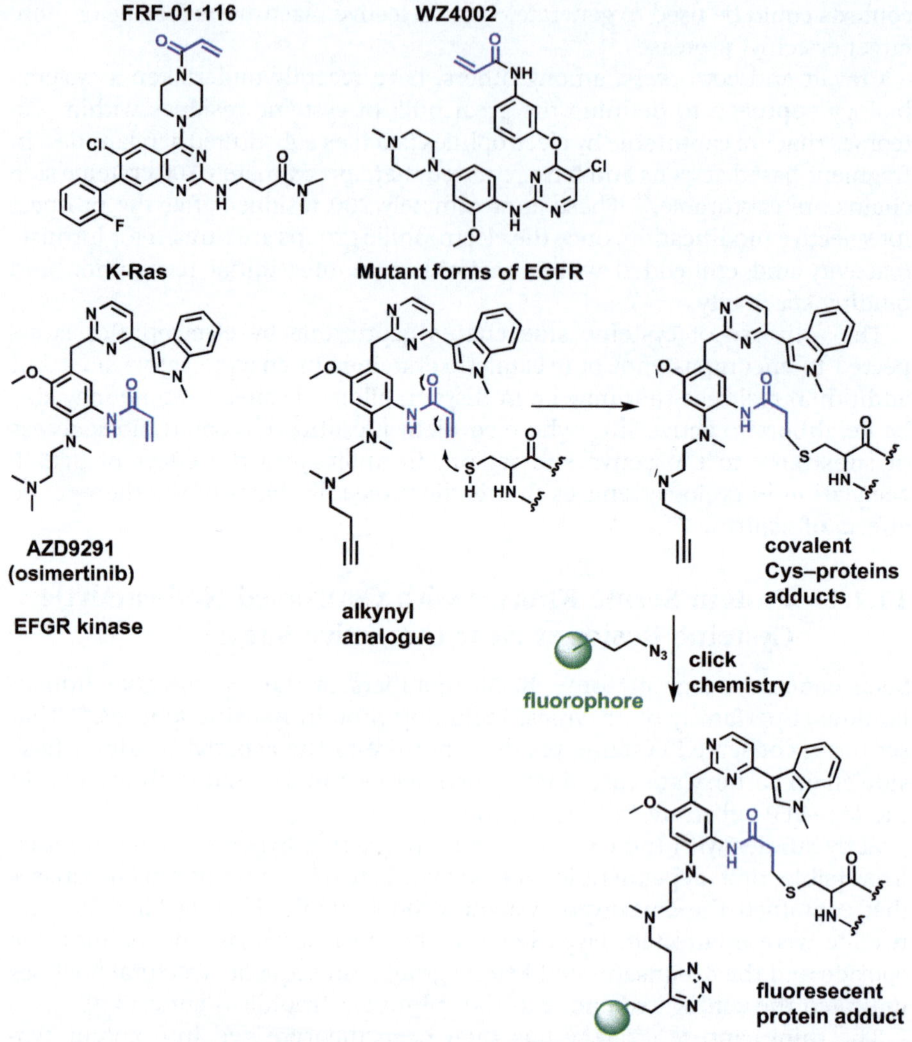

Figure 14.10 Three optimized protein kinase inhibitors with electrophilic acrylamide moieties to covalently modify a cysteine residue within the active site region of targeted protein kinases. The alkynyl probe of AZD9291 allowed evaluation of specificity and promiscuity for off-target protein labeling after click chemistry with an azide-containing fluorophore.

alkyne retained potency and selectivity by activity loss assays and allowed subsequent click chemistry with an azide-containing fluorophore to detect covalently modified off-target proteins.[33] Among other conclusions, such proteome-wide screens could be used to optimize on-target *vs.* off-target structure–activity relationships for inactivators in this 40–50-member subset of therapeutically relevant protein kinases with a reactive cysteine in or near the active site.[34]

References

1. P. Arnison, M. Bibb, G. Bulaj, G. Bierbaum, J. A. Cmarero, D. J. Camopiano, J. Clardy, P. D. Cotter, D. J. Craik, E. Dittman, S. Donadio, P. B. Dorrestein, M. A. Fischbach, U. Goransson, C. Gruber, D. H. Haft, T. K. Hemschidt, C. Hertwick, C. Hil, M. Jaspers, W. L. Kelly, O. Kuipers, A. J. Link, W. Liu, M. A. Marahiel, D. A. Mitchell, G. N. Moll, B. S. Moore, I. F. Nes, G. E. Norris, B. M. Olivera, H. Onaka, M. L. Patchett, S. Rbuffat, P. R. Ross, H.-G. Sahl, E. W. Schmidt, M. E. Selsted, K. Severinov, B. Shen, K. Sivonen, R. E. Sussmuth, J. R. Tagg, W. A. van de Donk, J. C. Vederas, C. T. Walsh, J. D. Walton and J. M. Wiley, Ribosomally synthesized and posttranslationally modified natural products, *Nat. Prod. Rep.*, 2012, **30**, 108–160.
2. N. M. Okeley and W. van der Donk, Novel cofactors via post-translational modifications of enzyme active sites, *Cell Chem. Biol.*, 2000, **7**, R159–R171.
3. Q. Zhang, J. Doroghazi, X. Zhao, M. C. Waler and W. A. van der Donk, Expanded natural product diversity revealed by analysis of lanthipeptide-like gene clusters in Actinobacteria, *Appl. Environ. Microbiol.*, 2015, **81**, 4339–4349.
4. L. M. Repka, J. Chaikan, S. K. Nair and W. A. van der Donk, Mechanistic understanding of lanthipeptide biosynthetic pathways, *Chem. Rev.*, 2017, **117**, 5457–5520.
5. C. T. Walsh, S. Garneau-Tsodikova and G. J. Gatto, Protein posttranslational modifications: the chemistry of proteome diversifications, *Angew. Chem., Int. Ed.*, 2005, **44**, 7342–7372.
6. R. Y. Tsien, The green fluorescent protein, *Annu. Rev. Biochem.*, 1998, **67**, 509–544.
7. P. A. Beachy, M. Cooper, K. E. Young, D. P. von Kessler, W. J. Park, T. M. Hall, D. J. Lahey and J. A. Porter, Multiple roles of cholesterol in hedgehog protein biosynthesis and signaling, *Cold Spring Harb. Symp. Quant. Biol.*, 1997, **62**, 191–204.
8. J. L. Mullor, P. Sanchez and A. R. Altaba, Pathways and consequences: hedgehog signaling in human disese, *Trends Cell Biol.*, 2002, **12**, 562–569.
9. Y. Ye and M. Fortini, Proteolysis and developmental signal transduction, *Semin. Cell Dev. Biol.*, 2000, **11**, 211–221.
10. J. A. McIntosh, M. Donia and E. W. Schmidt, Ribosomal peptide natural products: bridging the ribosomal and noribosomal worlds, *Nat. Prod. Rep.*, 2009, **26**, 537–539.
11. M. I. Mclaughlin, N. Lanz, P. J. Goldman, K.-H. Lee, S. J. Booker and C. L. Drennan, Crystallographic snapshots of sufur insertion by lipoyl synthase, *Proc. Natl. Acad. Sci. U. S. A.*, 2016, **113**, 94346–94350.
12. C. T. Walsh, M. Acker and A. A. Bowers, Thiazolyl peptide antibiotic biosynthesis: a cascade of posttranslational modifications on ribosomal nascent proteins, *J. Biol. Chem.*, 2010, **285**, 27525–27531.
13. C. J. Schwalen, G. Hudson, B. Kille and D. A. Mitchell, Bioinformatic expansion and discovery of thiopeptide antibiotics, *J. Am. Chem. Soc.*, 2018, **140**, 9494–9501.

14. K. Y. Dunbar, J. Melby and D. A. Mitchell, YcaO domains use ATP to activate amide backbones during peptide cyclodehdrations, *Nat. Chem. Biol.*, 2012, **8**, 569–575.
15. Y. M. Li, J. Milne, L. L. Madison, R. Kolter and C. T. Walsh, From peptide precursors to oxazole and thiazole-containing peptide antibiotics: microcin B17 synthase, *Science*, 1996, **274**, 1188–1193.
16. F. B. Perler, Protein splicing mechansisms and applications, *IUBMB Life*, 2005, **57**, 469–476.
17. K. V. Mills, M. Johnson and F. B. Perler, Protein splicing: how inteins escape from precursor proteins, *J. Biol. Chem.*, 2014, **289**, 14498–14505.
18. A. Berk, Discovery of RNA splicing and genes in pieces, *Proc. Natl. Acad. Sci. U. S. A.*, 2016, **113**, 801–805.
19. D. S. Berkholz, H. Faber, S. N. Savvides and P. A. Karplus, Cataytic cycle of human glutathione reductase near 1 Agstrom resolution, *J. Mol. Biol.*, 2008, **382**, 371–384.
20. M. Vila-Perello and T. Muir, Biological applications of protein splicing, *Cell*, 2010, **143**, 191–200.
21. A. J. Stevens, Z. Brown, N. H. Shah, G. Sekar, D. Cowburn and T. W. Muir, Design of a split intein with exceptional protein splicing activity, *J. Am. Chem. Soc.*, 2016, **138**, 2162–2165.
22. C. A. Joseph and M. Maroney, Cysteine dioxygenase: structure and mechanism, *Chem. Commun.*, 2007, 3338–3349.
23. J. C. Powers, J. Asgian, O. D. Ekici and K. E. James, Irreversible inhibitors of serine, cysteine, and threonine proteases, *Chem. Rev.*, 2002, **102**, 4639–4750.
24. A. Schirmer, J. Kennedy, S. Murli, R. Reed and D. V. Santi, Targeted covalent inhibition of protein kinases by resorcylic acid lactone polyketides, *Proc. Natl. Acad. Sci. U. S. A.*, 2006, **103**, 4234–4239.
25. M. Gersch, J. Kreuzer and S. A. Sieber, Electrophilic natural products and their biological targets, *Nat. Prod. Rep.*, 2012, **29**, 659–682.
26. K. M. Backus, B. Correia, K. M. Lum, S. Forli, B. D. Horning, G. E. Gonzalo-Paez, S. Chatterjee, B. R. Lanning, J. R. Teijaro, A. J. Olson, D. W. Wolna and B. F. Cravatt, Proteome-wide covalent ligand discovery in natve biological systems, *Nature*, 2017, **534**, 570–574.
27. H. J. Kim, S. Ha, H. Y. Lee and K. J. Lee, ROSics: chemistry and proteomics of cysteine modifications in biology, *Mass Spectrom. Rev.*, 2015, **34**, 184–208.
28. M. Zeng, J. Liu, L. Li, F. Feru, C. Quan, T. W. Gero, S. B. Ficarro, Y. Xiong, C. Ambrogio, R. M. Paranal, M. Catalano, J. Shao, K. K. Wong, J. A. Marto, E. S. Fischer, P. A. Janne, D. A. Scott, K. D. Westver and N. S. Gray, Potent and Selective Quinazolne Inhibitors of KRAS G12C, *Cell Chem. Biol.*, 2017, **24**, 1005–1016.
29. W. Zhou, D. Ercan, L. Chen, C. H. Yun, D. Li, M. Capelletti, A. B. Cortot, R. E. Iacob, R. Padera, J. R. Engen, K. K. Wong, M. J. Eck, N. S. Gray and P. A. Janne, Novel mutant-selective EGFR kinase inhibitors against EFGR T790M, *Nature*, 2009, **462**, 1070–1074.

30. D. Kidd, T. Liu and B. F. Cravatt, Profiling serine hydrolase activities in complex proteomes, *Biochemistry*, 2001, **40**, 4005–4015.
31. D. A. Bachovchin and B. Cravatt, The pharmacological landscape and therapeutic potential of serine hydrolases, *Nat. Rev. Drug Discovery*, 2013, **11**, 52–68.
32. Z. Huang, D. Ogasawara, U. I. Seneviratne, A. B. Cognetta III, C. W. am Ende, D. M. Nason, K. Lapham, J. Litchfield, D. S. Johnson and B. F. Cravatt, Global portrait of protein targets of metabolites of the neurotoxic compound BIA 10-2472, *ACS Chem. Biol.*, 2019, **14**, 192–197.
33. S. Niessen, M. Dix, S. Barbas, Z. E. Potter, S. Lu, O. Brodsky, S. Plnken, D. Behenna, C. Almaden, K. S. Gajiwala, K. Ryan, R. Ferre, M. B. Lazear, M. M. Hayward, J. C. Kath and B. F. Cravatt, Proteome-wide map of targets of T790M EGFR-directed covalent inhibitors, *Cell Chem. Biol.*, 2017, **24**, 1388–1400.
34. N. J. Fowler, C. Blanford, S. P. de Visser and J. Warwicker, Features of reactive cysteines discovered through computation: from kinase inhibition to enrichment around protein degrons, *Sci. Rep.*, 2017, **7**, 16338.
35. K. V. Mills, in *Handbook of Proteolytic Enzymes*, ed. N. D. Rawlings and G. Salvesen, Academic Press, 2013, pp. 315–321.
36. N. Mahanta, D. M. Szantai-Kis, E. J. Petersson and D. A. Mitchell, *ACS Chem. Biol.*, 2019, **14**, 142–163.
37. S. Elleuche and S. Pöggeler, *Appl. Microbiol. Biotechnol.*, 2010, **87**, 479–489.

Part IV: Overview and Summary

This concluding part IV contains the final chapter. Chapter 15 summarizes several facets of sulfur chemical biology. Those roles of sulfur range through the redox versatility of the inorganic oxidation states between inorganic sulfate and inorganic sulfide ions, eight electrons apart in redox terms, with dramatically different metabolic niches as electrophiles and nucleophiles, respectively.

The enzymatic bridge from inorganic to organic sulfur in microbes and plants occurs at that most reduced state of inorganic sulfide as it is incorporated at carbon 3 of L-cysteine. The sulfur nucleophilicity of inorganic sulfide is preserved in the cysteine thiolate side chain.

The biological roles of the cysteine–cysteamine thiol in low molecular weight metabolites are most fully realized and functionally differentiated in two molecules: coenzyme A and the tripeptide glutathione. Coenzyme A is synonymous with acyl thioester metabolism. The combined thermodynamic activation and kinetic stability allow the acyl CoAs to function as diffusible packets of acyl transfer cellular currency.

Glutathione is vectored to a distinct set of enzyme-catalyzed cellular functions. Most notable are three such functions: reduction of cysteine sulfenates, free or protein-bound; reductive cleavage of hydrogen peroxide and alkyl peroxides by glutathione peroxidase; detoxification of electrophilic organic metabolites such as drugs and other xenobiotics. In the first two functions the oxidized glutathione disulfide products are reduced back to the pair of glutathione thiol forms by glutathione reductase action.

The iron–sulfur redox clusters, bound to hundreds of proteins with particular cysteine-rich sequence motifs, are most likely ancient inventions of one-electron redox agents, functioning in dozens of contexts, from electron transport to sensors. We have noted the specific functions of 4Fe–4S clusters in sulfite reductase, mitochondrial NADH dehydrogenase and the ever-expanding radical SAM enzyme family. Given the coordination of 2Fe–2S and 4Fe–4S clusters by a canonical set of four cysteine residue thiolate side chains in their host proteins, the Fe–S clusters represent a key nexus between inorganic and organic sulfur chemical biology and metabolism.

The biosynthesis of those inorganic sulfide molecules from cysteine follows two divergent paths from pyridoxal-phosphate (PLP)-containing lyases. The β-elimination route yields small quantities of H_2S (most likely as HS^-) along with pyruvate and ammonia. That H_2S may serve as signaling agent but is generally *not* the source of inorganic sulfide in Fe–S clusters. Instead, that arises when PLP-β-replacement cysteine lyases, release a nascent six-electron sulfur atom, (not the eight-electron HS^-) that is sponged up by active site cysteinyl residue. The result is a covalent persulfide (–Cys–S–SH) enzyme adduct. That sulfane –SH gets passed along by a bucket brigade of chaperone protein cysteine side chains to the partner proteins where 2Fe–2S centers are first assembled, dimerized, and/or passed on to the client proteins. In the course of assembly as S^{2-} ligands to two Fe^{3+} ions, the itinerant six-electron inorganic sulfur must be oxidized in controlled loci.

The behavior of thiols in proteomes in comparison to the low molecular weight metabolites is dominated by the cysteine side chain thiol chemistry. The modifications in some ways replicate the chemistry seen in free cysteine and its derivatives. These include oxidation to sulfenates and then to disulfides, *S*-nitrosylation, persulfide (–Cys–SSH) formations and transfers. A direct route of Cys-thiols to cystine disulfides, bypassing the intermediate sulfenic acid stage, appears dominant in protein disulfide isomerase-catalyzed folding processes in the endoplasmic reticular component of cell secretory pathways.

Cysteine side chain chemistry unique to certain protein classes, in their native folded states arises from autocatalytic attack of the cysteine thiolate side chain on its immediate upstream peptide carbonyl group. A plethora of fates from the resultant five-membered ring tetrahedral adducts, include auto-cleavage of the protein chain, protein splicing and cyclic thiazole formations.

Two final points about Part IV. Four molecular arrays involving sulfur play key, but vastly different, roles in cell metabolism. The biology follows the chemical properties of the sulfur atom in the specific molecular scaffolds.

On the inorganic side, the 4Fe–4S clusters are elegant self-assembled nano-machines that probably arose in anaerobic organisms to use one-electron transfers for a range of novel chemistries. The second molecule family is populated by 5′-adenylylsulfate (APS) and 3′phospho-5-adenylylsulfate (PAPS). The key functional group is the inorganic mixed anhydride between phosphoric acid and sulfuric acid. This is a thermodynamically activated form of sulfate, set up for the sulfuryl ($-SO_3^-$) group as an electrophile, akin to phosphoryl transfers to nucleophilic substrates. APS and PAPS have sufficient kinetic stability to be mobile substrates in cells. Remarkably, APS is used as a sulfuryl donor to start a redox cascade (sulfate to sulfite), while PAPS is used exclusively for non-redox, sulfuryl transfer roles in biology.

The remaining two sulfur metabolites, acyl CoAs (*e.g.* acetyl CoA) and *S*-adenosylmethionine are organosulfur molecules with two and three C–S bonds, respectively. They occupy vastly distinct chemical and metabolic spaces but have in common the same two principles of thermodynamic activation, the kinetic stability balance noted above in APS and PAPS. In these cases it is acyl thioester chemistry (acetyl CoA) *vs.* trigonalized sulfur cation chemistry (SAM), rather than mixed phosphoric–sulfuric anhydride chemistry. Clearly nature has invented three distinct strategies for effecting acyl, alkyl and sulfuryl group transfer chemistries, utilizing the different properties of sulfur as an activating group in three independent oxidation states.

The overview provided in Part IV emphasizes the chemical biology perspective presented in the previous three parts of this volume. Appreciation of the chemical diversity of inorganic and organic forms of sulfur accessible to organisms, how each is formed and the scope of their utilization gives insights into the life-giving roles of the approximately 120–150 g of sulfur in adult humans.

CHAPTER 15

A Traverse Through Sulfur Chemical Biology

15.1 Inorganic Sulfur Compounds Emphasize the Eight-electron Spread in Redox States of the Sulfur Atom

In the course of the preceding fourteen chapters we have examined three aspects of sulfur chemical biology. Figure 15.1 shows some of the inorganic and organic forms of sulfur that have been featured.

The first, and probably among the earliest, biological roles for sulfur are those of the predominant inorganic compounds of sulfur, including those with bonds to oxygen. The predominant current form is the dianion of sulfuric acid, sulfate ions, reaching up to 28 mM in global oceans from a combination of global weathering of rocks and anthropogenic production of sulfuric acid for industrial purposes.

Less abundant is the two-electron-reduced form, the sulfite monoanion. Sulfite, with its somewhat more nucleophilic sulfur, in contrast to the electrophilic sulfur in sulfate ions, has toxicity in cells, and organisms have evolved enzyme strategies both to reduce it and to oxidize it back to sulfate. Four electrons down from sulfite anion is elemental sulfur, S^0, which remarkably is actually redox-active in bacterial metabolism. Elemental sulfur is a yellow solid, mined extensively over the globe, historically and contemporaneously. It forms polysulfide allotropes, including the S^8 polygon.

Finally, two electrons down from elemental sulfur and a total of eight electrons down from sulfate is H_2S. The first pK_a is around 6 so dissolved hydrogen sulfide is in equilibrium with the nucleophilic HS^- hydrosulfide anion. When we examined the subject of the ubiquitously distributed, protein-bound 2Fe–2S and 4Fe–4S clusters as one-electron redox elements,

Chemical Biology No. 12
The Chemical Biology of Sulfur
By Christopher T. Walsh

Published by the Royal Society of Chemistry, www.rsc.org

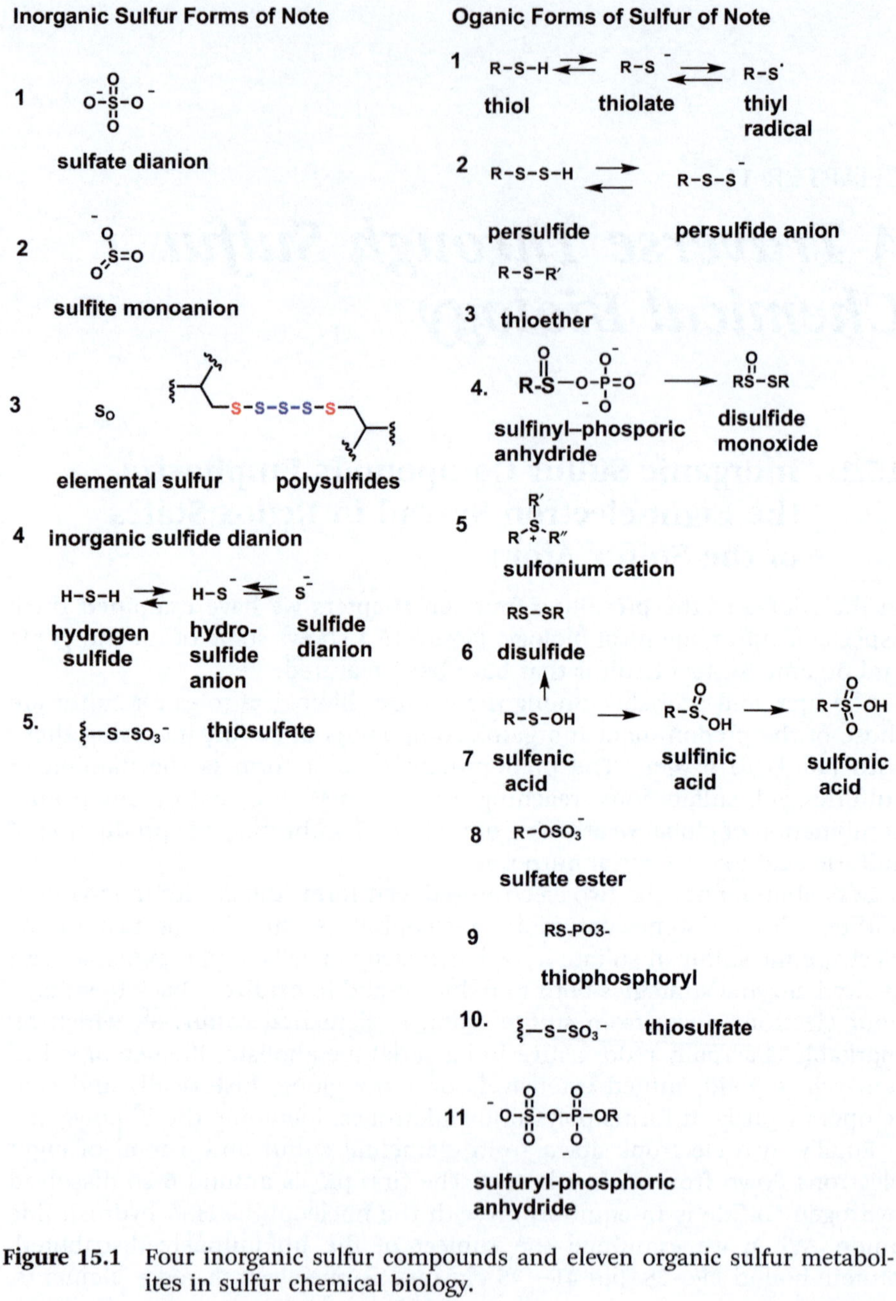

Figure 15.1 Four inorganic sulfur compounds and eleven organic sulfur metabolites in sulfur chemical biology.

we noted that the sulfide was incorporated as S^{2-}, even though its pK_a is around 20; there will be negligible amounts of free sulfide in biologic milieus. That limitation represents one of several challenges in construction of Fe–S clusters.

The ability of sulfur to exist in stable oxidation states at +6 (sulfate), +4 (sulfite), 0 (elemental sulfur), −2 (hydrogen sulfide) and some in between (sulfonium cations) is the overarching property that has made sulfur and its compounds so central to metabolic pathways and underlying chemical logic and reactivity.

During the first billion and more years of the emergence of life in anaerobic global atmospheres, low valent sulfur may have been crucial and the two-electron redox traverse between H_2S and S^0 perhaps central in the early organization of membrane electron transport chains and extraction of energy out of reduced inorganic and organic materials *via* oxidation and passage of electrons down potential drops. The affinity of thiols (sulfide, HS^-) for ferrous cations, zinc cations and others such as Hg^{2+} were probably early organizing elements for protein folding and construction of redox active nanomachinery.

As the planet became oxygenated, *e.g.* from cyanobacterial photosynthesis, the oxygenated forms of inorganic sulfur became more abundant and pathways to transfer electrons from sulfate down all the way to sulfide, continued to allow microbes to live in anaerobic niches. In complementary roles on the other side of the oxic–anoxic interface in water columns, the sulfide-oxidizing bacteria made a living ultimately transferring metabolic electrons to O_2 as a terminal acceptor.

In its redox versatility sulfur is almost unique among the six major elements of life (carbon, oxygen, nitrogen, hydrogen, sulfur and phosphorus). It offers different properties and reactivities in its most oxidized and most reduced oxidation states in biology. The ability of the valence electrons in the 3d shell to participate in bonding gives sulfur unique place in chemical biology.

In the $S = +6$ sulfate oxidation state, as the dianion, sulfate has functional analogies to the phosphate dianion ($P = +5$) oxidation state. Unlike sulfur, phosphate is rarely redox-active in biology. The phosphate group with electrophilic phosphorus surrounded by a phalanx of electron-rich oxygen substituents is almost exclusively the only form. With pK_{a2} of approximately 6, the phosphate dianion predominates at neutral pH, akin to the sulfate dianion. The sulfur is electrophilic in the +6 oxidation state and, as we have noted, reacts with nucleophiles for sulfuryl ($-SO_3^-$) group transfer in analogy to phosphoryl ($-PO_3^-$) transfers. One clear distinction is that unlike phosphate, the sulfate group can undergo two-electron enzymatic reduction to sulfite. The corresponding two-electron-reduced ($P = +3$) phosphite species is known, but very rare, in biology.

At the other end of the redox spectrum, electron rich sulfur at $S = -2$ oxidation state in H_2S and derived anions are divalent. They behave much more like divalent oxygen, in the periodic table below sulfur. They form thiols akin to alcohols. They form thioethers akin to oxygen ethers. On the other hand, the sulfur atom is large, the electrons are further from the nucleus and the dissociation energy of the S–H bond is lower than that of the O–H bond. Thiolate anions are much stronger nucleophiles than alkoxides and are approximately 10^6-fold more acidic. The thiolate anion is kinetically and thermodynamically accessible at physiological pH such that the thiolate anion dominates most of thiol biochemistry. H_2S is a gas, while H_2O is a liquid because of a much lower tendency of H_2S to hydrogen bond to neighboring molecules. H_2S may be one of three gaseous signaling agents in mammalian biochemistry (H_2S, NO and CO), but it is toxic, except at subnanomolar concentrations, in cells.

Finally, there is elemental sulfur, not only a solid under physiological conditions but an element that is both stable and a transformable metabolite. In this sense sulfur, S^0, differs from atomic hydrogen, oxygen and nitrogen which are stable only as dimers, H_2, O_2, N_2, albeit all subject to metabolic transfers.

In some respects though, sulfur redox chemical biology is most approximated by nitrogen. One can find nitrogen in the [−3] oxidation state (NH_3) and in the [+5] oxidation state in nitrate, a comparable eight-electron traverse. Some anaerobic bacteria that can reduce sulfate can also reduce nitrate all the way down to ammonia. That most reduced oxidation state of nitrogen has a vast role in metabolism on organification of NH_3 to amino groups, R–NH_2, even more extensive than organification of H_2S to cysteine. On the other had at the oxidized end, N = [+5] oxidation state, nitrate has a much more limited chemical biology than sulfate. There is the intriguing role for the $NO^•$ radical as mammalian signaling agent, perhaps analogized by gaseous H_2S.

15.2 AMP–Sulfate and Phosphoadenosine Phosphosulfate: Sulfuryl Transfer and Two-electron Redox Capacity

The sulfate dianion with its electrophilic sulfur and nucleophilic oxy anion substituents is utilized by ATP sulfurylase to attack the α-phosphoryl group of ATP. This generates a mixed anhydride of sulfuric acid and the substituted phosphoryl group of AMP as AMP–sulfate (Figure 15.2). This mixed anhydride can be used as a sulfuryl donor or can undergo further phosphorylation on the 3′-OH of the ribose moiety of APS to create PAPS (Figure 10.1). The second phosphorylation tends to segregate distinct metabolic roles for APS and PAPS.

The analogies with phosphoryl transfers are enzymatic sulfuryl transfers, reflecting the thermodynamic activation and kinetic stability that makes -PO_3 transfers so common to cosubstrate nucleophiles. While sulfuryl transfers are more restricted than phosphoryl transfers, there are both low molecular weight substrates and protein substrates that undergo sulfuryl transfer. The protein sub class that does react involves the tyrosine-phenol group as a nucleophile to give *O*-sulfotyrosine derivatives.

The ability of sulfate and sulfate esters or anhydrides to undergo enzyme-mediated reduction distinguishes sulfate from phosphate groups. The clearest

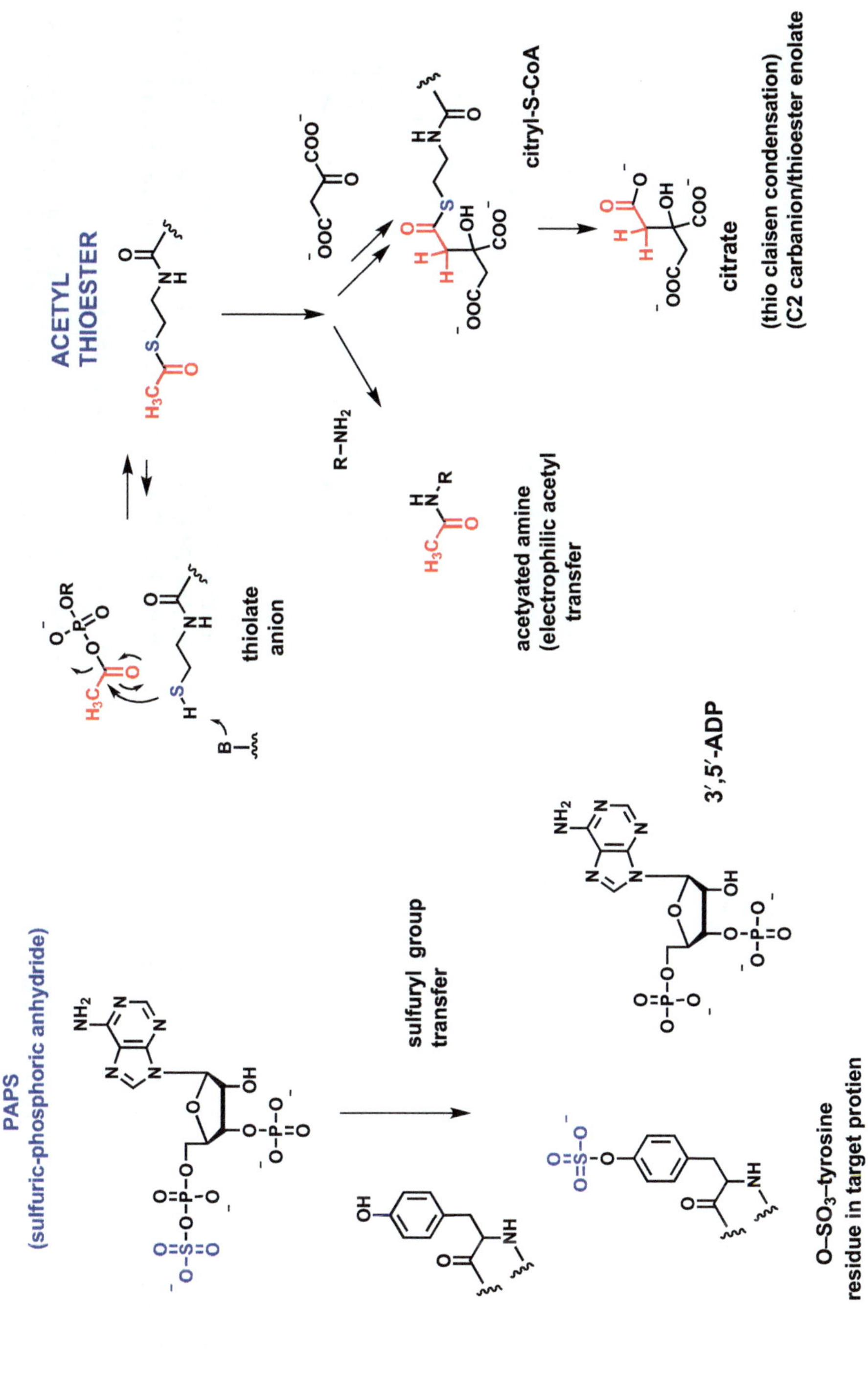

Figure 15.2 Continued on next page.

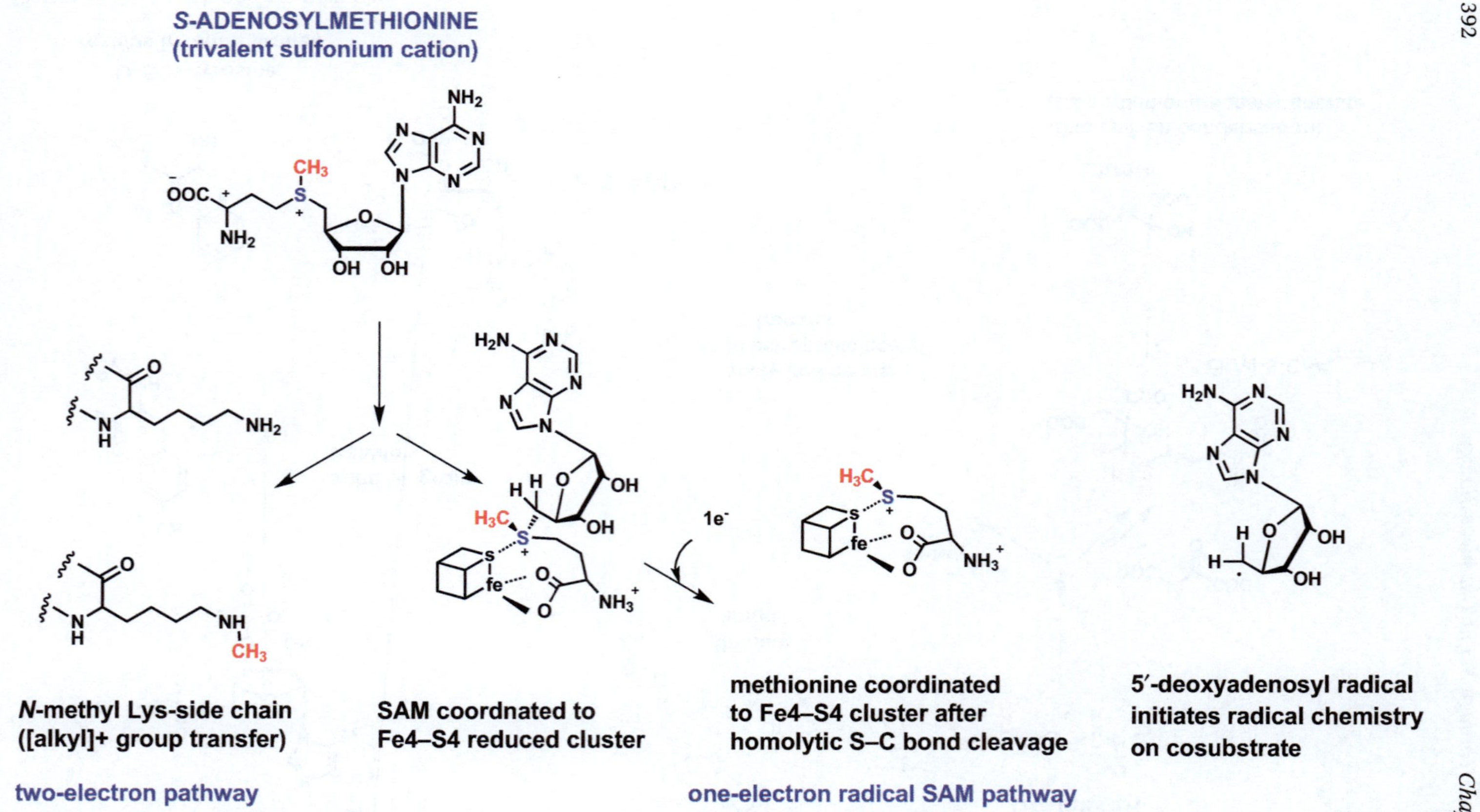

Figure 15.2 Three thermodynamically activated forms of organosulfur metabolites that donate electrophilic groups to cosubstrate nucleophiles. PAPS as sulfuryl donor; acyl CoAs as acyl donors; *S*-adenosylmethionine as alkyl (methyl) donor.

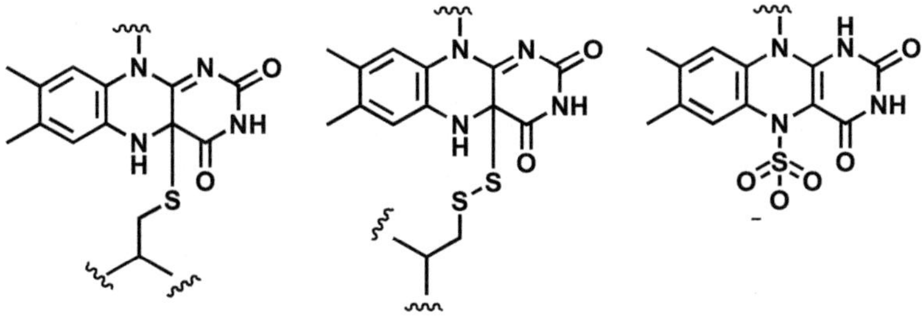

Figure 15.3 Three types of adducts between flavins and sulfur atoms in enzyme active sites.

indications are the two forms of APS reductases, the bacterial and the plant versions noted in Chapter 2. This also brings into focus the role of the flavin coenzyme as a redox active organic framework that acts as a conduit for electron transfer to and from sulfur. Figure 15.3 depicts three adducts between the flavin coenzyme and sulfur metabolites at distinct oxidation states. The figure highlights that covalent adducts mediate the electron transfer between sulfur and either N5 or C4a of the flavin and sulfur atoms. This is the first instance of a deep connectivity between the riboflavin-based vitamin B_2 coenzymes and sulfur redox metabolism. Their concurrent invention is essential for sulfur redox metabolism.

In the APS reductase case, starting the cascading eight-electron transfer pathway from sulfate to sulfide, AMP sulfate is reductively cleaved to sulfite and AMP. The actual reduction step is initiated by attack of N5 of the reduced flavin coenzyme on the sulfur in the sulfuric–phosphoric anhydride group. This creates an N5-flavin–sulfite adduct as AMP is released. Note that the relevant oxidation state is the dihydroflavin, which, in turn, was generated by NADPH reducing bound FAD. The genius of the FAD redox framework is then revealed when the N1 anion acts as a neighboring group to eject the sulfur as the two-electron-reduced sulfite anion as the flavin is converted to oxidized FAD.

An alternative strategy for reduction of the sulfur in the APS mixed anhydride has been implemented by plants. They replace the $FADH_2$ N5 nucleophile with a cysteine thiolate anion, generating an S–S bond ($-S-SO_3^-$) in which the two sulfurs are at highly distinct oxidant states. This is a thiosulfate linkage, well documented in nonenzymatic sulfur chemistry but quite rare in enzymatic transformations. The capture of the enzyme–thiosulfate on the thiol sulfur, by a second cysteine–SH side chain, creates a disulfide and the free sulfite. The redox step can be parsed either formally in formation of the thiosulfate but most obviously in the resolution step where two thiols in the reductase have been oxidized to disulfide and the sulfate reduced two electrons to sulfite.

The versatility of APS as a sulfuryl transfer reagent *or* a sulfate reduction donor is unmatched in ATP chemistry largely due to the resistance of phosphorus to undergo reduction from the +5 to the +3 state in biology. So why not make an adenosine trisulfate (ATS) as surrogate for adenosine triphosphate for sulfuryl transfers and have an expanded role for sulfate esters akin to phosphate esters in low molecular weight and macromolecular metabolism? To date ATS has not been detected as a metabolite in contemporary organisms.

15.3 Six-electron Reduction and Reoxidation Between Sulfite and Sulfide: Heme Connected to 4Fe–4S Cluster

Nowhere is the susceptibility of sulfur to undergo up to six electron inputs or outputs more evident than in the catalytic cycles of sulfite reductase. (Nitrate reductase logic and strategy is similar). We mostly consider sulfite to sulfide reduction but have noted that the enzyme runs in the other direction in sulfide-oxidizer metabolism so the machinery is reversible and driven by outside equilibria. The other six-electron oxidoreductase enzyme, often touted to be of larger global significance, is nitrogenase, reducing otherwise inert N_2 to two NH_3 molecules. Both are majestic in their architecture and machinery for solving the iterative input of six electrons and protons.

The key strategy in sulfite reductase couples two of the three common types of iron cofactors, heme and a 4Fe–4S cluster, *via* an intervening cysteine residue thiolate side chain (Figure 15.4). This machinery is the port of entry, *via* the Fe–S cluster, one electron at a time, for a total of six electron inputs. The SO_3^- to be reduced is ligated *via* an Fe–S in the top axial position of the siroheme, a heme biosynthetic variant that has its architecture and reduction potential tuned for the iterative one-electron inputs. The nature of the

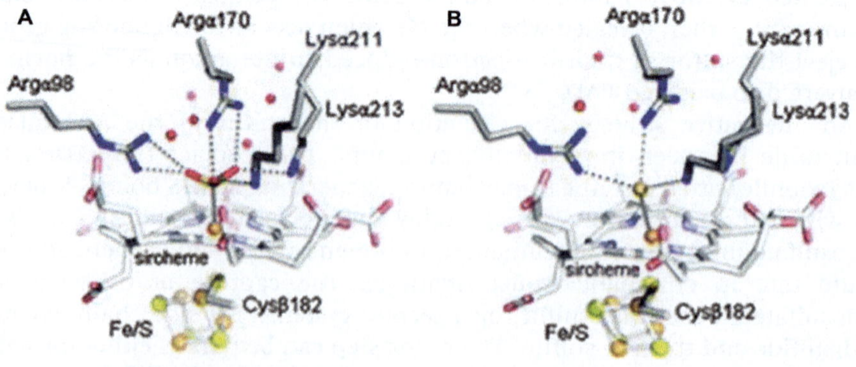

Figure 15.4 Representation of a synthetic model of the 4Fe–4S-thiol–heme cluster in sulfite reductase.
Reproduced from ref. 1 with permission from the Royal Society of Chemistry.

individual energy barriers for each electron input to flow through the Fe–S cluster and then the siroheme Fe into the coordinated sulfur substrate at each stage is not fully understood. There are X-ray structures of some of the partially reduced sulfo-heme intermediates. From an inventory point of view, as opposed to a detailed mechanistic landscape, each two electrons in to the reductase corresponds to subsequent cleavage of one of the three S–O bonds of the original sulfite anion substrate. Each of those oxygen atoms ends up as an H_2O, three in all. In the back direction of sulfide oxidation to sulfate, three water molecules give up an oxygen each to sulfur as it oxidizes.

One might wonder why and how the elemental sulfur oxidation state is not released but instead subject to further two-electron input to sulfide, or two-electron removal to a coordinated sulfur monoxide equivalent.

Going back up from hydrogen sulfide or hydrosulfide anion to elemental sulfur can be by reversal of the direction of electron flow in sulfite oxidase. On the other hand, there are bacteria that only carry out two-electron oxidations of sulfide to elemental sulfur.

Those sulfide to elemental sulfur oxidizers reveals a different aspect of biological strategy in sulfur redox chemistry. The responsible enzyme in these organisms, a sulfide quinone electron transferase, illuminates another facet of the equilibration of disulfides and inorganic sulfide oxidation machinery.

In this case, the inorganic HS^- substrate attacks an enzyme disulfide, cleaving it to a thiol and a persulfide (enzyme–Cys–S–SH). Now, the second variant of intersection between flavin coenzymes and sulfur occurs as the persulfide side chain dissociates to the persulfide anion (pK_a approximately 6.3) and then attacks the oxidized FAD coenzyme at C4a. (Figure 15.3). This covalent adduct, the FAD-4a-persulfide differs from the N5-sulfite adduct noted in the previous section. First, it is at C4a not N5, revealing two points of exit and entry of electrons between flavin and thiols. Second, the polarity is reversed. Here, an oxidized FAD is attacked by a persulfide anion as nucleophilic sulfur. In the N5-sulfite adduct it was the dihydroflavin attacking the electrophilic sulfate sulfur of the mixed anhydride in APS. Third, the oxidation state of the sulfur that becomes attached to the flavin ring system is different.

The fate of the persulfido-4a-FAD adduct is of interest. It is resolved by attack of a cysteine thiolate on the distal sulfur of the persulfide adduct. This cleaves the sulfur–C4a bond, releasing dihydroflavin that constitutes the redox step. The other product is the enzyme *trisulfide.* The middle sulfur is formally at the oxidation state of elemental sulfur, S^0. It is also known as a sulfane atom. In the cleavage of the persulfido-4a-flavin adduct, the electron pair in that bond goes into the flavin. The sulfur being attacked behaves as a six-electron sulfur (sulfane) as it yields the trisulfide.

Intriguingly this form of the enzyme can turn over multiple times. For example, another three catalytic cycles could yield the pentasulfide bridge. This can be cleaved by thiol attack and the $(S^0)_n$ chains deposited in accumulating S^0 globules in these bacteria. We will defer discussion of this '*ompolung*' behavior of persulfide side chains to a subsequent section of this final chapter.

Up to this point we have seen both thiosulfate (enzyme–Cys–S–SO_3^-) and persulfide (enzyme–Cys–S–SH) (Figure 15.1) forms of enzymes in these microbes that take sulfate up and down to low-valent sulfur compounds. We have seen the versatility of flavin coenzymes in both reduced and oxidized forms as covalent conduits for electron transfers to and from sulfur at different oxidation states. We have begun to appreciate the underlying logic and nano-machinery invented by sulfate reducers and sulfide oxidizers, as they cobbled together energy plans around available and accessible inorganic sulfur oxidation states.

15.4 Traverse from Inorganic Hydrogen Sulfide to Organic Thiol Group in Cysteine

Back in Chapter 1, Figure 1.9 illustrates the single enzymatic path that has evolved to connect the inorganic sulfur and organic sulfur metabolic universes. The incorporation of sulfur as a thiol group into the three-carbon framework of cysteine occurs at the level of H_2S, accounting for the presence in prokaryotes, lower eukaryotes and plants of the sulfate to sulfide reduction pathways. These serve an *assimilatory* purpose rather than acting as pathways for dumping electrons from reduced substrates as they get oxidized.

The strategy takes advantage of the nucleophilicity and kinetic accessibility of the hydrosulfide anion and uses a three-carbon amino acid–pyridoxal phosphate adduct as the appropriate electron sink for HS–C3 bond formation. The cosubstrate is an *O*-acylated serine molecule, acylated to lower the energy for cleaving the C3–O bond in the amino acid–PLP adduct. As in all amino acid–PLP adducts, the PLP aldehyde forms an imine linkage to the amino acid. That creates the electron sink that stabilizes subsequent carbanion intermediates on the amino acid framework.

In these *O*-acylserine lyase–cysteine synthase catalytic cycles, enzyme abstraction of the C2–H as a proton yields the delocalized C2-carbanion that can participate in ejection of the acyl substituent at C3 with departure of the oxygen atom as depicted in Figure 15.5A. The resultant enamino acid–PLP adduct is electrophilic at C3. This is a crucial contribution of the PLP vitamin B6 coenzyme. Absent prior formation of the PLP–aldimine adduct, the enzyme could not find a path to eliminate the C3–O acyl group, nor would the residual framework be electrophilic. Once the C3–S bond has formed the enzyme must reprotonate at C2 before cleaving the imine link to the PLP coenzyme (a net β-replacement of *O*-acyl by –SH). That releases L-cysteine and the metabolic diaspora of organic sulfur can begin (Figure 15.5).

There is no change is sulfur valence state ($S=[-2]$) as it traverses from inorganic H-SH to organic thiol R-SH. The thiol oxidation state is central to cysteine biology, both as a free metabolite and in proteomes, as noted in Chapters 12–14.

We will consider a separate PLP-enzyme working in the other direction on L-cysteine in Section 15.6 where the released sulfur atom is an electrophilic –SH (persulfide) rather than a nucleophilic one!! (Figure 15.5B).

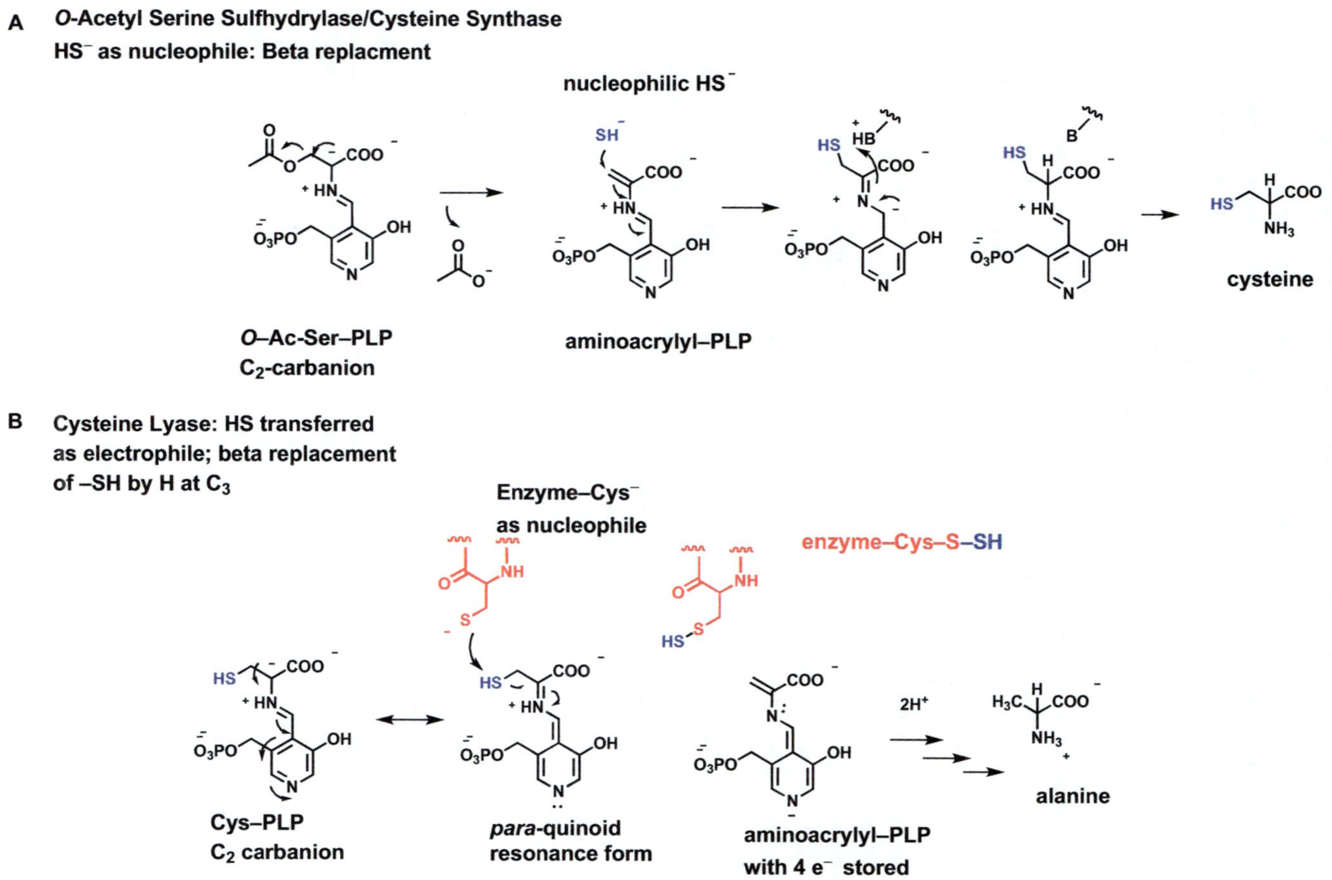

Figure 15.5 Two PLP enzymes that transfer "HS" with different polarities. (A) *O*-acetylserine sulfhydrylase generates L-cysteine using HS^- as a nucleophile. (B) Cysteine lyase releases HS as an electrophile, captured by an active site cysteine to form a persulfide adduct.

15.5 Coenzyme A (CoASH) and Glutathione (GSH): Acyl Thioesters *vs.* Electrophile Scavenging

Before taking up persulfides, we note that free cysteine can be toxic because of the side chain thiolate nucleophilicity so it is kept in quite low concentrations in cells. The flux is (a) to protein biosynthesis, (b) to taurine biosynthesis *via* oxidation to the sulfinate side chain, (c) to incorporation into the coenzyme A scaffold or (d) to incorporation into the tripeptide glutathione. The latter two fates preserve the nucleophilic thiol and lead to differential roles for coenzyme A *vs.* reduced glutathione.

Coenzyme A (Figure 15.2) is used primarily to attack activated acyl-AMPs or acyl-Ps. This creates kinetically stable acyl-*S*-thioesters that retain the thermodynamic activation for favored transfer to nucleophilic cosubstrates but gain enhanced kinetic stability over the acyl-phosphate precursors. Thus, acyl CoAs are the prominent diffusible acyl transfer currencies in all cells. The corresponding acyl oxoesters are at least 1000-fold less activated for acyl transfers.

Furthermore, we have noted in Chapter 5 that the acyl groups in the freely mobile CoA thioesters are orthogonally activated at C1 and C2 of those acyl groups. They are electrophilic at C1, but nucleophilic at C2, by virtue of stabilization of derived carbanions as thioester enolates. Hence, the acyl groups in acyl thioesters are ideal for C–C bond formation, carrying both electrophile and nucleophile required to make a C–C bond, within or between acyl chains. This is key chemical logic for construction of C–C bonds in central metabolites, such as citrate and acetoacetyl CoA.

While one could have imagined the evolution of mobile acyl-*S*-glutathione thioesters for equivalent purposes, the tripeptide with its nucleophilic thiolate side chain has been deployed largely for two distinct and complementary purposes. The first is to capture reactive oxygen species, notably peroxides, such as HOOH, but also alkyl peroxides. The resulting sulfenates (Figure 15.1) are intermediates on the way to disulfides. The change from thiol to sulfenate converts a nucleophilic thiol to an electrophilic sulfur in the thiolate, setting up the attack by a second GSH to form the disulfide.

This is a surveillance–protectant role for GSH. Cells then spend an NADPH molecule to reduce the GS–SG disulfide back to the two GSH molecules. In the disulfide form GSSG has no nucleophilic sulfurs with which to scavenge electrophiles. Thus, cells spend the diffusible electron currency NADPH to get back to the GSH forms. Typically, more than 95% of cellular glutathione is maintained in the surveillance-competent GSH state, by spending as many NADPH equivalents as necessary.

The second scavenging role for GSH and its nucleophilic thiol is to capture organic electrophiles, rather than the inorganic ones above. In Chapter 6 we noted several variants of cosubstrate–xenobiotic carbon electrophiles, from aldehydes, to epoxides, to allyl cations, to iminoquinones that are captured in thioether linkages. These adducts detoxify the electrophiles before they could covalently capture other cellular targets, such as DNA, RNA or proteins.

15.6 Cysteine to Fe–S Clusters: Role of Persulfide Side Chains and Electrophilic -SH

One other feature of free cysteine metabolism and its processing by PLP enzymes to note here is the action of the cysteine lyases that are the first steps in providing the inorganic sulfur that ends up as the protein-associated sulfide in the hundreds of protein-embedded 2Fe–2S, 4Fe–4S, and more complex Fe–S cluster architectures (Figure 15.5B).

The premise is that Fe–S clusters are ancient redox elements, set up for one-electron transfers, as one of the constituent Fe ions cycle between Fe^{3+} and Fe^{2+} states. Tracing back the inorganic sulfide in Fe–S clusters reveals they come from cysteine rather than directly from H_2S generated by the assimilatory sulfite reductases noted above. Perhaps, this is to avoid toxicity of free hydrosulfide anion or sulfide dianion.

In the event, the cysteine lyases process L-cysteine to L-alanine *via* PLP–aldimine adducts while cleaving the C3–SH bond. However, this is not just the reverse of the cysteine synthase reaction route of Figure 15.5A. Examination of the C3–S bond cleavage step shows capture of the itinerant sulfur atom by an active site Cys thiolate side chain. The polarity of transfer in Figure 15.5B shows the departing sulfur leaves the cysteine scaffold without the electron pair in the C3–S bond. That is an *electrophilic sulfur* in flight. This polarity reversal ("ompolung") is enabled by the unique ability of the PLP coenzyme to serve, transiently, not just as a two-electron sink, but also as a four-electron sink.

This insight identifies the resultant Cys–S–SH persulfide residues as carriers of electrophilic sulfur when that terminal-SH is protonated. That explains how chaperone proteins can use Cys–S- side chains to move the sulfur, without equilibration with solvent, and without the promiscuous nucleophilicity one might see it were behaving as an S^{2-} in flight. In Chapter 6 we also outlined the current thoughts on how Fe–S clusters are first built as 2Fe–2S clusters and then reductively dimerized to 4Fe–4S clusters. (In contrast, in Figure 14.3 we noted that the Cys–S–S^-, once dissociated to the anions were strong nucleophiles).

Lastly, we note that Fe–S clusters are actually a perfectly designed intersection of inorganic sulfur at the −2 oxidation state and the four cysteine thiolates that complete the iron ligand shells, also with S at the −2 oxidation state. The evolution of cluster assembly, back in the anaerobic eras needed a confluence of the two kinds of S =−2 sulfur atoms (Figure 15.6).

15.7 Methionine to *S*-Adenosylmethionine: Two-electron Transfers

By comparison to the thiol-containing cysteine, the other sulfur-containing amino acid, methionine, with its sulfur tied up in a thioether linkage, is very much the staid chemical partner. That is, until it meets up with ATP in the active site of *S*-adenosylmethionine synthetase. Remarkably, there, the

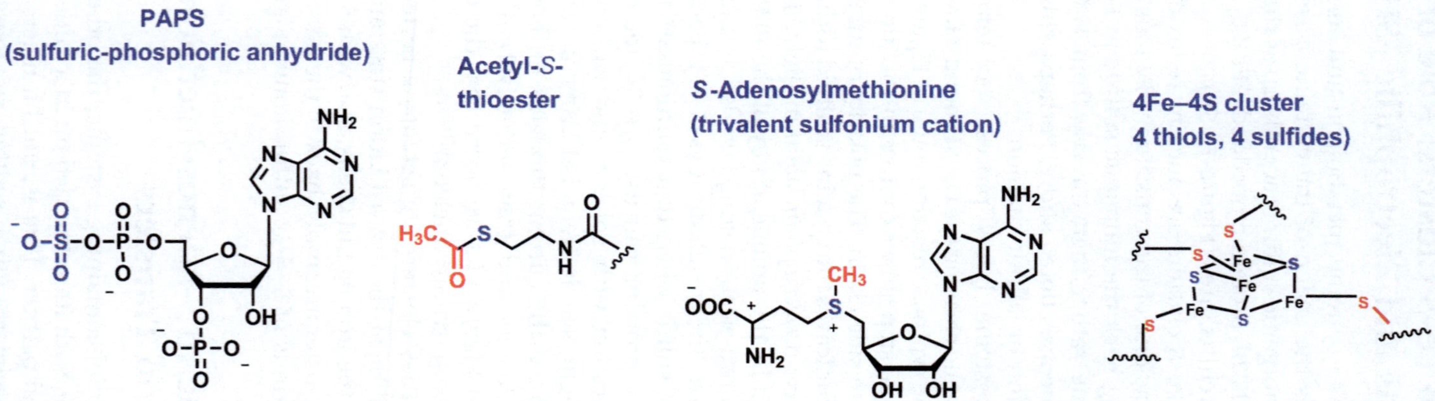

Figure 15.6 Four forms of activated sulfur in metabolism.

thioether sulfur is coaxed into attacking, not any of the three electrophilic phosphorus atoms in the triphosphoric anhydride side chain but, instead the unprepossessing C5′ carbon of the ribose moiety. The resultant cationic trivalent sulfonium center in *S*-adenosylmethionine allows SAM to spring into remarkable cellular metabolic prominence.

That prominence rivals, and in some ways, outshines the metabolic competencies of ATP, acetyl CoA and NAD(P)H. Thus, the permanent positive charge on the sulfonium sulfur activates all three of the substituents for capture by cosubstrate nucleophiles (Figure 15.2): a three for one center, creating neighboring electrophiles. Of course, by far the most important [R^+] group transfers are [CH_3^+] equivalent transfers to S, O, N and C nucleophiles in low and high molecular weight cosubstrates. The transfer of the one carbon methyl group electrophile, small as it is, has enormous ramifications whether in cell membrane phospholipids, DNA methylations, RNA methylations or histone and other protein methylations. SAM reveals the power of a single positive charge, trigonalizing the sulfur, to set in motion a tsunami of biological downstream effects.

The underlying chemical strategy that connects ATP, NADH and acetyl CoA to SAM is the constancy of chemically diverse frameworks that preserve thermodynamic activation with sufficient kinetic stability to be used as diffusible intracellular currencies. This is the case, even though distinct types of chemical scaffolds, functional groups and reactivities are at work in each molecule.

15.8 Radical SAM Enzymes; Fe–S–SAM Coordination: Prelude to Radical Chemistry

The most startling connection between Fe–S cluster one-electron transfer logic and *S*-adenosylmethionine over the past two decades has been the realization that many radical-based metabolic transformations originate from a subset of 4Fe–4S clusters that lack one cysteine ligand from the host protein. That protein thiolate is replaced by the carboxylate of SAM, such that the functional unit is the 4Fe–4S–SAM ligated cluster.

This is the very large family of enzymes known as radical SAM enzymes. Given that 4Fe–4S clusters are ports of entry and exit for only one electron at a time, this reaction manifold is imposed on the adjacent SAM molecule. In turn, one-electron transfer into the coordinated SAM molecule sets in motion radical fragmentation of its S–C5′-adenosine bond. This brings one-electron sulfur chemistry to the fore in over 1 000 000 predicted radical SAM open reading frames.

One-electron input to the SAM sulfur, from external reductant *via* passage through the Fe–S cluster, would give a transient SAM sulfur radical. As the S–C5′ adenosine bond homolyzes, one electron stays with the sulfur, creating an Fe–S cluster-ligated methionine, plus a single unpaired electron left on the 5′-deoxyadenosine (dA•) fragment. As we noted in Chapter 9, the dA• is a robust oxidant and in a given enzyme active site, typically abstracts an H• from a bound cosubstrate, often from an unactivated C–H center. This sets off many variants of radical-based metabolic transformations, unique in their ability to use one-electron pathways.

At the end of the radical-based catalytic cycles we noted that SAM can be regenerated and therefore function catalytically. Or, it can be cleaved as a stoichiometrically consumed substrate to form 5′-deoxyadenosine (dA-H) and methionine. This product pattern is distinct from the formation of *S*-adenosylhomocysteine after electrophilic methyl transfers, or from 5′-thiomethyl adenosine after aminobutyryl transfers to a cosubstrate nucleophile.

The product patterns clearly delineate two-electron *vs.* one-electron reaction manifolds from SAM. They reflect the different mechanisms for quenching the positive charge on the sulfonium ion center, as sulfur goes from trivalent cation to divalent thioether. While the primary contemporary cellular role of *S*-adenosylmethionine is typically viewed as a thermodynamically activated donor of the [CH_3^+] group to nucleophiles, that role may actually be dwarfed by the plethora of air-sensitive radical SAM enzymes, across many types of organisms.

15.9 The Broad Sweep of the Cysteinyl Proteomes: Covalent Cys–X Enzyme Intermediates and Covalent Posttranslational Modifications

Part III of this volume turned from low molecular weight inorganic and organic sulfur-containing metabolites to the roles of cysteine and methionine as protein residue side chains. Given the embedding of both amino and carboxy groups of these two amino acids in peptide linkages, any chemical biology is dominated by thiol or thioether group reactivity, respectively. Methionyl residue thioether reactivity is limited to both adventitious and purposeful oxidation and rescuing reduction at the sulfoxide, but not the sulfone, level.

Cysteine residue thiolate chemistry operates both in enzyme catalytic machinery and also as reactive groups in stable posttranslational modifications of protein substrates. Transient covalent enzyme adducts during catalytic cycles are found throughout many diverse metabolic pathways. Enzyme side chains that participate as nucleophiles include lysine (DNA ligase), histidine (bacterial sensor kinases), tyrosine (protein tyrosine kinases; type II topoisomerases), aspartate and glutamate (glycosyl transferases), serine (proteinases), along with cysteine *via* its powerfully nucleophilic thiolate side chain.

One might then have imagined that enzyme–cysteinyl thiolate nucleophiles would function to capture covalently most categories of electrophiles that undergo transfers: acyl groups, phosphoryl groups, glycosyl groups and alkyl groups. Three of the four strategies occur; the glycosyl transferases seem not to have evolved to use cysteine nucleophiles in mid catalytic cycles, although a variant of this strategy is found in plant glucosinolate pathways.

In Chapter 12 we noted the wide sweep of acyl-*S*-cysteinyl enzyme intermediates. They span diverse families of catalysts, from proteases, to the ubiquitin network both in enzymes 1, 2 and some E3s, and the six dozen deubiquitinases, as well as the family of glutaminase enzymes. We also noted *S*-phospho-cysteinyl enzyme intermediates only in the protein tyrosine phosphatase family of approximately 100 members.

The alkylation logic appears in two enzymes of central consequence in DNA metabolism: thymidylate synthase and DNA cytosine 5-methyltransferase. The prior addition of a cysteine thiolate side chain across the pyridine 5,6-double bond in a conjugate addition mode sets up the carbanions required for C–C bond formation. Perhaps most surprising of the covalent enzyme adducts during catalytic cycles is the formation of enzyme–SSH persulfide adducts. In generating that S–S bond, one sulfur (the enzyme cysteine) acts as a nucleophile while the other sulfur (the substrate cysteine–PLP adduct) acts as an electrophile, despite both being at the same oxidation state ($S = -2$).

Chapter 13 then switched from transient covalent cysteine residue adducts that get dissipated at the end of every catalytic cycle of an enzyme to cysteine covalent adducts that persist as protein substrates are converted to products. These constitute categories of posttranslational modifications (PTMs). The PTM universe is large and also includes covalent modifications at side chains other than cysteine. Notably these include serine phosphorylation, lysine methylation and acetylation, even tyrosine sulfation (two of which connect back, *via* SAM and APS, to sulfur chemical biology).

However, the cysteine thiolate provides some unique posttranslational modification outcomes. The most well-known is probably the dimerizing oxidation of pairs of cysteine thiol side chains to cystine disulfides. We noted this may have arisen as major rigidifying, stabilizing architectural responses in proteins produced by organisms as the atmosphere morphed from anaerobic to aerobic. Most proteins in intracellular compartments that are not part of secretory pathways have their complement of cysteine residues reduced, thanks in large part to active reduction of any disulfides that form adventitiously by high relative concentrations of reduced glutathione.

In the secretory pathway of eukaryotic cells, most proteins as they emerge from endoplasmic reticular membrane-associated ribosomes, start to form disulfides as they refold. These two-electron oxidations occur in two routes. One is *via* sulfenates as noted earlier in this chapter and in Chapter 13. The other, dominant route is catalysis of disulfide formation by the 15 protein disulfide isomerases in the ER and Golgi lumens that rise to approximately millimolar concentrations. Key to the oxidative folding, thiol–disulfide interchange process is the need to dump electrons removed from the nascent proteins undergoing net disulfide formation. This is the role of the ER resident ERO1 oxidase and QSOX oxidase, reducing O_2 to H_2O_2 (and thereby providing more HOOH for the sulfenate to disulfide oxidation route).

This mass oxidation of dithiols to disulfide bonds in proteins that move through cellular secretory pathways also re-illuminates the intimate chemical biology role of flavin coenzyme and cysteine thiols and disulfides: in these cases, to mediate net two-electron transfer from dithiols to O_2 by way of the 4a thiol flavin adducts illustrated in Figure 15.3.

Chapter 14, the third of the three chapters on reactions of the cysteinyl proteome examined a limited but remarkable set of structural and functional consequences when a specific cysteinyl side chain in a particular protein becomes oriented for productive attack on the immediate upstream carbonyl group of the peptide bond. This is surprising in many aspects, including the background that peptide bonds are highly stable covalent entities with lifetimes of *circa* 7 years.

The tetrahedral thioaminal adduct that results can lead to several fates. It could revert back to the starting protein in a traceless reversal. However, if it goes forward, one outcome is cleavage of the C–N peptide bond, as an acyl thioester linkage forms. The resonance-stabilized low energy amide bond in the peptide linkage has been replaced by a thermodynamically activated thioester, set up for acyl transfer to nucleophilic cosubstrates.

Two of the most intriguing subsequent biological acyl transfers are of the upstream protein acyl moiety to cholesterol in the plasma membrane or to a downstream cysteine thiolate side chain in the same protein chain (Figure 14.3). The first event tethers the approximately 170-residue N-terminal fragment of the signaling protein hedgehog to the 3-β-OH of a cholesterol molecule, restricting diffusion of that protein ester with its lipophilic tail to the plane of the plasma membrane. The second event is a key step in protein splicing, where upstream and downstream regions of a protein are fused together as the intervening sequence (the intein) is excised and discarded. This occurs in folded precursor proteins and was presumably an early experiment in protein diversification of function and structure. In the last step of protein splicing, the migrating or migrated acyl chain undergoes an S to N migration. This converts the acyl thioester back to the final peptide linkage. These *S*- to *N*-acyl transfers highlight the thermodynamically favored acyl chain migration from sulfur to amine at the newly revealed N-terminus of the spliced intermediate.

The other outcome from the initial tetrahedral thioaminal product would be C–O bond cleavage rather than C–N cleavage (Figure 15.3). Indeed, there is enzymatic machinery that routes the tetrahedral adduct in this direction by phosphorylation of that oxyanion. Now as a phosphate ester ($O–PO_3^-$) it is a lower energy leaving group than naked $–O^-$. The consequence of C–O bond cleavage is formation of a cyclic five-membered ring, a thiazoline. Thiazolines are hydrolytically labile to reversion, but a companion FMN enzyme can dehydrogenate the C2–C3 bond of thiazoline rings and produce the resonance-stabilized heteroaromatic thiazole ring. The net outcome has been to alter the backbone connectivity in this region of a protein or peptide. A protease-sensitive peptide linkage has been converted to the protease-resistant thiazole. This strategy has been adapted in hundreds of proteins that get morphed by these posttranslational steps into natural products transmogrified from their peptide roots.

15.10 The Versatility of Inorganic and Organic Sulfur Metabolites

Figure 15.1 may be a good way to end this tour through the chemical biology of sulfur. In the first column four compounds representing four different oxidation states of inorganic sulfur that figure in metabolism are the subjects of Chapters 2 and 3. Much of the remaining 11 chapters, dealing with low molecular weight sulfur organic metabolites and then cysteine proteomes have dealt with the 11 entries in the second column.

The sulfuryl–phosphoric and sulfinyl–phosphoric anhydrides reflect electrophilic sulfur metabolites at high oxidation states of sulfur. The sulfate esters are the products from sulfuryl–phosphoric anhydride–SO_3 transfers. The sulfenate, sulfinate and sulfonate oxidation states reflect the progressive oxygenation of sulfur most likely by one-electron reaction manifolds. The thiophosphoryl, sulfinyl–phosphoric anhydride and disulfide monoxide species all form in one remarkable rescue sequence as sulfiredoxins interact with over-oxidized peroxiredoxins to reduce sulfinates back to thiols in a four-electron reduction sequence.

The three staples of organosulfur functional groups in metabolism are organic thiols (cysteine), thioethers (methionine) and the cationic trivalent sulfonium *piece de resistance* in *S*-adenosylmethionine. Disulfides and persulfides are two-electron-oxidized forms that track the reversible oxidation of thiols. The persulfides in particular connect the potentially toxic inorganic hydrogen sulfide metabolite with the invention of the iron–sulfur clusters that also meld thiolate and inorganic sulfide chemistry into the one-electron transfer Fe–S clusters.

Nature has explored and exploited much or most of the versatility of sulfur in its many guises in chemical biology.

References

1. K. Parey, G. Fritz, U. Ermler and P. M. H. Kroneck, *Metallomics*, 2013, 5, 302.

Subject Index

β-elimination, cysteine to pyruvate, 83
β-replacement *vs.* β-elimination, 75
    cysteine lyases, 131–133

acetaminophen, 114, 115
acetyl *S*-coenzyme A (acetyl CoA), 96
2-acetylthiazolidine, 115
actin, 349–353
acyl-adenosine monophosphate (AMP), 95
acyl adenylates, 95–98
acyl CoA metabolites, 68
acyl phosphate, 95–98
acyl-*S*-Cys–enzyme intermediates, 314–319
acyl-*S*-cysteine, 318
acyl-*S*–enzyme cascade processes, 290
acyl thioesters, 95–98, 103
    vs. acyl oxoesters, 71
    carbon electrophiles, 98–103
    carbon nucleophiles, 98–103
    double headed reactivity of, 98–103
    vs. electrophile scavenging, 398
    hydrolysis, 67
acyl transfer, 373
adenosine-5′-phosphosulfate (APS), 233
adenosine trisulfate (ATS), 394
adenylyl sulfate reductases
    logic, 31–34
    mechanistic strategies for, 34
adventitious oxidation, 180
AMP–sulfate, 390–394
AMP–sulfate reductase, 27
anaerobic green algae, 173
APS reductase, 31
    mechanism, 33
Arabidopsis thaliana, 316
Archeoglobus, 12
Archeoglobus fulgidus reductase, 27
assimilatory sulfate reduction, 14
    nascent hydrogen sulfide to organic sulfur transition, 54–55
assimilatory sulfite reduction, 35
ATP sulfurylase, 26–27
ATP *vs.* 3′-phosphoadenosine-5′-phosphosulfate (PAPS), 234
autoproteolysis, 361–364

bacterial adenylyl sulfate reductase, 29
bacterial methionine-γ-lyase, 271
biotin synthase, 224

cannibalization, 221–228
carbamate nitrogen, 239
carbon electrophiles, 98–103
carbon nucleophiles, 98–103
carboxylic acid anions, 96
carboxymethyl transfers, 191–196
carboxy-SAM formation, 191–196
cellular electrophiles, 108–112
cellular metabolites, 68
cephalosporin, 208
cephem ring system, 208
chemokine receptor 5 (CCR5), 246, 247

cholic acid metabolism, 168
Chromatiacaea, 45
cinnabar (HgS), 8
citrate synthase, 101, 103
citric acid cycle, 101, 137
Cmo pathway, 196
cobalt, 180
coenzyme A (CoASH), 95–98, 398
coordination, iron and zinc, 279–283
 iron in proteins, forms, 280–281
 pervasive super acid/organizing ligand, zinc, 281–282
 zinc finger domains, 283
covalent Cys–X enzyme intermediates, 402–404
covalent enzyme–cysteinyl adducts, 307–310
covalent enzyme intermediates, 309
covalent posttranslational modifications, 402–404
Crithidia fasciculata, 123
C–S bond-forming strategies
 cysteine-persulfide anions, conjugate additions, 266
 disulfide to methylthioacetal, 262–265
 oxidative cleavage, disulfides, 266
 plant and microbial biosynthetic pathways, 262–272
 in primary metabolic pathways, 258–261
 thioglycosides and glucosinolates, 266–272
Cys_2 residues dioxygenation, 344–345
cystathionine, 78–83
cystathionine β syase, 79–81
cystathionine β-synthase, 84
cystathionine-γ-lyase, 82–83
cysteamine, 93–98
cysteine–cysteamine thiol, 383
cysteine lyases, 139
 persulfide-forming, 131–134
cysteine persulfides, 72
 residues, 345–347
cysteines, 17, 59, 66, 76–87, 92, 93, 115, 396–397, 399
 misprocessing to H_2S, 83–85
 oxidative dimerization of, 125
 S-alkylations and acylations, 116
 to sulfoxide to sulfenate to thiocarbonyl group, 170–174
cysteine side chain, 384
 attack by immediate upstream peptide carbonyl group, 361–374
 protein automodification, 363
cysteine *S*-oxygenation, 161–167
cysteine sulfenate, 169
cysteine sulfinate, 165, 166, 169
cysteine sulfinate dioxygenase, 163
cysteine sulfonate, 169
cysteine synthase, 14
 reaction, 72
cysteine thiolates, 277
cysteine–tyrosine thioether automodification, 374
cysteinyl proteomes, 279, 402–404
 acyl-*S*–enzyme intermediates, 284
 covalent changes in, 283–298
 deubiquitinases, 290–293
 fatty acid, polyketides and nonribosomal peptide, assembly lines, 295–298
 foreign electrophiles, 374–378
 glutaminase catalytic logic, 293–295
 nitrosylation of, 340–342
 posttranslational modifications of, 314
 reactions of, 285
 ubiquitinylated proteome, 286, 289–290
cysteinyl residue modifications, reactive oxygen species, 342–345
cysteinyl thiolate, 308

cystine disulfides, 225, 324–334
    cytoplasmic cysteinyl side chains, 325–327
    peroxiredoxins, 327–330
    sulfiredoxin, 330–334
cytochrome b5, 52
cytosine, catalysis, 300–304

Desulfovibrios, 12, 24
detoxification catalyst family, 122
deubiquitinases, 290–293
dianion of sulfuric acid, 7
dissimilatory sulfate reduction, 11
dissimilatory sulfite reductases, 35
disulfide bond formation, 324
DNA cytosine 5-methyltransferase, 403
DNA methyltransferase catalytic cycle, 303
D. vulgaris enzyme, 27

eight-electron spread, sulfur atom, 387–390
electronic configuration, sulfur, 6
electron transfer chemistry, 1
electrophilicity, 47
electrophilic –SH, 399
electrophilic sulfur, 85, 399
elemental sulfur, 8
enzymatic machinery
    S^0 to sulfate conversion, 50–54
    S^0 to sulfite conversion, 50–54
enzymatic transsulfuration, 78–83
enzyme catalysis, tandem glycyl and thiol radical intermediates, 201–205
enzyme cysteine active site nucleophile, 304–305
enzyme thiolate, covalent addition, 300–304
epidermal growth factor receptor (EGFR), 68
ergothioneine, 171, 173, 174
Escherichia coli iscU apo-protein, 145
estradiol-3-sulfate, 236
estrone-3-sulfate, 236
expandase (DAOCS) reaction, 208

F-actin, 349
Fe–S clusters, 129, 130, 399
2Fe–2S clusters, 221–228
4Fe–4S clusters, 221–231, 394–396
Fe–S–SAM coordination, 401–402
4Fe–4S-thiol–heme cluster, 394
flavin adenine dinucleotide (FAD), 27, 30
flavin–sulfur covalent adducts, 47
foreign electrophiles, 374–378
formylglycine (Fgly)-containing sulfatases, 249–252
four-electron sink, 140

G-actin, 349
galena, 8
geranyl-geranyl-*S*-glutathione (GGG), 112–113
glutaminase catalytic logic, 293–295
glutathione (GSH), 103–117, 161, 383, 398
    biosynthesis, 106
    with cellular electrophiles, 108–112
    with nitric oxide, 118
    nucleophilic cosubstrate, cellular reactions, 111
    S-alkylations and acylations, 116
    S-nitrosylation of, 117–119
    sulfenates and disulfides, 119–125
    S- *vs.* *N*-acylations, 106
    thiol oxidations, 119–125
    two-electron thiol reagent, 108
    with xenobiotic electrophiles, 114–115
glutathione peroxidase, 121
glutathione reductase, 121
glutathione *S*-transferases, 115
glutathione sulfenic acid (GS-OH), 110

glycosaminoglycans, 239–241, 243
Gly–Ser–Cys tripeptide sequence, 365
gram-negative bacteria, 24

heme, 394–396
HemN decarboxylase, 216–219
heparan sulfate tetramer, 240
homocysteine, 83, 177–182
homocysteine derives, four-carbon scaffold, 76–78
homocystinuria, 79
homolytic reaction modes, 198
homoserine, 76–87, 179
human sulfite oxidase, 51
Hunter's syndrome, 252
hydrogen peroxide, 21
hydrogen sulfide (H_2S), 10, 396–397
  assimilation of, 72–76
  signaling role in mammals, 87–89
  two electron oxidation of, 46–49
hypotaurine, 167–170
hypothemycin, 377

IcsS–IcsU tetramer, 142
inorganic sulfate, 13
  eight-electron traverse, 23–25
inorganic sulfide, 2
  atoms, 130
  eight-electron traverse, 23–25
inorganic sulfur compounds, 11–15, 387–390
  metabolism of, 56
inorganic sulfur metabolites, 405
inorganic sulfur redox biochemistry, 56
inorganic sulfur redox cycles, enzymatic strategies, 55–57
iron–sulfur (Fe–S) clusters, 383
  assembly, 134–145
  in biotin and lipoic acid sulfur insertions, 146–148
  variant, 145–148
isopenicillin *N* synthase (IPNS), 205

kinetoplastid-containing single-cell eukaryotes, 123

L-cysteine
  organic scaffold, 72–76
  transaminase action, 86
lead sulfide (PbS), 8
L-homocysteine, 78
L-iduronate sulfatase, 252
lipoic acid, 221–228
lipoic acid synthase, 223
lipoyl-*S*–ACP synthase, 226
long chain acyl-thioesters, 314–319
low molecular weight thiols, 103–117
lysine aminomutase, 216, 217
lysosomal sulfatase deficiencies, 252–255

mammalian metabolism, 76–87
3-mercaptopyruvate, 86
mercaptopyruvate sulfotransferse, 85
mercaptopyruvate transferase, 85–87
mesophilic sulfate-reducing bacteria, 12
metabolomics, 5
metalloprotein sulfatases, 249
methanethiol, 271
  methionine-γ-lyase activity, 272
methionine, 7, 17, 176, 399–401
  biosynthesis from homocysteine, 177–182
  biosynthesis, plants and microbes, 177
methionine residue enzymatic sulfidation, 349–353
methionine residue *S*-oxidations, 347–353
methionine sulfoxide, 349
methionine synthase, 180
methionyl aminopeptidases, 176
methylmalonyl CoA mutase, 180
methylthiolation, 228–231
  tRNA, 229

2-methylthio-*N*-6-isopentenyladenine, 228
5-methyluridine, 300
MiaA, 228
MiaB, 228
microbes, differential positioning, 56
mixed phosphoric–sulfuric anhydride, 27
molecule interacting with CasL (MICAL), 350, 353
molybdenum cofactor, 54
molybdopterin, 3
  cofactor, 52
multiple oxidation states, sulfur, 9–11
mycobacteria, 106, 189

net six-electron reductions, 35–42
nitrate-reducing bacteria, 55
N-methylbacillothiol, 104
N-5-methyl tetrahydrofolate, 177
N-myristoylation, 318
non-catalytic cysteine residues, 377–378
nonenzymatic methionyl residue *S*-oxidation, 347–349
N-sulfuryl transferase, 239
nucleophilicity, 47
nucleophilic thiol side chains, 374–378

O-acetyl-L-serine sulfhydrylase, 54, 55, 73, 129
one-electron enzyme chemistry, 198–208
one-electron transfer, 165
organic sulfur metabolic scaffolds, 15–19
organic sulfur metabolites, 405
organic thiol group, 396–397
organosulfur metabolites, 61, 392
  transformations, 59
organosulfur molecular scaffolds
  in biological systems, 65–72
O-sulfation, 237, 239
oxazole formation, 364–366
oxygen *vs.* sulfur chemical biology, 19–21

palmitoylation cycle, 319
Pectobacterium atrosepticum, 340
persulfide-forming cysteine lyases, 131–134
  persulfides *vs.* persulfide anions, differential reaction, 133–134
  six-electron *vs.* eight-electron sulfur atom products, 131–133
  β-replacement *vs.* β-elimination, 131–133
persulfides
  electrophilic/nucleophilic forms of sulfur, 154–157
  side chains, 399
persulfide S–S bond, 304–305
persulfides, sulfur donors, 151–154
  to pyrimidines, 148–151
persulfido–flavin adduct, 46
phosphatidylethanolamine, 189
phosphoadenosine phosphosulfate, 390–394
3′-phosphoadenosine-5′-phosphosulfate (PAPS), 16, 236
phosphopantetheinylated carrier protein domains, 100
phosphoryl *vs.* sulfuryl transfers, 234
polyubiqitinylation, 344–345
posttranslational lanthionine thioether, 359
post-translation lipidations
  of cysteinyl side chains, 314–324
  thioether adducts, 319–324
prenyl-*S*-protein transferases, 322
protein Cys–S–S–Cys disulfide bonds, 339–340
protein machinery, disulfide formation, 324–338
protein persulfides, sulfur donors, 151–154

protein prenylations, 319–324
protein serine kinase action, 360
protein serine kinases, 377–378
protein *S*-palmitoylation, 318
protein splicing, 368–374
protein sulfenates, 342–344
protein sulfotransferases, 243–248
protein tyrosine kinases (PTKs), 298
protein tyrosine phosphatase action, 298–299
protein tyrosine phosphatase receptors (PTPRs), 240
protein tyrosine phosphatases, 239–241
protein tyrosine sulfuryl transferase, 246
protein *vs.* proteoglycan substrates, sulfuryl transfer, 241–243
proteoglycan, 243
proteoglycan *O*- and *N*-sulfation, 242–243
Pseudomonas syringae, 196
purified dissimilatory sulfite reductases, 35
purple sulfur bacteria, 45
pyrite, FeS_2 ("fools gold"), 8
pyruvate formate lyase, 201, 204

radical SAM enzymes, 401–402
    5′-deoxyadenosyl radical, convergent routes, 209–211
radical SAM reactions
    ancient coenzymes, 219–221
    lysine aminomutase, 216, 217
    scope of, 211–216
ralfuranone D, 272
Ras proteins, 319
redox versatility sulfur, 11, 389
reduced glutathione. *see* glutathione
reductive oxygen metabolism, 19
ribonucleotide reductases, 199–201
Rieske type oxygenase, 239
RimO, 228

S-adenosylation, 65
S-adenosyl homocysteine (SAH), 76
S-adenosylmethionine (SAM), 7, 11, 15, 59, 62, 65, 176, 399–401
    consumption and products, 219
    5′-deoxyadenosine and methionine, 216–219
    electrophilic group transfer substrate, 183–191
    electrophilic one-carbon $[CH_3]^+$ fragment transfer, 184–191
    HemN decarboxylase, 216–219
    sulfonium sulfur, positive charge, 184
    thioether to trivalent sulfonium cation, 182–183
S-adenosylmethionine synthase, 183
S-adenosylmethionine synthetase, 65, 76
Salmonella typhimurium, 76
SAM-ylide formation, 191–196
Sanfilippo syndrome type A, 252
sequential arginylation, 344–345
SF_6, abiotic compound, 7, 8
simplified sulfur metabolic cycle, 14
six-electron sulfite reductase machinery, 2
S-methylation, 189
S-nitrosylation, 68
S-nitrosyl-glutathione (GSNO), 118, 119
S-oxygenation pathway, 162
S-phosphoryl–enzyme intermediates, 298–299
substrate electrophiles, 307–310
sulfatases, 248–252
    formylglycine (Fgly)-containing sulfatases, 249–252
    metalloprotein, 249
sulfated disaccharide, 239–241
sulfate reducers, 11, 13, 27–34
sulfate-reducing bacteria, 11
sulfate to sulfite microbial conversion, 26–27
sulfate uptake, by transport proteins, 25–26

sulfenamides, 342–344
sulfenic acids, 68
sulfhydrylase reaction, 72
sulfide oxidizers, 13, 45
sulfide oxidizing bacteria, 46
sulfide quinone oxidoreductase, 46, 47, 49, 87, 131
  mechanism, 69
sulfite oxidase, 50
  mechanistic proposal for, 53
sulfite reductase, 2, 39, 41, 52
  electron flow, 40
sulfite to hydrogen sulfide microbial reduction, 35–42
Sulfobales order, 45
sulfotransferases, 233–236
  nonprotein substrates for, 236–239
sulfur donors
  carrier protein acyl persulfides, 151–154
  persulfides, 148–154
  protein thiocarboxylates, 148–151
sulfur, electrophile, 148–154
sulfuric acid, 9
sulfur nucleophile, cysteine *vs.* glutathione, 115–117
sulfur nucleophilicity, 383
sulfur oxidation states, 6
sulfur redox equilibria, 12
sulfuryl transfer, 390–394

taurine, 167–170
taurine amide conjugation, 170
thiacoraline, 262
thiazole formation, 364–366
3-thio-Asp88, 231
thiocarbonyl group formation, 366–368
thioester intermediate, 361–364
thioethers, 358–361
thiohemiacetal adducts, 305–307
thiol aldehyde, 94
thiol chemical biology, 93–98
thiol radicals, 198–208
  bevy, in active site, 199–201
  β-lactam formation, 205–208
thioredoxin (TRX), 31, 32
thymidylate, catalysis, 300–304
thymidylate synthase, 300, 304
transsulfuration, 65
  pathway, 76–87
tripeptide glutathione, 16
trypanothione, 123
two-electron oxidation, 131
two-electron redox capacity, 390–394
two-electron transfers, 399–401
tyrosyl protein sulfuryl transferases (TSPTs), 247
tyrosyl residue *O*-sulfation, 313

ubiquitin–deubiquitin cascades, 275
ubiquitinylated proteome, 286, 289–290
ubiquitous dianion, sulfate (SO_4^{2-}), 1
universal liquid solvent *vs.* toxic gas, 20

vitamin B_{12}, organometallic forms, 210
vitamin B_6 (pyridoxine), 73

water *vs.* hydrogen sulfide, 20
Wittig reaction, 191

xenobiotic electrophiles, 114–115
xenobiotic molecules, 103

Yersinia pestis, 262

Zif 268, 283
zinc sulfide (ZnS), 8